Medicinal Plants

Promising Future for Health and New Drugs

Medicinal Plants

Promising Future for Health and New Drugs

Edited by
Parimelazhagan Thangaraj

CRC Press
Taylor & Francis Group
Boca Raton London New York

CRC Press is an imprint of the
Taylor & Francis Group, an **informa** business

CRC Press
Taylor & Francis Group
6000 Broken Sound Parkway NW, Suite 300
Boca Raton, FL 33487-2742

Printed on acid-free paper

International Standard Book Number-13: 978-0-8153-7052-9 (Hardback)

Library of Congress Cataloging-in-Publication Data

Names: Thangaraj, Parimelazhagan, editor.
Title: Medicinal plants : promising future for health and new drugs / editor, Thangaraj Parimelazhagan.
Description: Boca Raton : Taylor & Francis, 2018. | "A CRC title, part of the Taylor & Francis imprint, a member of the Taylor & Francis Group, the academic division of T&F Informa plc." | Includes bibliographical references and index.
Identifiers: LCCN 2017055054| ISBN 9780815370529 (hardback : alk. paper) | ISBN 9781351046510 (ebook)
Subjects: LCSH: Materia medica, Vegetable. | Medicinal plants.
Classification: LCC RS164 .M357 2018 | DDC 615.3/21--dc23
LC record available at https://lccn.loc.gov/2017055054

Visit the Taylor & Francis Web site at
http://www.taylorandfrancis.com

and the CRC Press Web site at
http://www.crcpress.com

Contents

Foreword ... ix
Preface ... xi
Contributors ... xiii

Chapter 1 Cabbage: A Storehouse of Nutraceuticals ... 1
Vandna Pandey, Abhishek Chura and Hemant Kr. Pandey

Chapter 2 Phytonutrients: Their Relevance to Human Health ... 17
Ranganathan Kumar, Sulaxana Kumari Chauhan, Subramanian Vijayalakshmi and Shanmugam Nadanasabapathi

Chapter 3 Lukoskin: An Effective Herbal Treatment of Leucoderma Developed by Defence Institute of Bio-Energy Research (DRDO), Haldwani ... 47
Hemant Kr. Pandey, Harsahay S. Meena, Sanjai Kr. Dwivedi and Madhu Bala

Chapter 4 Centipede Envenomation Effects on Human Beings and Scientific Research on Traditional Antivenom Agents: *Aristolochia* and *Piper* sp. ... 73
Dhivya Sivaraj, Revathi Ponnusamy, Rahul Chandran and Parimelazhagan Thangaraj

Chapter 5 Antidiabetic Herbal Formulations: An Effective Therapeutic Approach to Diabetes ... 89
Binu Thomas and Krishna M. Chinchu

Chapter 6 Antidiabetic Activity by the *In Vitro* α-Amylase and α-Glucosidase Inhibitory Action of Indian Ayurvedic Medicinal Plants: *Ficus amplissima* Smith ... 101
Karuppusamy Arunachalam, Puthanpura Sasidharan Sreeja, Domingos Tabajara de Oliveira Martins and Parimelazhagan Thangaraj

Chapter 7 Anti-Obesity Potential of Indian Traditional Medicinal Plants and Their Phytochemicals 111

Vellingiri Vadivel, Pichai Venkatalakshmi and Pemaiah Brindha

Chapter 8 Promise of Medicinal Plants for Neurodegenerative Disease: Alzheimer's, A Review 133

Kalidass Subramanaiam

Chapter 9 Plant Sources as Potential Therapeutics for Alzheimer's Disease... 161

Azhwar Raghunath, Kiruthika Sundarraj, Vishnu Vignesh Kanagaraj and Ekambaram Perumal

Chapter 10 *Schefflera* Genus (Araliaceae): A Review of Its Botanical Description, Ethnobotanical Uses, Phytochemistry, Pharmacology and Toxicology 193

Puthanpura Sasidharan Sreeja, Karuppusamy Arunachalam and Parimelazhagan Thangaraj

Chapter 11 Selective Chinese Viviparous Ferns, Their Bioactive Principles and Economical Values: A Review 223

Xavier Ravi Baskaran, Antony Varuvel Geo Vigila, Wenbo Liao and Zhang Shouzhou

Chapter 12 Prospects of Local Flora of Trans-Himalayan Region of Ladakh for Various Medicinal Uses 239

Gyan P. Mishra, Tsering Stobdan, Parimelazhagan Thangaraj, Tania Seth and Bijendra Singh

Chapter 13 Phytochemistry of *Muntingia calabura* L. Fruits 261

Kathirvel Preethi

Chapter 14 Pharmacognostical and Phytochemical Investigation on *Pterolobium hexapetalum* (Roth.) Sant. & Wagh. 269

Saikumar Sathyanarayanan, Rahul Chandran, Murugan Rajan and Parimelazhagan Thangaraj

Chapter 15 Antifungal and Anti-mycotoxigenic Activities of *Pogostemon mollis* Benth. against *Fusarium graminearum* 285

Kasipandi Muniyandi, Elizabeth George and Parimelazhagan Thangaraj

Chapter 16 *In Vitro* Calli Induction, Biomass Accumulation and Different Biological Activity of *Leucas aspera* (willd.) Linn. 297

Masilamani Sri Devi, Krishnamoorthy Vinothini, Blassan P. George, Sudharshan Sekar, Heidi Abrahamse, Bettine van Vuuren and Arjun Pandian

Chapter 17 Antioxidant Potential of Cup Saucer Plant: *Breynia retusa* (Dennst.) Alston 317

Prabu Jaganathan, Murugan Rajan, Saikumar Sathyanarayanan, Iniyavan Murugaiyan, Geetha Sree Subburaj Jeyalakshmi and Parimelazhagan Thangaraj

Chapter 18 The Promising Antiradical Potential of *Castanospermum australe* A. Cunn. and C. Fraser ex Hook. 331

Sajeesh Thankarajan, Rahul Chandran, Murugan Rajan, Saikumar Sathyanarayanan and Parimelazhagan Thangaraj

Chapter 19 *In Vitro* Plant Regeneration, Comparative Biochemical and Antioxidant Potential of Calli and Seeds of *Sesbania grandiflora* (L.) Poiret 355

Krishnamoorthy Vinothini, Masilamani Sri Devi, Sudharshan Sekar, Blassan P. George, Heidi Abrahamse, Bettine van Vuuren and Arjun Pandian

Index 379

Foreword

Medicine and health are central concerns for people everywhere. Mankind has used medicinal plants as indispensable cures for many ailments. Plants and other natural products play a significant role in traditional, botanical and pharmaceutical medicine. In many developing countries, including India, a large proportion of the population relies on traditional practitioners and medicinal plants in order to meet their healthcare needs. Although modern synthetic medicines are in demand by the ever-growing population, with such traditional practises, herbal medicines have often maintained their popularity for historical and cultural reasons. In recent years, there has been a recovery of the use of herbs due to the side effects of chemical drugs, lack of therapies for several chronic diseases (even with highly efficient modern medicines and microbial resistance), as well as the unprecedented investment in pharmaceutical research and development. With the emergence of the modern science and technology, global pharmaceutical companies have begun to search for novel sources of drugs from potential plants, and renewed their strategies in favour of natural product drug development and discovery. Herbal products with well-defined constituents and reliable preclinical and clinical trials could promote the better use of medicinal plants. This book is a very timely one. Dr. Parimelazhagan Thangaraj made an extraordinary effort in gathering valuable scientific information on medicinal plants and herbal drugs. I congratulate Prof. Thangaraj for his tremendous worth in bringing out this inclusive publication on phytomedicine. I hope the book will be widely read and used for development of herbal drug research, and hopefully stimulate the young minds of students worldwide.

Dr. Shashi Bala Singh, FNASc, FIAN, FAMS
Distinguished Scientist & Director General (LS)
Defence R & D Organisation (DRDO)
Min. of Defence, Govt of India
Rajaji Marg, New Delhi-110011

Preface

Plants are one of the most important resources of human foods and medicines. Rapidly increasing knowledge on nutrition and medicine has revolutionized our concepts about food, health and agriculture. Strong recommendations for the use of neutraceuticals, natural plant foods and phytotherapy have become progressively popular to improve health, and to prevent and treat certain diseases. Even though the health and disease-fighting benefits of phytochemicals have been widely published in many books, *Medicinal Plants: Promising Future for Health and New Drugs* has unique the contribution of including recent trends of phytomedicine on different aspects. This book explores the hidden medicinal potential of significant plants through 19 research and review chapters encircling the valuable works of eminent academicians and research scientists from different parts of the world. This book has attempted to encapsulate scientific information relevant to ethnomedicine, phytonutrients, pharmacognosy, phytochemistry and tissue cultures in a simple and clear way, so that students, researchers, scientists and industrialists can easily understand. The review processes of the articles have been carried out by experts from Universities and Research Institutes. The editor conveys his sincere thanks to Bharathiar University authorities for their boundless support and guidance. With great pleasure, the editor extends heartfelt thanks to all the authors for their excellent contributions and consistent cooperation. The support rendered by Professor and Head, Department of Botany, Bharathiar University is also acknowledged. The editor also thanks the bioprospecting research team, Dr. Thamburaj Suman, Dr. Rajan Murugan, Mrs. Puthanpura Sasidharan Sreeja, Ms. Sivaraj Dhivya, Mr. Muniyandi Kasipandi and Mr. Sathyanarayanan Saikumar for their involvement in the compilation of the manuscripts in an elegant manner. The editor would like to express a special gratitude for the publishers and their team for the technical support and continued encouragement in bringing out the book in time.

Parimelazhagan Thangaraj

Contributors

Heidi Abrahamse
Laser Research Centre
University of Johannesburg
Johannesburg, South Africa

Karuppusamy Arunachalam
Department of Basic Sciences in Health
Faculty of Medicine
Federal University of Mato Grosso
Mato Grosso, Brazil

Madhu Bala
Defence Institute of Bio-energy Research
Defence Research and Development Organization
Uttarakhand, India

Xavier Ravi Baskaran
State Key Laboratory of Biocontrol and Guangdong Key Laboratory of Plant Resources
School of Life Sciences
Sun Yat-sen University
Guangzhou, China

Pemaiah Brindha
Centre for Advanced Research in Indian System of Medicine
SASTRA University
Tamil Nadu, India

Rahul Chandran
Laser Research Centre
University of Johannesburg
Johannesburg, South Africa

Sulaxana Kumari Chauhan
ICAR: National Dairy Research Institute
Haryana, India

Krishna M. Chinchu
Post Graduate Department of Botany
Deva Matha College
Kerala, India

Abhishek Chura
Defence Institute of Bio-Energy Research
Defence Research and Development Organization
Uttrakhand, India

Sanjai Kr. Dwivedi
Defence Institute of Bio-energy Research
Defence Research and Development Organization
Uttarakhand, India

Antony Varuvel Geo Vigila
Department of Zoology
St. Xavier's College
Tamil Nadu, India

Blassan P. George
Laser Research Centre
University of Johannesburg
Johannesburg, South Africa

Elizabeth George
Bioprospecting Laboratory
Department of Botany
Bharathiar University
Tamil Nadu, India

Prabu Jaganathan
Bioprospecting Laboratory
Department of Botany
Bharathiar University
Tamil Nadu, India

Vishnu Vignesh Kanagaraj
Molecular Toxicology Laboratory
Department of Biotechnology
Bharathiar University
Tamil Nadu, India

Ranganathan Kumar
DRDO: Defence Food Research Laboratory
Karnataka, India

Wenbo Liao
State Key Laboratory of Biocontrol and Guangdong Key Laboratory of Plant Resources
School of Life Sciences
Sun Yat-Sen University
Guangzhou, People's Republic of China

Harsahay S. Meena
Defence Institute of Bio-Energy Research
Defence Research & Development Organization
Uttrakhand, India

Gyan P. Mishra
ICAR: Indian Agricultural Research Institute
New Delhi, India

Kasipandi Muniyandi
Bioprospecting Laboratory
Department of Botany
Bharathiar University
Tamil Nadu, India

Iniyavan Murugaiyan
Bioprospecting Laboratory
Department of Botany
Bharathiar University
Tamil Nadu, India

Shanmugam Nadanasabapathi
DRDO: Defence Food Research Laboratory
Karnataka, India

Domingos Tabajara de Oliveira Martins
Area of Pharmacology
Department of Basic Sciences in Health
Faculty of Medicine
Federal University of Mato Grosso (UFMT)
Cuiabá, Brazil

Hemant Kr. Pandey
Defence Institute of Bio-Energy Research
Defence Research and Development Organization
Uttarakhand, India

Vandna Pandey
Defence Institute of Bio-Energy Research
Defence Research and Development Organization
Uttrakhand, India

Arjun Pandian
Centre for Research and Development
PRIST University West Campus
Tamil Nadu, India

Parimelazhagan Thangaraj
Bioprospecting Laboratory
Department of Botany
Bharathiar University
Tamil Nadu, India

Ekambaram Perumal
Molecular Toxicology Laboratory
Department of Biotechnology
Bharathiar University
Tamil Nadu, India

Revathi Ponnusamy
Bioprospecting Laboratory
Department of Botany
Bharathiar University
Tamil Nadu, India

Kathirvel Preethi
Department of Microbial Biotechnology
Bharathiar University
Tamil Nadu, India

Azhwar Raghunath
Molecular Toxicology Laboratory
Department of Biotechnology
Bharathiar University
Tamil Nadu, India

Murugan Rajan
Bioprospecting Laboratory
Department of Botany
Bharathiar University
Tamil Nadu, India

Saikumar Sathyanarayanan
Bioprospecting Laboratory
Department of Botany
Bharathiar University
Tamil Nadu, India

Sudharshan Sekar
Centre for Ecological Genomics and Wildlife Conservation
Department of Zoology
University of Johannesburg
Johannesburg, South Africa

Tania Seth
ICAR: RCER, Research Centre,
Jharkhand, India

Zhang Shouzhou
Shenzhen Key Laboratory of Southern Subtropical Plant Diversity
Fairy Lake Botanical Garden
Shenzhen and Chinese Academy of Sciences
Shenzhen, People's Republic of China

Bijendra Singh
ICAR: Indian Institute of Vegetable Research
Uttar Pradesh, India

Dhivya Sivaraj
Bioprospecting Laboratory
Department of Botany
Bharathiar University
Tamil Nadu, India

Puthanpura Sasidharan Sreeja
Bioprospecting Laboratory
Department of Botany
Bharathiar University
Tamil Nadu, India

Masilamani Sri Devi
Department of Biotechnology
Jeppiaar Engineering College
Tamil Nadu, India

Tsering Stobdan
Defence Institute of High Altitude Research (Formerly FRL)
Defence Research and Development Organisation (DRDO)
Jammu and Kashmir, India

Geetha Sree Subburaj Jeyalakshmi
Bioprospecting Laboratory
Department of Botany
Bharathiar University
Tamil Nadu, India

Kalidass Subramanaiam
Department of Animal Science
Manonmaniam Sundaranar University
Tamil Nadu, India

Kiruthika Sundarraj
Molecular Toxicology Laboratory
Department of Biotechnology
Bharathiar University
Tamil Nadu, India

Sajeesh Thankarajan
Department of Botany
and Biotechnology
Mar Ivanios College
Kerala, India

Binu Thomas
Post Graduate Department of Botany
Deva Matha College
Kerala, India

Vellingiri Vadivel
Centre for Advanced Research in Indian
System of Medicine
SASTRA University
Tamil Nadu, India

Pichai Venkatalakshmi
PG and Research Department
of Biochemistry
Sengamala Thayaar Educational Trust
Women's College
Tamil Nadu, India

Subramanian Vijayalakshmi
DRDO: Defence Food Research
Laboratory
Karnataka, India

Krishnamoorthy Vinothini
Department of Biotechnology
Jeppiaar Engineering College
Tamil Nadu, India

Bettine van Vuuren
Centre for Ecological Genomics and
Wildlife Conservation
Department of Zoology
University of Johannesburg
Johannesburg, South Africa

1 Cabbage
A Storehouse of Nutraceuticals

Vandna Pandey, Abhishek Chura and Hemant Kr. Pandey

CONTENTS

1.1 Introduction 1
1.1.1 Health Benefits 2
1.1.2 Research and Development Work 3
1.1.3 Assessment of Antioxidant Phytochemicals in the Leaves of Green Cabbage Varieties 3
1.2 Material and Methods 4
1.2.1 Nutraceutical Evaluation 4
1.2.2 Estimation of Ascorbic Acid 4
1.2.3 Estimation of β-Carotene 4
1.2.4 Determination of Antioxidant Activity 5
1.2.5 Chlorophyll Estimation 5
1.2.6 Estimation of Iodine and Antioxidant Activity in Cabbage Hybrids 5
1.2.7 Estimation of Macro, Micro and Carotene in Cabbage Hybrids 6
1.3 Results and Discussion 6
1.3.1 Consumption Tips 12
1.4 Conclusion 14
References 15

1.1 INTRODUCTION

Vegetables are an intrinsic part of the human diet. They are the richest and cheapest source of minerals and vitamins. Besides being nutritious, vegetables also act as powerful medicines and offer preventive health benefits. The discovery of phytochemicals in vegetables and their strong antioxidant potential in scavenging free radicals has generated tremendous attention among scientists, horticulturists and plant breeders (Chu et al. 2002; Proteggente et al. 2002; Sun et al. 2002). Antioxidants are free radical scavengers; antioxidants are naturally formed inside the body as a byproduct of different metabolic reactions. Free radicals are molecules with unpaired electrons and are highly reactive. They can easily react with lipids, nucleic acid, enzymes and proteins, causing tissue injury (Thrombino et al. 2004; Ohr 2004). Vitamin C, carotenoids and phenolics are the major sources of antioxidants found in

White cabbage Red cabbage Savoy cabbage

FIGURE 1.1 Three categories of cabbage.

vegetables. Antioxidants are also called 'lifesaving' elements, which protect us from degenerative diseases such as cancer, aging and cardiovascular diseases (Hashimoto et al. 2002; Gosslau and Chen 2004; Kavanaugh et al. 2006; Lee et al. 2008).

Cabbage (*Brassica oleracia* L. var. *capitata*), as one of the most important vegetable crops, belongs to the Cruciferae or mustard family. Its edible portion is called the head. The growing point or terminal bud increases in size and becomes a storage area for essential nutrients (Dickson and Wallace 1996). During the early growth and development of the cabbage plants, the leaves expand, remain unfolded and form a frame around the plant. Next, the newly expanded leaves partially unfold, thus forming the shell of the head. Then the growing point increases in size and is filled with a number of thick, overlapping, smooth sessile leaves that form a compact head. Nieuwhof (1969) categorized three forms of cabbage viz. red cabbage, green cabbage and Savoy cabbage (Figure 1.1).

Nutritionally, it is a rich source of minerals and vitamins. Cabbage is one of the most popular vegetables for cancer prevention. This vegetable is nutrient packed and low in calories. It contains high content levels of calcium, iron, iodine, potassium, sulphur and phosphorus. It is loaded with vitamin A, B1, B2, B6, C, E, K and folic acid. Each layer of cabbage is packed with an abundance of natural antioxidants that help fight cancer. Cabbages possess both antioxidant and anticarcinogenic properties (Cohen et al. 2000; Chu et al. 2002).

The criteria for good quality cabbage are compactness, brightly colored outer leaves, freshness and crispiness. In the head, older leaves are arranged in the outermost and outer middle layers, and the newer leaves are arranged in the innermost and inner middle layers.

1.1.1 Health Benefits

Cabbage is well-known for its medicinal properties. In Ayurvedic medicine, cabbage leaves are prescribed for coughs, fevers, skin diseases, peptic ulcers, etc. Cabbage leaves are a good source of chlorophyll and cure anaemia by blood building. Fresh cabbage juice is reported to contain an antiseptic ulcer component, called vitamin U. Cabbage is one of the largest sources of sulphur-containing amino acids and is also reported to have significant anticancer activity. The benefits of cabbage have been proven by animal studies and epidemiological data found in humans. The antioxidant properties in cabbage make it a powerful healing food in combating cancer.

Iodine is one of the most important trace elements in the body, and iodine deficiency causes a number of dreaded diseases, like goiters, mental retardation, hypothyroidism and pregnancy-related problems. Iodine is not synthesized in the body, so it is always recommended to include iodine-rich foods in our diet. In developing nations, iodine deficiency is the main cause of hypothyroidism. When the body does not get enough iodine, the thyroid gland has trouble in synthesizing its hormones. The result is often a goiter, an enlargement of the thyroid. Human iodine intake is closely dependent on the iodine concentration of water, soil, iodine-rich salts and different foods. There are parts of the world where iodine is so scarce that the sight of a neck without a goiter is rare. According to a World Health Organization report, the number of people affected by an iodine deficiency is about 740 million. Normal adults and children need 150 micrograms and 200 micrograms of iodine per day, respectively. Without it, the thyroid gland cannot do its job. Table salt is one of the major sources of iodine in the body. In some parts of the world, the soil contains little or no iodine. In developing countries, where iodized salt may not be available, thyroid problems are very common and as a result many children suffer from mental and psychomotor retardation. Cabbage is a natural source of iodine. (Key et al. 1992; Appleby et al. 1999) and thus aids the proper function of the brain and nervous system.

1.1.2 Research and Development Work

Defence Institute of Bio Energy Research, one of the establishments of Defence Research & Development Organization, is actively engaged in the development and production of cabbage hybrids and varieties. A large number of hybrids and varieties have been evaluated (Figure 1.2) and assessed for their nutraceuticals.

1.1.3 Assessment of Antioxidant Phytochemicals in the Leaves of Green Cabbage Varieties

The main objective of the study was to assess the phytochemicals viz. ascorbic acid, β-carotene and the total chlorophyll and antioxidant activity in the leaves of the cabbages, so the cultivars with the maximum number of phytochemical antioxidants could be identified. Due to variation in the nutraceuticals in the leaves from outermost side to innermost side, the cultivars could guide the appropriate utilization of cabbage leaves in various food preparation techniques in order to maximize the nutritional benefits.

FIGURE 1.2 Variability in cabbage grown in the field.

1.2 MATERIAL AND METHODS

Nine genotypes of cabbage (namely C1, C2, C3, C5, C7, SEL 1, SEL 7, DARL 851 and DARL 852) were evaluated in a complete, randomized block design with three replications at the experimental field. The experimental site was located in Pithoragarh (Uttarakhand) at an altitude of 5500 ft above sea level. The site was situated in the western Himalayas, which extends from lat 29°29′ N to lat 30°49′ N and long 85°05′ E to long 81°31′ E. The annual rainfall is approximately 1250 mm, out of which 70–75% is received during the rainy season. The temperature of the place ranged from a maximum of 35°C in summer to a lower of –2°C during winter. The seedlings of cabbage genotypes were transplanted in an open field. The mature heads were selected in each replication. Leaves were separated individually from each head and grouped into outermost leaves (L1–L3), outer middle leaves (L4–L10), inner middle leaves (L11–L17) and innermost leaves (L18–L30). The leaves in the four groups were screened for ascorbic acid (mg/100 g), β-carotene (µg/100 g), total chlorophyll (mg/100 g) and antioxidant activity (IC_{50}).

1.2.1 Nutraceutical Evaluation

The chemical analysis of fresh heads included determination of ascorbic acid by the 2,6-di-chlorophenol indophenols method (AOAC 1990). The fresh fruit samples of cultivars were analysed for antioxidant activity using the DPPH (2,2-diphenyl-1-picryl–hydrazyl radical) method (Hatano et al. 1988). β-Carotene was estimated through a spectrophotometer, and the results were expressed as µg/100 g (AOAC 1980). Chlorophyll estimation was carried out by the method given by Witham et al. (1971).

1.2.2 Estimation of Ascorbic Acid

Ascorbic acid was estimated by the volumetric method. Three grams of the fresh sample were extracted with 4% oxalic acid and the volume was made up to 100 mL and centrifuged. Five milliliters of this supernatant was pipetted out, combined with 10 mL of 4% oxalic acid and the titration was done against the dye. Ascorbic acid reduces the 2,6-dichlorophenol dye to a colorless leuco base and gets oxidized to dehydroascorbic acid. It was estimated in mg/100 g.

1.2.3 Estimation of β-Carotene

For the estimation of β-carotene, 5 g of dried sample was taken in 150-mL glass stoppered Erlenmeyer flask and 40 mL water saturated butanol (WSB) was added. The contents of the flasks were mixed vigorously for 1 min and kept overnight (16–18 h) at room temperature in the dark for a complete extraction of β-carotene. The contents were shaken and filtered through the Whatman Filter Paper No. 1 into a 100-mL volumetric flask. The optical density of clear filtrate was measured at 440 nm using an ECIL, double-beam UV-VIS Spectrophotometer 5704SS. Pure WSB was used as a blank. The WSB was prepared by mixing n-butanol with distilled water in 8:2 ratios. The β-carotene contents were calculated by a calibration curve from the known amount of β-carotene and expressed as parts per million (ppm). The standard solution of β-carotene (Sigma) was prepared in WSB with a concentration of 5μg/mL. A calibration curve was made by known amounts of pure β-carotene from 0.25 μg/mL to 1.5 μg/mL. The results were expressed as μg/g.

1.2.4 Determination of Antioxidant Activity

The DPPH (2, 2-diphenyl-1-picrylhydrazyl) method was used for estimating free radical scavenging activity of the methanol extracts of samples. Two milliliters of methanol extract (4 mg/mL) were taken in a test tube and final volume of 3 mL was made with methanol. The absorbance of the mixture was measured after 40 min at 517 nm against methanol as a blank. Ascorbic acid was used as a standard. The free radical scavenging activities (%) of tested samples were evaluated by comparing with a control (2 mL DPPH and 1 mL of methanol). Each sample was then measured in triplicate and averaged. The free radical scavenging activity (FRSA) was calculated using the formula:

$$\text{FRSA} = \left[(\text{Ac} - \text{At})/\text{Ac} - \text{As}\right] \times 100,$$

where

Ac = Absorbance of the control,
As = Absorbance of the standard and
At = Absorbance of the test.

1.2.5 Chlorophyll Estimation

One gram of leaf samples were ground with the addition of 20 mL of 80% acetone. After centrifuging at 5000 rpm for 5 min, the supernatant was transferred to a 100-mL volumetric flask. The residue was further ground with 20 mL of 80% acetone, centrifuged and the supernatant was transferred to the same volumetric flask. This process was repeated until the residue become colorless. Volume was made up to 100 mL with 80% alcohol. Chlorophyll was extracted in 80% acetone and the absorption was read at 663 nm and 645 nm by a spectrophotometer. Using the absorption coefficients, the amount of chlorophyll was calculated.

1.2.6 Estimation of Iodine and Antioxidant Activity in Cabbage Hybrids

Owing to the significance of iodine (μg/100 g) in the human diet, the estimation of iodine was carried by the arsenic-cerium redox method (Brown and Hutchinson 1949). Seventeen hybrids viz. Cabbage Hy 1, Cabbage Hy 2, Cabbage Hy 3, Cabbage Hy 4, Cabbage Hy 5, Cabbage Hy 6, Quisto, Kranti, DARL-801, DARL-802, CH 21, Green Flash, SIR, CH 2200, Speed 50, Krishna and Varun were collected from different public and private seed companies and were grown in an open field in randomized block design in three replications. The marketable heads were selected for iodine estimation on a dry weight basis in three replications. The head of each hybrid was cut into small pieces and oven-dried at 40°C and then ground into a fine powder. Iodine estimation was based on the principal of quantitative determination of micro amounts of iodine on a catalytic reduction of microelements of ceric (Ce^{+4}) to cerous (Ce^{+3}) ions by iodine. A suitable amount of dried sample (usually containing 0.04 to 0.08 μg of the iodine) was taken in a test tube (15 × 125 mm). After digestion, incineration and extraction, the reduction of ceric to cerous ions is read in a spectrophotometer at 420 nm. A standard solution of KI (potassium iodide) containing 0.0 to 0.16 μg of iodine per milliliter was run simultaneously. A straight line response is obtained by plotting concentration of iodine in μg against reading on a spectrophotometer. Using this standard graph, the values for any unknown sample are read.

1.2.7 Estimation of Macro, Micro and Carotene in Cabbage Hybrids

The macro, micro and carotene studies were also undertaken in cabbage hybrids. Cabbage is a source of sodium and potassium, an essential mineral that helps to regulate the body's balance of fluid. Calcium is an essential mineral for the human body. It is a source of calcium needed for the formation of protein and bone. An iron deficiency leads to general anemia while copper causes cardiovascular malfunction and bone disorders. Biologically active minerals like sodium (Na), potassium (K) and calcium (Ca) were estimated through a flame photometer and zinc (Zn), manganese (Mn), copper (Cu) and iron (Fe) were estimated through a spectrophotometer. Cabbage contains β-carotene, which exhibits great antioxidant activity. β-Carotene was estimated through a spectrophotometer and the results were expressed as μg/100g, RE (retinol equivalents) and recommended daily allowance (RDA%). All qualitative estimations were on dry weight basis.

Statistical analysis was carried out according to Gomez and Gomez (1984). Significant variation in all parameters was observed among different genotypes of cabbage. The significance of the treatment effect was determined using a F-test and to determine the significance of the two treatments CD at 1% and 5% probability level was used.

1.3 RESULTS AND DISCUSSION

Ascorbic acid, also known as vitamin C, is an antiscorbutic and is very important for health. It is required for functional activities, such as the formation of collagen fibers and mucopolysaccharides of connective tissues, osteoid tissue and dentin (Fain 2004; Iqbal et al. 2004). More than 85% of vitamin C in the human diet is supplied by fruits and vegetables (Davey et al. 2000; Lee and Kader 2000). Ascorbic acid ranged from 40.00 to 75.83 mg/100 g in the outermost leaves, 35.83 to 71.66 mg/100 g in the outer middle leaves, 28.50 to 52.50 mg/100 g in the inner middle leaves and 19.33 to 45.33 mg/100 g in the innermost leaves (Table 1.1). Genotype C5 showed maximum ascorbic acid content (75.83 mg/100 g) in the outermost leaves. There was a decrease in the ascorbic acid content in the outer middle leaves (71.66 mg/100 g), inner middle leaves (52.50 mg/100 g) and innermost leaves (45.33 mg/100 g). All nine genotypes

TABLE 1.1
Ascorbic Acid (mg/100 g)

Genotype	Outermost	Outer Middle	Inner Middle	Innermost
C1	58.33	41.66	28.50	19.33
C3	70.73	52.50	32.50	30.00
C5	75.83	71.66	52.50	45.33
C7	75.00	51.24	50.00	45.00
SEL 1	66.42	57.14	46.66	42.85
SEL 7	39.93	41.89	45.70	50.09
DARL 851	46.18	36.18	31.42	30.47
DARL 852	27.50	35.00	35.83	40.00
DARL SEL	51.66	37.33	35.83	25.00
SEM	2.695	2.159	1.857	2.656
CD at 1%	11.13*	8.92*	7.67*	10.97*
CD at 5%	8.08*	6.47*	5.57*	7.96*
SD	4.56	3.21	3.74	4.66

*Significant at P = 0.01.

showed the similar trend of a decrease in the ascorbic acid content, with an increased internal positioning of leaves. The content of vitamin C varies significantly within cultivars of *Brassica* vegetables (Podsedek 2007). The effects of ascorbic acid on white cabbage was also studied (Singh et al. 2007).

β-Carotene is the dominant carotene in Brassicas. It is the precursor of vitamin A. It ranged from 52.30 to 192.95 mg/100 g in the outermost leaves, 17.42 to 189.77 mg/100 g in the outer middle leaves, 13.85 to 126.97 mg/100 g in the inner middle leaves and 9.76 to 79.00 mg/100 g in the innermost leaves of different cultivars. Genotype C7 exhibited maximum amounts of β-carotene. However, β-carotene decreased with the increased internal positioning of leaves in the cabbage head viz. 192.95 mg/100 g in the outermost leaves, 189.77 mg/100 g in the outer middle leaves, 126.97 mg/100 g in the inner middle leaves and 79.00 mg/100 g in the innermost leaves. All genotypes showed the similar trend of decreasing of β-carotene from the outer side to the inner side (Table 1.2). Singh et al. (2006) also studied β-carotene for white cabbage.

The total amount of chlorophyll also decreased from the outermost leaves to the innermost leaves in all genotypes. In the outermost leaves, it ranged from 1.50 to 14.60 mg/100 g. In the outer middle leaves, its range was 0.76 to 6.49 mg/100 g. In the inner middle leaves it ranged from 0.75 to 1.82 mg/100 g. The range of total chlorophyll was from 0.55 to 1.27 mg/100 g in the innermost leaves. Genotype C1 exhibited maximum contents of total chlorophyll (14.60 mg/100 g) in the outermost leaves, (6.49 mg/100 g) in the outer middle leaves, (1.82 mg/100 g) in the inner middle leaves and (1.27 mg/100 g) in the innermost leaves (Table 1.3).

The chlorophyll and β-carotene contents were highest in the outermost leaves of the cabbages. The reason behind it may be that the outermost leaves are more exposed to sunlight during photosynthesis aiding the formation of chlorophyll.

TABLE 1.2
β-Carotene (mg/100 g)

Genotype	Outermost	Outer Middle	Inner Middle	Innermost
C1	91.75	19.20	13.85	11.05
C3	68.00	27.25	17.38	10.45
C5	84.82	17.42	17.65	23.68
C7	192.95	189.77	126.97	79.00
SEL 1	63.13	25.25	24.56	12.27
SEL 7	48.37	37.34	23.15	22.40
DARL 851	87.96	34.83	23.30	15.81
DARL 852	52.30	30.00	14.93	13.36
DARL SEL	66.84	19.55	17.08	9.76
SEM	0.720	2.03	0.636	0.662
CD at 1%	2.97*	8.38*	2.62*	2.73*
CD at 5%	2.16*	6.08*	1.90*	1.98*
SD	1.24	3.51	1.10	1.14

*Significant at P = 0.01.

TABLE 1.3
Total Chlorophyll (mg/100 g)

Genotype	Outermost	Outer Middle	Inner Middle	Innermost
C1	14.60	6.49	1.82	1.27
C3	5.68	2.21	1.38	0.55
C5	7.46	3.24	2.16	1.98
C7	4.67	3.08	0.75	0.68
SEL 1	3.96	1.70	1.10	0.67
SEL 7	2.81	0.79	0.71	0.62
DARL 851	5.35	2.34	0.74	0.59
DARL 852	1.50	0.76	0.63	0.34
DARL SEL	2.06	1.80	1.60	1.51
SEM	0.01	0.06	0.00	0.02
CD at 1%	0.04*	0.25*	0.02*	0.08*
CD at 5%	0.03*	0.18*	0.02*	0.06*
SD	0.017	0.106	0.010	0.034

*Significant at P = 0.01.

Chlorophyll is recognised as the health-promoting phytochemical. The quantity of chlorophyll is directly related to the photosynthetic rate, which is important for all life on Earth. Carotene, also known as an accessory pigment, protects the chlorophyll from oxidation in the presence of sunlight and also absorbs some light and transfers it to chlorophyll for photosynthetic activity.

Free radicals are highly reactive and have deleterious effects on the body's organs. Vegetables, being a rich source of antioxidants, are responsible for many health benefits. Total antioxidant activity (IC_{50}) was ranged from 2.02 to 2.60 in the outermost leaves, 2.02 to 2.69 in the outer middle leaves, 2.20 to 3.47 in the inner middle leaves and 2.20 to 3.33 in the innermost leaves. Genotype C1 exhibited maximum antioxidant activity (IC_{50}) in the outermost leaves (2.03) (Table 1.4). IC_{50} value decreased with increasing external positioning of the leaves. Its value was highest in the innermost leaves (2.20). A low IC_{50} value is the sign of strong antioxidant activity. Hence, the outermost leaves showed the highest amount of antioxidant activity. Kim et al. (2004) studied the antioxidant potential often varieties of green cabbage and observed a significantly higher level of total phenolics and antioxidants in the 'Fresco' and 'Bobcat' varieties. Podsedek et al. (2006) measured the antioxidant potential of different crops of *Brassica oleracea*, including red, white and Savoy cabbage and Brussels sprouts. It was found that red cabbage and Brussels sprouts had higher antioxidant potential than white and Savoy cabbages. Cabbages from Belgium displayed the highest antioxidant potential, while the lowest ones were from Poland. Singh et al. (2007) studied the variability of carotenes, vitamin C, vitamin E and phenolics in *Brassica* vegetables. Results indicated that the cruciferous vegetables are a relatively good source of abundant antioxidants and there was a substantial and significant variation, both within and between the subspecies for the antioxidant phytochemicals. Eighteen different cultivars were analysed for the antioxidant phytochemicals in cabbage. The vitamin C content ranged from 5.70 to 23.50 mg/100 g. Cultivar Sprint Ball recorded maximum vitamin C content (23.5 mg/100 g) followed by Gungaless (12.9 mg/100 g). The β-carotene content in cabbage ranged from 0.01 to 0.12 mg/100 g. The maximum β-carotene was recorded in the cultivar Quisto.

TABLE 1.4
Antioxidant Activity (IC_{50} Value)

Genotype	Outermost	Outer Middle	Inner Middle	Innermost
C1	2.06	2.07	2.32	2.39
C3	2.60	2.81	2.96	3.27
C5	2.19	3.22	3.29	3.33
C7	2.27	3.02	3.47	2.80
SEL 1	2.03	2.03	2.20	2.20
SEL 7	2.03	2.44	2.45	2.45
DARL 851	1.99	2.44	2.44	2.49
DARL 852	2.02	2.18	2.21	2.40
DARL SEL	2.22	2.69	2.79	3.14
SEM	0.57	0.023	0.19	0.06
CD at 1%	0.23*	0.98*	0.79*	0.26*
CD at 5%	0.17*	0.77*	0.57*	0.19*
SD	0.112	0.334	0.041	0.019

*Significant at P = 0.01.

Fernandez-Leon et al. (2012) assessed the antioxidant phytochemicals in the Savoy cabbage. The contents of the main phytochemical compounds measured in Savoy cabbages were 49.06 mg/100 g fresh weight of vitamin C, 0.37 mg/100 g fresh weight of β-carotene and 102.71 mg/100 g fresh weight of total phenolics.

The iodine content of the different cabbage hybrids was cited in Table 1.5. There were significant differences in the iodine content among the hybrids grown in the Middle Hill climatic conditions of Uttarakhand Himalayas. The concentration range and mean of seventeen samples were 1.60 to 7.41 and 4.55, respectively. Antioxidant activity was also estimated and found 1.23 μg/100 g and 5.59 μg/100 g (mean values) respectively, on the basis of the iodine concentration hybrid Varun, which recorded 7.41 (IC_{50}), followed by Green Flash (6.40 IC_{50}). Cabbage hybrid 3 exhibited antioxidant activity of 1.20 IC_{50} value, followed by Cabbage Hy 4 (1.45 IC_{50}) (Figure 1.3).

By estimating the iodine concentration in soil and water, it will be helpful for us to judge whether this area is sufficient in iodine or not. Iodine concentration in water and soil reflects the environmental iodine distribution and is also an important index of the natural iodine intake of the surrounding human population. Iodine content in water and soil is directly associated with the content of iodine in food and affects both the mortality rate and quality of life of human populations. The iodine

TABLE 1.5
Iodine Concentration in Cabbage Hybrids

Cabbage Hybrids	Iodine (μg/100 g)	Antioxidant Activity (IC_{50}) Value
Cabbage hy 1	5.34	5.40
Cabbage hy 2	2.73	1.74
Cabbage hy 3	4.29	1.20
Cabbage hy 4	1.75	1.45
Cabbage hy 5	4.51	2.70
Cabbage hy 6	4.95	2.60
Quisto	4.78	2.31
Kranti	3.33	2.30
DARL 802	4.00	1.73
DARL 801	5.80	2.50
CH 21	5.94	2.37
Green Flash	6.40	2.94
SIR	6.20	2.37
CH 2200	6.41	2.08
Speed 50	2.91	2.93
Krishna	1.60	2.02
Varun	7.41	2.10
CD @ 1%	1.62	0.21*
CD @ 5%	1.21	0.84
SD	0.18	0.036

*Significant at P = 0.01.

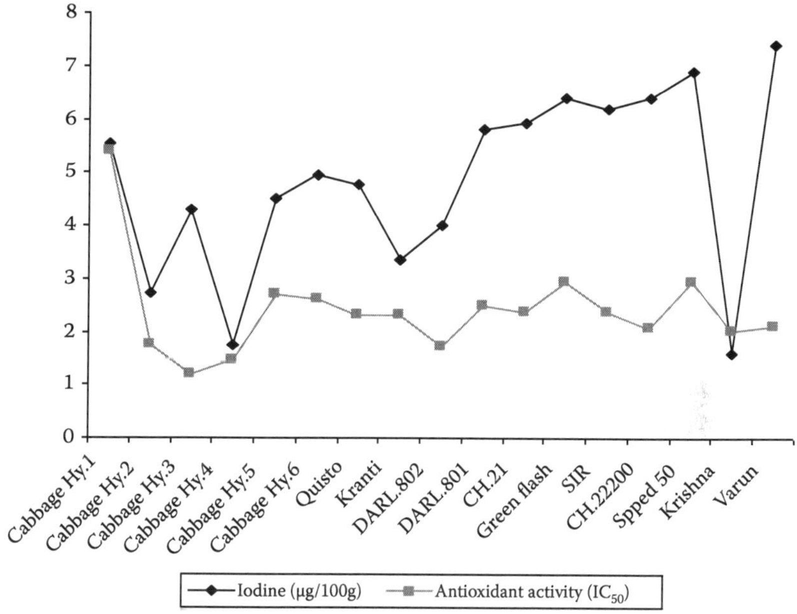

FIGURE 1.3 Iodine concentration (μg/100 g) and antioxidant activity (IC_{50}) in cabbage hybrids.

estimation of soil and water was undertaken on the assumption that the iodine content of the water was related in some way to the iodine content of the plant. Since cabbages consists of leaves, the soaked water is evaporated from the leaves of the plants and hence, the iodine is concentrated in the leaves. The range of the iodine content in water, soil and cabbages in the Middle Hill region was 1.10 to 1.36 μg/100 g, 3.46 to 7.72 μg/100 g and 1.59 to 7.41μg/100 g, respectively (Figure 1.4).

Food, soil and water are necessary for all forms of life. Some ingredients in soil and water play an important role in human development. Iodine has long been

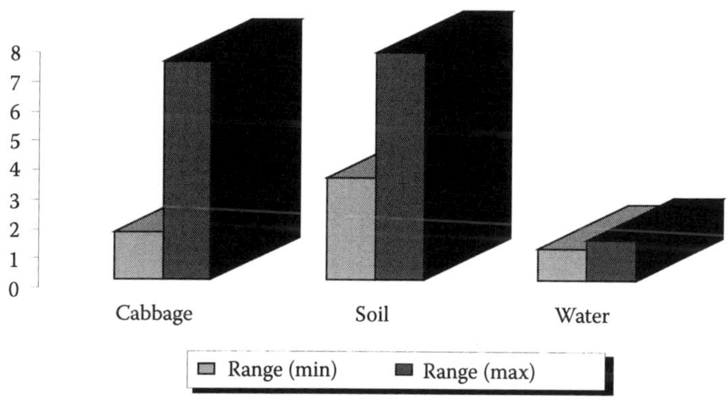

FIGURE 1.4 Iodine content in cabbage, soil and water.

recognised as an essential micronutrient for humans and livestock. Among the most common causes of iodine deficiency is the intake of iodine deficient food. Cabbage is rich in iodine. On the basis of this study, the above-cited iodine deficiency can be cured by introducing an appropriate amount of cabbage in regular diet. Besides the many iodine supplements available on the market, cabbage can be used as one of the cheapest and richest sources of iodine.

There were significant differences in all qualitative characters studied among the hybrids grown in the Middle Hill climatic conditions of the western Himalayas (Table 1.6). The β-carotene (μg/100 g) among hybrids ranged from 80.71 to 220.71. Hybrid Quisto exhibited the highest levels of β-carotene (220.71), while hybrid Krishna had the lowest levels (80.71). The provitamin A ranged from 14.42 to 37.52 RE/100 g. Provitamin A (RDA%) ranged from 1.37 to 3.75.

The value of sodium (mg/100 g) ranged from 19.53 to 70.21. Highest sodium was found in hybrid DARL-801 followed by Cabbage Hy 2 and Cabbage Hy 3. Potassium (mg/100 g) ranged lowest from 375.50 in Cabbage Hy 3 and highest in Cabbage Hy 1 (733.60). Mineral Calcium (mg/100 g) ranged from 71.42 to 214.26. Hybrid Speed 50 exhibited the maximum calcium content (214.26) followed by Cabbage Hy 2 (142.84).

The value of zinc ranged from 1.55 to 3.70 mg/100 g. Hybrid Cabbage Hy 4 exhibited the highest zinc content (5.75 mg/100 g) followed by SIR (3.70 mg/100 g). The maximum manganese content was exhibited by Cabbage Hy No. 1 (6.05 mg/100 g) followed by Cabbage Hy 3 (2.75 mg/100 g). Regarding copper content (mg/100 g), Cabbage Hy 2 was the highest-performing (6.85 mg/100 g) followed by DARL-801 (5.20 mg/100 g). Iron content varied from 4.50 to 28.55. DARL 801 exhibited the highest iron content (28.55 mg/100 g) followed by DARL 802 (22.25 mg/100 g).

1.3.1 Consumption Tips

1. Smaller size heads are tastier. It is advisable to choose from smaller varieties when buying cabbage.
2. Always buy cabbages with a complete head. Avoid buying precut cabbage that is either halved or precut. After cutting, it begins to lose its nutrient contents.
3. Keep the cabbage refrigerated in a perforated plastic bag to prevent the loss of vitamin C.
4. Cabbage consumption may cause gas formation or flatulence in some people, especially when cooked. In this case, add turmeric powder or season with cumin seeds.
5. Cabbage is commonly used to prepare sauerkraut, which is a fermented product.
6. Cabbage can be freeze-dried or canned after shredding, but these products have a lower quality than fresh products.
7. Cabbage soup is helpful for losing weight. A substance in cabbage inhibits the conversion of sugar and other carbohydrates into fat, which proffers a painless way of dieting.

TABLE 1.6
Quality Parameters in Cabbage Hybrids

Cabbage Hybrids	β Carotene (g/100 g)	Provitamin A (RE/100 g)	Provitamin A (RDA%)	Na (mg/100 g)	Ca (mg/100 g)	K (mg/100 g)	Zn (mg/100 g)	Mn (mg/100 g)	Cu (mg/100 g)	Fe (mg/100 g)
Cabbage hy 1	84.83	14.13	1.41	38.13	89.27	733.60	1.65	6.05	4.75	11.4
Cabbage hy 2	145.38	24.23	2.42	66.03	142.84	426.70	1.85	2.50	6.85	13.7
Cabbage hy 3	157.30	26.21	2.62	66.03	107.13	375.50	2.35	2.75	2.65	33.9
Cabbage hy 4	122.85	20.47	2.05	40.45	71.42	429.70	5.75	2.40	1.85	10.3
Cabbage hy 5	199.18	33.20	3.32	31.62	107.13	495.40	1.65	2.40	2.15	17.5
Cabbage hy 6	104.30	17.38	1.74	38.59	214.25	445.70	1.95	1.60	3.55	11.05
Quisto	220.71	37.52	3.75	47.43	124.98	400.40	2.10	2.30	3.40	13.6
Kranti	194.83	36.69	3.66	34.87	71.42	441.30	2.45	2.45	2.85	11.9
DARL 802	133.21	22.20	2.20	70.21	124.98	469.10	1.70	1.65	3.75	22.25
DARL 801	158.18	26.36	2.63	26.02	74.42	472.00	1.60	2.00	5.20	28.55
CH 21	135.70	22.62	2.26	35.80	71.42	425.25	1.55	1.95	5.15	10.90
Green Flash	147.00	24.50	2.45	30.69	89.27	420.90	2.00	2.05	4.70	19.10
SIR	142.81	23.80	2.38	46.96	160.69	369.70	3.70	2.15	4.50	27.45
CH 2200	129.29	21.54	2.15	19.53	71.42	406.20	2.10	1.70	4.50	4.50
Speed 50	126.45	21.08	2.11	141.30	214.26	412.10	2.65	1.40	4.35	6.20
Krishna	80.71	13.45	1.35	31.15	71.42	448.60	1.55	1.00	2.95	4.65
Varun	108.30	18.05	1.81	38.59	71.42	439.80	1.75	1.75	3.00	5.25
SEM±	3.87			0.58	10.45	8.71	0.571*	0.51*	0.67*	3.37*
CD @ 1%	15.01*			2.64*	40.50*	33.92*	2.21	2.01	2.62	13.08
CV	4.77			2.50	16.42	3.37	43.85	40.26	30.10	39.42
SD	6.33			1.12	17.12	14.25	0.024	0.89	1.03	6.38

Note: Provitamin A (RE/100g) based on 6 μg β-carotene = 1RE; Provitamin A (RDA%) based 1000 RE/100 g.
*Significant at P = 0.01.

8. Daily and frequent consumption of cabbage juice is effective in preventing and treating cancers of the breast, colon, liver, lung and ovary. The juice should be taken in small amounts of about 100 mL, three times a day, on an empty stomach.
9. Cabbage has slightly laxative effect, which is helpful in stimulating bowl movement. Glutamine amino acid in cabbage is gentle and cleansing on the digestive system.
10. Red cabbage is a rich source of anthocyanin pigments. This anthocyanin pigment acts as strong antioxidant. The red pigment is used as a natural food colorant. This color is recommended for use in beverages, chewing gums, candies, sherbets, dressings, yogurt and other fermented products.

1.4 CONCLUSION

The world's most urgent need is to increase the production and consumption of nutritious food, as billions of people suffering from severe malnutrition could potentially be saved. The production of good quality vegetables is the primary factor of commercial vegetable cultivation for better economic returns. It is chemical composition that plays the most crucial role in determining the quality of vegetables that is acceptable for consumption. The importance of a food-based approach for preventing micronutrient malnutrition has become widely accepted. Cabbage plays a very important role in the human diet, being the cornerstone of health and supplying us with a wealth of vitamins, minerals, fibres and carbohydrates. Therefore, it has assumed utmost importance after the discovery of phytochemicals and their strong antioxidant potential in scavenging free radicals. The antioxidant compounds viz. total chlorophyll, β-carotene, ascorbic acid and free radical scavenging activity in different leaf positions of white cabbage were studied. The outermost leaves had the highest total chlorophyll, β-carotene, ascorbic acid and free radical scavenging activity. Results indicated that white cabbage is a good source of antioxidants. There was significant variation among the nine cultivars studied. It may be due to many factors such as cultivar, maturity at harvest, growing conditions and the soil state. The knowledge of genetic variation will help devise improved breeding strategies for the production of high-quality cabbage cultivars with enhanced functional properties for the benefit of consumers and breeders. Most people remove the second or third outermost leaves of the cabbage and use the rest when cooking. As evident from this study, the outermost leaves possess the largest amount of nutraceuticals. So, the outermost leaves should not be removed from the head of cabbage. After thorough washing, they can be safely used for cooking. The nutritive value of cabbage is enhanced greatly by the presence of minerals and vitamins. Due to the presence of iodine in cabbage, this vegetable has become very important. Iodine is absent in the soil of many Himalayan and alluvial regions, when high land pressure has eroded the topsoil and the iodine content with it. There is potential for producing good-quality cabbage in the hills of western Himalayas. The productivity of cabbage is much higher; with an average yield potential of 500–600 q/ha, especially if they are grown from hybrid seeds and high-yielding varieties. It is therefore critical to increase its production through hybrids/varieties for global nutritional security.

REFERENCES

A.O.A.C. 1980. Official Method of Analysis. Washington, DC, USA.

A.O.A.C. 1990. Official Method of Analysis. Washington, DC, USA.

Appleby, P. N., Thorogood, M., Mann, J. I., and Key, T. J. 1999. The Oxford vegetarian study: An overview. *American Journal of Clinical Nutrition* 70 (3): 525S–531S.

Boyettte, M. D., Sanders, D. C., and Estes, E. A. 1999. Postharvest cooling and handling of cabbage and leafy vegetables. In: *Postharvest Commodity Series*, edited by North Carolina State University, 413–415. Department of Biological and Agricultural Engeneering Publication AG.

Brown, H. D. and Hutchinson, C. 1949. *Vegetable Science*. J. B. Lippincot Company, New York.

Chu, Y. F., Sun, J., Wu, X., and Liu, R. H. 2002. Antioxidant and antiproliferative activities of common vegetables. *Journal of Agricultural and Food Chemistry* 50, 6910–6916.

Cohen, J., Kristal, R., and Stanford, J. 2000. Fruit and vegetable intakes and prostate cancer. *Journal of the National Cancer Institute* 9, 61–68.

Davey, M. W., Van Motangu, M., Inze, D., Sanmartin, M., Kanellis, A., and Smirnoff, N. 2000. Plant L-ascorbic acid; chemistry, function bioavilability and effects of processing. *Journal of the Science of Food and Agriculture* 80, 825–860.

Dickson, M. H. and Wallace, D. H. 1996. Cabbage Breeding. In: *Breeding Vegetable Crops*. Ed. Mark J. Basset. The AVI Publishing Company, INC. Westport, Connecticut. 395–432.

Duffy, S. J., Gokse, N., Holbrook, M., Huang, A., Frei, B., Keaney, J. F. J., and Vita, J. A. 1999. Treatment of hypertension with ascorbic acid. *Lancet* 354, 2048–2049.

Fain, O. 2004. Vitamin C deficiency. *La Revue de Médecine Interne* 25, 872–880.

Fernandez-Leon, M. F., Fernandez-Leon, A. M., Lozano Ruiz, M., and Gonzalez-Gomez, D. 2012. Antioxidant phytochemicals in Savoy cabbage (*Brassica oleracea* L. var. *sabuda* L.) Proc. XXVIIIth IHC-IS on emerging health topics in Fruits and vegetables. *Acta Horticulturae* 939, ISHS. 295–300.

Gomez, K. A. and Gomez, A. A. 1984. *Statistical Procedure for Agricultural Research* (2nd Edition). John Wiley & Sons, New York.

Gosslau, A. and Chen, K. Y. 2004. Nutraceuticals, apoptosis and disease prevention. *Nutrition* 20, 95–102.

Hashimoto, K., Kawamata, S., Usui, N., Tanaka, A., and Uda, Y. 2002. *In vitro* induction of the anticarcinogenic marker enzyme, quinine reductase, in human hepatoma cells by food extracts. *CancerLit* 180, 1–5.

Hatano, T., Edamatsu, R., Hiramatsu, M., Mori, A., Fujita, Y., and Yashuhara, A. 1988. Effects of tannins and related polyphenols on superoxide anion radical and on DPPH radical Chem. *Chemical and Pharmaceutical Bulletin* 37, 2016–2021.

Iqbal, M. P., Kazim, S. F., and Ali, N. M. 2004. Biological significance of ascorbic acid (vitamin C) in human health: A review. *Pakistan Journal of Nutrition* 3, 5–13.

Jagdish, S., Upadhyaya, A. K., Singh, S., and Rai, M. 2009. Total phenolic content and free radical scavenging activity of Brassica vegetables. *Food Science and Technology* 46, 595–597.

Jeffery, E. H., Brown, A. F., Kurilich, A. C., Keek, A. S., Matusheski, N., Klein, B. P., and Juvik, J. A. 2003. Variation in content of bioactive components in Broccoli: Study review. *Journal of Food Composition and Analysis* 16, 323–330.

Kavanaugh, C. J., Trumbo, P. R., and Ellwood, K. C. 2006. The U.S. Food and Drug Administration's evidence-based review for qualified health claims: Tomato, lycopene and cancer. *Natural Cancer Institute* 99, 1074–1085.

Key, T. J., Thorogood, M., Keenan, J., and Long, A. 1992. Raised thyroid stimulating hormone associated with kelp intake in British vegan men. *Human Nutrition and Diet* 5, 323–326.

Khanzadi Fatima, K. 2011. Nutrient composition, phenolic content and free radical scavenging activity of some uncommon vegetables of Pakistan. *Pakistan Journal of Pharmaceutical Sciences* 24 (3), 277–283.

Kim, D. O., Padilla-Zakour, O. I., and Griffiths, P. D. 2004. Flavonoids and antioxidant capacity of various cabbage genotypes at juvenile stage. *Food Science* 69, C685–C689.

Kusznierewicz, B., Bartoszek, A., Wolska, L., Drzewiecki, J., Gorinstein, S., and Namiesnik, J. 2008. Partial characterization of white cabbages (*Brassica oleracea* var *capitata*) from different regions by glucosinolates, bioactive compounds, total antioxidant activities and proteins. *LWT-Food Science and Technology* 41, 1–9.

Lee, S. K. and Kadar, A. A. 2000. Preharvest and postharvest factors influencing vitamin C content of horticultural crops. *Postharvest Biology and Technology* 20, 207–220.

Lee, Y., Lee, H. J., Lee, H. S., Jang, Y. A., and Kim, C. 2008. Analytical dietary fiber database for the national health and nutrition survey in Korea. *Journal of Food Composition and Analysis* 21, S35–S42.

Murkovic, M., Gams, K., Draxi, X., and Pfannhauser, W. 2000. Development of an Austrian carotenoid database. *Journal of Food Composition and Analysis* 13, 435–440.

Nieuwhof, M. 1969. Cole Crops: Botany, Cultivation and Utilization. *World Crop Series.* Leonard Hill. London, UK, PP353.

Ohr, L. M. 2004. Dietary antioxidants. *Food Technology* 58, 67–74.

Podsedek, A., Sosnowska, D., and Redzynia, M., Anders, B. 2006. Antioxidant capacity and content of *Brassica oleracea* dietary antioxidants. *International Journal of Food Science and Technology* 41, 49–58.

Podsedek, A. 2007. Natural antioxidant capacity of *Brassica* vegetables: A review. *LWT-Food Science and Technology* 40, 1–11.

Puupponen-Pimia, R., Hakkinen, S. T., Aarni, M., Suortti, Lampi, A. M., Eurola, M., Piironen, V., Nuutila, A. M., and Oksman-Caldenteye, K. M. 2003. Blanching and long-term freezing affect various bioactive compounds of vegetables in different ways. *Journal of the Science of Food and Agriculture* 83, 1389–1402.

Proteggente, A. R., Pannala, A. S., Paganga G., Van Buren, L., Wagner, E., Wiseman, S., Van de Put, F., Dacombe, C., and Rice-Evans, C. A. 2002. The antioxidant activity of regularly consumed fruits and vegetables reflects their phenolic and vitamin C composition. *Free Radical Research* 36, 217–233.

Singh, J., Upadhyaya, A. K., Bahadur, A., Singh, B., Singh, K. P., and Mathura, R. 2006. Antioxidant phytochemicals in cabbage. (*Brassica oleracea* L. var *capitata*) *Scientia Horticulturae* 108, 233–237.

Singh, J., Upadhyaya, A. K., Prasad K., Bahadur, A., and Rai, M. 2007. Variability of carotenes, vitamin C, E and phenolics in *Brassica* vegetables. *Journal of Food Composition Analysis* 20, 449–459.

Sun, J., Chu, Y. F., Wu, X., and Liu, R. H. 2002. Antioxidant and antiproliferative activities of common fruits. *Journal of Agricultural and Food Chemistry* 50, 7449–7454.

Thrombino, S., Serini, S., Di Nicuolo, F., Celleno, L., Andò, S., Picci, N., Calviello, G., and Palozza, P. 2004. Antioxidant effect of ferulic acid in isolated membranes and intact cells; Synergistic interactions with β-tocopherol, β-carotene and ascorbic acid. *Journal of Agricultural and Food Chemistry* 52, 2411–2420.

Van der Berg, H., Faulks, R., Granado, F. H., Hirschberg, Olmedilla, and Sandmann, B. 2000. The potential for the improvement of carotenoid levels in foods and the likely systematic effects. *Journal of Agricultural and Food Chemistry* 80, 880–912.

Witham, F. H., Blaydes, D. F., and Develin, R. M. 1971. *Experiments in Plant Physiology.* Van Nostrand, New York. 245.

2 Phytonutrients
Their Relevance to Human Health

Ranganathan Kumar, Sulaxana Kumari Chauhan, Subramanian Vijayalakshmi and Shanmugam Nadanasabapathi

CONTENTS

2.1 Introduction 18
2.2 Classification of Phytonutrients 19
  2.2.1 Carotenoids 20
  2.2.2 Phenolic Compounds/Polyphenols 21
  2.2.3 Glucosinolates 24
  2.2.4 Organosulphur Compounds 24
  2.2.5 Nondigestible Carbohydrates 25
  2.2.6 Terpenes 25
  2.2.7 Betalains 27
  2.2.8 Alkaloids 27
  2.2.9 Capsaicinoids 28
  2.2.10 Chlorophyll 28
2.3 Sources of Phytochemicals 28
  2.3.1 Broccoli and Other Cruciferous Vegetables 28
  2.3.2 Oats 29
  2.3.3 Flaxseed (Linseed) 30
  2.3.4 Tomatoes 30
  2.3.5 Soybeans 31
  2.3.6 Citrus 32
  2.3.7 Berries 33
  2.3.8 Tea 33
  2.3.9 Grapes and Wine 34
  2.3.10 Garlic 35
2.4 Prospect for Developing Countries 36
References 37

2.1 INTRODUCTION

An assortment of foods plays a vital role in maintaining the normal function of the human body by providing adequate and necessary nutrients. With the advancement in recent years in the medical and nutrition sciences, natural products and health-promoting foods have received widespread interest from both health professionals and the general population. The new concept of functional foods raises apprehension in relevance to issues of food security and diet design, beyond the conventional benefit of providing adequate energy, essential fats, protein, vitamins and minerals. In recent times, it has been recognized that foods not only provide basic nutrition, but also help in preventing diseases, assuring good health and longevity. Almost every nation has traditional folk medicines or folk remediation with medicinal plants. Plants used as natural medicines to benefit humans have a long history, particularly in countries such as China, India, Egypt and Greece. The idea in these concepts goes back three thousand years ago. Hippocrates, well-known as the father of modern medicine, stated 'Let food be thy medicine and medicine be thy food' (Lucock 2004, p. 214), signifying the pivotal relationship between appropriate foods for health and their therapeutic benefits. The truth in this saying is widely recognized today. Thousands of years of using plants have accumulated abundant information worldwide about use of medicinal plants and their functions or toxicities. The secondary metabolites from plant origin provide humans with numerous biologically active products, which have been used extensively as food additives, preservatives, flavours, colours, insecticides, drugs, fragrances and other fine chemicals. These plant secondary metabolites include several classes like terpenoids, flavonoids and alkaloids, which have a diverse chemical structures and biological activities, and also exist widely in different crops such as rice, wheat, maize and soybeans. Therefore, these natural compounds consumed as dietary components have a considerable impact on human health. Plant foods rich in dietary fibre provide well-addressed benefits to humans, such as improving digestive system health. New concepts like nutraceuticals, nutritional therapy, phytonutrients and phytotherapy have appeared with this trend. These phytonutrients (functional foods) or phytomedicines (medicinal foods) play positive roles in maintaining well-being, enhancing health and modulating immune response to avert specific diseases. They also hold great promise in clinical therapy due to their potential in reducing side effects associated with chemotherapy or radiotherapy and significant advantages in reducing healthcare cost.

Ever since the establishment of the industrial era, the lifestyles of human beings have dramatically changed. The increasing pace of life and work, with longer working schedules and the amalgamation of various psychological cultures, has resulted in an increased intake of instant and tasty meals, but with decreased quantity and quality in nutrients. At the same time, industrialisation has caused air and water pollution, which has contaminated food and soil, because of the extensive use of various chemicals, heavy metals, electromagnetic waves and other potentially harmful man-made products. These issues have eventually led to an increased occurrence of diabetes, obesity, physiological problems, various cancers and vascular diseases, as well as other degenerative diseases. The raised demands for healthcare have dramatically augmented the medical care expense. Now, people have realized that a healthy body is more important than money or anything else in life. Hence, people have aimed to achieve a better quality of life by eating more fruits and vegetables, taking dietary supplements or nutraceuticals or using nutritional therapy or phytotherapy to replace chemotherapy or radiotherapy. Scientific studies also have expanded to these areas and have given support to many of nutraceutical foods. Furthermore, plant biotechnologists have put lots of effort to engineer plants and crops in order to improve their nutritional value.

Various terms have been used interchangeably to designate foods for disease deterrence and health enhancement. 'designer foods', coined in 1989, is used to describe foods that are present naturally or enriched with nonnutritive and biologically active chemical components of plants that are effective in reducing cancer risk. 'Nutraceuticals' was introduced in 1989 by the U.S. Foundation for Innovation in Medicine to refer to "any substance that is a food or a part of a food which provides medical or health benefits, including the prevention and treatment of disease" (Brower 1998, p. 728). The U.S. Institute of Medicine's Food and Nutrition Board defined *functional foods* in 1994 as 'any food or food ingredient that may possibly provide a health benefit beyond the traditional nutrients it contains' (Hasler 2002, p. 3773).

Phytonutrients literally mean 'plant nutrients'. These phytonutrients are a large group of plant-derived compounds with particular biological activities in supporting human health, hypothesized to be responsible for much of the disease protection conferred from diets high in fruits and vegetables. Scientists have identified thousands of different phytonutrients, found in vegetables, fruits, beans, whole grains, nuts and seeds. This concept primarily refers to the phytonutrients which act as modifiers of physiological function. When compared with concepts like functional foods, dietary supplements and nutraceuticals, phytonutrients emphasise the natural bioactive compounds from plants that provide general health benefits to humans more specifically, which become closer to or even overlap with phytomedicines. The wide application of phytonutrients indicates that nutrition science has advanced beyond the treatment of deficiency syndromes to reduce disease risk. No longer are food nutrients evaluated only in terms of macronutrient and micronutrient levels, as contents of some biologically active compounds are becoming more essential.

Several factors – scientific advances, consumer demand, increasing health care expenditure, an aging population, technical advances in the food industry and changing regulatory norms – have inspired the field of functional foods. Scientific investigations have resulted in the accretion of scientific substantiation supporting the vital role of diet in overall health and well-being. For instance, six of the ten leading causes of death in world – cancer, coronary heart disease, stroke, diabetes, atherosclerosis and liver disease – have been increasingly revealed to be related to diet.

2.2 CLASSIFICATION OF PHYTONUTRIENTS

Hundreds of Phytonutrients, with several different biological functions, have been identified in plant-based foods in the last decade. The evidence from epidemiological studies, biological, experimental studies and clinical intervention trials have demonstrated that a plant-based diet can reduce the risk of degenerative diseases, especially cardiovascular disease and cancer (Block et al. 1992; Ames et al. 1993; Steinmetz and Potter 1996; World Cancer Research Fund 1997; Ness and Powles 1997; Law and Morris 1998; Kaur and Kapoor 2001). It is estimated that plant-based diets prevent 20–50% of all types of cancer (Steinmetz and Potter 1996; WCRF 1997). Thus, dietary recommendations for the prevention of cancer and other chronic diseases have always emphasized the consumption of a variety of plant foods. The single composite approach has given way to the concept that overall protection against disease is provided by a range of phytonutrients contained in foods.

2.2.1 Carotenoids

Among the phytochemicals, carotenoids have been the widely studied ones. Carotenoids are tetraterpenoids responsible for the yellow, orange and red colour of many fruits. Red, orange and green fruits and vegetables (including broccoli, leafy greens, carrots, sweet potatoes, cooked tomatoes, winter squash, apricots, cantaloupe, oranges and watermelon), a few roots, egg yolks, fish (like salmon and trout) and crustaceans are a rich source of carotenoids. They are synthesized by plants, algae, fungi, yeasts and bacteria, but they are merely accumulated from the diet consumed, unchanged or slightly modified, in some animals.

In foods, about a hundred of carotenoids have been established. Normally a food would have one to five major carotenoids with a series of minor carotenoids in trace amounts. The principal carotenoids encountered in human blood and are the most investigated in terms of human health benefits are: β-carotene, α-carotene, β-cryptoxanthin, lutein and lycopene. These are the carotenoids which are most commonly found in foods (Rodriguez-Amaya 1993, 1999).

The provitamin A activity of some carotenoids, such as β-carotene, α-carotene and β-cryptoxanthin, have been known for a long time. In recent times, carotenoids, whether they are provitamin A or not, have been recognized with other health-promoting effects: immunity enhancement and reduction of the risk of emergence of degenerative diseases like cancer, cardiovascular diseases (CVD), cataracts and macular degeneration (Gaziano and Hennekens 1993; Krinsky 1993; Mayne 1996; Olson 1999; Krinsky and Johnson 2005). The physiological activities of carotenoids were attributed to its antioxidant property, that is, the ability to quench singlet oxygen and interact with free reactive radicals (Palozza and Krinsky 1992; Palace et al. 1999). However, other mechanisms of action against chronic diseases involve the modulation of carcinogen metabolism, regulation of cell growth, inhibition of cell proliferation, cell differentiation enhancement, stimulation of cell-to-cell gap junction communication and the retinoid-dependent signal and filtering of blue light (Astorg 1997; Olson 1999; Stahl et al. 2002; Stahl and Sies 2005; Krinsky and Johnson 2005).

Many retrospective and prospective epidemiological studies performed in various countries have consistently and strongly revealed that the dietary intake of β-carotene or its serum level was inversely associated with the occurrence of cancer, especially lung cancer (Ziegler 1991; Block et al. 1992; Ziegler et al. 1996). This inverse relation was also observed with CVD (Gaziano and Hennekens 1993; Kohlmeier and Hasting 1995; Mayne 1996). β-carotene, though, faced disgrace when intervention studies unexpected found that this carotenoid, which was given in capsules to smokers and asbestos workers, augmented rather than reduced the occurrence of lung cancer (ATBC Study Group 1994). Later, it was recognized that β-carotene was administered alone in the intervention studies at 20–30 mg doses, much higher than the optimal daily intake in the epidemiological studies (about 4 mg), in which the diets comprised of other carotenoids and food constituents act synergistically with β-carotene (CARIG 1996). With these considerations, carotenoids regained their eminence, however the current emphasis is on carotenoids other than β-carotene.

Lycopene. Lycopene's possible role in the deterrence of cancer has enticed considerable attention (Gerster 1997; Clinton 1998; Sies and Stahl 1998; Rao and Agarwal 1999; Giovannucci 1999; Rissanen et al. 2002), with special emphasis on prostate cancer (Hadley et al. 2002; Giovannucci et al. 2002; Wertz et al. 2004; Stacewicz-Sapuntzakis and Bowen 2005). This carotenoid has also been associated with the prevention of CVD (Kohlmeier and Hastings 1995). A case-control study of 1379 men from 10 European countries (EURAMIC Study) revealed that a higher lycopene concentration in body fat was correlated with a lower risk of heart attack (Kohlmeier et al. 1997a).

Lutein and Zeaxanthin. Lutein and Zeaxanthin contribute to the yellow pigment in the macula of the human retina (Bone et al. 1988; Handelman et al. 1988) and also have been reported to be associated with the reduced risk for macular degeneration (EDCC 1993; Seddon et al. 1994), the primary cause of irrevocable blindness in the elderly. These two carotenoids have also been consistently associated with lowering the risk of cataracts (Moeller et al. 2000).

Although numerous studies support the protective outcome of carotenoids against chronic diseases, there have been some inconsistencies in the results, requiring more well-designed studies to be performed. The consumption of carotenoid-rich foods is widely recommended, but caution and more investigations are also advised to evaluate the benefits and risks of supplementation (Mayne 1996; Granado et al. 2003; Krinsky and Johnson 2005).

Limonoids. Limonoids are terpenes present in citrus fruit. Limonoids appear to provide chemotherapeutic activity by impeding the phase I enzymes and inducing phase II detoxification enzymes in the liver. D-Limonene, the most common monocyclic monoterpene, is found within orange peel oil, which inhibits pancreatic carcinogenesis induced in hamsters by N-nitrosobis(2-oxopropyl)amine and gastric carcinogenesis induced in Wistar rats by N-methyl-N′-nitro-N-nitrosoguanidine. Limonoids may also provide protection to lung tissue (Nakaizumi et al. 1997; Uedo et al. 1999).

2.2.2 Phenolic Compounds/Polyphenols

Phenolic compounds are the most abundant and widely-distributed groups of substances in the plant kingdom, with more than 8000 phenolic structures presently known (Bravo 1998). These are secondary plant metabolites which act as the most abundant antioxidants in the human diet. These compounds are designed with an aromatic ring, carrying one or more hydroxyl moieties. According to the number of phenol rings and the structural elements that bind to these rings, they can be classified under several classes. Traditionally, they were categorized under two major groups of polyphenols, termed flavonoids and nonflavonoids.

The flavonoid group encompasses compounds with a C6-C3-C6 structure: flavanones, flavones, flavonols, flavan-3-ols, dihydroflavonols, anthocyanidins, isoflavones and proanthocyanidins. The nonflavonoids group is comprised of xanthones, chalcones, lignans, stilbenes and secoiridoids; and are classified according to the number of carbons and comprises the following subgroups, such as simple phenols, benzoic acids, hydrolyzeable tannins, acetophenones and phenylacetic acids, benzophenones, coumarins, cinnamic acids, chalcones, xanthones, stilbenes, lignans and secoiridoids.

Phenolic acids, such as caffeic, ellagic and ferulic acids, are commonly found in fruits, vegetables, tea and wine; many are present in foods in the form of glycosides. Ellagic acid is generally found in high concentrations in fruits and nuts, specifically blackberries, raspberries, strawberries, pecans and walnuts (Hollman and Venema 1993). Green tea, grapes, wine, berries, citrus fruits, apples, whole grains and peanuts are rich source of polyphenols.

The most common group of plant phenolics are the flavonoids, which are named based on the structure of flavones, comprised of two benzene rings linked through a three-carbon γ-pyrone ring. The common classes of flavonoids include flavones, isoflavones, flavonols, anthocyanins, catechins (flavanols) and flavanones. More than 4000 different types of flavonoids have been reported; and except for catechins (Aherne and O'Brien 2002), most flavonoids occur as glycosides in nature.

Flavonoids. Flavonoids are the most diverse group of phytochemicals. Research suggests that flavonoids, a vital phytochemical group, may contribute to the actions such as inhibiting xanthine oxidase and arachidonic acid metabolism. These are found in fruits, vegetables, coffee, tea and wine. The flavonols quercetin, kaempferol and myricetin are widely available in fruits and vegetables. Berries, tomatoes, potatoes, broad beans, broccoli, Italian squash, apples, kale and onions are rich in quercetin (Hollman and Artz 2000; Aherne and O'Brien 2002). Vegetables such as radishes, horseradishes, endive and kale are relatively high in kaempferol (Hertog et al. 1993). Citrus fruits are comprised of various flavanoids such as naringenin (a flavanone), rutin (a flavonol glycoside) and tangeretin (a methylated flavone).

The health-promoting properties of flavonoids are believed to be based on their antioxidant activity wherein they act as the hydrogen donating free radical scavengers (Rice-Evans et al. 1996; Prior and Cao 2000). The primary target of free radicals are proteins (including enzymes), lipids (related to the induction of heart disease), DNA (responsible for the occurrence of cancer) and RNA. Though, the most frequently occurring oxidative event inside the body is the oxidation of the unsaturated fatty acid components of the phospholipid cell membranes forming lipid peroxides. Many researchers have shown that the lipid peroxides and reactive oxygen species formed during oxidation are held responsible for the development of a variety of diseases, including cancer, atherosclerosis, heart disease, kidney damage and also accelerated aging (Ames et al. 1993; Yu 1996). Flavonoids also acts as metal chelators that binds to metals, such as copper and iron, which catalyze lipid oxidation.

Yang et al. (2001) reviewed the carcinogenesis inhibition by dietary polyphenolic compounds and questioned the association between the antioxidative and anticarcinogenic properties, asserting that polyphenols may inhibit carcinogenesis by impinging the molecular events at the initiation, promotion and progression stages of cancer. Beyond their antioxidative properties, flavonoids possess many beneficial roles, such as deactivating carcinogens, inhibiting the expression of mutated genes, deactivating the carcinogenesis-promoting enzymes and also promoting detoxification of xenobiotics (Kris-Etherton et al. 2002).

Isoflavones, lignans and stilbenes are phytoestrogens, a group of nonsteroid plant constituents that elicit an oestrogen-like biological response (Murphy and Hendrich 2002).

Isoflavones. Isoflavones are found in just a few botanical families. Although present in several other legumes, soybeans are well-known to be a principal dietary source. Isoflavones, genistein and daidzein, with their β-glucosides, are present in up to 3 mg/g of soybean (Price and Fenwick 1985).

Anthocyanidins. Anthocyanidins are water-soluble flavonoids that are aglycones of anthocyanins. The naturally occurring, principal anthocyanidins are pelargonidin, cyanidin, paeonidin, petunidin, delphinidin and malvidin. These naturally occurring compounds are among the principal pigments in fruits and flowers. The colours of these pigments are greatly influenced by the pH and metal ion complexes. Like other flavonoids, anthocyanidins act as antioxidants in *in vitro* conditions and are also expected to have antioxidative and anti-mutagenic properties *in vivo*. However, they have potent antioxidant activity in isolated anthocyanidins (aglycons and glycosides) extracts from plants (Haslem and Lilley 1998).

Catechins and Gallic Acids. The major sources of catechins are grapes, berries, cocoa and green tea. Tea contains considerable quantity of gallic acid esters, like epicatechin, epicatechin gallate and epigallocatechin gallate. Numerous studies have suggested that the protective benefits provided by these components were due to their free radical scavenging ability (Hanasaki et al. 1994), their inhibition of eicosanoid synthesis (Moroney et al. 1988) and platelet aggregation (Pace-Asciak et al. 1995). On the basis of experimental evidence in cell culture systems, animal models, as well as epidemiological evidence, Gupta et al. (1999) support the possible use of tea, especially green tea, for prevention of prostate cancer. However, Hollman et al. (1999) regard the question of flavonol protection against cardiovascular disease and cancer as remaining open. In wines, catechins and procyanidins are involved in the astringency sensation (Haslam and Lilley 1988). Catechin is one of the prime phenolics present in grapes and red wines, and is considered to be responsible for the protective effect of red wine against atherosclerotic cardiovascular disease. Donovan et al. (1999) measured catechin and catechin metabolites in the plasma of human subjects following consumption of both alcoholised and dealcoholised red wine. Sulfate and sulfate± glucuronide conjugates, but little free catechin, were present. The metabolites/conjugates of catechin were eliminated from blood with a half-time of approximately 4 h. These data confirmed that the grape polyphenolic flavonoid catechin is well absorbed, but rapidly metabolized and conjugated. The results suggest that the physiological properties of the metabolites in blood and tissues may be more important than the parent compounds found in plants.

Stilbenes. Stilbenes are 1,2-diaryl ethenes that are biosynthesized from cinnamic acid derivatives. They are commonly present in liverworts and plants with higher monomeric, dimeric, trimeric and polymeric (viniferins) forms (Gorham 1989). Amongst the monomeric form of stilbenes, the major active compound is trans-resveratrol and most of the physiological studies conducted have been targeted on its activity (Cassidy et al. 2000). The main dietary sources of stilbenes are grapes, peanuts and peanut products. Trans-resveratrol is a phytoallexin that protects grapevines from fungal infection.

2.2.3 Glucosinolates

Glucosinolates are sulphur-containing glucosides prevalent in the cruciferous family of vegetables, especially *Brassicas* (e.g. cabbage, cauliflower, broccoli and Brussels sprouts) and are also present at relatively high levels in oilseeds (such as rapeseed) and also in condiments (like mustard seed). Over a hundred different glucosinolates have been recognized in the plant kingdom, but about ten are only present in cruciferous vegetables (Stoewsand 1995). Although glucosinolates are structurally diverse, based on the side-chain structure, there are only three principal groups: aliphatic, aromatic and heteroaromatic (indolyl) glucosinolates (Mithen et al. 2000). All the glucosinolates have a β-D-thioglucose, a sulphonated oxime moiety and a variable side-chain derived from methionine, tryptophan, phenylalanine or branched-chain amino acids (Fenwick et al. 1983).

During food preparation or chewing, the plant tissue is damaged, leading to the contact of the glucosinolates with myrosinase, the endogenous enzyme which hydrolyzes the former, to yield a complex mixture of products, primarily isothiocyanates, nitriles and thiocyanates. The glucosinolate breakdown of products exerts a range of antinutritional and toxic effects in animals higher up in the animal kingdom, the most thoroughly studied of which is the goitrogenic effect of some products. There is little or no epidemiological evidence for this goitrogenic effect, resulting in disease in humans, at present.

Finley (2005) on the other hand, summarized that *in vitro* and *in vivo* studies have reported that isothiocyanates intrude on many steps of cancer development, including modulation of phase I and II of detoxifying enzymes, acting as a direct or indirect antioxidant by phase II enzyme induction, modulating cell signalling, induction of apoptosis, control of the cell cycle and reduction of *Helicobacter* infections. Apoptosis and alteration of phase I and phase II detoxification pathways have been considered the most significant mechanisms by which the glucosinolate/isothiocyanates inhibit carcinogenesis (Mithen et al. 2000; Talalay and Fahey 2001; Finley 2005). The bioactivity of glucosinolates is enhanced only when they have been enzymatically hydrolysed to the associated isothiocyanates.

2.2.4 Organosulphur Compounds

Allium compounds are organosulphur compounds (OSCs) found in *Allium* vegetables such as garlic, onions, shallots, scallions, chives and leeks, which are responsible for their distinctive flavour and aroma. In addition, as many authors reported, these compounds also contribute to the medicinal effects of these above-listed vegetables. The OSCs in *Allium* vegetables have been reported to exert various physiological activities, including antimicrobial activity, lipid-lowering effects, hypocholesteremic activity, hypoglycemic activity, antithrombic effects, inhibition of platelet aggregation and lipoxygenase and tumor inhibition (Huang et al. 1994).

2.2.5 Nondigestible Carbohydrates

Several definitions have been given to the term 'dietary fibre'. The most widely accepted is a physiological definition, wherein dietary fibre were considered as vegetable wall residues that were resistant to enzymatic hydrolysis in the small intestine. A chemical description indicates dietary fibre as nonstarch polysaccharides, although, more recently, some authors classify it under resistant starch as a new class. The most commonly used definition was 'oligosaccharides, polysaccharides and their hydrophilic derivatives that cannot be digested by the human digestive enzymes to absorbable products in the upper alimentary tract; this definition includes lignin' (Thebaudin et al. 1997, p. 42). There are three main forms of carbohydrates that are indigestible in the human small intestine: nonstarch polyssacharides (NSP), nondigestible oligosaccharides (NDOs) and resistant starch (RS) (Voragen 1998).

NDOs occur naturally in food raw materials and food products. The most well-known example are fructans (e.g. inulin), which occur in the edible parts of various plant foods like onions, artichokes, bananas, rye, chicory, leek, garlic, barley and yacon (Voragen 1998). Galactosyl sucroses (raffinose and stachyose) are present in soybeans and other leguminous seeds, while xylooligosaccharides occur in bamboo shoots. NDOs are generated during processing in some foods. The NDOs have been described as prebiotics (Crittenden and Playne 1996). Prebiotics are nondigestible food ingredients which positively affect the host by selectively stimulating the growth and/or activity of one or a few number of beneficial bacterial species already resident in the colon (Gibson and Roberfroid 1995). This is why, the resistant starches, although considered as fibre, are not necessarily prebiotics.

Dietary fibre have several established physiological effects: modulation of glucose absorption, regulation of gastrointestinal passage time, faecal bulking, acidification of colonic material and control of cholesterol bioavailability. Prebiotics, on the other hand, specifically alter the colonic microbiota and modulate hepatic lipogenesis (Crittenden and Playne 1996; Roberfroid 1996). A balanced intestinal microflora permits improved bowel regularity, reduction of bacteria, protection against a range of toxins and increased absorption of nutrients.

Lignans are diphenols that are minor components associated with dietary fibre. Sources of dietary lignans are oilseeds, cereal grains, vegetables, fruits and legumes. Flaxseed and sesame seeds have been recognized as the richest sources of these compounds (Thompson et al. 1991; Coulman et al. 2005). Plant lignans, once ingested, are converted into enterolactone and enterodiol by bacteria in the large intestine, which are known as mammalian lignans, as they have been found only in mammals (Crosby 2005). Mammalian lignans are associated with the reduced risk of CVD and cancer.

2.2.6 Terpenes

The term terpenes originate from turpentine (*Balsamum terebinthinae*). Turpentine, also known as the 'resin of pine trees', is the viscous, pleasantly smelling balsam that flows upon cutting or whittling the bark and the new wood of various pine tree species (*Pinaceae*). Turpentine contains 'resin acids' and some hydrocarbons, which were initially termed as terpenes. Traditionally, all those natural compounds which are built up from isoprene subunits and, mostly, originating from plants are denoted as terpenes.

All living organisms produce terpenes for certain vital physiological functions and therefore have the potential to produce terpene natural products. Terpenes have a unique antioxidant activity in their interaction with free radicals. Terpenes react with free radicals by means of partitioning themselves into fatty membranes by virtue their long carbon side chain. The most-studied terpene antioxidants are tocotrienols and tocopherols.

Tocotrienols. Tocotrienols are unsaturated analogues of tocopherol, i.e. vitamin E. They are present in the unsaponifiable fraction of vegetable oils; rice bran oil and palm oil are the most noteworthy sources (Eitenmiller 1997). A number of plant foods, ranging from kale and broccoli, to cereal grains and nuts, were also found to have tocotrienols (Piironen et al. 1986).

There are at least four known forms of tocotrienol, with γ-tocotrienol being the foremost potent cholesterol-lowering form (Parker et al. 1993). The cholesterol-lowering nature of tocotrienols was attributed to their ability to inhibit hydroxymethylglutaryl-CoA (HMG-CoA) reductase, known as the rate-limiting enzyme in the cholesterol synthesis pathway (Parker et al. 1993). It has been reported that α-tocopherol aids the inhibitory effect of γ-tocotrienol (Qureshi et al. 1996). Tocotrienols have also been demonstrated to have the vitamin E activity (Tan 1989), antioxidant activity (Serbinova et al. 1993) and antitumor properties (Komiyama et al. 1992). It was theorized that the inhibition of HMG-CoA reductase by tocotrienols also results in suppression of tumour growth (Elson and Qureshi 1995).

Phytoestrogens. Phytoestrogens can compete with steroid hormones for various enzymes and receptors, thereby kindling the production of sex hormone-binding globulin in the liver. Thus, they may alter steroid hormone metabolism and may hinder the growth and proliferation of hormone-dependent cancer cells. Like other phenolic compounds, phytoestrogens possess antioxidant activity, and like oestrogens, they can be capable of influencing the lipoprotein metabolism and enhancing vascular reactivity. Hence the phytoestrogens are proven to have potential protective effects against CVD.

Plant sterols or phytosterols are structurally alike and functionally equivalent to the animal sterol, cholesterol. Phytosterols are triterpenoids occurring in both free and esterified form. Among more than 40 identified phytosterols, β-sitosterol, stigmasterol and campesterol are the most abundant, and are predominantly provided by vegetable oils (Piironen et al. 2000; Hicks and Moreau 2001), which are affluent sources of sterol esters. Ferulic acid esters of phytosterols, commonly known as oryzanol, occur in rice bran oil (Kaneko and Tsuchiya 1954). Nuts, cereals, grains and vegetables are also sources of sterols, albeit of lesser significance.

Stanols or Phytostanols. Plant stanols or phytostanols are the less abundant class of related compounds, which are the saturated forms of phytosterols. Phytostanols are obtained primarily from corn, wheat, rye and rice (Hicks and Moreau 2001). Corn fiber oil is said to be unique, not only because it contains fatty acid and phenolic esters of phytosterols, but also because it appears to be the most abundant source of stanols and stanol esters (Moreau et al. 1996). Phytosterols and phytostanols hinder the intestinal absorption of cholesterol. The cholesterol-lowering property of these compounds has been established some decades ago. For example, β-sitosterol has been used from 1950s as a supplement and as a drug (Cytellin, marketed by Eli Lilly) for reducing the serum cholesterol levels in hypercholesterolemic individuals (Hicks and Moreau 2001).

It is also equally effective in reducing plasma cholesterol and plant sterol esters, unlike stanols, which increase their own absorption. The resulting increased serum sterol levels would result in values seen as in phytosterolemia, a strongly atherogenic hereditary metabolic abnormality.

As of September 2000, the U.S. FDA has allowed a health claim for reducing the risk of coronary heart disease for foods products, such as spreads and salad dressings, containing phytosterol and phytostanol esters (Jones and Raeini-Sarjaz 2001).

2.2.7 Betalains

The name 'betalain' is designated from the Latin term of the common beet (*Beta vulgaris*), from which betalains were first extracted. The deep red colour of beets, amaranth, bougainvillea and many cacti results from the presence of pigments. Betalains are a class of red and yellow indole-derived pigments present in plants of the *Caryophyllales*, as well as in some higher-order fungi, where they replace anthocyanin pigments (Strack et al. 2003). There are two categories of betalains: Betacyanins comprise the reddish to violet betalain pigments and betaxanthins that are those betalain pigments that impart yellow to orange colour.

Amongst the betaxanthins, various types of pigments are present in plants, which include vulgaxanthin, miraxanthin, portulaxanthin and indicaxanthin (Tiwari et al. 2013). The few edible known sources of betalains are red and yellow beetroot (*Beta vulgaris* L.ssp. *vulgaris*), coloured Swiss chard (*Beta vulgaris* L. ssp. *cicla*), grain or leafy amaranth (*Amaranthus* sp.) and cactus fruits like those of *Opuntia* and *Hylocereus genera* (Azeredo 2009).

2.2.8 Alkaloids

The term 'alkaloid' was coined in 1819 by Carl Friedrich Wilhelm Meissner, the German pharmacist, to refer to plant natural products, which was found to show basic properties similar to that of the inorganic alkalis. The ending '-oid' is still in practise even today, which suggests the similarity of their structure or activity, as is evident in modern names such as peptoid, terpenoid or vanilloid (Hesse 2002). Among the secondary metabolites that are produced by plants, the alkaloids act as a very prominent class of defence compounds. Over 21,000 alkaloids have been identified, hence they constitute the largest group among the nitrogen-containing secondary metabolites (besides 700 nonprotein amino acids, 100 amines, 150 alkylamides, 100 glucosinolates and 60 cyanogenic glycosides). Alkaloids are usually present as a mixture of a few major and several minor alkaloids of a particular biosynthetic unit, which differ in functional groups (Wink et al. 2005).

2.2.9 Capsaicinoids

The nitrogenous compounds produced in pepper fruit, which are responsible for the burning sensation, are known as capsaicinoids. These capsaicinoids are purported to have antimicrobial effects for food preservation (Tiwari et al. 2013), and their most medically relevant use is as an analgesic. Capsaicinoids have been used successfully to treat a wide range of painful conditions including arthritis, cluster headaches and neuropathic pain. The analgesic actions of the capsaicinoids are dose-dependent and specific for polymodal nociceptors. The gene for the capsaicinoid receptor has been cloned (TRPV1) and this receptor transduces multiple pain-producing stimuli. Capsaicin (trans-8-N-vamllyl-6-nonenamide) is an acrid, volatile alkaloid accountable for the hotness in peppers.

2.2.10 Chlorophyll

Chlorophyll is a green pigment produced in almost all plants, algae and cyanobacteria. The term was originated from the Greek words *chloros* ('green') and *phyllon* ('leaf'). Chlorophyll is an extremely important biomolecule, and vital for the photosynthesis process, wherein it allows the plants to obtain energy from light. Chlorophyll absorbs light most strongly in the blue region of the electromagnetic spectrum, followed by the red portion. However, it is a poor absorber of green and near-green regions of the spectrum; thereby imparting green colour to the chlorophyll-containing tissues. Chlorophyll was first isolated in 1817 by Joseph Bienaime Caventou and Pierre Joseph Pelletier (Pelletier and Caventou 1951).

In peppers, the colours of unripe fruit can differ from ivory, green or yellow. The green colour derives from accumulation of chlorophyll in the chloroplast while the ivory represents chlorophyll degradation as the fruit ripens (Wang et al. 2005). The stable presence of chlorophyll in fruit ripening was to accumulate other pigments like carotenoids or anthocyanins results in brown or black mature fruit colours. Chlorophyll in black pepper fruit is 14-fold higher compared to violet fruit (Lightbourn et al. 2008).

2.3 SOURCES OF PHYTOCHEMICALS

The following are plant foods that are established or emerging functional foods, each with a body of scientific proof on the positive effect on health that the constituent nutrients and phytochemicals confer.

2.3.1 Broccoli and Other Cruciferous Vegetables

Cruciferous vegetables contain little fat, are low in energy and are sources of fibre, micronutrients (vitamins A, C and E, selenium and folic acid) and other phytochemicals (carotenoids, coumarins, glucosinolates, flavonoids and other phenolic compounds) (Steinmetz and Potter 1996; Delaquis and Mazza 1998).

A review of 87 case-control studies were reported by Verhoeven et al. (1996) and revealed that an inverse relation between consumption of cruciferous vegetables and cancer risk was observed. This epidemiological evidence is supported by a host of experimental studies, which have indicated that glucosinolate breakdown products exerted anticarcinogenic activity in both experimental and animal models (Verhoeven et al. 1997).

Stoewsand (1995) reported that the cancer chemopreventive effects of *Brassica* vegetables were attributed by the two types of phytochemicals such as certain glucosinolates and *S*-methyl cysteine sulfoxide present in it.

The indolyl glucosinolate glucobracissin, found in higher levels in Brussels sprouts, was hydrolyzed by myrosinase to give indole-3-carbinol (I3C). This indole was suspected to be responsible for its chemopreventive property and this relation has been under investigation, especially toward the mammary gland. It may reduce the risk of cancer by increasing 2-hydroxylation over 16-hydroxylation of oestrogen – this shift in hydroxylation represents a reduction in the activity of oestrogens and may protect against oestrogen-related cancers (Hasler 1998).

In a human clinical trial, women were given a dose of 500 mg I3C (roughly 50 times the estimated average daily intake in the United States) daily for a week, which resulted in a significant increase in 2-hydroxylation of estradiol (Michnovicz and Bradlow 1991). The result recommended that I3C may be a new approach for minimizing the risk of breast cancer. However, animal studies have shown that I3C and other indoles induced the production of both phase I and II enzymes. Induction of phase I enzymes could activate or deactivate the carcinogens, whereas the induction of phase II enzymes leads to detoxification. Thus, caution was urged before going on with an extensive clinical trial (Dashwood 1998); although similar phase I clinical trials to the one described above are being carried out (Wong et al. 1998).

Three-day-old sprouts of cultivars of certain crucifers, including broccoli and cauliflower, contain 10 to 100 times higher amounts of the aliphatic glucosinolate glucoraphanin than the corresponding mature plants (Fahey et al. 1997). Myrosinase hydrolyzes glucoraphanin to yield sulforaphane, an isothiocyanate, which induces a phase II enzyme (Zhang et al. 1992). Hence, broccoli sprouts have more desirable anticancer properties than matured vegetables (Fahey et al. 1997; Nestle 1998).

2.3.2 Oats

Among the food grains, oats are the most concentrated source of β-glucan, a soluble non-starch polysaccharide recognized to reduce the risk of coronary heart disease (Bell et al. 1999). Compounds such as phytates, phenolics, vitamins and minerals, which confer other physiological benefits, are also present.

Oats were the first food to be permitted to have a health claim under the U.S. Nutrition Labeling and Education Act (Hasler 1998). The FDA approved the claim 'soluble fiber from oatmeal, as part of a low saturated fat, low cholesterol diet and may possibly reduce the risk of heart disease'. The FDA has acknowledged that β-glucan is the main active ingredient accountable for this health claim (Oomah and Mazza 1999).

Several clinical studies were conducted to find the impact of oats on serum lipids (de Groot et al. 1963). The most conclusive study was a meta-analysis for the effect of oat products, wherein a clinical trial on free-living subjects was performed and summarized, so that out of 20 trials conducted, 12 trials became the bases of analyses (Ripsin et al. 1992). This report provided the strongest substantiation to the FDA, so that about 3 g per day of soluble fibre from oat products can achieve a clinically relevant lowering effect of cholesterol in serum, and that the reduction was also greater in individuals with higher initial blood cholesterol levels.

To explain the cholesterol-lowering effect of β-glucan from oats, four mechanisms have been proposed by Bell et al. (1999). First, it has been hypothesized that the soluble fibre binds to bile acids in the intestinal lumen, resulting in a reduced bile acid pool flowing back to the liver. This binding activity stimulates the production of more bile acids from cholesterol, thus reducing the serum cholesterol concentration (Lia et al. 1997). A second mechanism centres on the short-chain fatty acids (like acetic, butyric and propionic acids) that are formed from the fermentation of soluble fibre in the large bowel by native intestinal microflora (Glore et al. 1994). These fatty acids are absorbed through the portal vein, inhibiting the HMG-CoA reductase activity or increasing the catabolism of LDL cholesterol. A third mechanism involves the delay of gastric emptying by oat soluble fibre – this minimizes post-prandial serum insulin concentrations (Inks and Mathews 1997), which in turn diminishes hepatic cholesterol production through mediation of HMG-CoA reductase. Lastly, the increase in intestinal viscosity was induced by oat soluble fibre that may interfere with the absorption of dietary fat, including cholesterol (Inks and Mathews 1997).

2.3.3 Flaxseed (Linseed)

Flaxseed or linseed is known as an abundant source of omega 3-fatty acid, α-linoleic acid (ALA), mammalian lignan precursors and viscous fibre components (Oomah and Mazza 1999). The high concentrations of ALA, dietary fibre (polysaccharide gum or mucilage) and lignans have been related with the many potential health benefits of flaxseed.

The first meta-analysis investigated the relationship between intake of flaxseed or its components and risk reduction of disease in humans was expolited by Oomah (2001). Out of the 24 clinical studies identified, only 12 (6 with flaxseed and 6 with flaxseed oil), comprising a total of 208 subjects, were observed to meet all the criteria of well-designed clinical trials. Four of these studies substantiated the protective effect of ALA of flaxseed oil against cardiovascular disease. Three studies reported that utilization of raw or defatted flaxseed reduced total and LDL cholesterol. Five studies showed the role of flaxseed in promoting bone health and its phytoestrogenic and therapeutic effect in reducing the risk of hormone-associated cancers in women.

2.3.4 Tomatoes

Tomatoes and tomato products have been the focus of intense investigation in recent years, particularly for their relation to prostate cancer (Giovannucci et al. 2002; Hadley et al. 2002; Campbell et al. 2004; Stacewicz-Sapuntzakis and Bowen 2005). Giovannucci (1999) studied the epidemiological literature on the relationship between intake of tomatoes and tomato-based products, the plasma levels of lycopene and the risks of various cancers. Among 72 studies identified, 57 of them reported an inverse relation between tomato intake or blood lycopene level and the risk of cancer at defined anatomical sites and 35 of those inverse relations were statistically significant. No study indicated that higher tomato consumption or blood lycopene level augmented the risk of cancer at any of the sites investigated. Evidence for a benefit was greatest for the lung, stomach and prostate gland cancers. Data were also suggestive of a benefit towards the oral cavity, esophagus, breast, cervix, pancreas, colon and rectum cancers.

Tomatoes and tomato-based products are the main contributors for lycopene in the diet of many countries and lycopene has been considered the primary phytochemical responsible for the reduction in the risk of prostate cancer. Tomatoes, however, is also a rich source of nutrients like folate, vitamins C and E and other potentially beneficial phytochemicals including phytosterols, phenolic acids and flavonoids (Beecher 1998). Hence, it is more probable that the combination of these compounds is responsible for the effect on prostrate carcinogenesis (Hadley et al. 2002; Campbell et al. 2004; Stacewicz-Sapuntzakis and Bowen 2005).

2.3.5 Soybeans

Soybeans have been grown and consumed as food in Asia for over 5000 years. But this ancient bean was cultivated in abundance throughout the world only during the 20*th* century and scientific interest on its health benefits started even much later. Soybeans are known as a source of high quality proteins is also a source of phytosterols, isoflavones, phenolic acids, saponins and phytic acid (Messina and Barnes 1991).

Soybeans have been known to have a protective role in women's health, in particular in the alleviation of menopausal symptoms and promotion of bone health. In a clinical study of 66 postmenopausal women, it was reported that a daily intake of 40 g isolated soy protein (ISP), containing 90 mg total isoflavones, significantly improved both bone mineral content and density in the lumbar spine after a consumption for six months (Erdman and Potter 1997). This finding proposed that soybeans may have a protective role in osteoporosis. Asian women were found to have significantly lower levels of hot flashes and night sweats compared to Western women. A clinical study showed that the daily intake of 60 g ISP for three months decreased the hot flashes by 45% in 104 postmenopausal women (Albertazzi et al. 1998). Human ecological observations support a cancer-protective effect of soybeans. Vegetarians and population groups (e.g. Japanese women) who relatively consume greater amounts of soy products, have a lower risk of certain cancers, including breast cancer (WRCF 1997). Various classes of anticarcinogenic phytochemicals have been identified in soybeans, among which the isoflavones, genistein and daidzein are noteworthy because soybeans are the only significant dietary sources for these compounds. The epidemiological data on soy intake and cancer risk are not consistent (Kris-Etherton et al. 2002). Though, a number of experimental studies have indicated a protective role of soybeans and their components in cancer.

Population studies indicate that countries where the diets contain a high amount of soy products have the lowest rates of CVD. An inverse relation between soy food product intake and cholesterol level has been observed in Japanese men and women (Nagata et al. 1998). A meta-analysis of 38 controlled clinical trials with a total of 730 subjects reported that the daily consumption of 47 g of soy protein resulted in significant decreases in total cholesterol (9%), LDL cholesterol (13%), triglycerides (11%) and also in an increase in HDL cholesterol (2%) (Anderson et al. 1995).

The well-documented physiological effect of soybeans is their cholesterol-lowering effect. Investigations on the specific components responsible for this role of soybeans have targeted the isoflavones. In two studies, however, isoflavone supplements were found not to be effective in reducing cholesterol in humans (Hodgson et al. 1998). On the other hand, Crouse et al. (1999) reported that naturally occurring isoflavones (62 mg) isolated with soy protein reduced the concentrations of total and LDL cholesterol in plasma, without affecting concentrations of triacylglycerols or HDL cholesterol in marginally hypercholesterolemic individuals. Ethanol-extracted soy protein (with only 3 mg isoflavones) was ineffective.

Animal studies indicated that the cardioprotective role of soybeans goes beyond cholesterol-lowering (Potter 1998), such as reducing the atherosclerotic lesion and thrombus formation and in atherosclerotic plaque. In 1999, the U.S. Food and Drug Administration permitted a health claim for soy protein in reducing the risk of heart disease (Department of Health and Human Services 1999).

2.3.6 Citrus

Citrus fruits are prime sources of vitamin C, folate, fibre, flavonoids, phenolic acids, monoterpenes, carotenoids and limonoids (Girard and Mazza 1998). The health benefits of citrus fruits have been attributed to the antioxidant activity of their essential flavonoids, such as flavanones, flavones, flavonols and anthocyanins (Benavente-Garcia et al. 1997). In biological studies, flavonoids from citrus have established anticarcinogenic (antimutagenic and antiproliferative effects and inhibition of the invasion of the carcinogenic cell) and cardiovascular (impact on capillary fragility, platelet aggregation and coronary heart disease) properties. Citrus flavonoids were also observed to have anti-inflammatory, anti-allergic and antiviral activities.

The monoterpene D-limonene, the main component of the oil from citrus peel, was also found to protect against cancer; it produces glutathione transferases, a family of phase II detoxification enzymes (Crowell 1997; Gould 1997). D-limonene was given a GRAS (generally regarded as safe) status for its use as a flavouring agent. D-limonene was proved to be a good candidate for human clinical chemoprevention trial evaluation possessing no toxicity in humans. Perillyl alcohol, a metabolite of limonene, has passed through phase I clinical trials in patients with advanced malignancies (Ripple et al. 1998).

Citrus fruits are also found to be a rich source for another class of phytochemicals, the highly oxidized triterpenes called limonoids. Recent researches have proposed that these compounds may have substantial anticancer activity. However, these studies have been conducted in *in vitro* and animal models, facilitating the need for further human studies to confirm such action (Ejaz et al. 2006).

2.3.7 Berries

Berries are not only low energy delicious food, but are also rich sources of fibre, antioxidant vitamins and various phenolic compounds comprising of flavonoids and phenolic acids. The main classes of flavonoids in berries are anthocyanins, proanthocyanidins, catechins and flavonols (Wang et al. 1997). Phenolic acids present in berries are hydroxylated benzoic and cinnamic acid derivatives (Wang et al. 1997). Various potential health benefits have been attributed to flavonoids and phenolic acids present in berries.

Berries of *Vaccinium* sp. have been reported to possess a range of biological activities. Cranberries and wild blueberries have been shown to prevent urinary tract infections. This defensive action has been credited to the condensed tannins or proanthocyanidins, which are said to act as anti-adhesive agents hindering bacterial colonization (Ofek et al. 1996; Howell et al. 1998). In wild blueberry, proanthocyanidins have been recognized as the active agents inhibiting the *in vitro* promotion of chemically-induced carcinogenesis (Bomser et al. 1996).

Fruit extract of wild bilberry repressed LDL oxidation (Laplaud et al. 1997), exhibited astringent and antiseptic characteristics, diminished the permeability and fragility of capillaries, inhibited platelet aggregation, inhibited urinary tract infection and intensified the collagen matrices via cross linkages (Morazzoni and Bombardelli 1996; Morazzoni and Magistretti 1990).

2.3.8 Tea

Tea first originated from China, where it has been consumed for its medicinal properties since 3000 BC (Balentine 1997). The three most important varieties of tea are green (unfermented), oolong (semi-fermented) and black (fully fermented). Green tea is known to contain significant amounts of catechins, which are easily extracted into hot water from the leaves as infusions. During the production of black tea, the dimeric theaflavins and thearubigens are formed by the enzyme catalytic oxidation reaction of catechins. Green tea contains catechins (90%) and flavonols (10%); black tea is comprised of catechins (30%), flavonols (10%), theaflavins (13%) and thearubigens (47%) (Tijburg et al. 1997).

Tea has been proved to possess antiviral, antibacterial and antifungal properties. Evidence has also suggested that drinking tea imparts protection against cancer and CVD and it is generally believed that flavonoids are mainly responsible for these biological activities. A study on native residents of Shizuoka, Japan, where green tea is produced and consumed, showed lower mortality rates due to stomach, lung and liver cancers than comparable populations in non-green tea consuming regions (Oguni et al. 1992). Epidemiological studies conducted showed that tea may reduce the risk for certain cancers (Blot et al. 1996; Kohlmeier et al. 1997b). However, the epidemiological evidence for a preventative effect of tea against cancer and CVD is still regarded as inconclusive (WCRF 1997; Kris-Etherton et al. 2002). On the contrary, research findings from animal and *in vitro* studies clearly supported a cancer chemopreventive effect for tea, which covers the entire process of carcinogenesis

(Dreosti 1996; Kuroda and Hara 1999; Wang et al. 2000). Likewise, the proposed protection against CVD had also rested heavily on cellular and animal experiments, conducted at conditions and/or concentrations implausible to occur in humans.

A study of green tea consumers from Japan (Kono et al. 1992) and another group in Norway (probably mainly black tea drinkers) (Stensvold et al. 1992) showed a significant inverse association between tea drinking and plasma cholesterol levels. A similar but not statistically significant trend was observed in presumably black tea drinkers from Israel (Green and Harari 1992). In addition, proof for a protective effect of tea against CVD – the lowering of blood pressure and a trend towards decreased coronary heart disease mortality – was observed in the Norwegian study. These protective effects were also observed in a study of a population of elderly men in the Netherlands, although the association was seen with above-average intake of dietary flavonoids such as quercetin, kaempferol, myricetin, apigenin and luteolin, which were largely derived from tea.

2.3.9 Grapes and Wine

Grapes, and wine derived from grapes, contain large amounts of phenolic compounds, including flavonoids like catechins, epicatechin, quercetin and anthocyanidins; phenolic acids (hydroxycinnamates); and tannins (Kaur and Kapoor 2001). The phenolic compounds in wine primarily originate from grapes, but the phenolic profile of wine is not the same as that present in fresh grapes or grape juice which is due to the significant changes that take place during wine making.

In certain regions of France, coronary heart disease mortality is low despite diets high in dairy fat (Renaud and de Lorgeril 1992). This phenomenon, was referred to as the 'French Paradox' is attributed to a high intake of red wine and has been partly explained by the association of moderate alcohol consumption with a decreased risk of CVD. More recent examinations have targeted the nonalcoholic components, particularly flavonoids and other phenolics; the phenolic content of red wine is 20 to 50 times greater that of white wine. Red wine was shown to inhibit oxidation of human LDL *in vitro* (Frankel et al. 1993; Teissedre et al. 1996), this property being attributed to wine phenolics. Collectively, as reviewed by Rotondo and Gaetano (2000) and Wollin and Jones (2001), phenolic compounds appear to have anti-thrombic effects as a result of the altered susceptibility of platelet aggregation, reduced synthesis of prothrombotic and proinflammatory intermediates and reduced expression of adhesion molecules and tissue factor activity. The inhibition of platelet-mediated thrombosis was reported with grape juice consumption (Freedman et al. 2001).

Aside from its antioxidant activity, resveratrol, found in grape skin and red wines, persuades the quinone reductase, a phase II detoxifying enzyme. It possesses anti-inflammatory activity and also hinders the hydroperoxidase activity of cyclooxygenase enzymes, thereby inhibiting the arachidonic pathway that produces prostaglandins that encourage tumor cell growth (Cassidy et al. 2000). In a dose-dependent manner, resveratrol inhibits the growth of preneoplastic lesions, thereby slowing down the progression of carcinogenesis (Jang et al. 1997). Other reports propose that resveratrol hinders key enzymes involved in DNA duplication and synthesis (Fontecave et al. 1998; Sun et al. 1998). As stated by Della Ragione et al.

(1998), resveratrol seems a very attractive molecule for the improvement of anticancer treatments and for inhibiting lymphocyte proliferation during immunosuppressive therapies.

2.3.10 Garlic

Garlic is one of the earliest of cultivated spices and foods and the most widely quoted in the literature for its medicinal properties and health benefits. Few of the epidemiological studies conducted have shown that garlic consumption is correlated to decreased cancer risk. An ecological study reported that Shandong Province, China, an area with a larger amount of garlic consumption, had the lowest national mortality rate for stomach cancer (Mei et al. 1982). You et al. (1989) found that consumption of more than 1.5 kg/year of garlic resulted in a significantly lower stomach cancer risk. The Iowa Women's Health Study illustrated a decreased colon cancer risk by almost 50% in over 40,000 women who consumed garlic more than once a week (Steinmeitz et al. 1994). An experiment of 20 epidemiological studies by Ernst (1997) suggested that allium vegetables, like garlic, may provide a protective effect against cancer of the gastrointestinal tract. In a more recent meta-analysis, a constant inverse relation between raw and cooked garlic consumption and stomach and colorectal cancers was observed (Fleischauer et al. 2000). In contrary, garlic supplement consumption in one case-control study of prostate cancer and four other studies performed in Netherlands (cohort of colorectal, stomach, lung and breast cancers) did not appear to be connected with cancer risk.

Composition of garlic has been demonstrated to inhibit carcinogenesis in several experimental models. These studies have suggested that allyl sulphur compounds in garlic act primarily on the initiation stage of carcinogenesis, hindering the development of chemically-induced tumours in various sites through the initiation of phase II detoxification enzymes and inhibition of P450 E1, the enzyme responsible for the metabolic activation of carcinogens. The capability of garlic to inhibit the synthesis of *N*-nitroso compounds (Mei et al. 1989) and its antibacterial action against *Helicobacter pylori*, a risk factor in stomach cancer (Sivam et al. 1997), are two other possible mechanisms. Studies have shown that the anticancer activities related with garlic are not limited to a particular tissue. Both the lipid-soluble and water-soluble allyl sulphur compounds are effective, supporting the possibility of multiple mechanisms (Milner 2006).

Substantiation from several experimental studies shows that garlic protects against CVD by bringing about lipid normalization, improved fibrinolytic activity, subdued platelet aggregation and reduced blood pressure (Petesch and Sumiyoshi 1999). These experimental studies are defended by ecological observations of lowered cardiovascular incidence in populations like those in the Mediterranean region and some parts of Asia known for consuming high amounts of garlic, compared to populations who have a different diet and lifestyle (Lin 1994). Many clinical trials have been conducted to investigate the cardioprotective effects of garlic. A clinical study on the effect of garlic supplementation on the endpoint of cardiovascular events like myocardial infarction or death, was investigated in 432 cardiac patients (Bordia and Verma 1990). Supplementation reduced the mortality rate by 50% in the second year, and by around 66% in the third year, and reduced the rate of reinfarction by 30% and 60% in the consecutive second and third year, respectively. A meta-analysis by Warshafsky et al. (1993) summarized the results of five randomized placebo-controlled clinical trials involving 410 patients. It was shown that an average of 900 mg of garlic a day (one-half to one clove of garlic) could decrease total serum cholesterol levels by approximately 9%. Another meta-analysis including 16 trials showed that 800 mg garlic per day reduced total cholesterol by 12% (Silagy and Neil 1994). Alternatively, a multi-centre, randomized, placebo-controlled trial showed that 12 weeks of garlic intake were unproductive in reducing cholesterol levels in hypercholesterolemic subjects (Isaacsohn et al. 1998). The contradictory results in these studies may be due to methodological shortcomings, differences in garlic preparation/formulation used (i.e. loss of active compounds during processing or inhibition of release of the active components) and insufficient duration of the studies. Rahman and Lowe (2006) analysed *in vitro* and *in vivo* studies published since 1993, concluding that although garlic appears to hold promise in reducing parameters associated with cardiovascular disease, more in-depth and appropriate studies are required.

2.4 PROSPECT FOR DEVELOPING COUNTRIES

In order to obtain the full benefit of the phytochemicals/functional foods, a holistic, concerted, multidisciplinary approach is essential, including the workers in diverse fields such as nutrition, medical sciences, epidemiology, statistics, analytical and organic chemistry, biochemistry, immunology, biology, agriculture, food science, food technology and engineering.

The prospect of functional foods depends on the explicit demonstration of their efficacy in promoting health. Epidemiological and intervention studies, on the other hand, may require large investigating teams of researchers from different areas, infrastructure to manage the large numbers of subjects and substantial financing to support the different activities for an extended period. Animal model and cell/tissue studies are more acquiescent to conditions in developing countries, but the results need to be endorsed in humans.

A pragmatic approach for developing countries would be to keep abreast of advances in human health research in developed countries and target the identification and substantiation of local sources of phytochemicals. Accurate quantification of phytochemicals, including monitoring and enhancing of their quantity all through the food chain, could establish product development for the domestic and international markets.

Every country has the responsibility to identify and promote their own sources, like what has been done by Canada (Oomah and Mazza 1999), Japan (Namiki 1994) and China (Dai and Luo 1996). Many developing countries have a diversity of yet unexplored or underutilized foods which can have higher levels of bioactive phytochemicals than those found in developed countries. The significance of databases on these compounds cannot be overemphasized. In fact, reliable data on food composition is a required for a successful epidemiological study so that reliable intakes could be calculated. A country can forcefully endorse its products only if their compositions are reliably known. In addition, exploitation of the favourable effects of the phytochemicals depends on how fully we understand their behaviour at different phases of the food chain.

The phytochemical composition can vary markedly as a function of factors such as cultivar, degree of maturity at harvest, climatic or geographic effects, soil composition, cultivation practises and the part of the plant used. Agronomic and post-harvest handling and processing measures can be taken to insure high levels of these compounds in the diet consumed. In fact, it is now progressively recognized that optimization of the nutrient/phytochemical composition and profiles of foods, by conservative plant breeding and agronomic practises or genetic manipulation, is a feasible approach (Mithen et al. 2000; Parr and Bolwell 2000; Schneeman 2000; Van den Berg et al. 2000; Miettinen 2001). The probability that manipulation of phytochemical metabolism by expensive molecular techniques may not substantially enhance phytochemical contents ahead of what can be accomplished by traditional agricultural procedures (Parr and Bolwell 2000) should not be unnoticed.

The majority of phytochemicals are prone to degradation during processing and storage. Therefore, food production should be intertwined with optimum food processing and storage, which can profoundly affect the health-promoting potential of functional foods. In fact, as discussed earlier, processing is necessary to transform some phytochemicals into their active form. As per the carotenoids are concerned, food processing can increase its bioavailability by easing the release of the bioactive compounds from the food matrix, the first step in the human digestive process (Gartner et al. 1997; Rock et al. 1998). Hence, processing conditions should be optimized, such that appreciable losses of the phytochemical are avoided whilst their bioavailability is enhanced.

REFERENCES

Aherne, S. A. and O'Brien, N. M. 2002. Dietary flavonols: Chemistry, food content and metabolism. *Nutrition* 18, 75–81.

Albertazzi, P., Pansini, F., Bonaccorsi, G., Zanotti, L., Forini, E., and de Aloysio, D. 1998. The effect of dietary soy supplementation on hot flashes. *Obstetrics and Gynecology* 1, 6–11.

Ames, B. N., Shigenga, M., and Hagen, T. M. 1993. Oxidants, antioxidants and degenerative diseases of aging. *Proceedings of the National Academy of Sciences of the United States of America* 90, 7915–22.

Anderson, J. W., Johnstone, B. M., and Cook-Newell, M. E. 1995. Meta-analysis of the effects of soy protein intake on serum lipids. *New England Journal of Medicine* 333, 276–82.

Astorg, P. 1997. Food carotenoids and cancer prevention: An overview of current research. *Trends in Food Science and Technology* 8, 406–13.

ATBC (Alpha-Tocopherol, Beta-carotene) Cancer Prevention Study Group. 1994. The effect of vitamin E and beta-carotene on the incidence of lung cancer and other cancers in smokers. *New England Journal of Medicine* 330, 1029–35.

Azeredo, H. M., Pereira, A. C., de Souza, A. C. R., Gouveia, S. T., and Mendes, K. C. B. 2009. Study on efficiency of betacyanin extraction from red beetroots. *International Journal of Food Science and Technology* 44 (12), 2464–9.

Balentine, D. 1997. Tea and health. Critical Review in Food Science and Nutrition. 37, 691–2.

Beecher, G. R. 1998. Nutrient content of tomatoes and tomato products. *Proceedings of the Society for Experimental Biology and Medicine* 218, 98–100.

Bell, S., Goldman, V. M., Bistrian, B. R., Arnold, A. H., Ostoff, G., and Forse, R. A. 1999. Effect of β-glucan from oats and yeast on serum lipids. *Critical Reviews in Food Science and Nutrition* 39, 189–202.

Benavente-Garcia, O., Castillo, J., Marin, F. R., Ortuno, A., and del Rio, J. A. 1997. Uses and properties of *Citrus* flavonoids. *Journal of Agricultural and Food Chemistry* 45, 4505–15.

Block, G., Patterson, B., and Subar, A. 1992. Fruit, vegetables, and cancer prevention: A review of the epidemiological evidence. *Nutrition and Cancer* 18, 1–29.

Blot, W. J., Chow, W. H., and Maclaughlin, J. K. 1996. Tea and cancer: A review of the epidemiological evidence. *European Journal of Cancer Prevention* 5, 425–38.

Bomser, J., Madhavi, D. L., Singletary, K., and Smith, M. A. L. 1996. *In vitro* anticancer activity of fruit extracts from *Vaccinium* species. *Planta Medica.* 62, 212–6.

Bone, R. A., Landrum, J. T., Fernandez, L., and Tarsist, S. L. 1988. Analysis of the macular pigment by HPLC: Retinal distribution and age study. *Investigative Ophthalmology and Visual Science* 29, 843–9.

Bordia, A. K. and Verma, S. K. 1990. Effect of three years treatment with garlic on the rate of reinfarction and mortality in patients with coronary artery disease. In: *Garlic in Biology and Medicine:* Proceedings of the First World Congress on the Health Significance of Garlic and Garlic Constituents.

Bravo, L. 1998. Polyphenols: Chemistry, dietary sources, metabolism, and nutritional significance. *Nutrition Reviews* 56, 317–33.

Brower, V. 1998. Nutraceuticals: Poised for a healthy slice of the healthcare market? *Nature Biotechnology* 16, 728–731.

Campbell, J. K., Canene-Adams, K., Lindshield, B. L., Boileau, T. W. M., Clinton, S. K., and Erdman, J. W. 2004. Tomato phytochemicals and prostate cancer risk. *Journal of Nutrition* 134, 3486S–92S.

CARIG. 1996. Beta-carotene and the carotenoids: Beyond the intervention trials. *Nutrition Reviews* 54, 185–8.

Cassidy A., Hanley, B., and Lamuela-Raventus, R. M. 2000. Isoflavones, lignans and stilbenes – Origins, metabolism and potential importance to human health. *Journal of the Science of Food and Agriculture* 80, 1044–62.

Clinton, S. K. 1998. Lycopene: Chemistry, biology, and implications for human health and disease. *Nutrition Reviews* 56, 35–51.

Clydesdale, F. M. 1996. What scientific data are necessary? *Nutrition Reviews* 54, S195–S198.

Coulman, K. D., Liu, Z., Hum, W. Q., Michaelides, J., and Thompson, L. U. 2005. Whole sesame seed is as rich a source of mammalian lignan precursors as whole flaxseed. *Nutrition and Cancer* 52, 156–165.

Crittenden, R. G. and Playne, M. J. 1996. Production, properties and applications of food-grade oligosaccharides. *Trends in Food Science and Technology* 7, 353–61.

Crosby, G. A. 2005. Lignans in food and nutrition. *Food Technology* 59, 32–6.

Crouse, J. R., Morgan, T., Terry, J. G., Ellis, J., Vitolins, M., and Burke, G. L. 1999. A randomized trial comparing the effect of casein with that of soy protein containing varying amounts of isoflavones on plasma concentrations of lipids and lipoproteins. *Archives of Internal Medicine* 159, 2070–6.

Crowell, P. L. 1997. Monoterpenes in breast cancer chemoprevention. *Breast Cancer Research Treatment.* 46, 191–7.

Dai, Y. and Luo, X. 1996. Functional food in China. *Nutrition Reviews* 54, S21–S23.

Dashwood, R. H. 1998. Indole-3-carbinol – Anticarcinogen or tumor promoter in *Brassica* vegetables. *Chemico-Biological Interactions* 110, 1–5.

de Groot, A. P., Luyken, R., and Pikaar, N. A. 1963. Cholesterol lowering effect of rolled oats. *Lancet* 2, 303–4.

Delaquis, P. and Mazza, G. 1998. Functional vegetable products. In: *Functional foods*. Ed. G. Mazza. Biochemical and processing aspects. Lancaster: Technomic Publishing Company, Inc. 193–233.

Della Ragione, F., Cucciolla, V., Borriello, A., Della Prieta, V., Racioppi, L., Soldati, G., Manna, C., Galletti, P., and Zappia, V. 1998. Resveratrol arrests the cell division cycle at S/G2 phase transition. *Biochemical and Biophysical Research Communications* 250, 53–58.

Department of Health and Human Services. 1999. Food labeling: Health claims, soy protein and coronary heart disease, final rule. Federal registers. FDA. 64, 57700–33.

Donovan, J. L., Bell, J. R., Kasim-Karakas, S., German, J. B., Walzem, R. L., Hansen, R. J., and Waterhouse, A. L. 1999. Catechin is present as metabolites in human plasma after consumption of red wine. *The Journal of nutrition* 129 (9), 1662–1668.

Dreosti, I. E. 1996. Bioactive ingredients: Antioxidants and polyphenols in tea. *Nutrition Reviews* 54, S51–S58.

EDCC (Eye Disease Case-Control) Study Group. 1993. Antioxidant status and neovascular age-related macular degeneration. *Archives of Ophthalmology* 111, 104–9.

Eitenmiller, R. R. 1997. Vitamin E content of fats and oils: Nutritional implications. *Food Technology* 51, 78–81.

Ejaz, S., Ejaz, A., Matsuda, K., and Lim, C. W. 2006. Limonoids as cancer chemopreventive agents. *Journal of the Science of Food and Agriculture* 86, 339–45.

Elson, C. E. and Qureshi, A. A. 1995. Coupling of cholesterol and tumor-suppressive actions of palm oil to the impact of its minor constituents on 3-hydroxy 3-methylglutaryl coenzyme: A reductase activity. *Prostaglandins Leukot Essent Fatty Acids* 52, 205–8.

Erdman, J. W. and Potter Jr., S. M. 1997. Soy and bone health. *The Soy Connection* 5, 1–4.

Ernst, E. 1997. Can allium vegetables prevent cancer? *Phytomedicine* 4, 79–83.

Fahey, J. W., Zhang, Y., and Talalay, P. 1997. Broccoli sprouts: An exceptionally rich source of inducers of enzymes that protect against chemical carcinogens. *Proceedings of the National Academy of Sciences of the United States of America* 94, 10366–72.

Fenwick, G. R., Heaney, R. K., and Mullin, W. J. 1983. Glucosinolates and their breakdown products in food and food plants. Critical *Reviews in Food Science and Nutrition* 18, 123–201.

Finley, J. W. 2005. Proposed criteria for assessing the efficacy of cancer reduction by plant foods enriched in carotenoids, glucosinolates, polyphenols and selenocompounds. *Annals of Botany* 95, 1075–96.

Fleischauer, A. T., Poole, C., and Arab, L. 2000. Garlic consumption and cancer prevention: Meta-analysis of colorectal and stomach cancers. *American Journal of Clinical Nutrition* 72, 1047–52.

Fontecave, M., Lepoivre, M., Elleingand, E., Gerez, C., and Guiter, O. 1998. Resveratrol, a remarkable inhibitor of ribonucleotide reductase. *FEBS Letters* 421, 277–9.

Frankel, E. N., Kanner, J., German, J. B., Parks, E., and Kinsella, J. E. 1993. Inhibition of oxidation of human low-density lipoprotein by phenolic substances in red wine. *Lancet* 341, 454–7.

Freedman, J. E., Parker, C.III., Li, L., Perlman, J. A., Frei, B., Ivanov, V., Deak, L. R., Iafrati, M. D., and Folts, J. D. 2001. Select flavonoids and whole juice from purple grapes inhibit platelet function and enhance nitric oxide release. *Circulation* 103, 2792–8.

Gartner, C., Stahl, W., and Sies, H. 1997. Lycopene is more bioavailable from tomato paste than from fresh tomatoes. *American Journal of Clinical Nutrition* 66, 116–22.

Gaziano, J. M. and Hennekens, C. H. 1993. The role of beta-carotene in the prevention of cardiovascular disease. *Annals of the New York Academy of Sciences* 691, 148–55.

Gerster, H. 1997. The potential role of lycopene for human health. *Journal of the American College of Nutrition* 16, 109–26.

Gibson, G. R. and Roberfroid, M. B. 1995. Dietary modulation of the human colonic microbiota: Introducing the concept of prebiotics. *Journal of Nutrition* 125, 1401–12.

Giovannucci, E. 1999. Tomatoes, tomato-based products, lycopene, and cancer: Review of the epidemiologic literature. *Journal of the National Cancer Institute* 91, 317–31.

Giovannucci, E., Rimm, E. B., Liu, Y., Stampfer, M. J., and Willet, W. C. 2002. A prospective study of tomato products, lycopene, and prostate cancer risk. *Journal of the National Cancer Institute* 94, 391–8.

Girard, B. and Mazza, G. 1998. Functional grape and citrus products. In: *Functional foods.* Ed. G. Mazza. Biochemical and processing aspects. Lancaster: Technomic Publishing Company, Inc. 139–91.

Glore, S. R., van Treeck, D., Knehan, A. W., and Guild, M. 1994. Soluble fiber and serum lipids: A literature review. *Journal of the American Dietetic Association* 94, 425–36.

Gorham, J. 1989. Stilbenes and phenanthrenes. Methods in plant biochemistry, vol. 1. New York: Academic Press 159–89.

Gould, M. N. 1997. Cancer chemoprevention and therapy by monoterpenes. *Environmental Health Perspectives* 105, 977–9.

Granado, F., Olmedilla, B., and Blanco, I. 2003. Nutricional and clinical relevance of lutein in human health. *British Journal of Nutrition* 90, 487–502.

Green, M. S. and Harari, G. 1992. Association of serum lipoproteins and health-related habits with coffee and tea consumption in free-living subjects examined in the Israeli CORDIS study. *Preventive Medicine* 21, 532–45.

Gupta, S., Ahmad, N., and Mukhtar, H. 1999. Prostate cancer chemoprevention by green tea. *Seminars in Urologic Oncology* 17, 70–76

Hadley, C. W., Miller, E. C., Schwartz, S. J., and Clinton, S. K. 2002. Tomatoes, lycopene, and prostate cancer: Progress and promise. *Experimental Biology and Medicine* 227, 869–80.

Hanasaki, Y., Ogawa, S., and Fukui, S. 1994. The correlation between active oxygens scavenging and antioxidative effects of flavonoids. *Free Radical Biology and Medicine* 16, 845–850.

Handelman, G. J., Dratz, E. A., Reay, C. C., and van Kujik, F. J. G. M. 1988. Carotenoids in the human macula and whole retina. *Investigative ophthalmology and visual science* 29, 850–85.

Haslam, E. and Lilley, T. H. 1988. Natural astringency of foodstuffs – A molecular interpretation. *Critical Reviews in Food Science and Nutrition* 27, 1–40.

Hasler, C. M. 1998. Functional foods: Their role in disease prevention and health promotion. *Food Technology* 52, 63–70.

Hasler, C. M. 2002. Functional foods: Benefits, concerns and challenges – A position paper from the American Council on Science and Health. *Journal of Nutrition* 132 (12), 3772–3781.

Hesse, M. 2002. Textbook Alkaloids: Nature's curse or blessing? John Wiley & Sons.

Hicks, K. B., and Moreau, R. A. 2001. Phytosterols and phytostanols: Functional food cholesterol busters. *Food Technology* 55, 63–7.

Hodgson, J. M., Puddey, I. B., Beilin, L. J., Mori, T. A., and Croft, K. D. 1998. Supplementation with isoflavonoid phytoestrogens does not alter serum lipid concentrations randomized controlled trial in humans. *Journal of Nutrition* 128, 728–32.

Hollman, P. C., Feskens, E. J., and Katan, M. B. 1999. Tea flavonols in cardiovascular disease and cancer epidemiology. *Proceedings of the Society for Experimental Biology and Medicine* 220, 198–202.

Hollman, P. C. H. and Artz, I. C. W. 2000. Flavonols, flavones and flavanols – Nature, occurrence and dietary burden. *Journal of the Science of Food and Agriculture* 80, 1081–93.

Hollman, P. C. H. and Venema, D. P. 1993. The content of the potentially anticarcinogenic ellagic acid in plant foods. In: *Food and cancer prevention: Chemical and biological aspects.* Ed. K.W. Waldron, I. T. Johnson, G. R. Fenwick, 202–8. Fenwick Cambridge, England: The Royal Society of Chemistry.

Howell, A. B., Versa, N., Der Marderosian, A., and Foo, L. Y. 1998. Inhibition of the adherence of P-fimbriated *Escherichia coli* to uroepithelial-cell surfaces by proanthocyanindin extracts from cranberries. *New England Journal of Medicine* 339, 1085–6.

Huang, M. T., Ferraro, T., and Ho, C. T. 1994. Cancer chemoprevention by phytochemicals in fruits and vegetables: An overview. In: *Food phytochemicals for cancer prevention I. Fruits and vegetables.* Ed. M. T. Huang, T. Osawa, C. T. Ho, and R. T. Rosen, 2–16. Washington, DC: American Chemical Society.

Inks, S. and Mathews, R. 1997. Oatmeal and oat-bran: Heart healthy benefits and more. In: *New Technologies for healthy foods and nutraceuticals.* Ed. M. Yalpani. Shrewsbury, Massachusetts: ATL Press.

Isaachson, J. L., Moser, M., Stein, E. A., Dudley, K., Davey, J. A., Liskov, E., and Black, H. R. 1998. Garlic powder and plasma lipids and lipoproteins: A multicenter, randomized, placebo-controlled trial. *Archives of Internal Medicine* 158, 1189–94.

Jang, M., Cai, L., Udeani, G. O., Slowing, K. V., Thomas, C. F., Beecher, C. W. W., Fong, H. H. S., Farnsworth, N. R., Kinghorn, A. D., Mehta, R. G., Moon, R. C., and Pezzuto, J. M. 1997. Cancer chemopreventive activity of resveratrol, a natural product derived from grapes. *Science* 275, 218–20.

Jones, P. J. and Raeini-Sarjaz, M. 2001. Plant sterols and derivatives: The current spread of results. *Nutrition Review* 59, 21–4.

Kaneko, R. and Tsuchiya, T. 1954. New compound in rice bran and germ oils. *Journal of the Chemical Society* 57, 526.

Kaur, C. and Kapoor, H. C. 2001. Antioxidants in fruits and vegetables – The millennium's health. *International Journal of Food Science & Technology* 36, 703–25.

Kohlmeier, L. and Hastings, S. B. 1995. Epidemiologic evidence of a role of carotenoids in cardiovascular disease prevention. *American Journal of Clinical Nutrition* 62, 1370S–6S.

Kohlmeier, L., Kark, J. D., Gomez-Garcia, E., Martin, B. C., Steck, S. E., Kardinaal, A. F. M., Ringstad, J., Thamm, M., Masaev, V., Riemersma, R., Martin-Moreno, J. M., Huttunen, J. K., and Kok, F. J. 1997a. Lycopene and myocardial infarction risk in the EURAMIC Study. *American Journal of Epidemiology* 146, 618–26.

Kohlmeier, L., Weterings, K. G. C., Steck, S., and Kok, F. J. 1997b. Tea and cancer prevention: An evaluation of the epidemiologic literature. *Nutrition Cancer* 27, 1–13.

Komiyama, K., Hayashi, M., Cha, S., and Yamaoka, M. 1992. Antitumor and antioxidant activity of tocotrienols. In: *Lipid-soluble antioxidants: Biochemistry and clinical applications.* Ed. A. S. H. Ong and L. Packer, 152–9. Basel. Switzerland: Birkhauser Verlag.

Kono, S., Shinchi, K., Ikeda, N., Yanai, F., and Imanishi, K. 1992. Green tea consumption and serum lipid profiles: A cross-sectional study in Northern Kyushu, Japan. *Preventive Medicine* 21, 526–31.

Krinsky, N. I. 1993. Actions of carotenoids in biological systems. *Annual Review of Nutrition* 13, 561–87.

Krinsky, N. I. and Johnson, E. J. 2005. Carotenoid actions and their relation to health and disease. *Molecular Aspects of Medicine* 26, 459–516.

Kris-Etherton, P. M., Hecker, K. D., Bonanome, A., Coval, S. M., Binkoski, A. E., Hilpert, K. F., Griel, A. E., and Etherton, T. D. 2002. Bioactive compounds in foods: Their role in the prevention of cardiovascular disease and cancer. *American Journal of Medicine* 113, 71S–88S.

Kuroda, Y. and Hara, Y. 1999. Antimutagenic and anticarcinogenic activity of tea polyphenols. *Mutation Reserach* 436, 69–97.

Laplaud, P. M., Lelubre, A., and Chapman, M. J. 1997. Antioxidant action of *Vaccinium myrtillus* extract on human low-density lipoproteins *in vitro*: Initial observations. *Fundamental and Clinical Pharmacology* 11, 35–40.

Law, M. R. and Morris, J. K. 1998. By how much does fruit and vegetable consumption reduce the risk of ischaemic heart disease? *European Journal of Clinical Nutrition* 52, 549–56.

Lia, A., Andersson, H., Mekki, N., Juhel, C., Senft, M., and Lairon, D. 1997. Postprandial lipemia in relation to sterol and fat excretion in ileostomy subjects given oat-bran and wheat with meals. *American Journal of Clinical Nutrition* 66, 357–65.

Lightbourn, G. J., Griesbach, R. J., Novotny, J. A., Clevidence, B. A., Rao, D. D., and Stommel, J. R. 2008. Effects of anthocyanin and carotenoid combinations on foliage and immature fruit color of *Capsicum annuum* L. *Journal of heredity.* 99 (2), 105–111.

Lin, X. Y., Lin, J. Z., and Milner, J. A. 1994. Dietary garlic suppresses DNA adducts caused by N-nitroso compounds. *Carcinogenesis.* 15, 349–52.

Lucock, M. 2004. Is folic acid the ultimate functional food component for disease prevention? *BMJ: British Medical Journal* 328 (7433), 211–214.

Mayne, S. T. 1996. Beta-carotene, carotenoids, and disease prevention in humans. *FASEB Journal* 10, 690–701.

Mei, X., Lin, X., Liu, J., Song, P., Hu, J., and Liong, X. 1989. The blocking effect of garlic on the formation of N-nitrosoproline in humans. *Acta Nutrimenta Sinica* 11, 141–5.

Mei, X., Wang, M. C., Xu, H. X., Pan, P., Gao, C. Y., Han, N., and Fu, M. Y. 1982. Garlic and gastric cancer – The effect of garlic on nitrite and nitrate in gastric juice. *Acta Nutrimenta Sinica* 4, 53–8.

Messina, M. and Barnes, S. 1991. The role of soy products in reducing risk of cancer. *Journal of the National Cancer Institute* 83, 541–6.

Michnovicz, J. J., and Bradlow, H. L. 1991. Altered estrogen metabolism and excretion in humans following consumption of indole carbinol. *Nutrition and Cancer* 16, 59–66.

Miettinen, T. A. 2001. Phytosterols – What plant breeders should focus on. *Journal of the Science of Food and Agriculture* 81, 895–903.

Milner, J. A. 2006. Preclinical perspectives on garlic and cancer. *Journal Nutrition* 136, 827S–31S.

Mithen, R. F., Dekker, M., Verkerk, R., Rabot, S., and Johnson, I. T. 2000. The nutritional significance, biosynthesis and bioavailability of glucosinolates in human foods. *Journal of the Science of Food and Agriculture* 80, 967–84.

Moeller, S. M., Jacques, P. F., and Blumberg, J. B. 2000. The potential role of dietary xanthophylls in cataract and age-related macular degeneration. *Journal of the American College of Nutrition* 19, 522S–7S.

Morazzoni, P. and Bombardelli, E. 1996. *Vaccinium myrtillus. Fitoterapia* 67, 3–30.

Morazzoni, P. and Magistretti, M. J. 1990. Activity of myrtocyan, anthocyanoside complex from *Vaccinium myrtillus* (VMA), on platelet aggregation and adhesiveness. *Fitoterapi.* 61, 13–21.

Moreau, R. A., Powell, M. J., and Hicks, K. B. 1996. Extraction and quantitative analysis of oil from commercial corn fiber. *Journal of Agricultural and Food Chemistry* 44, 214954.

Moroney, M. A., Alcaraz, M. J., Forder, R. A., Carey, F., and Hoult, J. R. S. 1988. Selectivity of neutrophil 5-lipoxygenase inhibition by an antiinflammatory flavonoid glycoside and related aglycone flavonoids. *Journal of Pharmacy and Pharmacology* 40, 787–792.

Murphy, P. A. and Hendrich, S. 2002. Phytoestrogens in foods. *Advances in Food and Nutrition Research* 44, 195–246.

Nagata, C., Takatsuka, N., Kurisu, Y., and Shimizu, H. 1998. Decreased serum total cholesterol is associated with high intake of soy products in Japanese men and women. *Journal of Nutrition* 128, 209–13.

Nakaizumi, A., Baba, M., Uehara, H., Iishi, H., and Tatsuta, M. 1997. DLimonene inhibits N-itrosobis (2-oxopropyl) amine induced hamster pancreatic carcinogenesis. *Cancer Letter* 117, 99–103.

Namiki, M. 1994. Antimutagen and anticarcinogen research in Japan. In: *Food Phytochemicals for Cancer Prevention I. Fruits and Vegetables.* Ed. M. T. Huang, T. Osawa, C. T. Ho, and R. T. Rosen, 65–81. Washington, DC: American Chemical Society.

Ness, A. R., and Powles, J. W. 1997. Fruit and vegetables, and cardiovascular disease: A review. *International Journal of Epidemiology* 26, 1–13.

Nestle, M. 1998. Broccoli sprouts in cancer prevention. *Nutrition Reviews* 56, 127–30.

Ofek, I., Goldhar, J., and Sharon, N. 1996. Anti-*Escherichia coli* adhesion activity of cranberry and blueberry juices. In: *Toward Anti-Adhesion Therapy for Microbial.* Ed. S. Kahane and I. Ofek.

Oguni, I., Chen, S. J., Lin, P. Z., and Hara, Y. 1992. Protection against cancer risk by Japanese green tea. *Preventive Medicine* 21, 332–3.

Olson, J. A. 1999. Carotenoids and human health. *Arch Latinoamer Nutr.* 49, 7S–11S.

Oomah, B. D. 2001. Flaxseed as a functional food source. *Journal of the Science of Food and Agriculture* 81, 889–94.

Oomah, B. D., and Mazza, G. 1999. Health benefits of phytochemicals from selected Canadians crops. *Trends Food Science and Technology* 10, 193–8.

Ottaway, P. B. 2008. *Food Fortification and Supplementation – Technological, Safety and Regulatory Aspects.* CRC Press, 1st Edition, ISBN 978-0849364099.

Pace-Asciak, C. R., Hahn, S., Diamandis, E. P., Soleas, G., and Goldberg, D. M. 1995. The red wine phenolics trans-resveratrol and quercetin block human platelet aggregation and eicosanoic synthesis: Implications for protection against coronary heart disease. *Clinica. Chimica. Acta.* 235, 207–219.

Palace, V. P., Khaper, N., Qin, Q., and Singal, P. K. 1999. Antioxidant potentials of vitamin A and carotenoids and their relevance to heart disease. *Free Radical Biology and Medicine* 26, 746–61.

Palozza, P. and Krinsky, N. I. 1992. Antioxidant effects of carotenoids *in vivo* and *in vitro*: An overview. *Methods in Enzymology* 213, 403–20.

Parker, R. A., Pearce, B. C., Clark, R. W., Gordon, D. A., and Wright, J. J. K. 1993. Tocotrienols regulate cholesterol production in mammalian cells by post-transcriptional suppression of 3-hydroxy-3-methylglutaryl-coenzyme: A reductase. *Journal of Biological Chemistry* 268, 11230–8.

Parr, A. J., and Bolwell, G. P. 2000. Phenols in the plant and in man. The potential of possible nutritional enhancement of the diet by modifying the phenol content or profile. *Journal of the Science of Food and Agriculture* 80, 985–1012.

Pelletier, J. and Caventou, J. 1951. Marcel Delepine. *Journal of Chemical Education* (28), 454–61.

Petesch, B. L. and Sumiyoshi, H. 1999. Recent advances on the nutritional benefits accompanying the use of garlic as a supplement. *Trends in Food Science and Technology* 9, 415–8.

Piironen, G. W., Price, K. R., Rhodes, M. J. C., and Williamson, G. 1986. Tocopherols and tocotrienols in Finnish foods: Vegetables, fruits and berries. *Journal of Agricultural and Food Chemistry* 34, 742–6.

Piironen, V., Lindsay, D. G., Miettinen, T. A., Toivo, J., and Lampi, A. M. 2000. Plant sterols: Biosynthesis, biological function and their importance to human health. *Journal of the Science of Food and Agriculture* 80, 939–66.

Potter, S. M. 1998. Soy protein and cardiovascular disease: The impact of bioactive components in soy. *Nutrition Reviews* 56, 231–5.

Price, K. R. and Fenwick, G. R. 1985. Naturally occurring oestrogens in foods – A review. *Food Addit. Contam.* 2, 73–106.

Prior, R. L. and Cao, G. 2000. Antioxidant phytochemicals in fruits and vegetables: Diet and health implications. *Horticultural Science* 35, 588–92.

Qureshi, A. A., Pearce, B. C., Nor, R. M., Gapor, A., Peterson, D. M., and Elson, C. E. 1996. α-Tocopherol attenuates the impact of tocotrienol on hepatic 3-hydroxy-3-methylglutaryl coenzyme: A reductase activity. *Journal of Nutrition* 126, 389–94.

Rahman, K. and Lowe, G. M. 2006. Garlic and cardiovascular disease: A critical review. *Journal for Nutrition* 136, 736S–740S.

Rao, A. V. and Agarwal, S. 1999. Role of lycopene as antioxidant carotenoid in the prevention of chronic diseases: A review. *Nutrition Research* 19, 305–23.

Renaud, S. and de Lorgeril, M. 1992. Wine, alcohol, platelets, and the French paradox for coronary heart disease. *Lancet* 339, 1523–6.

Rice-Evans, C. A., Miller, N. J., and Paganga, G. 1996. Structure–antioxidant activity relationship of flavonoids and phenolic acids. *Free Radical Biology and Medicine* 20, 933–56.

Ripple, G. H., Gould, M. N., Stewart, J. A., Tutsch, K. D., Arzoomanian, R. Z., Alberti, D., Feierabend, C., Pomplun, M., Wilding, G., and Bailey, H. H. 1998. Phase I clinical trial of perillyl alcohol administered daily. *Clinical Cancer Research* 4, 1159–64.

Ripsin, C. M., Keenan, J. M., Jacobs, D. R., Elmer, P. J., and Welch, R. R. 1992. Oat products and lipid lowering. *Journal of the American Medical Association* 267, 3317–25.

Rissanen, T., Voutilainen, S., Nyyssönen, K., and Salonen, J. T. 2002. Lycopene, atherosclerosis and coronary heart disease. *Experimental Biology and Medicine* 227, 900–7.

Roberfroid, M. B. 1996. Functional effects of food components and the gastrointestinal system: Chicory fructooligosaccharides. *Nutrition Reviews* 54, S38–S42.

Rock, C. L., Lovalvo, J. L., Emenhiser, C., Ruffin, M. T., Flatt, S. W., and Schwart, S. J. 1998. Bioavailability of β-carotene is lower in raw than in processed carrots and spinach. *Journal of Nutrition* 128 (5), 913–6.

Rodriguez-Amaya, D. B. 1993. Nature and distribution of carotenoids in foods. In: *Shelf-life Studies of Foods and Beverages. Chemical, Biological, Physical and Nutritional Aspects.* Ed. G. Charalambous. 547–89. Amsterdam: Elsevier Science Publishers.

Rodriguez-Amaya, D. B. 1999. Latin American food sources of carotenoids. *Arch Latinoamer Nutr.* 49, 74S–84S.

Rotondo, S. and de Gaetano, G. 2000. Protection from cardiovascular disease by wine and its derived products. Epidemiological evidence and biological mechanisms. *World Review of Nutrition and Dietetics* 87, 90–113.

Schneeman, B. O. 2000. Linking agricultural production and human nutrition. *Journal of the Science of Food and Agriculture* 81, 3–9.

Seddon, J. M., Ajani, U. A., Sperduto, R. D., Hiller, R., Blair, N., Burton, T. C., Ferber, M. D., Gragoudas, E. S., Haller, J., Miller, D. T., Yannuzzi, L. A., and Willet, W. 1994. Dietary carotenoids, vitamins A, C, and E, and advanced age-related macular degeneration. *Journal of the American Medical Association* 272, 1413–20.

Serbinova, E. A., Tsuchiya, M., Goth, S., Kagan, V. E., and Packer, L. 1993. Antioxidant action of α-tocopherol and α-tocotrienol in membranes. In: *Vitamin E in Health and Disease.* Ed. L. Packer and J. Fuchs, 235–43. New York: Marcel Dekker.

Sies, H. and Stahl, W. 1998. Lycopene: Antioxidant and biological effects and its bioavailability in the human. *Proceedings of the Society for Experimental Biology and Medicine* 218, 121–4.

Silagy, C. and Neil, A. 1994. Garlic as a lipid-lowering agent – A meta-analysis. *Journal of the Royal College of Physicians of London* 28, 39–45.

Sivam, G. P., Lampe, J. W., Ulness, B., Swanzy, S. R., and Potter, J. D. 1997. *Helicobacter pylori – In vitro* susceptibility to garlic (*Allium sativum*) extract. *Nutrition and Cancer* 27, 118–21.

Stacewicz-Sapuntzakis, M. and Bowen, P. E. 2005. Role of lycopene and tomato products in prostate health. *Biochimica et Biophysica Acta* 1740, 202–5.

Stahl, W., Ale-Agha, N., and Polidori, M. C. 2002. Non-antioxidant properties of carotenoids. *Biological Chemistry* 383, 553–8.

Stahl, W. and Sies, H. 2005. Bioactivity and protective effects of natural carotenoids. *Biochimica et Biophysica Acta* 1740, 101–7.

Standing Committee on the Scientific Evaluation of Dietary Reference Intakes and Its Panel on Dietary Antioxidants and Related Compounds, Dietary Reference Intakes: Proposed Definition and Plan for Review of Dietary Antioxidants and Related Compounds, in Dietary Reference Intakes, N.A. Press, Ed. 1–13. (1998) Institute of Medicine: Washington, DC.

Steinmetz, K. A., Kushi, H., Bostick, R. M., Folsom, A. R., and Potter, J. D. 1994. Vegetables, fruit, and colon cancer in the Iowa Women's Health Study. *American Journal of Epidemiology* 139, 1–15.

Steinmetz, K. A. and Potter, J.D. 1996 Vegetables, fruit, and cancer prevention: A review. *Journal of the American Dietetic Association* 96, 1027–39.

Stensvold, I., Tverdal, A., Solvoll, K., and Foss, O. P. 1992. Tea consumption, relationship to cholesterol, blood pressure and coronary artery disease mortality. *Preventive Medicine* 21, 546–53.

Stoewsand, G. S. 1995. Bioactive organosulfur phytochemicals in *Brassica oleracea* vegetable – A review. *Food and Chemical Toxicology* 33, 537–43.

Strack, D., Vogt, T., and Schliemann, W. 2003. Recent advances in betalain research. *Phytochemistry* 62(3), 247–69.

Sun, N. J., Woo, S. H., Cassady, J. M., and Snapka, R. M. 1998. DNA-polymerase and topoisomerase-II inhibitors from *Psoralea coryfolia. Journal of Natural Products* 61, 362–6.

Talalay, P. and Fahey, J. W. 2001. Phytochemicals from cruciferous plants protect against cancer by modulating carcinogen metabolism. *Journal of Nutrition* 131, 3027S–33S.

Tan, B. 1989. Palm carotenoids, tocopherols and tocotrienols. *Journal of the American Oil Chemists' Society* 66, 770–6.

Teissedre, P. L., Frankel, E. N., Waterhouse, A. L., Pelez, E., and German, J. B. 1996. Inhibition of *in vitro* human LDL oxidase by phenolic antioxidants from grapes and wines. *Journal of the Science of Food and Agriculture* 70, 55–61.

Thebaudin, J. Y., Febvre, A. C., Harrington, M., and Bourgeois, C. M. 1997. Dietary fibres: Nutritional and technological interest. *Trends in Food Science and Technology* 8, 41–8.

Thomas, P. R. and Earl, R. 1994. Enhancing the food supply. In: *Opportunities in the Nutrition and Food Sciences.* Washington, DC: National Academy Press, 98–142.

Thompson, L. U., Robb, P., Serraino, M., and Cheung, E. 1991. Mammalian lignan production from various foods. *Nutrition and Cancer* 16, 43–52.

Tijburg, L. B. M., Mattern, T., Folts, J. D., Weisgerber, U. M., Tiwari, B., Brunton, N., and Brennan, C. B. 2013. *Handbook of Plant Food Phytochemicals: Sources, Stability and Extraction.* Publishers John Wiley & Sons, Ltd. Wiley-Blackwell.

Torronen, R. 2000. Flavonols and ellagic acid in berries. *Recent Advance in Agricultural and Food Chemistry* 1, 31–7.

Uedo, N., Tatsuta, M., Iishi, H., Baba, M., Sakai, N., Yano, H., and Otani, T. 1999. Inhibition by D-limonene of gastric carcinogenesis induced by N-methyl-N′-nitro-N-nitrosoguanidine in Wistar rats. *Cancer Letter* 137, 131–136.

Van den Berg, H., Faulks, R., Fernando Granado, H., Hirschberg, J., Olmedila, B., Sandmann, G., Southon, S., and Stahl, W. 2000. The potential for the improvement of carotenoid levels in foods and the likely systemic effects. *Journal of the Science of Food and Agriculture* 80, 880–912.

Vansoest, P. J. 1994. Nutritional ecology of the ruminant, 2nd ed. Cornell University Press. Ithaca, NY, 476. 33.

Verhoeven, D. T. H., Goldbohm, R. A., van Poppel, G., Verhagen, H., and van den Brandt, P. A. 1996. Epidemiological studies on brassica vegetables and cancer risk. *Cancer Epidemiol Biomark Prev.* 5, 733–48.

Verhoeven, D. T. H., Verhagen, H., Goldbohm, R. A., van den Brandt, P. A., and van Poppel, G. 1997. A review of mechanisms underlying carcinogenicity by brassica vegetables. *Chemico-Biological Interactions* 103, 79–129.

Voragen, A. G. J. 1998. Technological aspects of functional food-related carbohydrates. *Trends in Food Science and Technology* 9, 328–35.

Wang, H., Cao, G., and Prior, R. L. 1997. Oxygen radical absorbing capacity of anthocyanin. *Journal of Agriculture and Food Chemistry* 45, 3425–31

Wang, H. C., Huang, X. M., Hu, G. B., Yang, Z. Y., Huang, H. B. 2005. A comparative study of chlorophyll loss and its related mechanism during fruit maturation in the pericarp of fast-and slow-degreening litchi pericarp. *Scientia Horticulturae* 106 (2), 247–257.

Wang, H., Provan, G. J., and Helliwell, K. 2000. Tea flavonoids: Their functions, utilisation and analysis. *Trends in Food Science and Technology* 11, 152–60.

Warshafsky, S., Kamer, R. S., and Sivak, S. L. 1993. Effect of garlic on total serum cholesterol. A meta-analysis. *Annals of Internal Medicine* 119, 599–605.

WCRF. 1997. Food, nutrition and the prevention of cancer: A global perspective. Washington, D. C.: American Institute for Cancer Research.

Wertz, K., Siler, U., and Goralczyk, R. 2004. Lycopene: modes of action to promote prostate health. *Archives of Biochemistry and Biophysics* 430, 127–34.

Wink, M., Alfermann, A. W., Franke, R., Wetterauer, B., Distl, M., Windhovel, J., Krohn, O., Fuss, E., Garden, H., and Mohagheghzadeh, A. 2005. Sustainable bioproduction of phytochemicals by plant *in vitro* cultures: Anticancer agents. *Plant Genetic Resource: Characterization and Utilization* 3, 90–100.

Wollin, S. D., and Jones, P. J. H. 2001. Alcohol, red wine and cardiovascular disease. *Journal of Nutrition* 131, 1401–4.

Wong, G. Y. C., Bradlow, L., Sepkovic, D., Mehl, S., Mailman, J., and Osborne, M. P. 1998. Dose-ranging study of indole-3-carbinol for breast cancer prevention. *Journal of Cellular Biochemistry* 22 (28–29), 111–6.

Yang, C. S., Landau, J. M., Huang, M. T., and Newmark, H. L. 2001. Inhibition of carcinogenesis by dietary polyphenolic compounds. *Annual Review of Nutrition* 21, 381–406.

You, W. C., Blot, W. J., Chang, Y. S., Ershow, A. G., Yang, Z. T., An, Q., Henderson, B. E., Fraumeni, J. F., and Wang, T. G. 1989. *Allium* vegetables and reduced risk of stomach cancer. *Journal of National Cancer Institute* 81, 162–4.

Yu, B. P. 1996. Aging and oxidative stress: Modulation by dietary restriction. *Free Radical Biology and Medicine* 21, 651–68.

Zhang, Y., Talalay, P., Cho, C. G., and Posner, G. H. 1992. A major inducer of anticarcinogenic protective enzymes from broccoli: Isolation and elucidation of structure. *Proceedings of the National Academy of Sciences of the United States of America* 89, 2399–403.

Ziegler, R. G. 1991. Vegetables, fruits, and carotenoids and the risk of cancer. *American Journal Clinical Nutrition* 53, 251S–9S.

Ziegler, R. G., Colavito, E. A., Hartge, P., McAdams, M. J., Schoenberg, J. B., Mason, T. J., and Fraumeni Jr., J. F. 1996. Importance of alpha-carotene, beta-carotene, and other phytochemicals in the etiology of lung cancer. *Journal of National Cancer Institute* 88, 612–5.

3 Lukoskin
An Effective Herbal Treatment of Leucoderma Developed by Defence Institute of Bio-Energy Research (DRDO), Haldwani

Hemant Kr. Pandey, Harsahay S. Meena, Sanjai Kr. Dwivedi and Madhu Bala

CONTENTS

3.1 Introduction 49
3.1.1 Incidence of Leucoderma 50
3.1.2 Clinical Features of Leucoderma 50
3.2 Types of Leucoderma 51
3.2.1 Leucoderma *Localisata* (Localized) 51
3.2.2 Leucoderma *Generalista* (Generalized) 51
3.2.3 Leucoderma *Universalis* (Universalized) 51
3.3 Etiology of Leucoderma 52
3.3.1 Some Triggering Factors Which also Precipitate the Disease 52
3.3.2 Autoimmune Disease 52
3.3.3 Autoimmune Hypothesis 52
3.3.4 Neurogenic Hypothesis 53
3.3.5 Self-Destruction Theory Lerner 53
3.3.6 Dietary Factors 53
3.3.7 Endocrine Factors 53
3.3.8 Hermoral Factor (Bodily Fluids) 53
3.3.9 Hepatic Function 54
3.3.10 Gastric Acidity 54
3.3.11 Synthesis of Melanin 54
3.4 Available Treatment of Leucoderma 54
3.4.1 Medical Therapies 55
3.4.1.1 Topical Steroid Therapy 55
3.4.1.2 Psoralen Photochemotherapy 55
3.4.1.3 Topical Psoralen Photochemotherapy 56

3.4.1.4 Oral Psoralen Photochemotherapy ... 56
3.4.1.5 Depigmentation ... 56
3.4.2 Surgical Therapies ... 57
3.4.2.1 Autologous Skin Grafts ... 57
3.4.2.2 Skin Grafts using Blisters ... 57
3.4.2.3 Micropigmentation (Tattooing) ... 57
3.4.2.4 Autologous Melanocyte Transplants ... 57
3.4.3 Adjunctive Therapies ... 57
3.4.3.1 Sunscreens ... 57
3.4.3.2 Cosmetics ... 58
3.5 Drawbacks of Existing Treatment Modalities ... 58
3.5.1 Medical Therapies ... 58
3.5.1.1 Topical Steroid Therapy ... 58
3.5.1.2 Psoralen Photochemotherapy ... 58
3.5.1.3 Depigmentation Therapy ... 58
3.5.2 Surgical Therapies ... 59
3.5.2.1 Autologous Skin Graft ... 59
3.5.2.2 Skin Grafts Using Blisters ... 59
3.5.2.3 Micropigmentation (Tattooing) ... 59
3.5.2.4 Autologous Melanocyte Transplants ... 59
3.6 DIBER–Invented Herbal Therapy of Leucoderma ... 59
3.6.1 Summary of Invention ... 59
3.6.1.1 Ointment ... 60
3.6.1.2 Oral Dose ... 60
3.7 Toxicological Studies ... 60
3.7.1 Materials ... 61
3.7.2 Acute Oral Toxicity ... 61
3.7.3 Primary Skin Irritation Test ... 62
3.7.4 Toxicological Studies of Ointment ... 62
3.7.5 Biochemical Studies in Blood Plasma of Leucoderma Patients ... 64
3.7.5.1 Screening of Tyrosinase Enzyme ... 64
3.7.5.2 Screening of Copper and Zinc in the Blood Plasma of Leucoderma Patients ... 64
3.8 Clinical Trials ... 67
3.9 Protocol for Use ... 68
3.9.1 Ointment ... 68
3.9.2 Oral Dose ... 69
3.9.3 Restriction to be Followed by the Patients ... 69
3.10 Recommendation ... 69
3.11 Launching of Anti-Leucoderma Herbal Product (Lukoskin) in the Market ... 69
References ... 69

3.1 INTRODUCTION

Leucoderma, or vitiligo, commonly known as Sweta Kustha, Kodh or Phuleri and is an idiopathic acquired disorder of the skin that results from the selective destruction of melanocyte cells (Agrawal et al. 2004; Koranne et al. 1986). Patients with leucoderma develop white spots on the skin of various sizes and in different locations. It is considered as a big social stigma in India, as people confuse it with leprosy. The affected individuals suffer from constant depression and social ostrcization. It is considered a medical disqualification for recruitment in the military and paramilitary. The etiological factors for leucoderma are not specific but considered, mainly due to the disturbed melanin synthesis owing to an inactivation of tyrosinase enzymes; a deficiency of copper, zinc and iron; the absence of certain aromatic amino acids like tyrosine, phenylalanine, di-hydroxy phenylalanine; and a lack of vitamins like folic acid, pentothenic acid and vitamin B_{12} in the diet (Antoniou et al. 1992; Behl et al. 1972). The disturbed function of nerve cells; destruction of melanocyte cells; external injuries; worm infestation; abnormal thyroid function (Bolognia et al. 1988; Nordlund et al. 1993); constant disturbed digestion; emotional and psychological stress; neurotoxins; and constant exposure to chemicals, paints and dyes are some additional causes responsible for its occurrence.

There are many existing remedies of this disorder viz. allopathic, surgical and adjunctive therapy, none of which has a satisfactory cure for this disorder. These are either costly or single component–based, with very low level of efficacy. Patients will often develop blisters, edema and skin irritation, and as a result, most of the patients discontinue the treatment.

The quest to find an effective cure leucoderma has finally ended with the development of a new herbal product by extensive laboratory studies and clinical trials by scientists of the Defence Institute of Bio-energy Research, (DIBER), Haldwani, a laboratory of Defence Research & Development Organization, Ministry of Defence, Govt. of India. These scientists have discovered the unique properties of a number of herbs from the high attitudes of Himalayan ranges and came out with a formulation to cure leucoderma patients without any side effects. Although, leucoderma may not be considered to be a disease, the patients affected by leucoderma may have a severe inferiority complex resulting in depression, thereby adversely affecting their work performance, personality and aesthetics. Clinically the product is quite effective and helps not only in restoring the normal complexion in the affected areas, but also relieves the affected ones from mental stress, emotional and psychological trauma and thereby, enhancing their confidence and efficiency.

The formulation is available in the form of liquid oral dose for internal medication to remedy the nutritional deficiency and to restore the immune system, and an ointment for topical application on leucoderma lesions. The AIMIL Pharmaceuticals (India) Ltd., New Delhi, has been granted the transfer of technology to manufacture and market this research product of DIBER (DRDO) under the brand name 'Lukoskin' (Figure 3.1) based on their respective technological and marketing credentials.

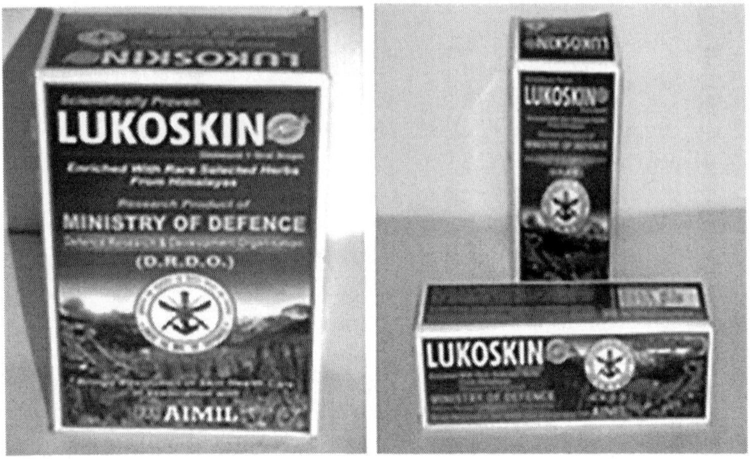

FIGURE 3.1 Lukoskin product.

3.1.1 Incidence of Leucoderma

Worldwide leucoderma incidence has been reported to be 1–2% of the population. In Japan it is 1.64%, 1.0% in the Soviet Union, 1.4% in the United States, 0.24%, in Great Britain 0.39% in Switzerland and 1.0% in Egypt. In India the incidence of disease is 3–4%, but in some parts of Rajasthan and Gujarat it is as high as 8–9%, in which the male/female ratio is reported to be 1:2.5.

Leucoderma could affect anybody, it is not at all an infectious disease. It cannot spread by any kind of contact, such as sharing food, linens or clothing. It is perfectly safe to touch, hug and kiss persons affected by this disease.

There is a 60–70% chance are that a child of one parent with leucoderma may not develop leucoderma at all. However, the chances of transmitting leucoderma to children become more prevalent in cases of marriage between cousins. If someone with leucoderma marries their first cousin, there is a 60–70% chance that the child will have normal skin (Hafez et al. 1983).

3.1.2 Clinical Features of Leucoderma

The disease is characterized by well-defined, varying sized macules of milk white appearance due to complete absence of melanin (Dutta et al. 1969). The following are the characteristics of leucoderma lesions:

- Localized or widespread: milk white, dead white or chalk white
- Scattered or confluent
- Small-sized macules; oval, round and irregular in shape, sometimes streaking and in larger patches
- Lesions may be unilateral, mostly bilateral, and not always symmetrical.

- Lesions may enlarge peripherally, may coalesce into extensive patches and gradually become generalized. In the summer, lesions become more conspicuous due to surrounding dark skin.
- In the menstrual cycle, white patches may turn pink-red and then return to normal after the cycle.
- The lesions normally have hyper-pigmented borders or the borders may be raised.
- The hairs may or may not be white on lesions, and the loss of pigment in hairs is called leucotrichia or achromotrichia.
- Skin, hair and mucus membranes are affected in this disease.
- Increased sweating in lesions, where there is vasoconstriction, less sensation in skin; patients become less tolerant to sunlight
- The leucoderma patches are most common in the hands, feet, arms, face, lips, armpits, groin, navel and genitals.
- In the case that the scalp is affected, the hair growing area may also be devoid of colour.

3.2 TYPES OF LEUCODERMA

According to extent of the involved areas, the lesions have been classified as

- Leucoderma *Localista*
- Leucoderma *Generalista*
- Leucoderma *Universalis*

3.2.1 Leucoderma *Localisata* (Localized)

This occurs only over one part of the body, e.g. face, hands and legs, scalp, genitalia, perineal area or on one side of the face. Localized types may be unilateral and cover the area of the segmental supply of the peripheral nerve, i.e. patches that occur along the directions of cutaneous nerves.

3.2.2 Leucoderma *Generalista* (Generalized)

This involves more than 50–60% of the body surface area. Small hypopigmented spots appear over exposed area like face, neck, body folds and external orifices, sites of recent trauma, umbilicus, genitals and mucus membranes. The borders of maculae appear convex, looking like an invasion on pigmented area, which shows the concave pigmented border area.

3.2.3 Leucoderma *Universalis* (Universalized)

The entire cutaneous surface is depigmented. Small hyperpigmented spots appear on the exposed part of the skin in response to unavoidable exposure to skin. It closely resembles albinism (pseudoalbinism) except the persistence of pigment remains in the eyes and hair.

3.3 ETIOLOGY OF LEUCODERMA

- It is more common in darker races, in the age group 10 to 30 years.
- Both sexes are considered equally affected, but others found the ratio 1:2.5 to male and female (young girls and boys) in India.
- The disease appears and progresses without any apparent cause.
- In some cases, the cause is suspected to be either local injury or psychological trauma. Thus only the constitution of the patients has been mainly held responsible.
- Predisposition to leucoderma may also be the cause:
  - i. 30–33% family history of disease
  - ii. It appears to be transmitted as autosomal dominant mendelian characteristics.
  - iii. Blood types B and AB are more predisposed.
  - iv. Alopecia areata and canities also contribute in family history.
  - v. Syphilis

3.3.1 Some Triggering Factors Which also Precipitate the Disease

- Mental stress
- Psychological trauma
- Emotional stress
- Worm infection
- Burns, injuries and itching friction
- Chemical exposure, particularly from rubber, rubber goods and phenol compounds

3.3.2 Autoimmune Disease

- Hyper thyroidism
- Pernicious anemia (low level of hemoglobin, defective absorption of vitamin B-12)
- Diabetes mellitus
- Adrenal insufficiency
- Prolonged gastrointestinal disturbances
- Adrenocortical insufficiency

3.3.3 Autoimmune Hypothesis

In addition to all the above predispositions, as well as precipitating and triggering factors contribute to leucoderma, a disturbed autoimmune system in which the destruction of melanocytes cells takes place by lymphocytes is also responsible (Cape et al. 2003; Kemp et al. 2002; Ongenae et al. 2003). The reason for the loss of melanocyte cells or their abnormal function, however, is not exactly clear. The theories (Harning et al. 1991) include:

- Poisoning of melanocyte by body's own immune system (autoimmunity)
- Poisoning of melanocytes by abnormal nerve cells
- Self-destruction of melanocyte
- Genetic defects which predispose melanocytes' injury

3.3.4 Neurogenic Hypothesis

- A compound released by peripheral nerve endings is toxic to melanocytes and inhibits melanogenesis (Liu et al. 1996).
- In dermatomal variants the affected area shows sympathetic nerve disfunction. Catechol neurotransmitters probably destroy melanacytes. Compounds like melantoin, noradranelin and hydroguinidole also found to have same destructive role (Shajil et al. 2005).

3.3.5 Self-Destruction Theory Lerner

Defects in melanin synthesis lead to the accumulation of toxic compounds (phenolic compounds), which destroy the melanocyte. This may be due to lack of copper and a defective tyrosinase enzyme system.

3.3.6 Dietary Factors

The deficiency of cuprominerals, vitamins (viz. folic acid, pentathenic acid and vitamin B_{12}), proteins and aromatic amino acids (viz. phenylalanine, dihydroxy phenylalanine, tyrosine etc.) could be contributing factors. An excess of vitamin C also has inhibitory action on melanogenesis.

3.3.7 Endocrine Factors

In most of the leucoderma cases, the endocrine glands have an incriminated tendency. The pituitary gland has been suspected, because of the high level of urinary melanotropic hormones and the absence of these hormones in affected areas. The thyroid gland is also suspected, because of association of leucoderma with Grave's disease. The specific thyroglobulin antibodies were associated with leucoderma and thyroid disease. The gonads also have been blamed for similar action. Higher amounts of adrenaline has also been noticed in leucoderma.

3.3.8 Hermoral Factor (Bodiiy Fluids)

The sulphydril compounds in epidermis have more affinity to Cu^{++} binding, which acts as a neutral inhibitor of tyrosinase activity. This may be reversed by SH-inhibitors like Cu and parachloro mercuribenzoate. Sulphi hydril (SH-) group containing compounds have been found more in white patches of leucoderma patients.

3.3.9 Hepatic Function

The hepatic function is disturbed by toxins of intestinal parasitism. Liver extract could help cure leucoderma, due to presence of tyrosine and Cu.

3.3.10 Gastric Acidity

There was apparent achlorhydria in leucoderma patients. Gastric acidity in 50% of cases was noticed.

3.3.11 Synthesis of Melanin

In vertebrates, melanin synthesis or melanogenesis is a specialized function expressed only in the differentiation products of neural crest–derived cells called melanocytes, widely distributed in the skin, hair follicle roots and other places of the body, such as eyes and the ears (Blanpain et al. 2004; Steingrímsson et al. 2004). Melanogenesis occurs through an enzymatic process, catalysed by tyrosinase (albino locus) and tyrosinase related protein (TRP)-I (b-locus protein/gp75) as well as TRP-2 (DCT/slaty locus protein), which convert tyrosine to melanin pigments (Cape et al. 2002; Huang et al. 2002; Kemp et al. 1998; Vance et al. 2004). The melanocytes in the skin localize the basal layers of the epidermis and hair bulbs, where they contact many adjacent keratinocytes through their dendritic processes (Caixia et al. 1999; Steingrímsson et al. 2005; Zhang et al. 2005). Melanin pigment is initially synthesized and contained in subcellular organelles (melanosomes) in melanocytes and then transferred to keratinocytes, where they shield the nucleus from damaging ultraviolet light. In this position, melanin effectively absorbs ultraviolet light (UV), penetrating the skin and preventing consequential DNA damage. The loss of skin pigmentation can result in various types of hypopigmentary disorders in human skin, which leads to psychological, social and interpersonal problems. The degree of visible skin pigmentation determines the amount, size and types of melanin produced by melanocytes residing at the junction of the epidermis and dermis, and the subsequent distribution of that pigment by keratinocytes to the surface. A defect at any stage of the complex procedure of proliferation and/or function of melonocytes and/or keratinocytes can contribute to the pigmentary disorder. Visible pigmentation in humans results from the synthesis and distribution of melanin in organs such as the skin, hair follicles and eyes. The pigment melanin plays a crucial role in the absorption of free radicals generated in the cytoplasm, shielding the host from various types of ionizing radiation, including UV light. Cellular melanisation is a process, which has been described as a fine stress regulatory mechanism linked to growth arrest involving a signalling cascade of stress-activated protein kinase (SAPK).

3.4 AVAILABLE TREATMENT OF LEUCODERMA

The existing treatment modalities of leucoderma are as follows.

1. Medical Therapies
    - Topical steroid therapy
    - Topical psoralen photochemotherapy
    - Oral psoralen photochemotherapy
    - Depigmentation

2. Surgical Therapies
    - Skin grafts from a person's own tissues (autologous)
    - Skin grafts using blisters
    - Micropigmentation (tattooing)
    - Autologous melanocyte transplants

3. Adjunctive Therapies
    - Sunscreens
    - Cosmetics
    - Counselling and support
    - General motivation

3.4.1 Medical Therapies

3.4.1.1 Topical Steroid Therapy

Steroids may be helpful in repigmenting the skin (returning the colour on white patches), particularly if started early in the disease. Corticosteroids are a group of drugs similar to the hormones produced by the adrenal glands (such as cortisone). Doctors often prescribe a mild topical corticosteroid cream for children under 10 years old and a stronger one for adults (Nagai et al. 2000). Patients must apply the cream to the white patches on their skin for at least three months before seeing any results. It is the simplest and safest treatment, but not as effective as psoralen photochemotherapy. There are different levels of potency of corticosteroids in use, ranging from low potency drugs, such as desonide (0.5% dexamethasone; 1% hydrocortisone; 0.25%, 0.5%, 1% & 2.5% methylprednisolone; 0.5% & 1% prednisoline) medium potency drugs like betamethasone valerate (1% sillorasone diacetate; 0.05%, fluocinol acetoxide; 0.1%, 0.25% & 1% triaminocinlone actinide) to high-potency corticosteroid like betamethasone dipropionate (0.05% cholesterol; 0.05% propionate; 0.05% fluocinonide and 0.1% halcinonide). The topical steroid therapy is useful only at the initial stage and resticated to localized lesions on the face and eyelids, mostly in children.

3.4.1.2 Psoralen Photochemotherapy

Psoralen photochemotherapy (psoralen and ultraviolet A therapy or PUVA) is probably the most beneficial treatment for leucoderma available (Abdel-Fattah et al. 1982). The goal of PUVA therapy is to repigment the white patches. However, it is time-consuming and care must be taken to avoid side effects, which can sometimes be severe. Psoralens are drugs that contain chemicals that react with ultraviolet light and cause darkening of the skin. The treatment involves taking psoralen (orally) or applying it to the skin (topically). This is followed by carefully timed exposure to ultraviolet A (UVA) 320–400 nm light from a special lamp or sunlight.

3.4.1.3 Topical Psoralen Photochemotherapy

Topical psoralen photochemotherapy often is used for people with a small number of depigmented patches (affecting less than 20% of the body) (Orecchia et al. 1992). It is also used for children 2 years old and older who have localized patches of leucoderma. Treatments are done in a doctor's office under an artificial UVA light once or twice a week. The doctor or nurse applies a thin coat of psoralen to the patient's depigmented patches about 30 min before UVA light exposure. The patient is then exposed to an amount of UVA light that turns the affected area pink. The doctor usually increases the dose of UVA light slowly over many weeks. Eventually, the pink areas fade and a more normal skin colour appears. After each treatment, the patient washes his or her skin with soap and water and applies sunscreen before leaving the doctor's office. In spite of backup, the therapy is also introduced in sunlight between 11 AM to 1 PM for a specific cure in topical application, and the natural as well as synthetic psoralen and 8-methoxy psoraline are used.

3.4.1.4 Oral Psoralen Photochemotherapy

Oral PUVA therapy is used for people with more extensive leucoderma (affecting greater than 20% of the body) or for people who do not respond to topical PUVA therapy. Oral psoralen is not recommended for children under 10 years of age because of an increased risk of damage to the eyes, such as cataracts. For oral PUVA therapy, the patient takes a prescribed dose of psoralen by mouth about 2 h before exposure to artificial UVA light or sunlight. The doctor adjusts the dose of light until the skin areas being treated, become pink. Treatments are usually given two or three times a week, but never for 2 d in a row.

For patients who cannot go to a PUVA facility, the doctor may prescribe psoralen to be used with natural sunlight exposure. The doctor will give the patient careful instructions on carrying out the treatment at home and monitor the patient during scheduled checkups. Injection of trimethyl psoralen 0.6–0.9 mg/kg or 8.0 methoxy psoralen 0.3–0.6 mg/kg body weight 2 h before exposure to UVA, three times a week in an alternate day, is recommended. The light source may be artificial or natural, with the dose and exposure time altered to achieve required pigmentation, 24–48 h after each treatment. Patients should also wear protective UVA sunglasses for 18–24 h after each treatment to avoid eye damage, particularly cataracts.

3.4.1.5 Depigmentation

Depigmentation involves fading the rest of the skin on the body to match the already white areas. For people who have leucoderma more than 50% of their bodies, depigmentation may be the best treatment option. Patients apply the drug monobenzylether of hydroquinone (monobenzone or Benoquin) twice a day on pigmented areas until they match with the already depigmented areas. Patients must avoid direct skin to skin contact with other people for at least 2 h after applying the drug.

Depigmentation is permanent and cannot be reversed. In addition, a person who undergoes depigmentation will always be abnormally sensitive to sunlight.

3.4.2 Surgical Therapies

All surgical therapies are viewed as experimental because their effectiveness and side effects remain to be fully defined yet.

3.4.2.1 Autologous Skin Grafts

In an autologous (use of a person's own tissues) skin graft, the doctor removes skin from one area of a patient's body and attaches it to another area. This type of skin grafting is sometimes used for patients with small patches of leucoderma. The doctor removes sections of the normal, pigmented skin (donor sites) and places them on the depigmented area (recipient sites) (Falabella 1988). The autologous grafting is successful only on localized or non-progressive leucoderma spots. If the skin is unresponsive to typical corticosteroid and Psoralen therapy, minigrafting 3–4 mm punches from the pigment donor side to leucoderma patches is done by cutting normal skin and then transplanting it on the white patches.

3.4.2.2 Skin Grafts using Blisters

In this procedure, the doctor creates blisters on the patient's pigmented skin by using heat, suction or freezing with liquid nitrogen. The blister roof at recipient site is discarded and sister roof from the donor site is transplanted onto it. The tops of the blisters are then cut out and transplanted to a depigmented skin area.

3.4.2.3 Micropigmentation (Tattooing)

Tattooing implants pigment into the skin with a special surgical instrument (Halder et al. 1989). This procedure works best for the lip area, particularly in people with dark skin; however, it is difficult for the doctor to match perfectly the colour of the skin of the surrounding area.

3.4.2.4 Autologous Melanocyte Transplants

In this procedure, the doctor takes a sample of the patient's normal pigmented skin and places it in a laboratory dish containing a special cell culture solution to grow melanocytes. When the melanocytes in the culture solution have multiplied, the doctor transplants them to the patient's depigmented skin patches (Halder et al. 1989). In another type of practise the melanocytes and keratinocytes are grown together and then transplanted.

3.4.3 Adjunctive Therapies

3.4.3.1 Sunscreens

People who have leucoderma, particularly those with fair skin, should use a sunscreen that provides protection from both the UVA and UVB forms of ultraviolet light. Sunscreen helps to protect the skin from sunburn and long-term damage. Sunscreen also minimizes tanning, which makes the contrast between normal and depigmented skin less noticeable.

3.4.3.2 Cosmetics

Some patients with leucoderma cover depigmented patches with stains, makeup or self-tanning lotions. These cosmetic products can be particularly effective for people whose leucoderma is limited to exposed areas of the body (e.g. Dermablend).

3.5 DRAWBACKS OF EXISTING TREATMENT MODALITIES

The existing remedy of leucoderma, as described in earlier paragraphs, is a multifaceted approach. But none of these are 100% successful. There are a number of formulations, both synthetic and natural, for chemotherapy, surgical and cosmetic modulation, but all have limitations. The drawback of these treatments are discussed as follows:

3.5.1 Medical Therapies

3.5.1.1 Topical Steroid Therapy

The main drawback of this therapy is that this is restricted to localized lesions at the initial stage of leucoderma on eyelids and face and treatment in children, thus it is not effective for all types of leucoderma. The success rate is very low and there are chances of side effects (i.e. skin shrinkage, skin strain and also problems of higher pigmentation). The another disadvantage is that if it is unsuccessful, then other modes of therapy will not work.

3.5.1.2 Psoralen Photochemotherapy

The major potential side effects of topical PUVA are:

- Severe sunburn and blistering
- Vesiculation of skin
- Too much repigmentation and darkening of the treated lesion (hyperpigmentation)
- The blister and vasiculation cause pain and itching, and as a result patients discontinue the treatment.
- Hyperkeratosis of treated lesions

Oral PUVA is also not safe because it is also not free from the side effects after treatment. The main drawbacks of this therapy are that it causes sunburn, nausea, vomiting, abnormal hair growth, pruritus erythema/oedema, xeroses, hyper-pruritus and hyperpigmentation.

3.5.1.3 Depigmentation Therapy

The major disadvantage of existing depigmentation therapy is inflammation with redness of the skin. Patients may experience itching, dry skin and the abnormal darkening of the membrane that covers the white of the eye. Depigmentation is permanent and irreversible, and furthermore, the person who undergoes depigmentation will become sensitive to sunlight.

3.5.2 Surgical Therapies

3.5.2.1 Autologous Skin Graft

There are several complications and disadvantages for this therapy. Infection may occur at the donor and recipient sites. The recipient and donor sites may develop scarring, a cobblestone appearance, a spotty pigmentation or they may fail to repigment at all. The other disadvantage is that it is successful only for localized lesions. It is also time-consuming and costly, which most patients found unacceptable.

3.5.2.2 Skin Grafts Using Blisters

The disadvantages of this therapy include the development of a cobblestone appearance, scarring and a lack of repigmentation, but this type of graft is produces less scarring than others. It is also costly.

3.5.2.3 Micropigmentation (Tattooing)

Tattooing tends to fade over a time, and in addition, this treatment restricted to the lips only. It may lead to blister outbreaks caused by the herpes simplex virus.

3.5.2.4 Autologous Melanocyte Transplants

All the surgical models of leucoderma are of experimental significance only, since these treatments are largely impractical for routine care of leucoderma.

The overall drawback of these models for leucoderma therapy is that either they are costly, single component–based, ineffective over a broad spectrum and the overall recovery rate in any mode of existing therapies is not satisfactory.

3.6 DIBER–INVENTED HERBAL THERAPY OF LEUCODERMA

Reviewing the limitations, drawbacks and lower level of efficacy of the available treatments, the poly component herbal formulation, both for topical and oral therapy of leukoderma, has been formulated by the institute with following objectives:

- To provide an effective multicomponent herbal formulation for the treatment of leucoderma and also to provide a process for preparation of this formula
- To provide a herbal formulation which does not cause any side effects like blisters, vesiculation, oedema formation, vomiting, nausea and other complications which are faced in existing topical and oral therapies of PUVA
- To provide a herbal-based formulation which is very effective and provides pigmentation at faster rate and more satisfactory level of recovery, irrespective of the type and extent of the disease
- To provide formulation which has poly herbal ingredients, which, due to the formulation has a broad spectrum of effectiveness

3.6.1 Summary of Invention

An effective and very cheap poly component herbal formulation for the cure of leucoderma for both topical and oral application has been developed by DIBER (DRDO).

The existing therapies as described earlier have many limitations and disadvantages. In the present formulation, attempts have been made to remove most of these disadvantages to a greater extent. This has been done by making the formulation a poly component product in which the plants viz. *Psoralea corylifolia, Ammi majus, Calotropis gigantia, Aloe vera, Ocimum basilicum, Arnebia benthamii*, petroleum ether extract of *Sesamum indicum, Jasmine arboresece*, ash of *Ammi majus* and *Psoralea corylifolia* have been used. The first two plants are the source of the active ingredient 'Furano-coumarins', providing the necessary amino acids for pigmentation, and the adequate amounts of copper, zinc and iron needed for tyrosinase activity and autoimmune modulation. For oral formulation plants viz. *Ocimum basilicum, Centella asiatica, Ammi majus, Psoralea corylifolia, Mucuna puriens* and *Aloe vera* along with ash of *Psoralea Corylifolia*; *Ammi majus* and *Mucuna pruriens* have been used. This anti-leucoderma herbal formulation has a two-component ointment and oral dose. The brief information of ointment and oral formulation is as follows:

3.6.1.1 Ointment

In the ointment ingredients, two plants *Psoralea corylifolia* and *Ammi majus*, which are rich in furano-coumarins, have been used as skin photosensitizer for initiation of erythema on the spots of leucoderma by the exposure to UVA radiation. This reaction is a free radical mechanism which creates heat sensation and blister formation. This activity has been subsided by the addition of two plants, *Calotropis gigantia* and *Ocimum basilicum*, ingredients which have anti-blister formation and anti-irritation properties. The fifth plant, *Aloe vera*, has antiseptic properties and checks the hyper pigmentation of skin. The ash of the first two plants is rich in copper, which has been added to make up the deficiency of copper needed for tyrosinase activity. The sixth plant has been added to impart the colour to the ointment and also to synergize the wound-healing activity. The ingredients of seventh plant impart the odor to the ointment and also improve the texture of skin.

3.6.1.2 Oral Dose

To check the emergence of new spots, an oral dose has been formulated, which contains the ingredients of six plants. The first two plants, *Psoralea corylifolia* and *Ammi majus*, have furano-coumarins, as in the case of the ointment. The third plant *Centella asiatica* has CNS–stimulating activity, which also helps in improving the psychological and emotional trauma of the patients. The fourth plant, *Aloe vera*, has properties to removes skin disorder of parasitic origin. Both the third and fourth plants have liver restorative properties which gradually improve the autoimmune system. The oral dose has been augmented with ash of first two plants, which are rich in copper, and ingredients of the fifth plant, *Mucuna puriens*, is rich in dihydroxyphenylalanine, which makes up the deficiency of copper and DOPA.

3.7 TOXICOLOGICAL STUDIES

The toxicological studies of this herbal formulation have been carried out at Industrial Toxicological Research Centre (ITRC), Lucknow, for acute oral toxicity, subacute oral toxicity, the mucous membrane irritation test, the skin sensitization test etc. and the formultation is found safe for human application. LD_{50} of oral dose is >2000 mg/kg of body weight.

The toxicity of the ethanol extract of the plants taken into ointment preparation was completed at DRDE, Gwalior and the ointment and oral dose at ITRC, Lucknow. The tests were carried out for LD_{50} and primary skin irritation. The results for toxicity evaluation from both the institutes are described as under.

3.7.1 Materials

Randomly bred Wister male rats (175–250 g body weight) were used. Animals were housed in polypropylene cages, four animals per cage, on dust-free rice husk. The animals were provided with pelleted diet (*Amrut Feed India Ltd.*) and water *ad libitum.* Female albiono rabbits (2.4 ± 0.2 kg) were taken for a skin irritation test. They were acclimatized in a well-ventilated room (room temperature 25–27°C and room humidity 55–70%) with regular lighting conditions for 3 to 4 d. They were provided with a pelleted diet (*Amrut Feed India Ltd.*) and water *ad libitum.*

3.7.2 Acute Oral Toxicity

The rats were fasted overnight (about 18 h), with only water allowed. Doses of 800, 1600, 3200 and 6400 mg/kg body weight of the alcoholic extract were fed through an 18-gauge, oral feeding cannula. For each dose of extract, four rats were used and they were observed for 14 d. Observation on general behaviour, weight changes, mortality and feed and water intake were recorded.

There was no mortality up to 3200 mg/kg body weight dose. Only one rat in 6400 mg/kg dose died within 6 h. The LD_{50} was observed above 5.0 g/kg body weight. The general behaviour of the rats was normal. No significant differences in water and feed intake was observed. The body weight was converted as percent before treatment. The percent change of body weight was recorded and analysed by a two-way analysis of variance with Dunnett's multiple comparisons test (Table 3.1).

TABLE 3.1
Percent Bodyweight Change after Oral Administration of Alcoholic Extract of Medicinal Plants Used for Leucoderma Treatment

		Percent Weight		
Group	**Dose (mg/kg)**	**24 h**	**7 d**	**14 d**
Control	–	101.8±0.5	110.6±2.1	121.2±2.3
Alcohol Extract	800	103.2±0.2	111.7±1.4	113.1±2.4
	1600	104.1±2.4	112.2±3.1	114.5±4.3
	3200	102.9±0.1	108.4±1.4	113.1±1.1
	6400	97.5±2.2	97.9±3.1	105.3±1.7

Note: Values are mean ±SEM.

TABLE 3.2
Computation of Primary Irritation Index (P11)

No.	Dose (ml)	Erythema 4 h	24 h	72 h	7 d	Oedema 4 h	24 h	72 h	7 d
1	0.5	+	0	0	0	0	0	0	0
2	0.5	+	0	0	0	0	0	0	0
3	0.5	+	0	0	0	0	0	0	0

Note: PII- of Extract = 0.17; For computation of PII three times readings, i.e. 4 h, 24 h and 72 h were taken; PII = Primary irritation tests, i.e. total average (types of skin sites × times of readings).

There was no significant change in the percent of weight gain when compared to a control, except at a dose of 6400 mg/kg of alcoholic extract.

3.7.3 Primary Skin Irritation Test

The rabbit's hair on the back was closely clipped. Four areas (each area = 2.5 cm^2) on the skin, two on the right and two on left site were marked. The left side was used as a control. On the right upper side, 0.5 ml of ethanol plant extract was applied. The skin sites were covered with a gauze paper and impervious soft sheet and held in place by means of adhesive tape. They were immobilized using rabbit restrainer for 4 h. Three rabbits were used for the primary skin irritation tests. After 4 h, the areas were cleaned inspected for the presence of erythema, escar and oedema. The animals were observed 24 h to 72 h and 7 d, post application. The scoring procedure described by Draize was used, and a primary irritation index was computed (Table 3.2).

3.7.4 Toxicological Studies of Ointment

The final product was evaluated for following toxicological tests at ITRC Lucknow:

- Acute dermal toxicity in rats
- Subacute (2–28 d) dermal toxicity test in rats
- Mucus membrane irritation test in rabbit vagina
- Skin sensitization in Guinea pigs
- Subchronic and chronic toxicity tests in rats

The acute dermal toxicity results indicated LD_{50} > 2000 mg/kg body weight. The herbal formulation at limit test dose, i.e. 2000 mg/kg body weight, did not produce any sign of toxicity or death in male rats. An autopsy of the surviving animals at the end of observation period did not indicate any gross pathological changes in vital organs (Table 3.3). The female rats also showed the same level of LD_{50}, i.e. a limit test dose showed a similar result. Results of subacute, dermal toxicity indicated that repeated dermal toxicity in animals treated with the herbal ointment (1000 mg/kg/d)

TABLE 3.3
Organ Body Weight (g) Ratio of Rats Dermal Exposed to SP-7 Anti-leucoderma Ointment for 21 d

Treatment (mg/kg/d)	Liver	Brain	Kidney	Spleen	Adrenal	Testis/ Ovary	Svessida/ Uterus	Prostate	Epidsdy-mis
Male control	3.53±0.11	0.70±0.02	0.73±0.03	0.58±0.03	6.02±0.00	1.00±0.03	0.49±0.01	0.35±0.04	0.39±0.05
Product (1000 mg/kg/d)	3.46±0.13	0.75±0.01	0.66±0.04	0.69±0.16	0.02±0.00	1.16±0.05	0.53±0.04	0.29±0.02	0.41±0.01
Female control	3.62±0.22	0.83±0.02	0.66±0.03	0.49±0.06	0.01±0.00	0.02±0.01	0.34±0.15	–	–
Product (1000 mg/kg/d	3.74±0.19	0.85±0.01	0.69±0.02	0.46±0.03	0.01±0.00	0.01±0.00	0.13±0.03	–	–

did not show any toxicity in animals throughout 28 d of exposure and therefore showed no morbidity and mortality. The body weight also did not show any significant variation. Similarly the organ:body weight ratio (Table 3.3) of treated animals did not show any significant variation. No gross significant changes were observed in the morphological characters and no pathological lesions were detected in any vital tissues like liver, kidney, testis, ovary and skin. Also, no significant change in hematological parameters was observed (Table 3.4).

The biochemical studies on the activity of GOT, GPT, ALP and proteins of the control and product-treated (1000 mg/kg/d) animals are shown in Table 3.5. No significance changes in the serum, protein and BUN between the control and experimental animals were observed. However, the activity of serum GAT and Alp were altered in both male and female rats. The serum GOT activity was significantly altered in male rats only.

The mucus membrane irritation test in female rabbits did not show any sign of erythema or oedema in the vaginal mucous membrane of rabbits. It was found to be non-irritant. The skin sensitization test in guinea pigs also indicated that the product was found to be unresponsive in guinea pigs.

3.7.5 Biochemical Studies in Blood Plasma of Leucoderma Patients

3.7.5.1 Screening of Tyrosinase Enzyme

The tyrosinase enzyme level was studied in the blood plasma of some of the leucoderma patients and was compared with healthy persons taken as a control. Standardization of the tyrosinase estimation methods was based on the conversion of dihydroxy-phenylalanine to red coloured oxidized product DOPA chrome. The absorbance of DOPA chrome was measured at 475 nm in spectrophotometer at two stages viz. before and after incubation. The change in absorbance at two levels is proportional to the enzyme concentration. The method standardization was done taking the blood plasma of white, variegated and black cows which showed respective variation, i.e. the highest levels in black and lowest in white cattle.

The comparative level of enzyme gave the following observation:

- Enzyme tyrosinase was found to be absent or beyond detectable limit in 15% of patients.
- In 60% of patients, the tyrosinase level was found to be lower as compared to healthy persons.
- In the rest of the the patients, the level of tyrosinase was found to be almost equal to that of healthy persons, but there was a severe deficiency of copper in their blood plasma.

3.7.5.2 Screening of Copper and Zinc in the Blood Plasma of Leucoderma Patients

With a view to standardize the anti-leucoderma ointment and its oral dose, screening of copper and zinc was carried out in the blood plasma of the patients. Copper acts as prosthetic group of enzyme tyrosinase, without which the enzyme becomes inactive. The estimation of both copper and zinc was done with the help of the Atomic Absorption Spectrophotometer. The following observation were recorded (Table 3.6):

TABLE 3.4
Hematological Changes in Rats, Dermal Exposed to Herbal Product for 28 d

Treatment (mg/kg/d)	Hb%	RBC 10^6 mm^3	WBL X mm^3	Leucocytes %	Polymorph %	Eosinophils %	Monocytes %
Male control	16.16±0.36	7.88±0.27	7090±145.65	81±1.63	17.8±1.39	0.5±0.25	0.6±0.24
Product (1000 mg/kg/d)	15.96±0.36	7.19±0.25	6600±235.69	76.8±0.86	22.2±0.86	0.8±0.20	0.2±0.20
Female control	15.80±0.30	7.99±0.44	7120±224.51	80.8±0.86	18.2±0.86	0.8±0.20	0.2±0.20
Product (1000 g/kg/d)	15.96±0.116	7.54±0.52	7140±205.73	75.8±0.80	23.2±0.66	0.4±0.24	0.6±0.245

Note: Volumes are the mean ±SE of five animals each.

TABLE 3.5
Serum Biochemical Changes in Rats, Dermally Exposed to Herbal Formulation for 28 d

Treatment mg/kg/d	GOT	GPT	Alp	BUN	Protein
Male control	0.264±0.0035	0.238±0.003	0.852±0.054	23.96±0.40	9.52±0.053
Product (1000 mg/kg/d)	0.223±0.003*	0.194±0.005*	0.629±0.025**	21.27±0.34**	9.598±0.112
Female control	0.241±0.005	0.173±0.001	0.568±0.21	22.69±0.64	9.83±0.067
Product (1000 g/kg/d)	0.234±0.001	0.199±0.002	0.47±0.009*	23.32±0.62	9.594±0.051

Note: Values are the mean ± SE g five animals. x - Mole-pyruvate liberated/ml/min; xx - Mole phenol liberated/ml/min;<> mg% g m%.
*P > 0.001; **P < 0.02.

TABLE 3.6
Tyrosinase, Zinc and Copper Level in the Blood Plasma of Leucoderma Patients

S. No	Name of Patient	Age	Tyrosinase Unit	Elements Mg/100 ml of Plasma	
				Copper	Zinc
1.	Sh. Vinod Kumar	30	9.382	0.040	0.00
2.	Sh. L.s. Waldia	68	12.839	0.00	0.00
3.	Sh. G. D. Nariyal	36	10.133	0.00	0.00
4.	Sh. Rajendra Singh	21	6.419	0.00	0.080
5.	Smt Shanti Devi	50	7.407	0.00	0.280
6.	Smt Usha	37	13.827	0.080	0.200
7.	Sh. Sushil Khatri	23	3.950	0.140	0.080
8.	Kum. Deepa Joshi	5	4.938	0.40	0.140
9.	Sh. Guddu	13	0.493	0.080	0.200
10.	Sh. Madan Mohan	37	ND[a]	0.820	0.480
11.	Sh. Naresh Kumar	37	ND[a]	0.860	0.220
12.	Ms. Namita Pant	14	2.200	0.960	0.280
13.	Smt Gpali Devi	48	4.938	0.900	0.160
14.	Sh. Harish Chandra	43	3.486	0.860	0.180
15.	Control (3 Healthy Persons)	40	13.456	0.940	3.150

[a] ND - Not Detected.

- In the majority of patients, the level of copper as well as zinc was found lower when compared to the control.
- Patients who had a copper level equal to the control group had drastically lower levels of tyrosinase enzymes.
- The copper and zinc levels were found to be inversely related with tyrosinase level of all the patients.
- Patients those who have higher level of copper had a lower level of zinc.

3.8 CLINICAL TRIALS

This herbal formulation has been evaluated in different age groups of patients with different ailments, ages and extent of disease under the supervision of Doctors of Govt. Civil and Ayurvedic Hospitals of District Pithoragarh. Two hundred patients are under regular treatment and placed under observation. The observations of two years revealed that 108 patients have recovered by 75–100% and 72 patients showed recovery up to 50%, while 20 patients either did not get any response from this medicine or, due to some other reason, they had discontinued the treatment. These patients were also identified on the basis of age and age of leucoderma disease (age of ailment), further categorized into three groups. Sarin et al. (1977) had also conducted a clinical study on vitiligo/leucoderma. In the first category, patients had white patches on the body and were 6 months to 1 year old. In the second category, patients had white patches and were 1 year to 5 years old and in third category, partients has white patches and were 6 to 12 years old. Out of the 200 patients taken in the study, 58 had leucoderma patches that were 6 months to 1 year old, 72 patients with white patches were 1 year to 5 years old and 50 patients who had white patches were 6 years to 12 years old. In first category, 46 patients had recovered by 75–100% and 12 patients observed up to 50% recovery (Figure 3.2).

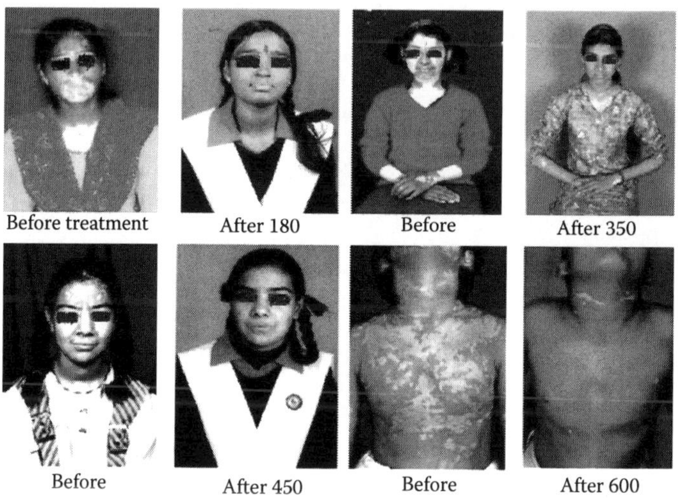

FIGURE 3.2 Result of clinical trials.

In the second category, 42 leucoderma patients showed 75–100% recovery and 30 showed recovery up to 50%. In the third category, 50 patients were taken and 20 patients got 75–100% recovery and rest showed up to 50% recovery. It is concluded from the study that the persons who have been diagnosed with leucoderma early on who started taking our herbal medicine recovered quicker/better than the patients with very old white patches. Results of clinical trials are cited in Table 3.7.

These patients were also classified on the basis of their sex (male/female) and age (i.e. below 40 years and above 40 years). The results of this study are cited in Table 3.8. It can be concluded from the data that persons (male or female) less than 40 years old responded better/quickly to the medicine than those who were more than 40 years of age.

It is evident from the results that the efficacy of anti-leucoderma herbal formulation developed by the institute is better than existing treatments.

TABLE 3.7
Recovery According to Age of Ailment

Ailment Age	No. of Patient	Recovery (%)		Time Taken (d)
		Up to 50	75–100	
6–12 months	58	12	46	90–300
1–5 years	72	30	42	120–450
6–15 years	50	30	20	350–750
No Response	20	–	–	–
Total	200	72	108	90–750

TABLE 3.8
Sex Wise Recovery of Leucoderma Patients

Sex/Age	No. of Patients	Recovery (%)		Time Taken (d)
		Up to 50	75–100	
Female				
Below 40	56	22	34	90–350
Above 40	20	10	10	120–480
Male				
Below 40	62	24	38	110–500
Above 40	42	16	26	250–750
No response	20	–	–	–
Total	200	72	108	90–750

3.9 PROTOCOL FOR USE

3.9.1 Ointment

The ointment has been recommended for use one or two times a day with gentle application on the spot of leucoderma. It has to be applied from the pigmented area to the non-pigmented area i.e. from outside to inside, because the process of spread of melanin is a three-way process viz. from the margin of spots, and from hair follicles and within the spots. If it is applied in a fashion from inside to outside i.e. from theaffected to non-affected area then there is a chance of spreading of the white spots over the normal pigmented area. After the application of the ointment, sun exposure between 11 AM to 1 PM should be done, starting from 6 min with an increment period of 2 to 3 min per day, reaching to 20–25 min exposure time.

3.9.2 Oral Dose

Oral dose has been recommended 2.50 ml for children and 5.00 ml for adults, twice a day with 20–25 ml of water.

3.9.3 Restriction to be Followed by the Patients

With the above drug, the patients have been advised not to eat any sour foods which contain organic acid, particularly ascorbic acid. Fast and preserved foods that contain organic acids and use salts as preservatives should be avoided. In addition to this, a non-vegetable diet has been restricted because it is devoid of aromatic amino acids, i.e. phenylalanine tyrosine and DOPA, which function as precursors of melanin.

3.10 RECOMMENDATION

The patients have been advised to use copper-rich water after storing it in a copper vessel at least for 12 h. Vegetable and fruits rich in vitamin B-Complex, aromatic amino acids, and pulses rich in protein have been recommended for their daily diet.

3.11 LAUNCHING OF ANTI-LEUCODERMA HERBAL PRODUCT (LUKOSKIN) IN THE MARKET

The anti-leucoderma herbal product developed by the institute had been launched by AIMIL Pharmaceuticals (India) Ltd., New Delhi, on September 7, 2011 at DRDO Bhawan, New Delhi by the brand name of Lukoskin. This herbal product is available in the form of ointment and oral liquid in separate packs as well as in a combo pack. This product is becoming a boon for leucoderma patients. It is available in combo pack form at chemist shops.

REFERENCES

Abdel-Fattah, A., Aboul-Enein, M. N., Wassel, G. M., and El-Menshawi, B. S. 1982. An approach to the treatment of leucoderma by khellin. *Dermatologica* 165(2): 136–40.

Agrawal, D., Shajil, E. M., Marfatia, Y. S., and Begum, R. 2004. Study of the antioxidant status of vitiligo patients of different age groups in Baroda. *Pigment Cell Research* 17: 289–94.

Antoniou, C. and Katsambas, A. 1992. Guidelines for the treatment of leucoderma. *Drugs* 43 (4): 490–98.

Behl, P. N. and Bhatia, R. K. 1972. 400 cases of Vitiligo. A clinico-therapeutic analysis. *Indian Journal of Dermatology* 17: 51–55.

Blanpain, C., Lowry, W. E., Geoghegan, A., Polak, L., and Fuchs, E. 2004. Self-renewal, multipotency, and the existence of two cell populations within an epithelial stem cell niche. *Cell* 118(5): 635–48.

Bolognia, J. L. and Pawelek, J. M. 1988. Biology of hypopigmentation. *Journal of the American Academy of Dermatology* 19: 217–55.

Caixia, T., Hongwen, F., and Xiran, L. 1999. Levels of soluble interleukin-2 receptor in the sera and skin tissue fluids of patients with vitiligo. *Journal of Dermatological Science* 21: 59.

Cape, C. B., She, J. X., and McCormack, W. T. 2002. Genetic association of the catalase gene (CAT) with vitiligo Susceptibility. *Pigment Cell Research* 15: 62.

Cape, C. B., She, J. X., and McCormack, W. T. 2003. Genes of the LMP/TAP cluster are associated with the human autoimmune disease vitiligo. *Genes and Immunity* 4: 49.

Dutta, A. K. and Mandal, S. B. 1969. A clinical study of 650 cases of vitiligo and their classification. *Indian Journal of Dermatology* 14: 103–5.

Falabella, R. 1988. Treatment of localized leucoderma by autologous minigrafting. *Archives of Dermatology* 124(11): 1649–55.

Hafez, M., Sharaf, L., and Abd El-Nabi, S. M. 1983. The genetics of vitiligo. *Acta Derm Venereol (Stokh)* 63: 249–51.

Halder, R. M., Pham, H. N., Breadon, J. Y., and Johnson, B. A. 1989. Micropigmentation for the treatment of leucoderma. *Journal of Dermatologic Surgery and Oncology* 15(10): 1092–8.

Harning, R., Cui, J., and Bystryn, J. C. 1991. Relation between the incidence and level of pigment cell antibodies and disease activity in vitiligo. *Journal of Investigative Dermatology* 97: 1078.

Huang, C. L., Nordlund, J. J., and Boissy, R. 2002. Vitiligo: A manifestation of apoptosis. *American Journal of Clinical Dermatology* 3: 301–306.

Kemp, E. H., Waterman, E. A., Gawkrodger, D. J., Watson, P. F., and Weetman, A. P. 1998. Autoantibodies to tyrosinase-related protein-1 detected in the sera of vitiligo patients using a quantitative radiobinding assay. *British journal of Dermatology* 139: 798.

Kemp, E. H., Waterman, E. A., Hawes, B. E., O'Neill, K., Gottumukkala, R. V., Gawkrodger, D. J., Weetman, A. P., and Watson, P. F. 2002. The melanin concentrating hormone receptor 1, a novel target of autoantibody responses in vitiligo. *Journal of Clinical Investigation* 109: 923.

Koranne, R. V., Sehgal, V. N., and Sachdeva, K. G. 1986. Clinical profile of vitiligo in North India. *Indian Journal of Dermatology, Venereology and Leprology* 52: 81–82.

Liu, P. Y., Bondesson, L., Loentz, W., and Tohansson, O. 1996. The occurrence of cutaneous nerve ending and neuropeptides in vitiligo vulgaris: A case control study. *Archives of Dermatological Research* 288: 670.

Nagai, K., Ichimiya, M., Yokoyama, K., Hamamoto, Y., Muto., 2000. Successful treatment of non-segmental vitiligo: Systemic therapy with sex hormone thyroid powder mixture. *Hormone Research* 54: 316.

Nordlund, J. J., Halder, R. M., and Grimes, P. 1993. Management of leucoderma. *Dermatologic Clinics* 11(1): 27–33.

Ongenae, K., van Geel, N., and Naeyaert, J. M. 2003. Evidence for an autoimmune pathogensis of vitiligo. *Pigment Cell & Melanoma Research* 2: 90.

Orecchia, G. and Perfetti, L. 1992. Photochemotherapy with topical khellin and sunlight in leucoderma. Journal of *Dermatology* 184 (2): 120–23.

Sarin, R. C. and Kumar, A. S. 1977. A clinical study of vitiligo. *Indian Journal of Dermatology, Venereology and Leprology* 43: 300–14.

Shajil, E. M., Sahani, M., Mistry, N., Sheril, A., and Begum, R. 2005. Evaluation of oxidative stress, autoimmune and neuro-chemical hypothesis in Baroda vitiligo patients, poster presented to the first conference of the Asian Society for Pigment Cell Research, New Delhi, India, 1–2.

Steingrímsson, E., Copeland, N. G., and Jenkins, N. A. 2004. Melanocytes and the microphthalmia transcription factor network. *Annual Review of Genetics*. 38: 365–411.

Steingrímsson, N. A. 2005. Melanocyte stem cell maintenance and hair graying. *Cell* 8, 121(1): 9–12.

Vance, K. W. and Goding, C. R. 2004. The transcription network regulating melanocyte development and melanoma. *Pigment Cell Research* 17(4): 318–25.

Zhang, X. J., Chen, J. J., and Liu, J. B. 2005. The genetic concept of vitiligo. *Journal of Dermatological Science* 39: 137.

4 Centipede Envenomation Effects on Human Beings and Scientific Research on Traditional Antivenom Agents

Aristolochia and *Piper* sp.

Dhivya Sivaraj, Revathi Ponnusamy, Rahul Chandran and Parimelazhagan Thangaraj

CONTENTS

4.1 Introduction 74
4.2 Venomous Centipede *Scolopendra* 74
4.3 Severity of Hypersensitivity/Delayed Hypersensitivity 75
4.4 Resemblance to Snake Envenomation Effects 77
4.5 Venom Components and Their Effects 77
  4.5.1 Metalloprotease 78
  4.5.2 γ-Glutamyl Transpeptidases (GGTs) 78
  4.5.3 Glycoside Hydrolases 78
  4.5.4 Hyaluronidases 78
  4.5.5 Phospholipase A2 (PLA2) 78
  4.5.6 Cardiotoxin, Neurotoxin, Myotoxin and Cytotoxin Effects 79
  4.5.7 CAP Proteins CRiSP (Cysteine Rich Proteins), Allergen (Ag-5) and Pathogenesis-Related (PR-1) 79
4.6 Current Treatment against Hypersensitivity 79
4.7 Ethnobotanical Uses of *Aristolochia* sp. 80
4.8 Ethnobotanical Uses of *Piper* sp. 80
4.9 Recent Research on Medicinal Plants against Venom 81
4.10 Requirements in Venom Research 82
4.11 Conclusion 83
References 83

4.1 INTRODUCTION

Throughout the world, rural people are suffering from venomous bites and go for acute therapeutic treatment (Kunjam et al. 2013). The sufferers experience mortality, local pathology and systematic pathology. Primarily, arthropod bites produce effects on the skin by means of the assortment of mechanisms (Bagnall and Rook 1977; Stawiski 1985; Walton 1988). Arthropod envenomation causes effects like local tissue damage, erythema, swelling, ulceration, itching, etc (Eivind et al. 2011).

As literature surveys reported, rual people who practise folk medicine often live in synchronization with nature, have a rich knowledge on the healing value of therapeutic plants against venomous bites and related problems (Revathi and Parimelazhagan 2010) and have evolved a unique system of unrecorded medicinal practises to cure various ailments (Perumal, Samy, and Ignacimuthu 2000) which are known as traditional systems of medicine (Martz 1992; Mors et al. 2000; Soares et al. 2004, 2005, 2009). People depend on medicinal plants to supplement emergency care (Kundal 2009). Plant-based traditional knowledge has been a recognised tool in the search for new sources of drug and nutraceuticals (Sharma and Mujumdar 2003). Several scientific studies are in progress using modern scientific tools based on the folklore and ethnomedicine for formulating new pharmaceuticals (Pradeeps and Poyyamoli 2013).

4.2 VENOMOUS CENTIPEDE *SCOLOPENDRA*

Centipedes belong to Class Chilopoda and have Myriapod lineages. There are about 3500 species worldwide within five orders: Scutigeromorpha, Lithobiomorpha, Craterostigmomorpha, Geophilomorpha and Scolopendromorpha; all except Geophilomorpha assault humans (Eivind et al. 2011). The Scolopendromorpha includes the genus Scolopendra, which is the largest and most dangerous (Bush et al. 2001) and is found to be cosmopolitan in India (Vinod 2006). Centipede bites are a major health problem among the rural community, and in 2014, the Tamil Nadu centipede bite rate was 17.8 per 1000 members of the population (Murugan et al. 2014).

The venom composition of *Scolopendra* has some similarities to snakes venom like *Bungarus caeruleus* (Krait), *Ophiophagus hannah* (Cobra) and *Bothrops jararacussu* (Viper) (Table 4.1 and Figure 4.1) (Eivind et al. 2011; Leijiane et al. 2013; Mirajkar et al. 2006; Rajan et al. 2013b; Singh et al. 2001a; Tan and Ponnudurai 1990; Tan et al. 2015; Warrell 1993; Zhang et al. 2002). The enzymes phospholipase A2 and metalloproteases and non-enzymes cardiotoxin and myotoxin are found commonly among centipede, krait, cobra and viper venom. These are responsible for local pathology, hypersensitivity and delayed hypersensitivity, which may lead to life-threatening effects for the victims.

TABLE 4.1
Similarity between Centipede and Snake Venom

Venom Components	*Scolopendra* sp.	*Bungarus caeruleus*	*Ophiophagus hannah*	*Bothrops jararacussu*
Alkaline phosphatases	+	+	–	–
Phospholipase A2	+	+	+	+
Esterases	+	+	–	–
Hyaluronidase	+	+	–	+
Metalloproteases	+	–	+	+
Cardiotoxin	+	+	+	+
Disintegrins	+	–	–	+
Myotoxins	+	–	+	+
Neurotoxins	+	+	+	+

FIGURE 4.1 The venomous centipede *Scolopendra morsitans.*

4.3 SEVERITY OF HYPERSENSITIVITY/DELAYED HYPERSENSITIVITY

Envenomation is responsible for several clinical complications of severe, systemic and local pathology. Immediate hypersensitivity triggered by venom may have direct toxic effects, inducing local reaction and clinical symptoms. The hypersensitivity is of four types based on their type of immune response and effector mechanism responsible for cell and tissue injury (Uzzaman and Cho 2012). Immediate or anaphylactic hypersensitivity (type I) is mediated by immunoglobulin E and results in eczema or skin damage. The primary cellular component of this type of hypersensitivity is mast cell and basinophills. The mast cell releases the histamines, platelet activation factors, vasoactive amines and allergens that result in itching and could cause erythema and local inflammation.

The eosinophil chemotactic factor of anaphylaxis and neutrophil chemotactic factors induce the release of hydrolytic enzymes and that causes necrosis.

An immediate hypersensitivity reaction was observed among 48 human beings in Australia during the year 2000–2002 as pain (29%), erythema (53%), edema (43%) and itchiness (14%). *Scolopendra* sp. and *Ethmostigmus* sp. caused more severe effects (Balit et al. 2004). Similarly, in Hong Kong, 46 cases were reported with venom effects (2010) like pain (100%), erythema (80%), edema (65%), numbness (22%), systemic symptoms (11%), bruising (7%) and itchiness (2%). Among them, one victim suffered from necrosis and five of them were reattending for the treatment for delayed inflammation and pruritus, these severities are due to the hypersensitive reaction that takes place at the time or much later (Fung et al. 2011). Centipede victims suffered from systemic urticaria, fatigue and dyspnea. Skin prick tests with centipede venom were positive in three patients with systemic allergic reactions to their bite (Harada et al. 2005).

Type II hypersensitivity involves in immune suppression activity, which is initiated by centipede venom components metalloproteases (immunosuppressive enzyme). Type III is mediated by immunoglobulin G and M which induces pathology in organs like skin (systemic lupus erythematosus) and kidney (lupus nephritis). This hypersensitivity reaction was reported in India with symptoms like localised pain, swelling, sweating, nausea, vomiting, acute myocardial infarction, multisystemic leukocytoclastic vasculitis and rhabdomyolysis with renal failure (Soylu et al. 2005). An 80-year old retired civil servant was affected with avascular necrosis on the right head of femur and severe pain due to centipede bite (Nwokeukwu et al. 2015).

Type IV hypersensitivity is cell-mediated which includes the T lymphocytes and monocytes/macrophages, and was observed in a 21-year old girl, who died in the Philippines (2004). A species of Scolopendra bit a 7-year old girl in the Philippines on the head, killing her after 29 h (1950). In the United States, centipede bites were responsible for six deaths from the year 1979 to 2001 (Eivind et al. 2011).

On the whole, centipede envenomation produces different effects on victims. This is due to the complexity of venom, and also depends on amount of venom injected; the age, size and gender of the afflicted; as well as the geographical areas and seasons at bite time (Vorono et al. 1999). It is also influenced by the immune power of victims. Eventually, many victims were observed have skin-related troubles like inflammation, erythema, itching and necrosis; myocardial infarction and renal failure are uncommon cases.

The immune system produces antibodies in response to antigens (venom) and different immune responses include the release of the alleregic mediators, like histamaine from the mast cell, which regulates the allergic reaction or anaphylactic hypersensitivity. Pathetically, centipede venom itself contains the histamine and serotonin (Eivind et al. 2011). So, the histamine range in the victim's blood has increased and causes enhanced allergic/hypersensitivity reaction.

4.4 RESEMBLANCE TO SNAKE ENVENOMATION EFFECTS

Nearly 70% of centipede venom composition is alike to snake venom. The literature shows that a number of effects common for centipede envenomation (like swelling, blistering, necrosis, pain, paralysis, rapid respiratory depression, hypertension, etc.) had similarities to viper, krait and cobra snake envenomation.

Kunjam et al. (2013) reported that viper bites cause local tissue swelling, hemorrhagic blisters from dermo-epidermal separation and ecchymoses. *E. ocellatus* envenomation causes swelling, blistering, necrosis and hemorrhages (Harrison et al. 2003) while *Naja nigricollis* envenoming leads to local necrosis, hemorrhage and respiratory arrest or paralysis (Schwersenski and Beatty 1982). Cobra victims experience severe pain, skin inflammation and ecchymoses around the bite site. Consequently, tense blebs, massive skin damage and subcutaneous tissues have developed that result in massive, untreatable ulcers. Envenomation causes death because of cardiogenic shock due to myocardial infarction. Rapid respiratory depression and hematotoxic effects were also seen (Girish 2006; Rajith and Ramachandran 2010; Shanmugam et al. 2012).

Envenomation from kraits can cause renal failure (Makhija and Khamar 2010), neuroparalysis (Basha and Sudarsanam 2012), hyponatremia (Vijaya 2013), hyperkalemia myocytolysis, myocardial damage with lethal arrhythmias, pulmonary edema and hypertension (Bawaskar 2004; Singh et al. 2001). It also causes respiratory muscle weakness and dysphasia. The severity of snake venom over centipede might be due to the low percentage of dissimilarities over venom composition and the concentration of each of them.

4.5 VENOM COMPONENTS AND THEIR EFFECTS

The centipede venom produced by merocrine, which contains multi-antigenic components, is responsible for its effects (Eivind et al. 2011). Table 4.2 explains the enzyme present in the family of Scolopendridae and their functions.

TABLE 4.2
Scolopendridae Venom Enzymes and Their Functions

Centipede	Enzyme	Function
Scolopendridae	Zinc metalloendopeptidase	Potential spreading factor
	Serine protease	Potentially involved in activation of toxins
	Glycoside hydrolase family 18	Unknown
	Glycoside hydrolase family 22	Potential antimicrobial component
	Glycoside hydrolase family 56	Degrades glycosaminoglycans, potentially facilitating the spread of venom components
	Phospholipase A2	Venom PLA2 can be myotoxic, inflammatory and neurotoxic
	CAP protein	CaV channel antagonist (KC144967) and trypsin inhibitor (KC144061)

Source: Eivind, A. B. et al., *Toxins*, 7(3), 679–704, 2015.

4.5.1 Metalloprotease

The literature reports that blister formation, hemorrhages, necrosis, itching (pruritus) and skin damage are due to metalloprotease enzyme. It is also responsible for vasodilatation by degrading the extracellular matrix, preventing blood clot formation (Gutierrez and Rucavado 2000), sturdy myotoxicity (Hamza et al. 2010) and hyperalgesic, oedematogenic activities (Malta et al. 2008). Therefore, it appears to be in the dominant form of proteases in centipede venom and serine proteases too play a role in toxin processing (Knapp et al. 2010; Siezen et al. 1997). In addition, to metalloproetase, there is serine protease (S1 and S8) found in Scolopendrid venom (Malta et al. 2008). They are involved in smooth muscle contraction, anticoagulation and immunosuppression (Fry et al. 2009; Low et al. 2013; Ma et al. 2012; Weston et al. 2013). The structure of metalloprotease resembles astacin metalloendoprotease (MEROPS family M12, subfamily A) (Undheim et al. 2014).

4.5.2 γ-Glutamyl Transpeptidases (GGTs)

GGTs are found abundantly in Scolopendridae than other centipedes (Undheim et al. 2014). They induce aggregation of platelets and haemolysis (Liu et al. 2012) and oxidative stress (Courtay et al. 1992).

4.5.3 Glycoside Hydrolases

Glycoside hydrolase enzymes (namely chitinase, lysozyme and hyaluronidase) hydrolyse the glycosidic bonds of carbohydrates. They are found in Scolopendridae (Gonzalez-Morales et al. 2014; Malta et al. 2008; Undheim et al. 2014).

4.5.4 Hyaluronidases

Among the three glycoside hydrolase enzymes, hyaluronidases are considered to be spreading factors that increase the pathological impact of venom components, such as hemorrhagins and neurotoxins (Girish et al. 2004; Kuhn-Nentwig et al. 2003; Long-Rowe and Burnett 1994). They also degrades the hyaluronic acid, which induces inflammation and tumor formation.

4.5.5 Phospholipase A2 (PLA2)

Enzyme PLA2 is found only in Scolopendrid venom (Gonzàlez-Morales et al. 2009; Liu et al. 2012; Malta et al. 2008; Undheim et al. 2014). PLA2 hydrolyse glycerophospholipids at the sn-2 position release lysophospholipids and fatty acids, such as arachidonic acids, thereby inducing skin diseases and prostrate cancer (Eividin et al. 2011). However, neofunctionalisation of snake venom PLA2 often removes the ability to catalyse the reaction (Fry et al. 2009; Low et al. 2013) and this may also be the case for PLA2 in scolopendrid venom. Local necrosis, hemorrhages and respiratory arrest or paralysis are due to PLA_2 (an anticoagulant enzyme which inhibits the prothrombinase complex by its binding to coagulation factor Xa) and cardiotoxin (Fry et al. 2009; King and Hardy 2013).

4.5.6 Cardiotoxin, Neurotoxin, Myotoxin and Cytotoxin Effects

Myocardial infarction, due to a surge of catecholamines, causes sinus bradycardia, A-V block and hypotension due to the cardio-depressant action of venom and sudden respiratory arrest without any other neurological manifestations, which can occur, resulting in anoxic cardiac arrest and acute abdominal pain, which is due to cholecystokinin release. The cardiotoxin leads to a vasoconstriction increase in capillary permeability, slow smooth muscle contraction and haemolysis. The neurotoxin interacts with the lysophosphatidic receptors and causes peripheral pain. The myotoxin is responsible for subcutaneous skin damage. The cytotoxin affects the cells and release histamine and causes edema, itching, pain and even intense headaches (Eividin et al. 2011).

4.5.7 CAP Proteins CRiSP (cysteine rich proteins), Allergen (Ag-5) and Pathogenesis-Related (PR-1)

CAP proteins are among the principal allergens in vespid and fire ant (*Solenopsis* sp.) venoms (Hoffman 2006). The relatively frequent allergic reactions observed after centipede envenomation might be due to the plentiful centiCAP proteins, because they act as ion channel modulators, vasodilators and myotoxins (Fry et al. 2009; Moran et al. 2013; Peichoto et al. 2009). CentiCAP1 have found only in *Scutigerid T. longicornis* and *Scolopendrid E. rubripes*, while centiCAP3 have been reported only in *S. morsitans* (Undheim et al. 2014). CentiCAP2 are the dominant form in Scolopendrinae, where they have diversified into multiple subtypes and undergone neofunctionalisation to include inhibitors of trypsin and voltage-gated calcium (CaV) channels (González-Morales et al. 2014; Liu et al. 2012; Rates et al. 2007; Undheim et al. 2014). The activities of centiCAP1 centiCAP2, centiCAP3 are most remaining to be determined but they might be of clinical relevance by virtue of their high abundance.

Apart from this enzyme, the most novel enzyme centipede peptidyl arginine deiminase (centiPAD) found in centipede venom has not been reported from any other animal venom, but several isoforms were detected scutigerid *T. longicornis* venom (Undheim et al. 2014). The function of CentiPAD in the venom remains to be determined, but they might be involved in post translational modification of toxin arginine residues.

4.6 CURRENT TREATMENT AGAINST HYPERSENSITIVITY

Centipede envenomation is not as simple as individuals take it and cannot be left as untreated. The victims were usually treated with analgesia, antihistamines, antibiotics (Fung et al. 2011), tabs of zerodol (100 mg bd x 10/7), tabs of neurobion T (bdx 2/52), IM diclofenac (50 mg bd x 3/7), caps of astyfer T (xdly 2/52) (Nwokeukwu et al. 2015) and TT injection. These drugs cause immediate relief for victims (Supakthanasiri et al. 2004; Crohn's and Colitis 2015).

These drugs offer temporary relief, since antihistamines may counteract the histamine, but can't counteract the other antigenic venom enzymes. The steroids hold back the action of the immune system and reduce the production of free radicals from the defence system, aiding the reduction of tissue damage. Though it suppresses the effects induced by antigens, these drugs' side effects are dose dependent (Web 1). TT injection can prevent secondary infection and neutralise of the venom (metalloprotease and other). Overall, these drugs will give only temporary suppression, since these drugs will interact with histamines, tetanus infections, analgesics and inflammatory mediators. The question is whether the drug will interact with the venom to neutralise its complex nature.

Apart from above reports, many centipede bites were not documented, since most of the victims or sufferers depend on traditional local practises (medicinal plants) for their treatment. The ethnobotanical documents contain the knowledge of medicinal plants which were passed on by different peoples. Many ethnobotanists reported that members of both Aristolochiaceae and Piperaceae have potential action against venomous bites as well as skin diseases. The ethnobotanical uses of the genus Aristolochia and Piper for venomous bites are described below.

4.7 ETHNOBOTANICAL USES OF *ARISTOLOCHIA* sp.

A. tagala Cham. roots are taken orally by the tribes of Tirunelveli hills in Tamil Nadu to treat poisonous snakebites (Ayyanar and Ignacimuthu 2005). The fresh roots of *A. tagala Cham.* are grounded along with roots of *Rauvolfia serpentine*, mixed in water and taken twice daily for 3 d to treat snakebites, which was used by the Kruichayas tribe of the Kannur district (Rajith and Ramachandran 2010). Roots of *A. indica* L. are taken orally in liquid form by the tribal community of Paschim Medinipur district of West Bengal, the Sugli tribes of Yerramalais of the Kurnool district, the Andhra Pradesh (Basha and Sudarsanam 2012) and the Malayali tribes of the Yercaud hills to treat centipede bite effects (Rekka et al. 2014). Leaf juice of *A. indica* L. is taken orally for snakebite (Shanmugam et al. 2012). *A. bracteolate* L., root powder taken about a teaspoonful internally for 3 d and whole plant decoction is used as an antidote (Vijaya et al. 2013). The leaf infusion of the species has been utilised by various tribes (Bhil, Meena, Garasia, Sahariya, Damor and Kathodia) of Rajasthan as an antidote. The fresh root extract of *A. indica* L., mixed with the root of *Rauvolfia serpentine* (L) Nenlt. and *Croton rozburghii* Balak., is consumed by the tribes of Bihar as an antidote.

4.8 ETHNOBOTANICAL USES OF *PIPER* sp.

The leaves and fruits of *P. nigrum* L. are used by people in Kalahandi district of Odisha, India to treat snakebite (Mund and Satapathy 2011) and its seeds are used against the effects of snakebite (Vijaya et al. 2013). The flower paste of *P. nigrum* L. is consumed along with ghee (Makhija and Khamar 2010) as antivenom. *P. longum L.* fruits have been traditionally used against snakebites in the northeastern and southern regions of India (Shenoy et al. 2013). The leaves, stems and branches of *P. pulchurm* C. DC. and the leaves and branches of *P. arboretum* Aub. are used against poisonous snakebites.

4.9 RECENT RESEARCH ON MEDICINAL PLANTS AGAINST VENOM

Antivenoms are basically of two types: monovalent and polyvalent. Antivenoms are polyclonal antibodies derived from horses or sheep for the treatment of venomous bites, which were used in the recent past. But from folk medicine, medicinal plants are used for the treatment of poisonous bites. Now, many research reports proves scientifically that medicinal plants can significantly inhibit lethality, cardiotoxicity, neurotoxicity, nephrotoxicity, myotoxicity, hemorrhaging, local pathology and respiratory paralysis. The plants or their extracts may also complement the benefits of conventional anti-serum treatment.

In general, the invention of antivenomic active principles from plant sources are based on *in vitro* assays, like procoagulant, PLA_2 and metalloprotease inhibition, proteolytic, fibrinolytic, defibrinogenolytic, haemolytic, inhibition of edema, inhibition of adenosine diphosphate, inhibition of induced platelet aggregation (Gutierrez 2009). Immunoassays correlate with the neutralization of lethality (Maria et al. 2001; Rial et al. 2006; Theakston and Reid 1979). The preclinical evaluation of antivenom engages with experimental animals, mostly with mice and rats. In view of this fact, venom that induces pain and other effects in these animals could be used in the evaluation of antivenom. Pharmacological studies have revealed that the extracts and fractions of some plants used in traditional medicine are able to antagonise the activity of various crude venoms and purified toxins (Biondo et al. 2003, 2004; Borges et al. 2000, 2001; da Silva et al. 2005; Januário et al. 2004; Maiorano et al. 2005; Oliveira et al. 2005; Otero et al. 2000; Ticli et al. 2005).

As above mentioned, the medicinal plants of both family Aristolochiaceae and Piperaceae showed the ability to inhibit the venom enzyme and their antivenomics properties, which are in Tables 4.3 and 4.4.

Hypothetically, some literature reports that the antivenom action of plants may be due to the active components that may block the active site of venom or receptor-structure prone to chemical attack or plant components that may inhibit venom metalloprotease enzyme action, and this could be due to the metal chelating potential of the components in the plants. The above hypothetical statement may endow us with some clues to identify the exact action between the venom and active component interaction or the inhibition or neutralisation action.

TABLE 4.3
Anitvenom Activity of Aristolochiaceae

Binomial Name	Plant Parts	Compound Name/Extract	Function	References
A. indica L.	Whole plant	Sesquiterpenes/ ethanol extract	Anti-scorpion venom Anti-phospholiphase A2	Attarde and Apte, (2013)
	Root	Methanol and ethanol	Anti-hyaluronidase	Girish and Kemparaju (2006)
	Whole plant	Methanol extract	Neutralizing the toxic effects of venom	Meenatchisundaram et al. (2009)
A. bracteolata L.	Leaf and root	Aristolochic acids	Anti-phospholiphase A2, anti-inflammatory	Anita et al. (2011)
A. elegans	Aerial parts and roots	Methanol extract	Scorpion antivenom	Izquierdo et al. (2010)
A. radix	Whole plant	Aristolochic acid	Anti-snake venom	Vishwanath et al. (1987)
A. albida	Whole plant	–	Anticoagulant, anti-hemolytic, anti-PLA_2	Abubakar et al. (2006)
A. odoratissima L.	Leaf	Aqueous extract	Against *Bothrops atrox* venom	Usubillaga et al. (2005)
A. shimadai	Whole plant	–	Anti-PLA_2	Kumar et al. (1998)

TABLE 4.4
Anitvenom Activity of Piperaceae

Binomial Name	Plant Parts	Compound Name/ Extract	Function	References
P. longum L.	Fruits	Piperine/ethanol extract	Anti-snake venom	Shenoy et al. (2013)
P. umbellatum L.	Branches	4-nerolidylcatechol	Anti-PLA_2	Núñez et al. (2005)
P. peltatum L.	Branches	4-nerolidylcatechol	Anti- PLA_2	
P. nigrum L.	Fruits	4- nerolidycatechol	Anti-myotoxicity Anti- PLA_2	

4.10 REQUIREMENTS IN VENOM RESEARCH

As per the previous studies, most of the victims were treated with the antihistamine, analgesic, anti-inflammatory and tetanus toxin injections. Those drugs are not ideal for the treatment of venom-induced effects. The researcher should focus on the invention of active principles from the medicinal plants and their mechanisms of interaction on venom to avoid the side effects displayed during conventional treatment. Each and every enzyme found in venom has its own peculiar role. So, the invention of a forthcoming drug might be based on the interaction of the active component with the enzymes of venom or at least with the dominant enzyme metalloprotease or metalloendopeptidase. The metal chelating agents are the inhibitors of metalloendopentidase enzyme (MEROPS database). Eventually, the metal chelating agents might be the best drug for treating envenomation.

4.11 CONCLUSION

Though centipede venom resembles snake venom, there is no specific treatment for centipede envenomation in modern pharmaceuticals. The specific drug that neutralizes or inhibits venom action optimistically could be obtained from medicinal plants with the antivenom properties *Aristolochia* and *Piper* sp., since both played an important role in the treatment of venomous bites traditionally.

REFERENCES

Abubakar, M. S., Balogun, E., Abdurahman, E. M., Nok, A. J., Shok, M., Mohammed, A., and Garba, M. 2006. Ethnomedical Treatment of Poisonous Snakebites: Plant Extract Neutralized *Naja nigricollis* Venom. *Pharmaceutical Biology* 44(5): 343–348.

Anita, J., Katewa, S. S., Sharma, S. K., Praveen, G., and Vartika, J. 2011. Snakelore and indigenous snakebite remedies practiced by some tribals of Rajasthan. *Indian Journal of Traditional Knowledge* 10(2): 258–268.

Attarde, S. S. and Apte, K. G. 2013. Studies on antivenom activity of *Aristolochia indica* plant extract against red scorpion venom by *in vivo* and *in vitro* methods. *International Journal of Pharmacognosy and Phytochemical Research* 5(3): 168–72.

Ayyanar, M. and Ignacimuthu, S. 2005. Medicinal plants used by the tribals of Tirunelveli hills, Tamil Naduto treat poisonous bites and skin diseases. *Indian Journal of Traditional Knowledge* 4(3): 229–236.

Bagnall, B. and Rook, A. 1977. Arthropods and the skin. Ed. A. Rook. In: *R.Ad. in Derma.* Edinburgh: Churchill Livingstone 4: 59–90.

Balit, C. R., Harvey, M. S., Waldock, J. M., and Isbister, G. K. 2004. Prospective study of centipede bites in Australia. *Journal of Toxicology: Clinical Toxicology* 42(1): 41–48.

Basha, S. K. and Sudarsanam, G. 2012. Traditional use of plants against snakebite in Sugali tribes of Yerramalais of Kurnool district, Andhra Pradesh, India. *Asian Pacific Journal of Tropical Biomedicine* 2(2): S575–S579.

Bawaskar, H. S. 2004. Snake venoms and antivenoms: Critical supply issues. *Journal – Association of Physicians of India* 52: 11–13.

Biondo, R., Pereira, A. M. S., Marcussi, S., Pereira, P. S., França, S. C., and Soares, A. M. 2003. Inhibition of enzymatic and pharmacological activities of some snake venoms and toxins by *Mandevilla velutina* (Apocynaceae) aqueous extract. *Biochimie* 85(10): 1017–1025.

Biondo, R., Soares, A. M., Bertoni, B. W., França, S. C., and Pereira, A. M. S. 2004. Direct organogenesis of Mandevilla illustris (Vell) Woodson and effects of its aqueous extract on the enzymatic and toxic activities of *Crotalus durissus terrificus* snake venom. *Plant Cell Reports* 22(8): 549–552.

Borges, M. H., Soares, A. M., Rodrigues, V. M., Andriao-Escarso, S. H., Diniz, H., Hamaguchi, A., Quintero, A., Lizano, S., Gutiérrez, J. M., Giglio, J. R., and Homsi-Brandeburgo, M. I. 2000. Effects of aqueous extract of *Casearia sylvestris* (Flacourtiaceae) on actions of snake and bee venoms and on activity of phospholipases A 2. *Comparative Biochemistry and Physiology Part B: Biochemistry and Molecular Biology* 127(1): 21–30.

Borges, M. H., Soares, A. M., Rodrigues, V. M., Oliveira, F., Fransheschi, A. M., Rucavado, A., Giglio, J. R., and Homsi-Brandeburgo, M. I. 2001. Neutralization of proteases from Bothrops snake venoms by the aqueous extract from *Casearia sylvestris* (Flacourtiaceae). *Toxicon* 39(12): 1863–1869.

Bush, S. P., King, B. O., Norris, R. L., and Stockwell, S. A. 2001. Centipede envenomation. *Wilderness Environmental Medicine* 12: 93–9.

Courtay, C., Oster, T., Michelet, F., Visvikis, A., Diederich, M., Wellman, M., and Siest, G. 1992. y-glutamyltransferase: Nucleotide sequence of the human pancreatic cDNA. Evidence for a ubiquitous y-glutamyltransferase polypeptide in human tissues. *Biochemical Pharmacology* 43: 2527–2533.

Crohn's & Colitis Foundation of America. Corticosteroids. Accessed 3/30/2015.

Da Silva, J. O., Coppede, J. S., Fernandes, V. C., Sant'ana, C. D., Ticli, F. K., Mazzi, M. V., Giglio, J. R., Pereira, P. S., Soares, A. M., and Sampaio, S. V. 2005. Antihemorrhagic, antinucleolytic and other antiophidian properties of the aqueous extract from *Pentaclethra macroloba. Journal of Ethnopharmacology* 100(1–2): 145–152.

Eivind, A. B., Undheim, G., Glenn, and King, F. 2011. Review on the venom system of centipedes (Chilopoda), a neglected group of venomous animals. *Toxicon* 57: 512–524.

Eivind, A. B., Undheim Bryan, G., and Glenn King, F. 2015. Centipede Venom: Recent Discoveries and Current State of Knowledge. *Toxins* 7(3): 679–704.

Fry, B. G., Roelants, K., Champagne, D. E., Scheib, H., Tyndall, J. D. A., King, G. F., Nevalainen, T. J., Norman, J. A., Lewis, R. J., and Norton, R.S. 2009. The toxicogenomic multiverse: Convergent recruitment of proteins into animal venoms. *Annu. Rev. Genom. Human Genet.* 10: 483–511.

Fung, H. T., Lam, S. K., and Wong, O. F. 2011. Centipede bite victims: A review of patients presenting to two emergency departments in Hong Kong. *Hong Kong Medical Journal* 17(38): 1–5.

Girish, K. S. and Kemparaju, K. 2006. Inhibition of Naja naja venom hyaluronidase: Role in the management of poisonous bite. *Life Sciences* 78(13): 1433–1440.

Girish, K. S., Shashidharamurthy, R., Nagaraju, S., Gowda, T. V., and Kemparaju, K. 2004. Isolation and characterization of hyaluronidase a "spreading factor" from Indian cobra (*Naja naja*) venom. *Biochimie* 86: 193–202.

González-Morales, L., Diego-García, E., Segovia, L., Gutierrez, M. D. C., and Possani, L. D. 2009. Venom from the centipede *Scolopendra viridis* Say: Purification, gene cloning and phylogenetic analysis of a phospholipase A2. *Toxicon* 54: 8–15.

González-Morales, L., Pedraza-Escalona, M., Diego-García, E., Restano-Cassulini, R., Batista, C. V. F., Gutierrez, M. C., and Possani, L. D. 2014. Proteomic characterization of the venom and transcriptomic analysis of the venomous gland from the Mexican centipede. *Scolopendra viridis. Journal of Proteomics* 111: 224–237.

Gutierrez, J. M., Lomonte, B., Leon, G., Alape-Girón, A., Flores Diaz, M., and Sanz, L. 2009. Snake venomics and antivenomics: Proteomic tools in the design and control of antivenoms for the treatment of snakebite envenoming. *Journal of Proteomics* 72: 165–182.

Gutierrez, J. M., and Rucavado, A. 2000. Snake venom metalloproteinases: Their role in the pathogenesis of local tissue damage. *Biochimie* 82(9–10): 841–850.

Hamza, L., Girardi, T., Castelli, S., Gargioli, C., Cannata, S., Patamia, M., Luly, P., Laraba-Djebari, F., Petruzzelli, R., and Rufini, S. 2010. Isolation and characterization of a myotoxin from the venom of *Macrovipera lebetina* transmediterranea. *Toxicon* 56: 381–390.

Harada, S., Yoshizaki, Y., Natsuaki, M., Shimizu, H., Fukuda, H., Nagai, H., and Ikeda, T. 2005. Three cases of centipede allergy – Analysis of cross reactivity with bee allergy [abstract]. *Arerugi* 54: 1279–1284.

Harrison, R. A., Oliver, J., Hasson, S. S., Bharati, K., and Theakston, R. D. G. 2003. Novel sequences encoding venom C-type lectins are conserved in phylogenetically and geographically distinct Echis and Bitis viper species. *Gene* 315(1–2): 95–102.

Hoffman, D. R. 2006. Hymenoptera venom allergens. *Clinical Reviews in Allergy & Immunology* 30: 109–128.

Izquierdo, A. M., Zapata, E. V., Jiménez-Ferrer, J. E., Muñnoz, C. B., Aparicio, A. J., Torres, K. B., and Torres, L. O. 2010. Scorpion antivenom effect of micropropagated *Aristolochia elegans. Pharmaceutical Biology* 48(8): 891–896.

Januario, A. H., Santos, S. L., Marcussi, S., Mazzi, M. V., Pietro, R. C. L. R., Sato, D. N., Ellena, J., Sampaio, S. V., França, S. C., and Soares, A. M. 2004. Neo-clerodane diterpenoid, a new metalloprotease snake venom inhibitor from *Baccharis trimera* (Asteraceae): Anti-proteolytic and anti-hemorrhagic properties. *Chemico-Biological Interactions* 150(3): 243–251.

King, G. F. and Hardy, M. C. 2013. Spider-venom peptides: Structure, pharmacology, and potential for control of insect pests. *Annual Review of Entomology* 58: 475–496.

Knapp, O., Stiles, B., and Popoff, M. R. 2010. The aerolysin-like toxin family of cytolytic, pore-formingtoxins. *Open Toxinol J.* 3: 53–68.

Kuhn-Nentwig, L. 2003. Antimicrobial and cytolytic peptides of venomous arthropods. *Cellular and Molecular Life Sciences* 60: 2651–2668.

Kumar, K., Murthy, A. R., and Upadhyay, O. P. 1998. Plants used as antidotes by the tribals of Bihar. *Ancient Science of Life* 17(4): 268.

Kundal, Das. 2009. Medicinal plants for snakebite treatment – Future Focus. *Ethnobotanical Leaflets* 13: 508–521.

Kunjam, S. R., Jadhav, S. K., and Tiwari, K. L. 2013. Traditional herbal medicine for the treatment of snakebite and scorpion sting by the Tribes of South Surguja, Chhittisgarh, India. *Med. Aromat. Plants* 2: 17.

Liu, Z. C., Zhang, R., Zhao, F., Chen, Z. M., Liu, H. W., Wang, Y. J., Jiang, P., Zhang, Y., Wu, Y., and Ding, J. P. 2012. Venomic and transcriptomic analysis of centipede *Scolopendra subspinipes dehaani. Journal of Proteome Research* 11: 6197–6212.

Long-Rowe, K. O. and Burnett, J. W. 1994. Characteristics of hyaluronidase and hemolytic activity in fishing tentacle nematocyst venom of *Chrysaora quinquecirrha. Toxicon* 32: 165–174.

Low, D. H. W., Sunagar, K., Undheim, E. A. B., Ali, S. A., Alagon, A. C., Jones, A., Jackson, T. N. W., Gonzalez, S. P., King, G. F., and Antunes, A. 2013. Dracula's children: The proteomics, transcriptomics and molecular evolution of vampire bat venom proteins. *Journal of Proteomics* 89: 95–111.

Ma, Y., He, Y., Zhao, R., Wu, Y., Li, W., and Cao, Z. 2012. Extreme diversity of scorpion venom peptidesand proteins revealed by transcriptomic analysis: Implication for proteome evolution of scorpionvenom arsenal. *Journal of Proteomics* 75: 1563–1576.

Maiorano, V. A., Marcussi, S., Daher, M. A., Oliveira, C. Z., Couto, L. B., Gomes, O. A., França, S. C., Soares, A. M., and Pereira, P. S. 2005. Antiophidian properties of the aqueous extract of *Mikania glomerata. Journal of Ethnopharmacology* 102(3): 364–370.

Makhija, I. K. and Khamar, D. 2010. Anti-snake venom properties of medicinal plants. *Der Pharmacia Lettre* 2(5): 399–411.

Malta, M. B., Lira, M. S., Soares, S. L., Rocha, G. C., Knysa, I., Martins, R., Guizze, S. P. G., Santoro, M. L., and Barbaro, K. C. 2008. Toxic activities of Brazilian centipede venoms. *Toxicon* 52: 255–263.

Maria, W. S., Pacheco, B. G., Barbosa, C. F., Velarde, D. T., Chávez-Olórtegui, C. 2001. Determination of the neutralizing potency of horse antibothropic and anticrotalic antivenoms in blood samples collected on filter paper. *Toxicon* 39: 1607–1609.

Martz, W. 1992. Plant with a reputation against snakebite. *Toxicon* 30(10): 1131–1142.

Meenatchisundaram, S., Parameswari, G., and Michael, A. 2009. Studies on antivenom activity of *Andrographis paniculata* and *Aristolochia indica* plant extracts against *Daboia russelli* venom by *in vivo* and *in vitro* methods. *International Journal of Science & Technology* 2: 4.

Mirajkar, K. K., More, S., and Gadag, J. R. 2006. Preliminary studies with a neurotoxin obtained from *Bungarus caeruleus* venom. *Journal of Venomous Animals and Toxins Including Tropical Diseases* 12: 1.

Moran, Y., Praher, D., Schlesinger, A., Ayalon, A., Tal, Y., and Technau, U. 2013. Analysis of soluble protein contents from the nematocysts of a model sea anemone sheds light on venom evolution. *Marine Biotechnology* 15: 329–339.

Mors, W. B., Nascimento, M. C., Pereira, B. M. R., and Pereira, N. A. 2000. Plant natural products active against snakebite – The molecular approach. *Phytochemistry* 55(6): 627–642.

Mund, N. K. and Satapathy, K. B. 2011. Notes on the traditional use of plants to treat snakebite in Kalahandi district of Odisha. *Biohelica* 2: 57–66.

Murugan, V., Amol Dongre, R., and Ganapathy, K. 2014. An Epidemiological Study of Animal Bites and Envenomings in a Rural District of Tamilnadu, India. *Online J. Health Allied Scs.* 13: 4.

Núñez, V., Castro, V., Murillo, R., Ponce-Soto, L. A., Merfort, I., and Lomonte, B. 2005. Inhibitory effects of Piper umbellatum and Piper peltatum extracts towards myotoxic phospholipases A2 from Bothrops snake venoms: Isolation of 4-nerolidylcatechol as active principle. *Phytochemistry* 66: 1017–1025.

Nwokeukwu, Ugochukwu, O., Chidinma, A., Debra, I., and Igbani, U. 2015. Hyperesthesia Secondary to Centipede Bite: A Case Report Huldah I. *International Journal of Medicine and Pharmaceutical Sciences* 2(2): 22–26.

Oliveira, C. Z., Maiorano, V. A., Marcussi, S., Santana, C. D., Januário, A. H., Lourenço, M. V., Sampaio, S. V., França, S. C., Pereira, P. S., and Soares, A. M. 2005. Anticoagulant and antifibrinogenolytic properties of the aqueous extract from *Bauhinia forficata* against snake venoms. *Journal of Ethnopharmacology* 98(2): 213–216.

Otero, R., Nunez, V., Barona, J., Fonnegra, R., Jimenez, S. L., Osorio, R. G., Saldarriaga, M., and Diaz, A. 2000. Snakebites and ethnobotany in the northwest region of Colombia: Part III: Neutralization of the hemorrhagic effect of *Bothrops atrox* venom. *Journal of Ethnopharmacology* 73(2): 233–241.

Peichoto, M. E., Mackessy, S. P., Teibler, P., Tavares, F. L., Burckhardt, P. L., Breno, M. C., Acosta, O., and Santoro, M. L. 2009. Purification and characterization of a cysteine-rich secretory protein from *Philodryas patagoniensis* snake venom. *Comparative Biochemistry and Physiology* 150: 79–84.

Perumal Samy, R. and Ignacimuthu, S. 2000. Antibacterial activity of some of folklore medicinal plants used by Tribals in Western Ghats of India. *Journal of Ethnopharmacology* 69: 63–71.

Pradeeps, M. and Poyyamoli, G. 2013. Ethnobotany and utilization of plant resources in Irula villages (Sigur plateau, Nilgiri Biosphere Reserve, India). *Journal of Medicinal Plants Research* 7(6): 267–769.

Rajan, S. S., Bannikuppe, A. R., and Viswanath, S. 2013. Biochemical characterization of proteins and enzymes of *Bugarus caeruleus* venom. *Biomedicine and Preventive Nutrition* 3(2): 145–149.

Rajith, N. P. and Ramachandran, V. S. 2010. Ethnomedicines of Kurichyas, Kannur district, Western Ghats, Kerala. *International Journal of Natural Products Research* 1(2): 249–253.

Rates, B., Bemquerer, M. P., Richardson, M., Borges, M. H., Morales, R. A. V., de Lima, M. E., and Pimenta, A. 2007. Venomic analyses of *Scolopendra viridicornis nigra* and *Scolopendra angulata* (Centipede, Scolopendromorpha): Shedding light on venoms from a neglected group. *Toxicon* 49: 810–826.

Rekka, R., Murugesh, S., and Prabakaran, R. 2014. Ethnomedicinal study of plants used for the treatment of diabetetes and antidote for poisonous bite by Malayali tribes of Yercaud hills, Southern Eastern Ghats, Salem District, Tamil Nadu. *Life Sci. Leaf* 49: 89–96.

Revathi, P. and Parimelazhagan, T. 2010. Traditional Knowledge on Medicinal plants used by the Irula tribe of Hasanur Hill, Erode district, Tamil Nadu, India. *Ethnobotan Leaf* 14: 136–60.

Rial, A., Morais, V., Rossi, S., and Massaldi, H. 2006. A new ELISA or 72. determination of potency in snake antivenoms. *Toxicon* 4: 462–466.

Schwersenski, J. and Beatty, D. W. 1982. Unusaual features in a case of snakebite, presumably due to the cape cobra (Naja nigicolis). *South African Medical Journal* 61(16): 597–598.

Shanmugam, S., Rajendran, K., and Suresh, K. 2012. Traditional uses of medicinal plants among the rural people in Sivagangai District, Tamil Nadu, Southern India. *Asian Pacific Journal of Tropical Biomedicine* 429–434.

Sharma, P. P. and Mujumdar, A. M. 2003. Traditional Knowledge on plants from Toranmal plateau of Maharashtra. *Indian J. Trad. Knowledge* 2(3): 292–296.

Shenoy, P. A., Nipate, S. S., Sonpetkar, J. M., Salvi, N. C., Waghmare, A. B., and Chaudhari, P. D. 2013. Anti-snake venom activities of ethanolic extract of fruits of *Piper longum L.* (Piperaceae) against Russell's viper venom: Characterization of piperine as active principle. *Journal of Ethnopharmacology* 20, 147(2): 373–382.

Siezen, R. J. and Leunissen, J. A. M. 1997. Subtilases: The superfamily of subtilisin-like serine proteases. *Protein Science* 6: 501–523.

Singh, G., Gourinath, S., Sharma, S., Paramasivam, M., Srinivasan, A., and Singh, T. P. 2001. Sequence and crystal structure determination of a basic phospholipase A2 from common krait (Bungarus caeruleus) at 2.4 A resolution: Identification and characterization of its pharmacological sites. *Journal of Molecular Biology* 6, 307(4): 1049–1059.

Soares, A. M., Januario, A. H., Lourenço, M. V., Pereira, A. M., and Pereira, P. S. 2004. Neutralizing effects of Brazilian plants against snake venoms. *Drugs Fut.* 29(11): 1105–1117.

Soares, A. M., Marcussi, S., Fernandes, R. S., Menaldo, D. Z., Cost, T. R., Lourenço, A. H. J., and Pereira, P. S. 2009. Medicinal plant extracts and molecules as the source of new anti-snake venom drugs. Rahman, A., Reitz, A. B., and Choudhary, M.I. Eds. Karachi-Pakistan: Bentham Science Publishers. *Frontiers in Medicinal Chemistry* 4: 309–346.

Soares, A. M., Ticli, F. K., Marcussi, S., Lourenço, M. V., Januario, A. H., Sampaio, S. V., Giglio, J. R., Lomonte, B., and Pereira, P. S. 2005. Medicinal plants with inhibitory properties against snake venoms. *Current Medicinal Chemistry* 12(22): 2625–2641.

Soylu, A., Kavukçu, S., Erdur, B., Demir, K., and Türkmen, M. A. 2005. Multisystemic leukocytoclastic vasculitis affecting the central nervous system. *Pediatric Neurology* 33(4): 289–291.

Stawiski, M. A. 1985. Insect bites and stings. *Emergency Medicine Clinics of North America* 3: 785–808.

Supakthanasiri, P., Ruxrungtham, K., Klaewsongkram, J., and Chantaphakul, H. 2004. Anaphylaxis to centipede bite. *Journal Allergy Clinical Immunology* 113: 244.

Tan, N. H. and Ponnudurai, G. A. 1990. Comparative study of the biological properties of Krait (Genus: Bungarus) venoms. *Comparative Biochemistry and Physiology C* 95(1): 105–109.

Theakston, R. D. G. and Reid, H. A. 1979. Enzyme-linked immunosorbent 74. assay (ELISA) in assessing antivenom potency. *Toxicon* 17: 511–515.

Ticli, F. K., Hage, L. I., Cambraia, R. S., Pereira, P. S., Magro, A. J., Fontes, M. R., Stábeli, R. G., Giglio, J. R., França, S. C., Soares, A. M., and Sampaio, S. V. 2005. Rosmarinic acid, a new snake venom phospholipase A2 inhibitor from *Cordia verbenacea* (Boraginaceae): Antiserum action potentiation and molecular interaction. *Toxicon* 46(3): 318–327.

Undheim, E. A. B., Jones, A., Clauser, K. R., Holland, J. H., Pineda, S. S., King, G. F., and Fry, B. G. 2014. Clawing through evolution: Toxin diversification and convergence in the ancient lineage Chilopoda (Centipedes). *Molecular Biology and Evolution* 31: 2124–2148.

Usubillaga, N., Khouri, S., Celdilo-Vaz, and Yibrin, E. 2005. Anti-snake venom effect of *Aritolochia odoratissima* L. aqueous extract on mice. *Acta Horiculture. (ISHS)* 677: 85.

Uzzaman, A. and Cho, S. H. 2012. Classification of hypersensitivity reactions. *Allergy and Asthma Proceedings* 33(1): 96–99.

Vijaya, P., Ranjani, R., Rao, M. R., and Sudarsanam, G. 2013. Indentification of antidote medicinal plants against snake venom – A field survey. *International Journal of Medicine and Pharmaceutical Sciences* 3(5): 21–32.

Vinod, K. 2006. A checklist of centipedes (Chilopoda: Scolopendramorpha) from central India. *Zoos' Print Journal* 21(2): 2164–2166.

Vishwanath, B. S., Kini, R. M., and Gowda, T. V. 1987. Characterization of three edema-inducing phospholipase A2 enzymes from habu (*Trimeresurus flavoviridis*) venom and their interaction with the alkaloid aristolochic acid. *Toxicon* 25(5): 501–515.

Vorono, E., Apte, R. N., and Sofer, S. 1999. The systemic inflammatory response syndrome related to the release of cytokines following severe envenomation. *Journal of Venomous Animals and Toxins* 5(1).

Walton, G. S. 1988. Cutaneous responses to arthropods. In: *New Clinical Applications in Dermatology*. Ed. J. Verbov. Talking Points in Dermatology III. Dordrecht 103–124.

Warrell, D. A. 1993. Snake bite and sanke venoms. *Quarterly Journal of Medicine* 86(6): 351–353.

Web 1: www. everydayhealth.com

Weston, A. J., Chung, R., Dunlap, W. C., Morandini, A. C., Marques, A. C., Moura-da-Silva, A. M., Ward, M., Padilla, G., da Silva, L. F., and Andreakis, N. 2013. Proteomic characterisation of toxinsisolated from nematocysts of the South Atlantic jellyfish *Olindias sambaquiensis*. *Toxicon* 71: 11–17.

Wong, E. S., Morgenstern, D., Mofiz, E., Gombert, S., Morris, K. M., Temple-Smith, P., Renfree, M. B., Whittington, C. M., King, G. F., and Warren, W.C. 2012. Proteomics and deep sequencing comparison of seasonally active venom glands in the platypus reveals novel venom peptides and distinct expression profiles. *Molecular and Cellular Proteomics* 11: 1354–1364.

Wong, E. S., Papenfuss, A. T., Whittington, C. M., Warren, W. C., and Belov, K. A. 2012. Limited role for gene duplications in the evolution of platypus venom. *Molecular Biology and Evolution* 29: 167–177.

Zhang, H., Xu, S. J., Wang, Q. Y., Song, S. Y., Shu, Y. Y., and Lin, Z. J. 2002. Structure of a cardiotoxic phospholipase A(2) from Ophiophagus hannah with the "pancreatic loop". *Journal of Structural Biology* 138(3): 207–15.

5 Antidiabetic Herbal Formulations

An Effective Therapeutic Approach to Diabetes

Binu Thomas and Krishna M. Chinchu

CONTENTS

5.1 Introduction 89
5.1.1 Area of Present Study 90
5.2 Methodology 91
5.3 Results and Discussions 91
5.3.1 Diversity of Antidiabetic Plants in the Study Area 91
5.3.2 Life Forms Analysis 92
5.3.3 Efficacy of Herbal Formulations in Antidiabetic Treatments 92
5.3.4 Diabetic-Associated Health Problems 97
5.3.5 Popularization of Herbal Formulations in the Study Area 98
5.3.6 Major Threats to the Medicinal Flora of the Study Area 98
5.4 Conclusion 99
References 100

5.1 INTRODUCTION

Diabetes is a chronic disorder of carbohydrate, fat and protein metabolism characterized by increased fasting and postprandial blood sugar levels. It is a complex metabolic disorder resulting from either insulin insufficiency or insulin dysfunction (Ramachandran et al. 2002). The different types of diabetes have been identified and categorized as type I, II and gestational diabetes. Type I is also referred as insulin-dependent diabetes mellitus or juvenile diabetes (IDDM), which results from the body's failure to produce insulin and presently requires the person to inject insulin. Type II diabetes also referred as non-insulin-dependent diabetes mellitus or adult-onset diabetes (NIDDM). This result from insulin resistance, a condition in which cells fail to use insulin properly (Binu Thomas et al. 2012). The other type is gestational diabetes, which is observed in pregnant women who have never had diabetes before and who have a high blood glucose level during pregnancy. It may precede development of type II. The other forms of diabetes mellitus include congenital diabetes, which is due to the genetic defect of insulin secretion; cystic fibrosis–related diabetes; steroid diabetes, induced by high dose of glucocorticoids; and several forms of monogenic diabetes (Manisha et al. 2012).

India has a diabetic population of approximately about 18 million. The WHO (World Health Organization) has estimated that more than 80% of the world's population use botanical medicines to control this chronic disorder. A variety of plant preparations have been mentioned in Ayurveda and other indigenous systems of medicine (Yeh et al. 2003). These traditional medicines are derived mainly from plants and play major role in the management of diabetes mellitus (Shokeen et al. 2008). There are a lot of hypoglycaemic agents from synthetic sources that are available, and the cost of these medicines is a burden to the people. In addition to these, secondary complications continue to be a major health problem. In this scenario, traditional medicine is widely used for the treatment of diabetes in various developing countries (Saravanan and Pari 2008). The allopathic drugs used for the treatment of diabetes have their own side effects and adverse effects, like hypoglycaemia, nausea, vomiting, hyponatremia, flatulence, diarrhoea, constipation, alcohol flush, headache, weight gain, lactic acidosis, pernicious anemia, dyspepsia, dizziness and joint pain (Kokar and Mantha 1998).

Many indigenous Indian medicinal plants have been found to be useful to successfully manage diabetes. One of the great advantages of medicinal plants is that these are readily available and have very low side effects (Balamurugan and Ignacimuthu 2011). The active principles present in medicinal plants have been reported to possess pancreatic beta cells regenerating, releasing insulin and fighting the problem of insulin resistance. Plants have always been an exemplary source of drugs, and many of the currently available drugs have been derived directly or indirectly from these medicinal plants. The ethnobotanical information reports that, there are about more than 500 plants are being utilized for the treatment of diabetes by various indigenous people (Alarcon-Aguilara et al. 1998).

5.1.1 Area of Present Study

Meenachil Taluk of the Kottayam district, Kerala, lies in the northeastern region of the Kottayam district (9.36°N and 76.17°E). The area is blessed with diversified habitats such as lush paddy fields, hills and hillocks, highlands and different crop plantations. It has a tropical climate. The humidity is high and rises to about 90% during the rainy season. This area gets rain from two monsoon seasons, the southwest monsoon and the northeast monsoon. The average rainfall is around 3600 mm per year. The southwest monsoon starts in June and ends in September. The northeast monsoon season is from October to November. Pre-monsoon rains during March to May are accompanied by thunder and lightning; the highest rainfall during this period is in December. January and February are cooler, while March, April and May are warmer. The temperature ranges between 38.5°C and 15°C. Though food crops like paddy and tapioca are cultivated, the majority of the population depends on cash crops like rubber and black pepper for income (Reni et al. 2014) (Figure 5.1).

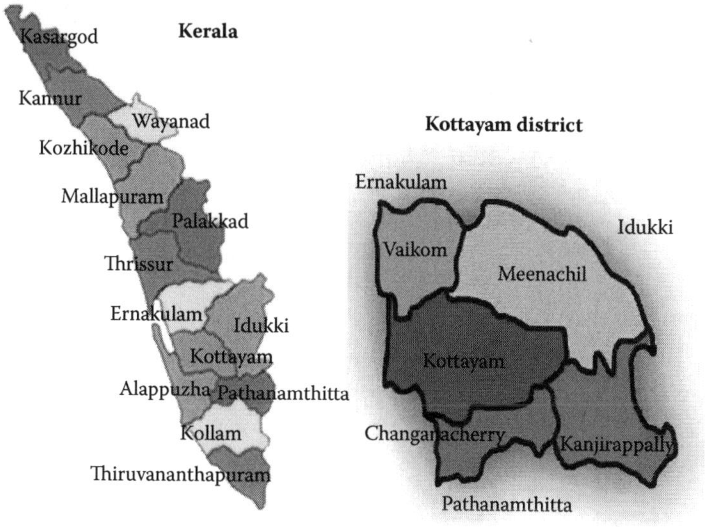

FIGURE 5.1 Map of Kerala showing Kottayam District and Meenachil Taluk.

5.2 METHODOLOGY

The present study was based on an extensive survey and field observations during the year 2014–2015. In this study, attempts were made to find out various antidiabetic plants, which were collected from Meenachil Taluk of the Kottayam district, Kerala. This documentation was mainly based on field observation, discussions with local peoples and herbal practitioners, as well as scrutinizing the literature review. The collected specimens were identified taxonomically with the help of available floras and literature (Gamble and Fischer 1915–1936; Sasidharan 2004). During the field visits, plant specimens were collected at different reproductive stages to prepare herbarium specimens. The specimens were processed for the preparation of herbarium by standard procedure (Santapau 1955). The voucher specimens were deposited in the herbaria of the PG Department of Botany, Deva Matha College, Kuravilangad, Kottayam, for future reference.

5.3 RESULTS AND DISCUSSIONS

5.3.1 Diversity of Antidiabetic Plants in the Study Area

The present study was undertaken to enumerate the diversity of potential antidiabetic plants in Meenachil Taluk of Kottayam district, Kerala. There are about 50 taxa which show antidiabetic properties and belong to 48 genera in 36 families. Besides these, the endangered species (such as *Humboldtia vahliana* Wight and vulnerable species like *Saraca asoca* Roxb.) are also found in the study area (Sasidharan 2004).

TABLE 5.1
Analysis of the Antidiabetic Plants in the Study Area

Analysis of Medicinal Plants		Families		Genera		Species	
Dicotyledons	Polypetalae	18	30	26	40	28	42
	Gamopetalae	8		9		9	
	Monochlamydae	4		5		5	
Monocotyledons		6		8		8	
Total		36		48		50	

Among the total 50 taxa of antidiabetic plants, the group dicots are dominating with 42 species in 30 families and monocots with eight species in six families. The analysis of total families documented from the study area reveals that, out of 36 families, Fabaceae are the first dominanat family with four species. Cucurbitaceae and Myrtaceae are the second dominant families with three species. All other families have two and one species, respectively (Tables 5.1 and 5.2; Figures 5.2 through 5.4).

The resulting dominant genera of antidiabetic plants which occured in the study area, revealed that, the genera like *Syzygium* and *Annona* are the dominant ones with two species, respectively. All other genera are represented with a single species each. Similarly, the analysis of various plant parts which are utilized for the preparation of different antidiabetic drugs are composed of leaves (16 Nos.), fruits (7 Nos.), roots (8 Nos.), whole plant (5 Nos.), seeds (2 Nos.), stems (2 Nos.), rhizomes (2 Nos.), bark (2 Nos.), tubers (2 Nos.), hard wood (1 No.), flower (1 No.), leaves and fruits (2 Nos.), respectively (Figure 5.5).

5.3.2 Life Forms Analysis

The total life forms of antidiabetic plants present in the study area revealed that trees are more dominant with 17 species, followed by herbs with 14 species, climbers with 10 species and shrubs with 9 species (Figure 5.6). The present results also indicate that the medicinal plants which are documented from the study area contribute more to the biodiversity of existing ecosystems.

5.3.3 Efficacy of Herbal Formulations in Antidiabetic Treatments

Diabetes mellitus is a group of metabolic diseases characterized by hyperglycemia resulting from defects in insulin secretion, insulin action or both. The chronic hyperglycemia of diabetes is associated with the long-term damage, dysfunction and failure of various organs, especially the eyes, kidneys, nerves, heart and blood vessels. Several pathogenic processes are involved in the development of diabetes. These range from autoimmune destruction of the β-cells of the pancreas with consequent insulin deficiency, to abnormalities that result in resistance to insulin action. The basis of the abnormalities in carbohydrate, fat and protein metabolism in diabetes is the deficient action of insulin on target tissues. Deficient insulin action results from inadequate insulin secretion or diminished tissue responses to insulin at one or more

TABLE 5.2
List of Antidiabetic Plants in the Study Area

S. No.	Botanical Name	Local Name	Family	Parts Used	Mode of Administration
1.	*Aegle marmelos* (L.) Corr.	Koovalam	Rutaceae	Leaves	Leaf decoction is consumed daily on an empty stomach.
2.	*Aerva lanata* (L.) Juss.	Cherula	Amarantaceae	Whole plant	Whole plant is boiled in water and consumed daily.
3.	*Aloe vera* (L.) Burm. f.	Kattarwasha	LIliaceae	Leaves	Oral administration of one tablespoon of *Aloe vera* juice, twice a day.
4.	*Andrographis paniculata* Brum.f.Wall	Kiriyath	Acanthaceae	Leaves	Oral administration of leaf juice usually used.
5.	*Annona muricata* L.	Mullatha	Annonaceae	Fruit	The juices of raw fruits are taken orally.
6.	*Annona squamosa* L.	Sithapazham	Annonaceae	Leaves	Leaf extract is taken orally.
7.	*Abrus precatorius* L.	Kunnikuru	Fabaceae	Leaves	Leaf juice is taken orally.
8.	*Artocarpus heterophyllus* Lam.	Plavu	Moraceae	Fruit	Mature fruits are eaten either raw or cooked.
9	*Asparagus recemosus* Wild.	Sathavary	Liliaceae	Rhizome	The rhizome of the plant is boiled and taken with water daily.
10.	*Azadirachta indica* A. Juss.	Veppu	Meliaceae	Leaves	One or two leaves are dipped in a glass of water for one night. This water is taken on an empty stomach the next morning.
11.	*Benincasa hispida* Thunb.	Kumbalam	Cucurbitaceae	Fruits	Fruit juice is taken daily.
12.	*Biophytum sensitivum* L.	Mukkutti	Oxalidaceae	Whole plant	The whole plant decoction is taken orally.
13.	*Boerhavia diffusa* (L.) DC.	Thazhuthama	Nyctaginaceae	Leaves	Leaf decoction is consumed daily.
14.	*Catharanthus roseus* (L.)G. Don.	Shavanari	Apocynaceae	Leaves	Leaf juice is taken twice a day.
15.	*Centella asiatica* L.	Kudakan	Apiaceae	Leaves	The paste prepared from the whole plant and is taken orally in the morning.

(*Continued*)

TABLE 5.2 (CONTINUED)
List of Antidiabetic Plants in the Study Area

S. No.	Botanical Name	Local Name	Family	Parts Used	Mode of Administration
16.	*Coccinia grandis* (L.) Wight.	Koval	Cucurbitaceae	Leaves and Fruits	Fruit and leaf juice taken daily.
17.	*Costus picatus* D.Don *ex* Lindl.	Insulinchedi	Zingiberaceae	Leaves	2–3 young leaves are taken daily.
18.	*Curcuma longa* L.	Manjal	Zingiberaceae	Rhizome	Mixture of juice of *Emblica officinalis* and *Curcuma longa* powder is taken on an empty stomach.
19.	*Cynodon dactylon* (L.) Pers. Syn.	Karuka pullu	Poaceae	Whole plant	The whole plant decoction is taken orally.
20.	*Desmodium gangeticum* (L.) DC	Orila	Fabaceae	Root	The roots of the plant along with the roots of *Pseudarthira viscid* and *Heliotropium indicum* are boiled in water and taken orally.
21.	*Dioscorea bulbifera* L.	Kattukachil, Adathappu	Dioscoreaceae	Tuber	Tuber is used to treat diabetes.
22.	*Ensete superba* (Roxb.) Chees.	Kalluvasha	Musaceae	Seed	Seed powder is mixed with milk and taken daily
23.	*Ficus racemosa* L.	Athi	Moraceae	Root	The root exudates are collected on a glass and taken orally
24.	*Gymnema sylvstris* R.Br.	Chakkarakolli	Asclepiadaceae	Leaves	2–3 leaves are consumed on an empty stomach.
25.	*Helicteres isora* L.	Idampiri-valampiri	Stericulaceae	Root	Root decoction is taken daily.
26.	*Heliotropium indicum* L.	Vayalchedi	Boraginaceae	Roots	The roots of the plant along with the roots of *Desmodium gangeticum* and *Pseudarthira viscida* are boiled in water and taken orally.
27.	*Hemidesmus indicus* (L.) R. Br.	Nannari	Asclepiadaceae	Root	Root decoction is taken internally for the treatment of diabetes.
28.	*Hibiscus rosa-sinensis* L.	Chemparathi	Malvaceae	Flower	Flower extract is taken dialy.

(*Continued*)

TABLE 5.2 (CONTINUED)
List of Antidiabetic Plants in the Study Area

S. No.	Botanical Name	Local Name	Family	Parts Used	Mode of Administration
29.	*Humboldtia vahlina* Wight.	Kurappunna	Cesalpinaceae	Root	A decoction of the root nodules in water is given twice a day.
30.	*Kyllinga nemoralis* Forst.	Peemuthanga	Cyperaceae	Tuber	Tuber juice is taken during the morning and evening.
31.	*Lagerstroemia speciosa* (L.) Pers.	Manimaruthu	Lythraceae	Bark powder	Dried bark powder is boiled and taken daily with water.
32.	*Lantana camera* L.	Kongini	Verbenaceae	Fruit	Fruits consumed in raw form.
33.	*Mangifera indica* L.	Mavu	Anacardiaceae	Leaves	Leaf extract is taken daily.
34.	*Mimosa pudica* L.	Thottavadi	Mimosaceae	Whole plant	Plant juice taken orally twice daily.
35.	*Momordica charantia* L.	Paval	Cucurbitaceae	Leaves and Fruits	The juice obtained from both leaf and fruits are taken orally in empty stomach in early morning.
36.	*Moringa oleifera* Beed.	Murringa	Moringaceae	Leaves	Leaf juice is taken orally
37.	*Murraya koenigii* (L.) Spreng.	Kariveppila	Rutaceae	Leaves	Leaf juice taken orally or leaves are boiled with water and taken on an empty stomach.
38.	*Ocimum tenuiiflorum* L.	Thulasi	Lamiaceae	Leaves	Leaf juice or extract is taken orally.
39.	*Passiflora edulis* Sim.	Passion fruit	Passifloraceae	Fruit	Fruit pulp is taken orally.
40.	*Phyllanthus emblica* L.	Nelly	Euphorbiaceae	Fruit	Mixture of fruit juice of *Phullanthus* and *Curcuma longa* powder is taken on an empty stomach.
41.	*Plumbago indica* L., Diss.	Chethikoduveli	Plumbaginaceae	Stem	Stem decoction of *Plumbago indica* and *Tinospora cordifolia* is given twice a day.

(Continued)

TABLE 5.2 (CONTINUED)
List of Antidiabetic Plants in the Study Area

S. No.	Botanical Name	Local Name	Family	Parts Used	Mode of Administration
42.	*Pseudarthria viscida* (L.) Wight.	Moovila	Fabaceae	Root	The roots of the plant along with the roots of *Desmodium gangeticum* and *Heliotropium indicum* are boiled in water and taken orally.
43.	*Psidium guajava* L.	Perakka	Myrtaceae	Leaves	Consume the leaves in raw form or boiled in the water and taken daily.
44.	*Pterocarpus marsupium* Roxb.	Venga	Fabaceae	Hard wood	Drinking of water extract of hard wood.
45.	*Salacia fruticosa* Heyne *ex* Lawson.	Korandi	Hippocrataceae	Root	Root boiled in water and this water is taken daily.
46.	*Saraca asoca* Roxb.	Asokam	Cesalpinaceae	Bark	Bark decoction is taken daily.
47.	*Scoparia dulcis* L.	Kallurukki	Scrophulariaceae	Whole plant	Consume whole plant decoction.
48.	*Syzygium aqueum* (Burm.f.) Alston, Ann.	Chamba	Myrtaceae	Fruit	Fruit juice is taken daily.
49.	*Syzygium cumini* (L.) Skeels, U.	Njaval	Myrtaceae	Seed	Seed powder is mixed with water and taken daily.
50.	*Tinospora cordifolia* (Willd.) Miers.	Amruthu	Menispermaceae	Stem	The juice is prepared from the stem of the plant is mixed with juice of *Emblica* and *Curcuma longa* powder is taken twice a day.

points in the complex pathways of hormone action. Impairment of insulin secretion and defects in insulin action frequently coexist in the same patient, and it is often unclear which abnormality, if either alone, is the primary cause of the hyperglycemia (Sullivan and Mahan 1964).

The results of the present survey also reveal that about 25% of diabetic patients were found in the study area. Most of them used various herbal formulations for the treatment of this chronic disorder, while some of them used both allopathic as well as herbal formulations for the treatments. Out of these two groups, the group of patients who used allopathy or modern medicine alone suffered from many health problems due to the side effects of modern medicine. Among the documented antidiabetic

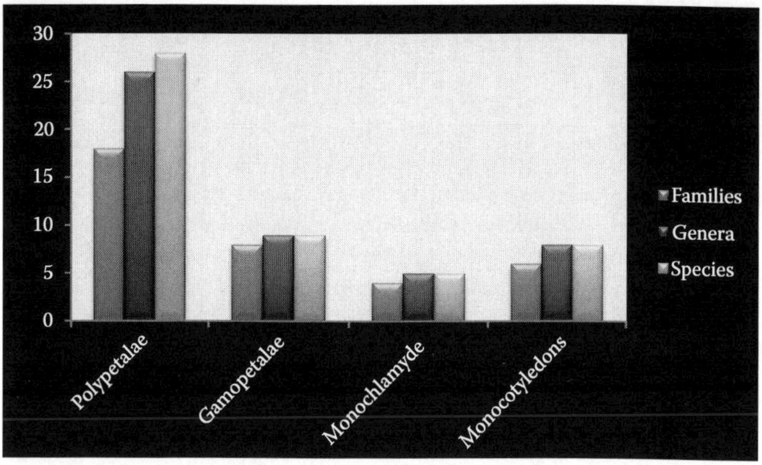

FIGURE 5.2 Analysis of antidiabetic plants in the study area.

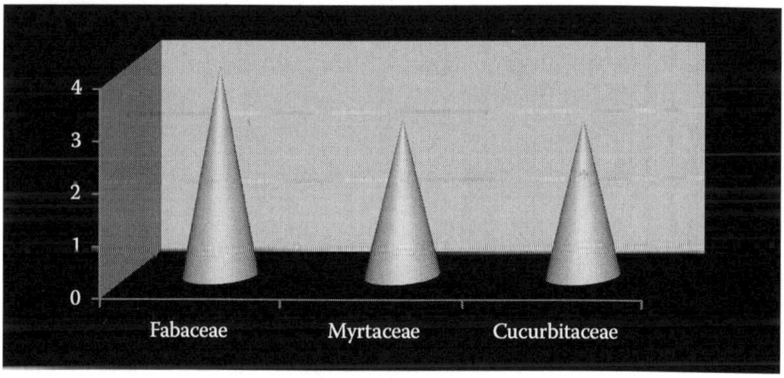

FIGURE 5.3 Analysis of dominant families with respect to the species.

plants in the study area, most of the patients were using the medicinal plants like *Azadirachta indica* A., *Coccinia grandis* (L.) Wight., *Curcuma longa* L., *Mimosa pudica* L., *Momordica charantia* L., *Phyllanthus emblica* L., *Aegle marmelos* (L.) Corr., etc.

5.3.4 Diabetic-Associated Health Problems

The present survey also noticed that there are several health problems associated with diabetic patients, some of them are exemplified here as potential loss of vision, renal failure, foot ulcers, amputations, Charcot joints, autonomic neuropathy, gastrointestinal, genitourinary, cardiovascular symptoms and sexual dysfunction. Patients with diabetes have an increased incidence of atherosclerotic cardiovascular, peripheral arterial and cerebrovascular disease. Moreover, the hypertension and abnormalities of lipoprotein metabolism are often found in people with diabetes.

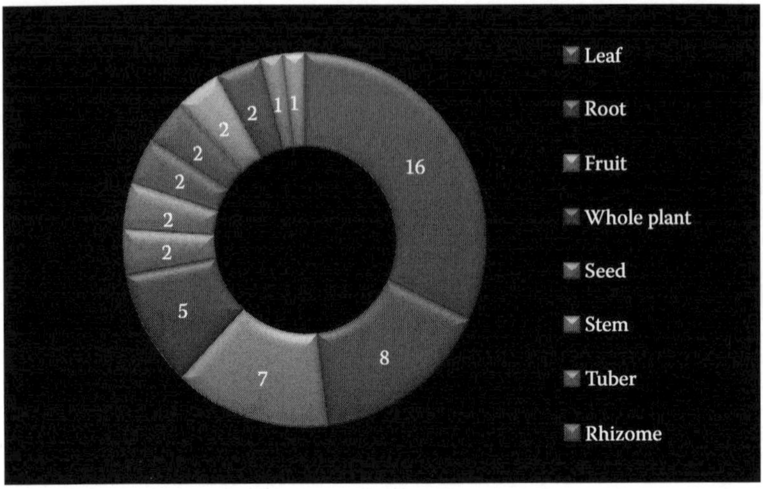

FIGURE 5.4 Analysis of various plant part(s) used to treat diabetes.

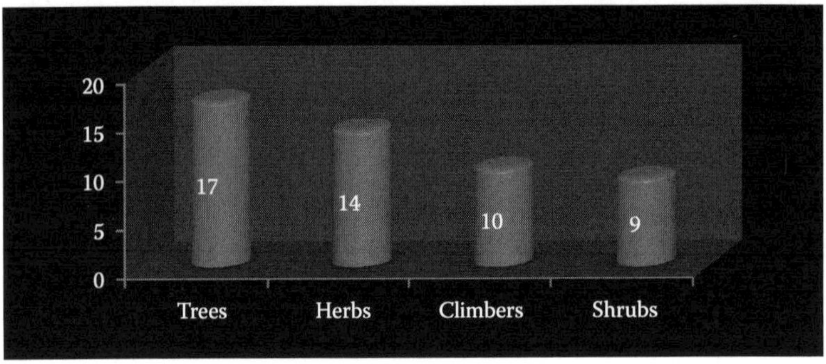

FIGURE 5.5 Life form analysis of antidiabetic plants.

5.3.5 Popularization of Herbal Formulations in the Study Area

The present survey also found that, many people in study area were not aware of the medicinal plant wealth of their surroundings. Moreover, the discussions with local inhabitants as well as herbal practioners of the study area reveals that the younger generations are not interested in using herbal drugs for even minor diseases. They always depend on the modern drugs. In this current scenario, the present study focused on popularizing many herbal formulations for the native and local inhabitants of the study area, by discussing antidiabetic medicinal plants with them based on our present observations, as well the previous literature.

5.3.6 Major Threats to the Medicinal Flora of the Study Area

The present study also noticed some threats in the study area, such as habitat destruction as well as habitat fragmentation of many medicinal plants due to anthropogenic

FIGURE 5.6 Photographs of collected plants. (a) *Annona muricata* L.; (b) *Annona squamosa* L.; (c) *Tinospora cordifolia* (willd.) miers.; (d) *Hibiscus rosa-sinensis* L.; (e) *Helicteres isora* L.; and (f) *Biophytum sensitivum* L.

activities. In addition to these, urbanization programs also affect the medicinal flora of the area. Some of the medicinal plants which occur in the study area were overexploited in the name of medicine. Moreover, the invasion of many exotic weeds also adversely affected the growth and survival of many native plant species. Hence, urgent attention is required to protect and conserve these valuable plant species for next generation.

5.4 CONCLUSION

The present study on the diversity of potential antidiabetic plants in Meenachil Taluk of the Kottayam District, Kerala reveals that there are about 50 taxa that show antidiabetic properties, which belong to 48 genera in 36 families. Out of the 36 families represented, Fabaceae is the first dominant family with four species. Cucurbitaceae and Myrtaceae are the second dominant families with three species, respectively. Among the total medicinal plant species documented, trees are the dominant growth form, followed by herbs, climbers and shrubs. The people in the study area used various parts and the whole plant of these medicinal plants (such as leaves, fruits, roots, seeds, stems, rhizomes, bark, tubers, hard wood, flowers, etc.) for the preparation of antidiabetic herbal formulations.

The present study also discussed the efficacy of herbal formulations in antidiabetic treatments and diabetic-associated health problems. In addition to these, the present study also focused on the popularization of many herbal formulations in the native and local inhabitants of the study area through discussions with them, based on our present observations as well as the previous literature.

The potential for antidiabetic plants, which are documented through the present study, are highly valuable for the treatment of diabetes. Due to the increasing demand of herbal drugs for diabetic treatments, these valuable plants should be conserved and protected within their natural ecosystems for future generations.

REFERENCES

Alarcon-Aguilara, F. J., Roman-Ramos, R., Perez-Gutierrez, S., Aguilar-Contreras, A., Contreras-Weber, C. C., and Flores-Saenz, J. L. 1998. Study of the anti-hyperglycemic effect of plants used asantidiabetics. *Journal of Ethnopharmacology* 61: 101–110.

Balamurugan, R. and Ignacimuthu, S. 2011. Antidiabetic and hypolipidemic effect of methanol extract of *Lippia nodiflora* L. in STZ induced diabetic rats. *Asian Pacific Journal of Tropical Biomedicine* 1: 30–36.

Binu Thomas, Rajendran, A., Aravindhan, V., and Maharajan, M. 2012. Antidiabetic Potential Plants of Southern Western Ghats in Kerala, India. *Advances Plant Science* 25 (1): 261–263.

Gamble, J. S. and Fischer, C. E. C. 1915–1936. *The Flora of Presidency of Madras*. Part 1-11 (Part 1–7 by Gamble and 8–11 by Fischer) (Repr. ed Vols. 1–3. 1957). Adlard and Sons Ltd., London.

Kokar, R. and Mantha, S. V. 1998. Increased oxidative stress in rat liver and pancreas during progression of streptozotosin induced diabetes. *Journal of Clinical Science* 19: 623–632.

Manisha Yadav, Khalid Kafeel Khan, and Beg, M. Z. 2012. Medicinal plants used for the treatment of diabetes by the *Baiga* tribe living in Rewa District, M.P. *Indian Journal of Life Science* 2(1): 99–102.

Ramachandran, A., Snehalatha, C., and Viswanathan, V. 2002. Burden of type 2 diabetes and its complications – The Indian scenario. *Current Science* 83: 1471–1476.

Reni, Scaria, Binu, Thomas, and Varghese, M. C. 2014. Medico-Potentiality of Leaves of Some Medicinal Plants from Meenachil taluk of Kottayam District, Kerala, India. *Int. J. Pharm. Integrat. Life Science*. 2(8): 28–46.

Santapau, H., 1955. *Botanical Collector's Mannual – A Handbook*. Govt. Press, New Delhi.

Saravanan, G. and Pari, L. 2008. Hypoglycaemic and antihyperglycaemic effect of *Syzygium cumini* bark in streptozotocin-induced diabetic rats. *Journal of Pharmacology and Toxicology* 3: 1–10.

Sasidharan, N. 2004. *Biodiversity documentation for Kerala, Part 6: Flowering plants*. Kerala Forest Research Institute (KFRI), Peechi.

Shokeen, P., Anand, P., Murali, Y. K., and Tandon, V. 2008. Antidiabetic activity of 50% ethanolic extract of *Ricinus communis*and its purified fractions. *Food and Chemical Toxicology* 46: 3458–3466.

Sullivan, J. B. and Mahan, C. M. 1964. Criteria for the oral glucose tolerance test in pregnancy. *Diabetes*. 13: 278–300.

Yeh, G. Y., Eisenberg, D. M., Kaptchuk, T. J., and Phillips, R. S. 2003. Systematic Review of Herbs and Dietary Supplements for Glycemic Control in Diabetes. *Diabetes Care*. 26: 1277–1294.

6 Antidiabetic Activity by the *In Vitro* α-Amylase and α-Glucosidase Inhibitory Action of Indian Ayurvedic Medicinal Plants

Ficus amplissima Smith

Karuppusamy Arunachalam,
Puthanpura Sasidharan Sreeja,
Domingos Tabajara de Oliveira Martins
and Parimelazhagan Thangaraj

CONTENTS

6.1 Introduction 101
6.2 Materials and Methods 104
 6.2.1 Plant Collection 104
 6.2.2 Extraction of Plant Material 104
 6.2.3 *In Vitro* Antidiabetic Activity 104
  6.2.3.1 Inhibition Assay for α-Glucosidase Activity 104
  6.2.3.2 Inhibition Assay for α-Amylase Activity 105
6.3 Results 105
 6.3.1 α-Glucosidase and α-Amylase Inhibitory Activity 105
6.4 Discussion 107
6.5 Conclusion 107
References 108

6.1 INTRODUCTION

Diabetes mellitus is a metabolic disorder characterized by hyperglycemia and disturbances of carbohydrate, protein and fat metabolisms, secondary to an absolute or relative lack of insulin (Fatima et al. 2010). The World Health Organization (WHO) estimates that more than 220 million people worldwide have diabetes and this number is liable to double by 2030 (WHO 2009). Hyperglycemia is known to produce reactive oxygen species (ROS), which play a central role in the complications of diabetes (Dewanjee et al. 2009). Antioxidants play a major role in the protection against molecular oxidative damage (Evans 2007). In addition to oxidative stress, insulin action is also impaired in diabetes, which leads to increased hepatic glucose production. Due to the side effects of the existing synthetic drugs, plant-derived food drugs are in great demand in industrially developed countries as an alternative approach to treating diabetes. The WHO expert committee has recommended that plants possessing hypoglycaemic activity may provide a utilizable source of new oral antidiabetic drugs for the development of pharmaceutical entities or may act as simple dietary adjuncts to the existing therapies (WHO 1980).

Until today, there is no treatment that can completely cure diabetes mellitus. In a study by Maria et al. (1999), it was found that insulin-dependent diabetes mellitus (IDDM) or type I, depends on insulin therapy to prevent the pathology that would arise from insulin deficiency due to a dysfunction of the β-cells. However, exogenous therapy, which is the main treatment, does not permit glycemic control as precise as that provided by natural secretion from functional islet β-cells, and acute decompositions and long term complications are always present (Larson 1988). It is therefore appropriate to look at possible alternative therapeutic strategies for IDDM. Type II diabetes, otherwise known as adult-onset diabetes, results from a combination of insulin resistance and an inadequate secretion of insulin (Reaven 1988). Management of type II diabetes is rarely straightforward. It requires rigorous control of blood glucose and special attention to a syndrome of associated vascular risk factors, including hypertension, dyslipidemia and abdominal obesity. The pharmacological agents currently used for the treatment of type II diabetes include sulfonylureas, biguanide, thiazolidinedione and α-glucosidase inhibitors.

One therapeutic approach to treat diabetes is to decrease the postprandial glycemia by the inhibition of enzymes that hydrolyze carbohydrates, such as α-amylase and α-glucosidase present in the digestive tract (Lebovitz 1997; Inzucchi 2002; Ali et al. 2006; Funke and Melzing 2006; Bhandari et al. 2008). α-Amylase is responsible for cleaving large maltooligosaccharides to maltose, which are converted to glucose by intestinal α-glucosidase (Whitcomb et al. 2007; Loizzo et al. 2008). Inhibitors of these enzymes delay the digestion of carbohydrates, producing a reduction in the rate of glucose absorption and consequently a reduction of the postprandial plasma glucose. α-Amylase and α-glucosidase inhibitors play a major role in managing postprandial hyperglycemia in diabetic patients; they inhibit the action of amylase and glucosidase enzymes, leading to a reduction in starch hydrolysis, which shows beneficial effects on the glycemic index control in diabetic patients. Examples of inhibitors used in therapy are acarbose, voglibose and miglitol. The main adverse effects of these inhibitors are gastrointestinal, including abdominal discomfort, flatulence and diarrhoea. Thus, to look for more efficacious agents with fewer side effects is necessary in order to expand the therapeutical options to control diabetes. The search for safer and more hypoglycemic pharmaceuticals remains an important area of research. Natural inhibitors from plants have shown an inhibitory effect against α-amylase and α-glucosidase activity and therefore can be potentially used as an effective therapy for postprandial hyperglycemia with minimal side effects (Grover et al. 2002; Gyemant et al. 2003; Cheng and Fantus 2005).

The genus *Ficus* (Moraceae) constitutes one of the largest genera of angiosperms, and includes more than 800 species and 2000 varieties. The genus *Ficus* occurs in the most tropical and subtropical forests; this plant is more commonly known as the fig (Hamed 2011). Sirisha et al. (2010) reported that all *Ficus* species possess latex-like material within their vasculatures that provide protection and self-healing from physical assaults. Various studies indicated that *Ficus* species are widely used in the management of various types of diseases like respiratory diseases, sexual disorders, central nervous system diseases (CNS), cardiovascular disorders (CVS), gastric problems, skin infections and diabetics etc. (Sirisha et al. 2010).

Fig fruits hold the highest levels of polyphenols, flavonoids, anthocyanins and exhibited the highest antioxidant capacity, which can be free of side effects versus those of synthetic antioxidants (Joseph and Raj 2011). Shukla et al. (2004) revealed the significant antioxidant effect of *F. bengalensis*. In addition, Duduku et al. (2007) reported that *F. microcarpa* bark contains highest free radical scavenging activity. Further, epidemiological studies have shown that many of these antioxidant compounds possess anti-inflammatory, analgesic, antimutagenic, anticarcinogenic, antibacterial and antiviral activities to a greater extent (Owen et al. 2000).

A large number of plants, plant extracts, decoctions or pastes are equally used by tribes and carried on in the folklore traditions in India for treatment of cuts, wounds and burns (Kumar et al. 2007). Besides, in Ayurvedic medicine, *F. racemosa* L. is used as a wound healing agent (Biswas and Mukherjee 2003) whereas, the aqueous extract of *F. deltoidea* was reported to have wound healing activity (Abdulla et al. 2010).

Medicinal plants play an important role in the cure of diabetes mellitus all over the world. A variety of ingredients present in medicinal plants are thought to act on a variety of targets by various modes and mechanisms. They have the potential to impart therapeutic effects in complex disorders like diabetes and its complications (Tiwari and Rao 2002). According to the Ayurvedic system of medicine, *F. bengalensis, F. carica, F. glomerata* (Rashid 2008; Singh et al. 2009), *F. exasperate* Vahl and *F. arnottiana* Miq. (Sonibare et al. 2006; Mazumdar et al. 2009; Sharma et al. 2010) are well-known in the treatment of diabetes. *F. carica* (Joseph and Raj 2011), *F. bengalensis* bark (Patil and Patil 2010) and *F. glomerata* were reported to have hypoglycemic effects.

Ficus amplissima Smith., commonly known as kal-itchchi, is endemic to India and commonly occurs in foothills of the Himalayas, Assam, Sikkim, Kerala, Tamil Nadu, Andhra Pradesh and Maharashtra. The ethnobotanical views on *F. amplissima* suggest that fruits are chewed for mouth ulcers, whereas leaf juice is applied externally for old chronic wounds. Bark is used against diabetes, which is reported in indigenous herbal formulations in India. In addition, some other reports on the medicinal properties of bark shows that, *F. amplissima* bark has been used for the ailments of the throat (Pankaj Oudhia 2012). Pulliah (2002) reported that boiled rice water and bark boiled with salt in equal proportions is used to treat colic. Leaves are used for jaundice, fruit is used to cure mental illness and latex is applied to wounds (Singh and Himadri 2005). Based on this ethnobotanical information, this study was planned to screen the leaves, bark and fruit of *F. amplissima* for new porcine pancreatic α-amylase and α-glucosidase inhibitors using *in vitro* assays, which may offer a new, effective and safe therapeutic approach in the management of diabetes.

6.2 MATERIALS AND METHODS

6.2.1 Plant Collection

Fresh plant materials were collected during the month of October 2009 from Sathyamangalam, Erode district, in Tamil Nadu, India. The taxonomic identity of the plant was confirmed by the Botanical Survey of India, Southern Circle, Coimbatore, Tamil Nadu (copy enclosed) and a voucher specimen (No: 006147) was deposited at the Botany Department Herbarium at Bharathiar University, Coimbatore, Tamil Nadu. The plant materials were washed under running tap water to remove the surface pollutants; and the different parts, such as leaves, bark and fruit, were separated mechanically. The separated plant parts were air-dried under shade. The dried samples were powdered and used for further studies.

6.2.2 Extraction of Plant Material

The powdered plant materials, such as leaves, bark and fruit, were packed in small thimbles separately and extracted successively with different solvents, such as petroleum ether, chloroform, acetone and methanol, in an increasing order of polarity using a Soxhlet apparatus. Before extraction with the next solvent, the thimble was air dried each time. Finally, the material was macerated using hot water with constant stirring for 24 h and the water extract was filtered. The different solvent extracts were concentrated by a rotary vacuum evaporator and then air-dried. The dried extract obtained with each solvent was weighed. The percentage of yield was calculated in terms of the air-dried weight of plant material.

The leaf (8.38%) and fruit (7.49%) methanol extract showed higher recovery percentage over other solvent extracts. The stock solution of the extract obtained was prepared (1 mg/mL of respective organic solvents) and used for further analysis.

6.2.3 *In Vitro* Antidiabetic Activity

6.2.3.1 Inhibition Assay for α-Glucosidase Activity

α-Glucosidase (0.075 units) was premixed with the extract at various concentrations (50–200 μg/mL). A substrate of 3 mM *p*-nitrophenyl glucopyranoside (pNPG) was added to the reaction mixture to start the reaction (Miller 1959). The reaction mixture was incubated at 37°C for 30 min and stopped by adding 2 mL of Na_2CO_3. The α-glucosidase activity was determined by measuring the *p*-nitrophenol release from pNPG at 400 nm. The IC_{50} value was defined as the concentration of α-glucosidase inhibitor to inhibit 50% of its activity under the assay conditions.

6.2.3.2 Inhibition Assay for α-Amylase Activity

α-Amylase was premixed with extract at various concentrations (50–200 μg/mL) and starch as a substrate was added (0.5% starch solution) to start the reaction. The reaction was carried out at 37°C for 5 min and terminated by the addition of 2 mL of DNS (3,5-dinitrosalicylic acid) reagent. The reaction mixture was heated for 15 min at 100°C and diluted with 10 mL of distilled water in an ice bath (Miller 1959). α-Amylase activity was determined by measuring the spectrum at 540 nm. The IC50 value was defined as the concentration of α-amylase inhibitor to inhibit 50% of its activity under the assay conditions.

6.3 RESULTS

6.3.1 α-Glucosidase and α-Amylase Inhibitory Activity

The IC_{50} values of α-glucosidase inhibition for *F. amplissima* leaf, bark and fruit extracts were indicated in Figures 6.1 through 6.3. All the extracts displayed potent α-glucosidase inhibitory activity at the primary screening concentration. Among the extracts, the methanol extract of leaves and bark and the acetone extract of fruit (IC_{50}: 150, 86 and 98 μg/mL, respectively) displayed strong α-glucosidase

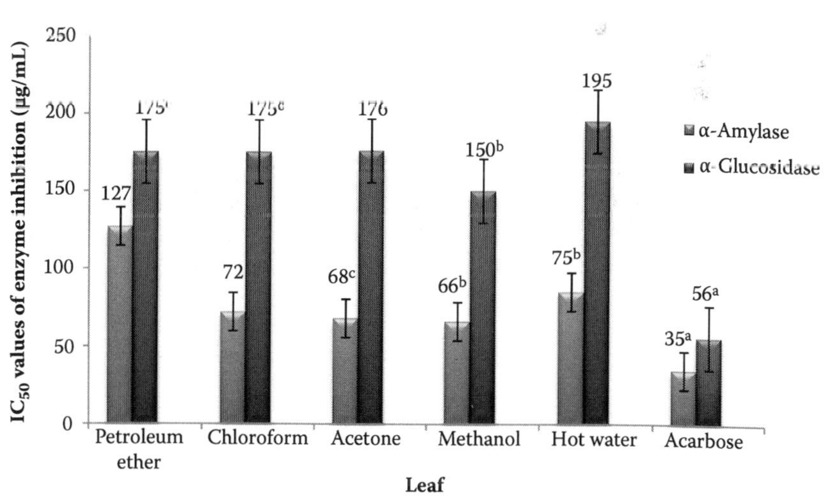

FIGURE 6.1 α-Amylase and α-glucosidase inhibitory activity of *F. amplissima* leaf extracts. Values are mean of triplicate determination (n = 3) ± standard deviation; Statistically significant at $p < 0.05$ where [a] > [b] > [c]; The IC_{50} value was defined as the concentration of enzyme inhibitor to inhibit 50% of its activity under the assay conditions.

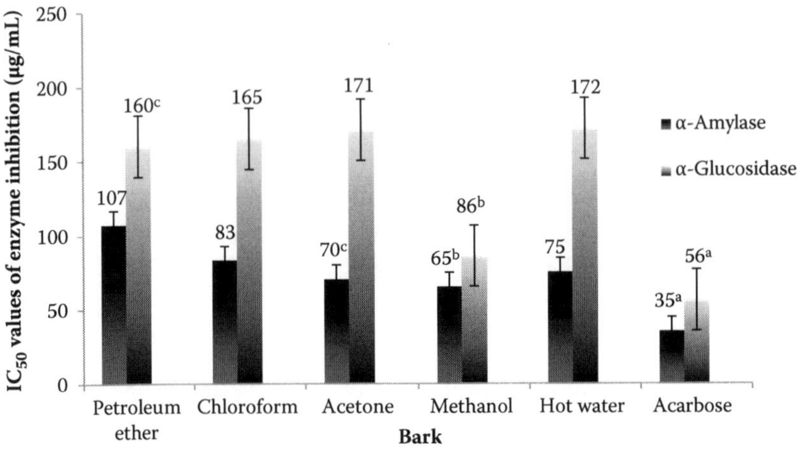

FIGURE 6.2 α-Amylase and α-glucosidase inhibitory activity of *F. amplissima* bark extracts. Values are mean of triplicate determination (n = 3) ± standard deviation; Statistically significant at $p < 0.05$ where $^a > ^b > ^c$; The IC_{50} value was defined as the concentration of enzyme inhibitor to inhibit 50% of its activity under the assay conditions.

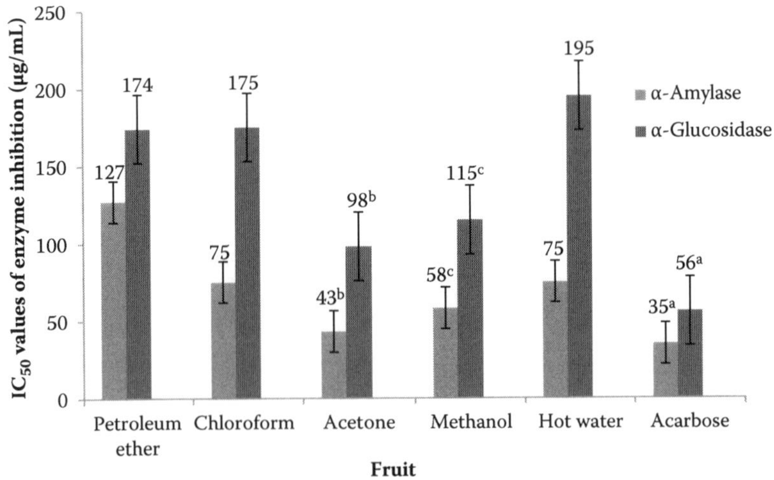

FIGURE 6.3 α-Amylase and α-glucosidase inhibitory activity of *F. amplissima* fruit extracts. Values are mean of triplicate determination (n = 3) ± standard deviation; Statistically significant at $p < 0.05$ where $^a > ^b > ^c$; The IC_{50} value was defined as the concentration of enzyme inhibitor to inhibit 50% of its activity under the assay conditions.

inhibitory activity. However, the petroleum ether, chloroform and hot water showed lower inhibitory activity compared to the standard acarbose (IC_{50}: 56 μg/mL). It appears, therefore, that the active components in the acetone and methanol extracts may be responsible for the significant inhibitory activity of the plant. α-Amylase inhibitory activity of the extracts are also shown in Figures 6.1 and 6.2. Among the extracts, the methanol extract of leaves and bark and the acetone extract of fruit (IC_{50} values 66, 65 and 43 μg/mL, respectively,) showed strong α-amylase inhibitory activity when compared to standard acarbose with an IC_{50} value of 35 μg/mL. However, petroleum ether extract was found to have a less inhibitory effect against α-amylase.

6.4 DISCUSSION

Plants play a major role in the discovery of new therapeutic agents and have received much attention as sources of biologically active substances, including antioxidants, hypoglycemic and hypolipidemic agents (Marles and Farnsworth 1995). Flavonoids and polyphenols are being used to treat diabetes and dyslipidemia (Martinello et al. 2006). This is based on the fact that, excessive oxidative stress is implicated in the pathology and complications of diabetes mellitus, and polyphenols with antioxidant properties exert beneficial antidiabetic effects by correcting the disturbed oxidative milieu in diabetic conditions (Abdelmoaty et al. 2010; Tiwari and Rao 2002). Between 2001 and 2005, four new drugs derived from natural products were introduced for the treatment of dyslipidemia and diabetes (Lam 2007).

α-Amylase inhibitory activity of *F. amplissima* leaf, bark and fruit extracts showed appreciable activity. Among the extracts, the methanol extract of bark and the acetone extract of fruit showed strong α-amylase inhibitory activity with an IC_{50} value of 65 and 43 μg/mL, respectively. Following the above extract, chloroform and hot water extracts exhibited better α-amylase inhibitory activity.

In the α-glucosidase inhibition of *F. amplissima*, all the extracts displayed potent α-glucosidase inhibitory activity at the primary screening concentration. Among the extracts, bark methanol and fruit acetone extracts displayed strong α-glucosidase inhibitory activity (IC_{50}: 86 and 98 μg/mL, respectively). It appears therefore that active components in acetone and methanol extracts may be responsible for significant inhibitory activity of the extracts. The methanol extract from leaves of *Adhatoda vasica* reported high α-glucosidase inhibitory activity (Hao et al. 2008).

Traditional systems of medicine used plants and herbal extracts as antidiabetic agents. Therefore, investigation of such agents from traditional medicinal plants has become more important, and researchers are competing to find new, effective and safe therapeutic agents for the treatment of diabetes. In this regard, *F. amplissima* was identified as a good α-glucosidase and α-amylase inhibitor, which will decrease the postprandial glucose level in the blood and can be used to manage diabetes.

6.5 CONCLUSION

The extracts of *F. amplissima* were screened for potential inhibitory activity on α-amylase and α-glucosidase. The results indicated that the methanol extract of bark had the highest activities on both digestive enzymes when compared to other plant part extracts. The target of this study was to provide *in vitro* evidence for potential α-amylase and α-glucosidase inhibitors so as to generate the stronger biochemical rationale for further *in vivo* researches and clinical studies on diabetes. Research has been conducted on the identification of the active principles derived from *F. amplissima* with inhibitory activity on α-amylase and α-glucosidase, and could be potentially useful in controlling diabetes.

REFERENCES

Abdelmoaty, M. A., Ibrahim, M. A., Ahmed, N. S., and Abdelaziz, M. A. 2010. Confirmatory studies on the antioxidant and antidiabetic effect of quercetin in rats. *Indian Journal of Clinical Biochemistry* 25: 188–192.

Abdulla, M. A., Ahmed, K. A., Abu-Luhoom, F. M., and Muhanid, M. 2010. Role of *Ficus deltoidea* extract in the enhancement of wound healing in experimental rats. *Biomedical Research* 21(3): 241–245.

Ali, H., Houghton, P. J., and Soumyanath, A. 2006. Alpha-Amylase inhibitory activity of some Malaysian plants used to treat diabetes; with particular reference to *Phyllanthus amarus*. *Journal of Ethnopharmacology* 107: 449–455.

Bhandari, M. R., Jong-Anurakkun, N., Hong, G., and Kawabata, J. 2008. Alpha-Glucosidase and alpha-amylase inhibitory activities of Nepalese medicinal herb Pakhanbhed (*Bergenia ciliata*, Haw.). *Food and Chemical Toxicology* 106: 247–252.

Biswas, T. K. and Mukherjee, B. 2003. Plant medicines of Indian origin for wound healing activity: A review. *International Journal of Lower Extremity Wounds* 2(1): 25–39.

Cheng, A. Y. and Fantus, I. G. 2005. Oral antihyperglycemic therapy for type 2 diabetes mellitus. *Canadian Medical Association Journal* 172: 213–226.

Dewanjee, S., Das, A. K., Sahu, R., and Gangopdhyay, M. 2009. Antidiabetic activity of Diospyros peregrine fruit: Effect on hyperglycemia, hyperlipidemia and augmented oxidative stress in experimental type 2 diabetes. *Food and Chemical Toxicology* 47: 2679–2685.

Duduku, K., Rosalam, S., and Leenah, N. J. 2007. Recovery of phytochemical components from various parts of *Morinda citrifolia* extracts by using membrane separator. *Journal of Applied Sciences* 7(15): 2093–2098.

Evans, J. L. 2007. Antioxidants: Do they have a role in the treatment of insulin resistance. *Indian Journal of Medical Research* 125: 355–372.

Fatima, S. S., Rajasekhar, M. D., Kumar, K. V., Kumar, M. T. S., Babu, K. R., and Rao, C. A. 2010. Antidiabetic and antihyperlipidemic activity of ethyl acetate:isopropanol (1:1) fraction of *Vernonia anthelmintica* seeds in streptozotocin induced diabetic rats. *Food and Chemical Toxicology* 48: 495–501.

Funke, I. and Melzing, M. F. 2006. Traditionally used plants in diabetes therapy – Phytotherapeutics as inhibitors of a-amylase activity. *Revista Farmacognosia* 16: 1–5.

Grover, J. K., Yadav, S., and Vats, V. 2002. Medicinal plants of India with antidiabetic potential. *Journal of Ethnopharmacology* 81: 81–100.

Gyemant, G., Kandra, L., Nagy, V., and Somsak, L. 2003. Inhibition of human salivary alpha-amylase by glucopyranosylidene-spiro-thiohydantoin. *Biochemical and Biophysical Research Communications* 312: 334–339.

Hamed, M. A. 2011. Beneficial effect of *Ficus religiosa* Linn. on highfat-induced hypercholesterolemia in rats. *Food Chemistry* 129: 162–170.

Hao, H., Huang, Y., Gao, B., Inagaki, C., and Kawabata, J. 2008. Inhibitory effect on alpha-glucosidase by *Adhatoda vasica* Nees. *Food Chemistry* 10: 965–972.

Inzucchi, S. E. 2002. Oral antihyperglycemic therapy for type 2 diabetes: Scientific review. *JAMA* 287: 360–372.

Joseph, B. and Raj, S. J. 2011. Pharmacognostic and phytochemical properties of *Ficus carica* Linn – An overview. *Internatonal Journal of PharmTech Research* 3(1): 8–12.

Kumar, B., Vijayakumar, M., Govindarajan, R., and Pushpangadan, P. 2007. Ethnopharmacological approaches to wound healing – Exploring medicinal plants of India. *Journal of Ethnopharmacology* 114: 103–113.

Lam, K. S. 2007. New aspects of natural products in drug discovery. *Trends in Microbiology* 15: 279–289.

Larson, R. A. 1998. The antioxidants of higher plants. *Photochem.* 27: 969–978.

Lebovitz, H. E. 1997. Alpha-Glucosidase inhibitors. *Endocrinology and Metabolism of Clinical Journal of North America* 26: 539–551.

Loizzo, M. R., Saab, A. M., Tundis, R., Menichini, F., Bonesi, M., Piccolo, V., Statti, G. A., de Cindio, B., and Houghton, P. J. 2008. *In vitro* inhibitory activities of plants used in Lebanon traditional medicine against angiotensin converting enzyme (ACE) and digestive enzymes related to diabetes. *Journal of Ethnopharmacology* 119: 109–116.

Maria, A. B., Assimina, G. T., Drakoulis, D., Alexandra, F., and Panayiotis, H. P. 1999. Salivary alterations in insulin-dependent diabetes mellitus. *International Journal of Paediatric Dentistry* 8(1): 29–33.

Marles, R. J. and Farnsworth, N. 1995. Antidiabetic plants and their active constituents. *Phytomedicine* 2(2): 137–189.

Martinello, F., Soares, S. M., Franco, J. J., Santos, A. C., Sugohara, A., Garcia, S. B., Curti, C., and Uyemura, S. A. 2006. Hypolipidemic and antioxidant activities from *Tamarindus indica* L. pulp fruit extract in hypercholesterolemic hamsters. *Food and Chemical Toxicology* 44: 810–818.

Mazumdar, P. M., Farswan, M., and Parcha, V. 2009. Hypoglycaemic effect of *Ficis arnottiana* Miq. bark extracts on streptozotocin induced diabetea in rats. *Natural Product Radiance* 8(5): 478–482.

Miller, G. L. 1959. Use of dinitrosalicylic acid reagent for determination of reducing sugar. *Analytical Chemistry* 31: 426–428.

Owen, R. W., Giacosa, A., Hull, W. E., Haubner, R., Spiegelhalder, B., and Bartsch, H. 2000. The antioxidant/anticancer potential of phenolic compounds isolated from olive oil. *Europesan Journal of Cancer* 36: 1235–1247.

Pankaj, O. 2012. Asi. Orlinseed (*Linum usitatissium* L.) and sanjeevani (*Sellaginella bryophytes* (L). Baker-Based herbal formulations in Indian traditional healing for diabeties complications: Pankaj Oudhia's Ethnobotanical Survey from year 1990–2012, India, 7.

Patil, V. V. and Patil, V. R. 2010. *Ficus Benghalensis* Linn.-an overview. *International Journal of Pharma. and Bio. Sciences* 6(2): 1–11.

Pulliah, T. 2002. *Biodiversity in India*. Vol. 1. Daya Publishing House, Delhi, India, 114–115.

Rashid, A. B. A. 2008. The chemical constituents from the stems of *Ficus deltoids*. B.Sc. Final Year Project Report, University, Teknologi Mara, Selangor.

Reaven, G. 1988. Role of insulin resistance in human disease. *Diabetes* 37: 1595–1607.

Sharma, V. K., Kumar, S., Patel, H. J., and Hugar, S. 2010. Hypoglycemic activity of *Ficus glomerata* in alloxan induced diabetic rats. *International Journal of Pharmaceutical Sciences Review and Research* 1(2): 18–22.

Shukla, R., Gupta, S., Gambhir, J. K., Prabhu, K. M., and Murthy, P. S. 2004. Antioxidant effect of aqueous extract of the bark of *Ficus bengalensis* in hypercholesterolaemic rabbits. *Journal of Ethnopharmacology* 92: 47–51.

Singh, M. P. and Himadri, P. 2005. *Medicinal herbs with their formulations*. Daya Publishing House, Delhi, Volume I: 158.

Singh, S., Gupta, S. K., Sabir, G., Gupta, M. K., and Seth, P. K. 2009. A database for antidiabetic plants with clinical/experimental trials. *Bioinformation* 4(6): 263–268.

Sirisha, N., Sreenivasulu, M., Sangeeta, K., and Chetty, C. M. 2010. Antioxidant Properties of Ficus species – A review. *International Journal of PharmTech Research*, 2174–2182.

Sonibare, M. O., Isiaka, A. O., Taruka, M. W., Williams, N. S., Soladoye, M., and Emmanuel, O. 2006. Constituents of *Ficus exasperata* leaves. *Natural Product Communications*, 23–26.

Tiwari, A. K. and Rao, M. J. 2002. Diabetes mellitus and multiple therapeutic approaches of phytochemicals: Present status and future prospects. *Current Science* 83: 30–8.

Whitcomb, D. C. and Lowe, M. E. 2007. Human pancreatic digestive enzymes. *Digestive Diseases and Sciences* 52: 1–17.

WHO (World Health Organization). 1980. Expert Committee on diabetes mellitus second report. Technical Report Series 646. World Health Organization, Geneva, 61.

WHO. 2009. Prevalence data of diabetes worldwide.

7 Anti-Obesity Potential of Indian Traditional Medicinal Plants and Their Phytochemicals

Vellingiri Vadivel, Pichai Venkatalakshmi and Pemaiah Brindha

CONTENTS

7.1 Introduction 111
7.2 Anti-Obesity Potential of Medicinal Plants 115
7.3 Polyphenols as Anti-Obesity Agents 116
7.4 Saponins as Anti-Obesity Agents 120
7.5 Triterpenes as Anti-Obesity Agents 123
7.6 Clinical Studies on Pancreatic Lipase Inhibitors 125
7.7 Conclusion 126
References 126

7.1 INTRODUCTION

Obesity is becoming one of the greatest threats to global health in this century, with more than 1.5 billion overweight adults and at least 400 million clinically obese subjects (Drew et al. 2007). Due to these increasing obesity rates, the World Health Organization (WHO) has prompted to consider it as the epidemic of 21st century and to promote strategies to prevent and control its progress (Brug and Crawford 2009). The development of obesity is characterized by a chronic imbalance between energy intake and energy expenditure (Schrauwen and Westerterp 2000), and it is often ascribed to changing lifestyles and inadequate dietary habits. Also, decreased energy expenditure is often associated with an inherited low basal metabolic rate, low physical activity and a low capacity for fat oxidation (Little et al. 2007). To reduce body weight and adiposity, a change in lifestyle habits is still the crucial cornerstone (Rubio et al. 2007). Physical activity might be helpful in the prevention of obesity by elevating the average daily metabolic rate and increasing energy expenditure. Unfortunately, this clinical approach is not long lasting, and weight regain is often seen.

Epidemiological studies have shown a direct relation between the incidence of overweight/obesity and dietary fat consumption (Little et al. 2007). Humans are frequently exposed to foods rich in fat, which are usually associated with a high-energy intake (Astrup 2001). Thus, those foods with a high-energy and dietary fat content are considered to promote body fat storage and weight gain in humans (Moreno et al. 2003). One explanation is that, in commercially available food items, the percentage of energy derived from fat is highly correlated with energy density. Given that fat contains 9 kcal/g compared with 4 kcal/g of carbohydrates and proteins, foods rich in fat are often high in energy density. Thus, when a similar volume of food is consumed, energy intake will be higher in high-fat diets compared with low-fat diets (Figure 7.1).

On the other hand, independently of an increased energy intake, specific dietary constituents may promote the development of obesity. This statement means that even when consuming an equal amount of energy, diet composition is important, especially concerning the balance between nutrients (Lomba et al. 2010). Thus, a macronutrient profile (a diet with high protein, carbohydrate and lipid contents) can affect diet-induced thermogenesis, the oxidation pathway, energy intake, gene expression or the level of some hormones (Hermsdorff et al. 2007). Following a high-fat diet, the diet-induced thermogenesis is lower than following high-protein and high-carbohydrate diets and also fat is more effectively absorbed from the gastrointestinal tract than carbohydrates, which translates into lower energy expenditure when following a high-fat diet. So, high-fat diets produce a metabolically more efficient state, at least in part because of the lower postprandial thermogenic effect of lipids in comparison with carbohydrates (Mobbs et al. 2010).

Furthermore, the consumption of a high-fat diet has the capacity to modulate the gastrointestinal responses to ingested fat and thereby, may lead to impairments in

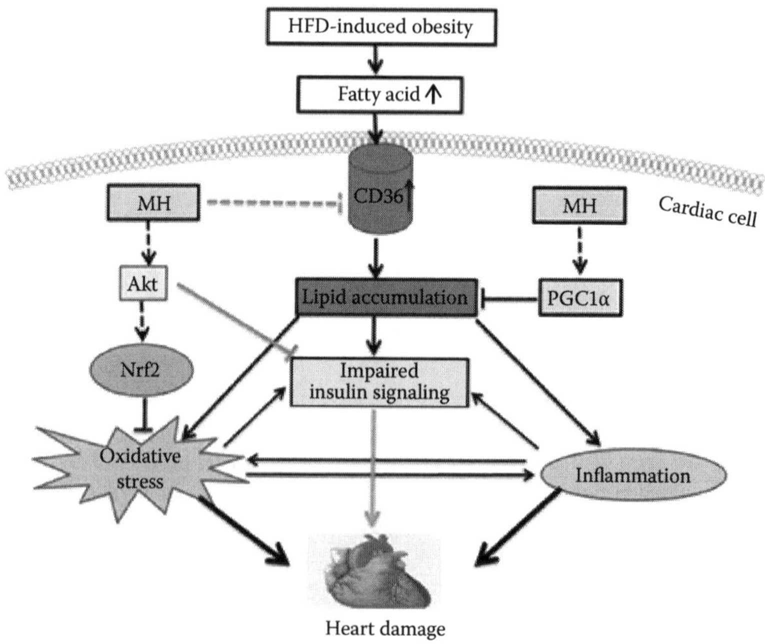

FIGURE 7.1 Consequences of high fat–diet-induced obesity.

appetite regulation that favor the development of obesity. Dietary fat usually implies an increase in energy consumption because it has a lower potential for inducing satiety than carbohydrates and protein. Hence, high-fat diets may play an important role in the increased prevalence of obesity and can be a triggering factor in the development of hyperglycaemia and hyperinsulinaemia (Jager et al. 2009). Moreover, the intake of dietary fats is usually accompanied by a higher intake of refined sweet carbohydrates (fast food and desserts), where the high intake of sucrose promotes weight gain, visceral adiposity and the development of diseases that are related to obesity, such as cardiovascular diseases (Stevenson et al. 2009). Therefore, low-fat diets often are prescribed in the prevention and treatment of obesity because a reduction in dietary lipids without restriction of total energy intake could cause weight loss.

Recent studies indicate that fat digestion is a prerequisite for the effects of fat on gastric emptying, gastrointestinal hormone secretion, appetite and energy intake. An increasing number of gastrointestinal enzymes involved in nutrient digestion are being identified and characterized, representing a rich pool of potential therapeutic targets for obesity and other metabolic disorders (Birari and Bhutani 2007). Enzymes that are especially related to dietary fat are interesting targets, which include preduodenal lipases (lingual and gastric lipases), pancreatic lipases, cholesterol-ester lipases and bile-salt stimulated lipases (Armand 2007). Most dietary fat is ingested as triglycerides (90–95%), and their hydrolysis starts in the mouth, then goes on through the stomach by an acid stable gastric lipase, and continues in the duodenum through the synergistic actions of gastric and colipase-dependent pancreatic lipases, leading to the formation of monoglycerides and free fatty acids (Figure 7.2). Free fatty acids are absorbed by the enterocyte to synthesize new triglyceride molecules, which are transported to the different organs via lipoproteins, especially chylomicrons, after a meal.

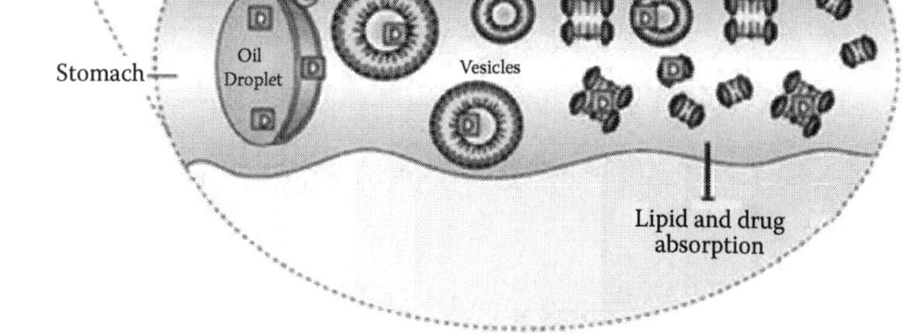

FIGURE 7.2 Role of pancreatic lipasema in fat digestion.

Pancreatic lipase plays a key role in the efficient digestion of triglycerides (Lowe 2002). It is secreted into the duodenum through the duct system of the pancreas and is responsible for the hydrolysis of 50–70% of total dietary fats. This enzyme has been widely used for the determination of the potential efficacy of natural products as anti-obese agents (Sugiyama et al. 2007). Orlistat is currently the only clinically approved drug for obesity management in Europe. This molecule acts by inhibiting pancreatic lipase activity and the reduction of triglyceride absorption, and its long-term administration accompanying an energy restricted diet results in weight loss (Neovius et al. 2008). Reduction on intestinal lipid digestion has been related to a decrease in the intra-abdominal fat content. Thus, this compound is associated with a small, but statistically significant weight loss of about 3% more than diet alone in overweight and obese people.

Since dietary lipids represent the major source of unwanted calories, the inhibition of fat digestion is an interesting approach for reducing fat absorption (Bray and Ryan 2007). Orlistat is the only authorized anti-obese drug in Europe and has been shown to act through the inhibition of pancreatic lipase, which is a key enzyme for the digestion of dietary triglycerides (McClendon et al. 2007). Orlistat is the saturated derivative of lipstatin (Figure 7.3), an inhibitor of pancreatic lipase isolated from the bacterium *Streptomyces toxytricini* (Weibel et al. 1987). This molecule exerts a modest weight-lowering effect when accompanying a suitable dietary advice. Thus, in a recent meta-analysis (Viner et al. 2010), the mean BMI change with Orlistat (120 mg three times daily) was a reduction of 0.83 kg m^{-2} (95% CI: 0.47–1.19) compared with placebo. Accompanying this anti-obesity action, Orlistat is also able to modestly reduce blood pressure, improve oral glucose tolerance and prevent the onset of type II diabetes (Heymsfield et al. 2000).

In addition to losing weight, Orlistat within a prescribed limit has been shown to be safe and more effective than diet alone in modifying some of the risks of coronary artery disease and other obesity-related comorbidities. The most commonly reported adverse effects of Orlistat are a range of gastrointestinal side effects, including steatorrhoea, bloating, oily spotting, fecal urgency and fecal incontinence, as well as hepatic adverse effects (Viner et al. 2010). These adverse effects are similar to those observed for other lipase inhibitors tested in phase II studies, such as Cetilistat (Kopelman et al. 2007). On the other hand, the inhibition of fat absorption could be accompanied by fat-soluble vitamin deficiencies, which could be prevented by the vitamin supplementation strategy, as other authors have recommended when vitamin deficiency

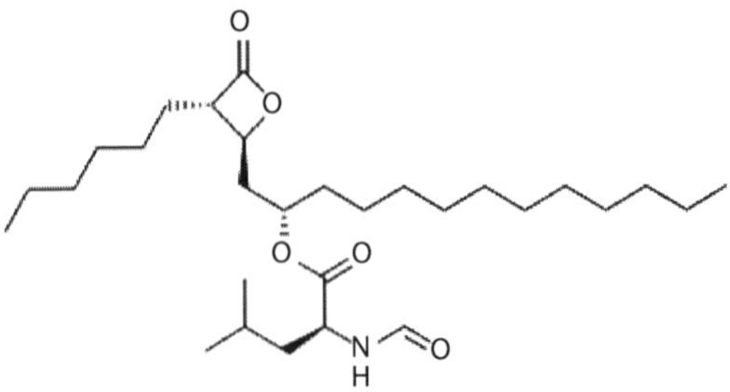

FIGURE 7.3 Structure of orlistat.

occurs in patients undergoing Orlistat therapy (Melia et al. 1996). Hence, researchers are interested in finding new natural substances that show potent inhibitory activity against pancreatic lipase, and have fewer side effects than the current ones.

Drugs that prevent weight regain appear necessary in obesity treatment. Thus, the development of natural products for the treatment of obesity is a challenging task, which can be launched faster and cheaper than conventional, single-entity pharmaceuticals (Moreno et al. 2006). Many medicinal plants may provide safe, natural and cost-effective alternatives to synthetic drugs. Currently, one of the most important strategies in the treatment of obesity includes development of inhibitors of nutrient digestion and absorption. For example, acarbose is an antidiabetic drug that inhibits carbohydrate digesting enzymes, thus preventing the digestion of complex carbohydrates and decreasing postprandial hyperglycaemia (Yamagishi et al. 2009). Similar compounds with α-amylase inhibiting activity that can be used for diabetes control are being isolated from different plants. The list includes valoneaic acid dilactone obtained from banaba (*Lagerstroemia speciosa*) (Hosoyama et al. 2003), the ethanol extract obtained from chestnut astringent skin (Tsujita et al. 2008) or the purified pancreatic α-amylase inhibitor isolated from white beans (*Phaseolus vulgar*), are able to reduce glycaemia in both non-diabetic and diabetic rats (Tormo et al. 2006). Now, extracts from hundreds of species of medicinal plants, vegetables and fruits (Slanc et al. 2009) are being screened for potential lipase inhibitory activity. Ideally, these treatments will be viewed as adjuncts to behavioural and lifestyle changes aimed at maintenance of weight loss and improved health.

Knowledge of herbs has been handed down from generation to generation for thousands of years. Herbal drugs constitute a major part in all traditional systems of medicines. Herbal medicine is a triumph of popular therapeutic diversity. Plants, above all other agents, have been used for medicine from time immemorial because they have fitted immediate personal needs, are easily accessible and inexpensive. In the recent past, there has been a tremendous increase in the use of plant-based health products in developing as well as developed, countries resulting in an exponential growth of herbal products globally.

7.2 ANTI-OBESITY POTENTIAL OF MEDICINAL PLANTS

Medicinal plants have been used as dietary supplements for body weight management and control in many countries. In this sense, the presence of pancreatic lipase inhibitors has been demonstrated in different plant species, although more research is needed for identifying and characterizing effective lipase inhibitors. Lipase inhibitors of plant origin include certain proteins, such as those from soybeans (Gargouri et al. 1984) and from wheat bran and wheat germ (Lairon et al. 1985). Other proteins that strongly inhibit hydrolysis of triglycerides are the basic protein protamine (Tsujita et al. 1996); ε-polylysine (Tsujita et al. 2003), which could act as several amphiphilic proteins like ovalbumin; and β-lactoglobulin by desorption of lipase from its substrate due to a change in interfacial quality (Gargouri et al. 1984).

Other lipase inhibitors from plant origins (Figure 7.4) are basic polysaccharides, especially chitosan oligosaccharides, water-soluble chitosan (46 kDa) and polydextrose when a basic group is introduced (Han et al. 1999); phytic acid and other

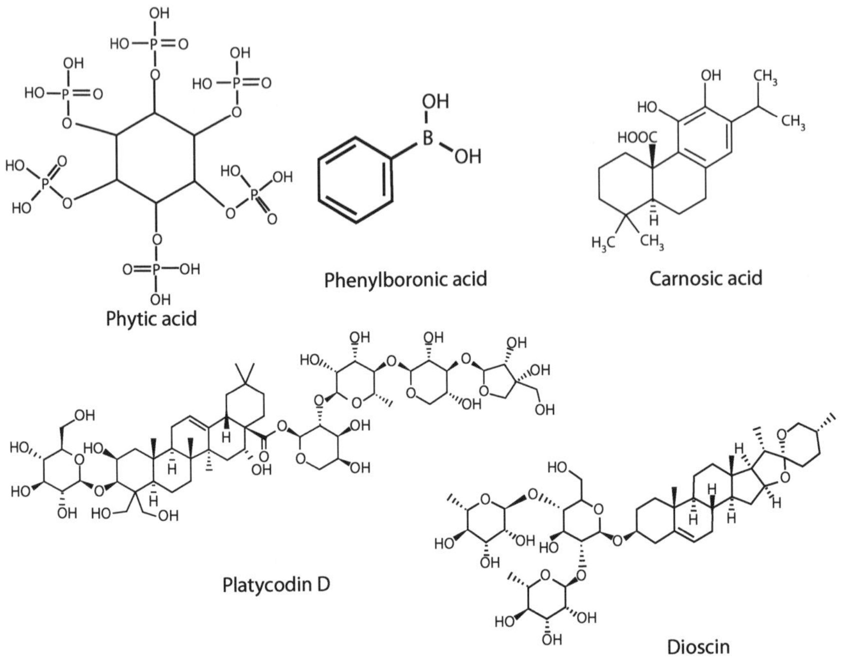

FIGURE 7.4 Structure of lipase inhibitors from plant origin.

myoinositol phosphate esters (Knuckles 1988); phenylboronic acid, a potent inhibitor of lipase from *Oryza sativa* (Raghavendra and Prakash 2002); and carnosic acid, a diterpene isolated from the methanolic extract of the leaves of sage (*Salvia officinalis*) and rosemary (Ninomiya et al. 2004). Korean and Chinese researchers have been worked on several new lipase inhibitors of herbal origin. The most promising compounds are platycodin D, isolated from the fresh roots of *Platycodon grandiflorum* (Zhao and Kim 2004); dioscin from *Dioscorea nipponica* (Kwon et al. 2003); licochalcone A from the roots of *Glycyrrhiza uralensis* (Won et al. 2007); phenolic constituents from the leaves of *Nelumbo nucifera* (Ono et al. 2006); the aqueous ethanol extracts of *Juniperus communis* or common juniper (bark) and *Illicium religiosum* (wood) (Kim and Kang 2005); the ethanol extract from stem, bark and leaves from mango tree (*Mangifera indica*); which is able to prevent weight gain induced by feeding a high-fat diet to Wistar rats (Moreno et al. 2006); a pomegranate leaf extract rich in ellagic acid and tannins (Lei et al. 2007); *Rhei rhizoma* (rhubarb) and the combinatorial drug Chunghyuldan (Yang et al. 2003); *Prunella vulgaris, Rheum palmatum* and other herbs (Zheng et al. 2007). Most of the common compounds that are found in different plant species are polyphenols, saponins and terpenes.

7.3 POLYPHENOLS AS ANTI-OBESITY AGENTS

A number of studies have revealed various health benefits of plant polyphenols and their importance in foods, beverages and natural medicine. In this context, polyphenols have some potential efficacy for preventing obesity. They inhibit enzymes related to fat metabolism including pancreatic lipase, lipoprotein lipase and glycerophosphate dehydrogenase (Yoshikawa et al. 2002). Polyphenol extracts are able to decrease the blood levels of glucose, triglycerides and LDL cholesterol, increase energy expenditure and fat oxidation and reduce body weight and adiposity (Terra et al. 2009). In fact, many polyphenols, including flavones, flavonols, tannins and chalcones, have shown an inhibitory activity of pancreatic lipase (Birari and Bhutani 2007).

Flavonoids are a group of plant secondary metabolites that are characterized as containing two or more aromatic rings, each bearing at least one aromatic hydroxyl and connected with a carbon bridge (García-Lafuente et al. 2009). Some of them are polymerized into large molecules, either by the plants themselves or as a result of food processing. These polymers are called tannins, and three subclasses (condensed tannins, derived tannins and hydrolysable tannins) exhibit a variety of beneficial effects on health (García-Lafuente et al. 2009). A flavonoid with pancreatic lipase inhibitory activity is hesperidin (Figure 7.5), obtained from the peels of *Citrus unshiu* (Kawaguchi et al. 1997).

Proanthocyanidins, also known as condensed tannins, are the most common group of flavonoids in the Western diet. They consist of monomeric units of flavans linked through carbon-carbon and ether linkages, which are considered the second most abundant group of natural phenolics after lignins (He et al. 2008). Proanthocyanidins can be found in such common foodstuffs as cereals, legumes, fruits, vegetables and beverages (red wine and tea in particular) (Quesada et al. 2009). They have a putative role as antioxidants, showing beneficial effects on inflammatory processes, cardiovascular diseases and other pathological conditions (Lee et al. 2005). For example, these compounds actively reduce plasma triglycerides by inhibiting the absorption of dietary lipids (Quesada et al. 2009) and possess inhibitory effects on different digestive enzymes, such as trypsin, amylase and lipase (Sugiyama et al. 2007).

Some examples of polyphenols with inhibitory action on pancreatic lipase are proanthocyanidins from edible herbs, such as those from *Cassia mimosoides* (Yamamoto et al. 2000) and tea catechins, especially (-)-catechin gallate and (-)-gallocatechin gallate, (Ikeda et al. 2005). Some of the most thoroughly-studied polyphenol extracts in relation to pancreatic lipase inhibition are the following:

Peanut (*Arachis hypogaea*) shells (hulls, seed coats), which are the by-products of the peanut industry, provide several compounds showing pancreatic lipase inhibitory

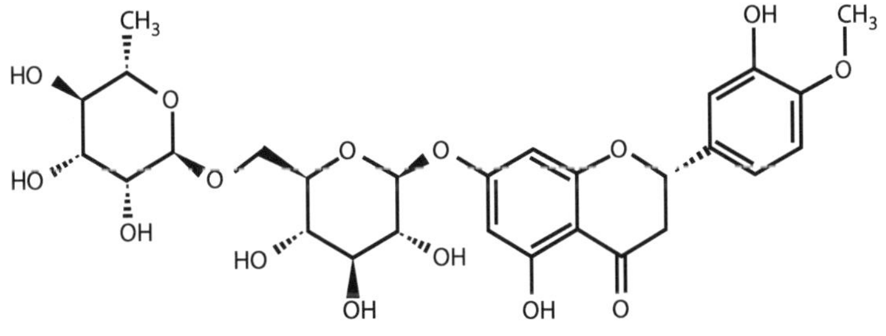

FIGURE 7.5 Structure of hesperidin.

activity in a dose dependent manner (1 mg/mL, 42% inhibitory effect) that are able to reduce body weight gain in rats fed a high-fat diet (Moreno et al. 2006). This plant contains several bioactive molecules, such as luteolin, certain fatty acids, caffeic, ferulic and benzoic acids, all of which are able to inhibit lipases (Birari and Bhutani 2007). Coumarin derivatses and phenolic acids were assumed to be the major active constituents. However, the authors have not examined the individual effects of each compound.

Camellia sinensis or tea plant (green tea, black tea or oolong tea) contains over 60 polyphenols, some of them with a potent pancreatic lipase inhibitory activity. It is likely the plant whose extracts have been more thoroughly used for searching new pancreatic lipase inhibitors. The major polyphenols are catechins which constitute about one-third of its total dry weight. A serving of tea is moderate to high in flavonoid and/or tannin content (Bose et al. 2008). Nakai et al. (2005) found that the polyphenols with more potent pancreatic lipase inhibitory effect were flavan-3-ol digallate esters isolated from oolong tea, such as (-)-epigallocatechin-3,5-digallate. Oolong tea-polymerized polyphenols reduced postprandial hypertriglyceridemia in olive oil-loaded rats and mice (Toyoda-Ono et al. 2007). Also (-)-epigallocatechin, abundant in green tea extract, is a weak inhibitor of pancreatic lipase and is able to decrease the postprandial hypertriglyceridemia in rodents (Ikeda et al. 2005).

The administration of black tea polyphenols suppressed postprandial hypertriglyceridemia in a dose-dependent manner in rats, with theaflavin-3,3'-digallate (Figure 7.6) as the most effective pancreatic lipase inhibitor (Kobayashi et al. 2009), whereas other authors point to the thearubigins (Kusano et al. 2008). These extracts are able to prevent increases in body weight and adiposity in mice fed a high-fat diet (Uchiyama et al. 2011). The pancreatic lipase inhibitory and hypotriglyceridemic effects of tea extracts were corroborated by Tanaka et al. (2010), who orally administered mixed fermented tea extracts and Loquat tea extracts to rats with a 10% soybean oil emulsion. Finally, cocoa tea extract (*Camellia sinensis* var. *ptilophylla*) is rich in polyphenols with pancreatic lipase inhibitory effect. A single oral administration of this extract produces an inhibition of plasma triglyceride levels in olive oil–loaded ICR mice and triolein-loaded rats (Kurihara et al. 2006).

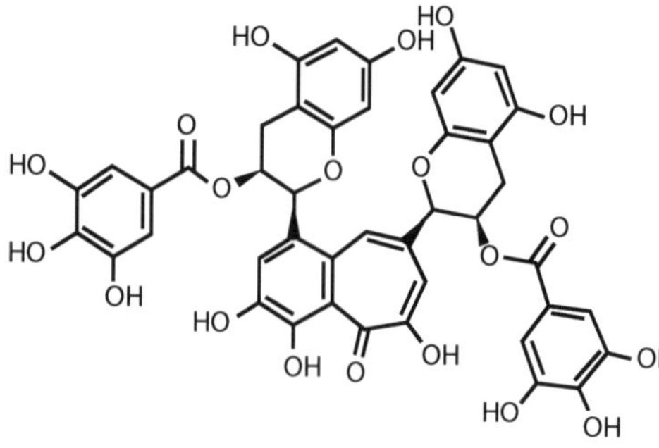

FIGURE 7.6 Structure of theaflavin-3,3'-digallate.

Daidzein belongs to the group of isoflavones and is produced almost exclusively by the members of the Fabaceae/Leguminosae (bean) family such as soybean. In one study, Guo et al. (2009) investigated the effects of daidzein on body weight, adipose tissue, blood and liver lipid levels in obese mice fed a high-fat diet, finding that daidzein reduced body and white adipose tissue weights in obese mice and ameliorated the hyperlipidemia induced by the high-fat diet. The authors attributed this effect to the inhibition of pancreatic lipase activity and fat digestion.

Yerba mate is a plant from the subtropical region of South America that is widely consumed in Brazil, Argentina, Paraguay and Uruguay. Yerba mate contains polyphenols, such as flavonoids (quercetin and rutin) and phenolic acids (chlorogenic and caffeic acids) and is also rich in caffeine and saponins (Martins et al. 2010). These substances act on the lipid metabolism by inhibiting pancreatic lipase activity in a concentration value of 1.5 mg/mL (Martins et al. 2010). Several triterpene saponins and monoterpene oligoglycosides from the leaves of yerba mate were found to exhibit potent inhibitory activity on porcine pancreatic lipase (Sugimoto et al. 2009).

Apples (*Malus domestica*) belong to the Rosaceae family whose fruits contain several phenolic substances (chlorogenic acid, catechin, epicatechin, phloridzin and procyanins). Procyanidins in apples are mainly composed of various polymerized catechins, with some of them showing a pancreatic lipase inhibitory activity and reducing triglyceride absorption (Sugiyama et al. 2007). In corn oil–loaded mice, a single oral administration of apple polyphenols reduced plasma triglyceride levels and a test diet containing 600 mg of apple polyphenols significantly inhibited triglyceride elevation at 6 h after ingestion, indicating an inhibition of triglyceride absorption (Sugiyama et al. 2007).

Salacia reticulata contains a high concentration of polyphenols, including catechins and condensed tannins. In hot water-soluble extract from the roots of *Salacia reticulata* the concentration is about 24% polyphenols (Yoshikawa et al. 2002). The polyphenols from *Salacia reticulata* inhibit enzymes related to fat metabolism, including pancreatic lipase, lipoprotein lipase and glycerophosphate dehydrogenase and are effective in preventing obesity (Yoshikawa et al. 2002). In fact, *Salacia* extract markedly improved metabolic syndrome symptoms (including body weight, adiposity, glucose intolerance, hypertension and peripheral neuropathy) in mice (Akase et al. 2009).

Dandelion (*Taraxacum officinale*) is a perennial herbaceous plant of the family Asteraceae that has been used as a phytomedicine due to its choleretic, antirhemetic, diuretic and anti-inflammatory properties (Zhank et al. 2008). Extracts from this plant have shown hypolipidemic effects and an inhibitory activity of pancreatic lipase, decreasing the area under curve for the postprandial triglyceride response curve (Zhank et al. 2008).

Grapevine (*Vitis vinifera*) has become a model plant for studying proanthocyanidin biosynthesis. Grapevine proanthocyanidins consist of two major flavan-3-ol monomers, catechin and epicatechin (Figure 7.7), that have inhibitory activity on pancreatic lipase (Zhao et al. 2010). Polyphenol-rich extracts from a range of berries, particularly cloudberry, are able to inhibit pancreatic lipase activity *in vitro*, which has been attributed to their content in ellagitannins and proanthocyanidins (McDougall et al. 2009).

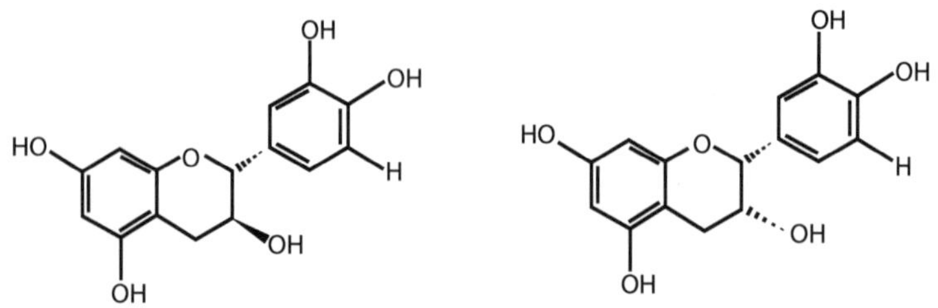

(+)-catechin (-)-epicatechin

FIGURE 7.7 Structure of catechin and epicatechin.

7.4 SAPONINS AS ANTI-OBESITY AGENTS

Saponins are a major family of secondary metabolites that occur in a wide range of plants species (Sparg et al. 2004). These compounds have been isolated from different parts of the plants, including the roots, rhizomes, stems, bark, leaves, seeds and fruit (Vincken et al. 2007). Saponins are categorized into two major classes, the triterpenoid and the steroid saponins, which are both derived from the 30 carbon atoms containing precursor oxidosqualene (Haralampidis et al. 2002). Some of the triterpene-rich plant materials are common foodstuffs consumed in large amounts in Mediterranean countries. Therefore, the correlation of a triterpene-rich diet and the beneficial effects of consuming a Mediterranean diet should be investigated in more detail (Jager et al. 2009). These types of plant secondary metabolites are found to inhibit pancreatic lipase and, thus, may represent potential effective treatments for obesity and related disorders (Slanc et al. 2009).

The Japanese horse chestnut (*Aesculus turbinata*) is a medicinal plant widely used in East Asia. The saponin mixture extracted from the seeds is called escins (Figure 7.8) and has a strong inhibitory activity on pancreatic lipase (Kimura et al.

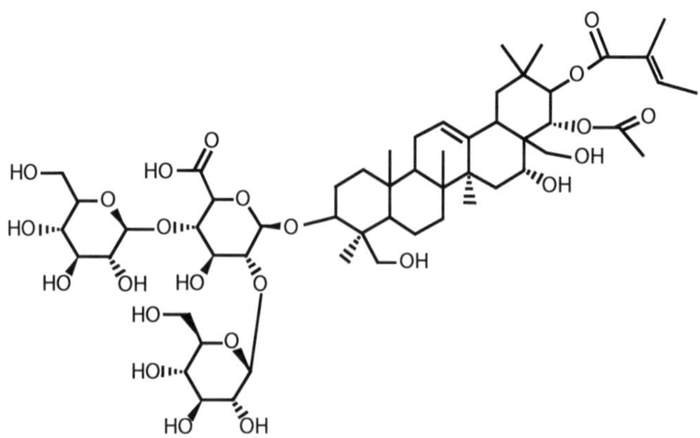

FIGURE 7.8 Structure of escin saponin.

2006). In mice fed a high-fat diet, the total amount of escins suppressed the increase in body weight, adiposity and liver fat and increased triglyceride level in the faeces, whereas it decreased plasma triglycerides after the oral administration of a lipid emulsion (Kimura et al. 2008).

The methanol extract of *Dioscorea nipponica* Makino powder has a potent inhibitory activity against porcine pancreatic lipase, with an IC_{50} value of 5–10 μg/mL (Kwon et al. 2003). In fact, the saponin dioscin and its aglycone, diosgenin, both suppressed the increase of blood triacylglycerols when orally injected with corn oil to mice. Rats fed a high-fat diet containing 5% *Dioscorea nipponica* Makino gained significantly less body weight and adipose tissue than control animals (Kwon et al. 2003), and a similar result has been observed after administering the aqueous extract of this rhizome to mice fed a high-fat diet (Song et al. 2009).

Eleutherococcus senticosus is a shrub, belonging to the family Araliaceae, which is commonly distributed in northeastern Asia. It is used as a traditional Chinese medicine against ischemic heart diseases, neurasthenia, hypertension, arthritis and tumours (Li et al. 2007). At least 15 triterpenoid saponins with *in vitro* pancreatic lipase inhibitory activity have been isolated from the fruits of *Eleutherococcus senticosus* (Jiang et al. 2006). The total saponin fraction obtained from the fruits of *Eleutherococcus senticosus* exhibits inhibitory activity on pancreatic lipase with an IC_{50} value of 3.63 mg/mL (Li et al. 2007).

Different lupine-type triterpene triglycosides isolated from a hot water extract of *Eleutherococcus sessiliflorus* leaves are able to inhibit pancreatic lipase activity *in vitro* and to suppress the body weight gain of mice fed a high-fat diet (Yoshizumi et al. 2006). Crocin is a glycosylated carotenoid extracted from the fruits of *Gardenia jasminoides*. *Gardeniae fructus* is used in Asian countries as a natural colorant, and in Chinese traditional medicine for its antioxidant, cytotoxic, antitumor and neuroprotective effects. Crocin and crocetin are effective hypolipidemic agents that act by reducing the absorption of fat and cholesterol through inhibition of pancreatic lipase activity (Lee et al. 2005). Sheng et al. (2006) demonstrated that crocin selectively inhibited the activity of pancreatic lipase as a competitive inhibitor.

Gypsophila oldhamiana (Caryophyllaceae) is a plant distributed in the north of China whose roots have high amounts of saponins, sterols and fatty acids. The extract from this plant shows a potent inhibitory activity of pancreatic lipase with an IC_{50} value of 0.54 mg/mL (Sheng et al. 2006) and different triterpenoid saponins, gypsosaponins A–C, as the more efficient compounds (Zheng et al. 2007).

Ginseng is one of the most popular medicinal herbs and is commonly consumed as powder, a beverage or a food supplement. Roots of *Panax ginseng* contain high levels of ginsenosides (Figure 7.9) which are steroidal saponins that show beneficial effects on lipid metabolism. Saponins from ginseng roots suppressed the expected increase in body weight and plasma triacylglycerols in mice following a high-fat diet, which was probably mediated by inhibiting pancreatic lipase with an IC_{50} value of 500 μg/mL (Karu et al. 2007).

The rhizomes of *Panax japonicus* (Japanese ginseng) are used in folk medicine for the treatment of arteriosclerosis, hyperlipidemia, hypertension and diabetes mellitus. Chikusetsu saponins prevented the increase in body weight and parametrial adipose tissue weight induced by a high-fat diet and inhibited the elevation of postprandial

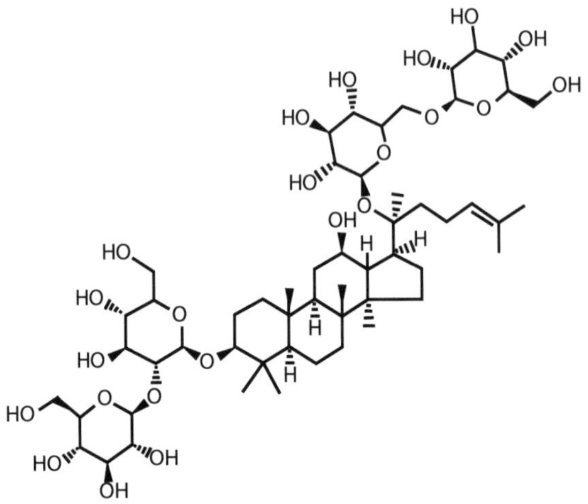

FIGURE 7.9 Structure of ginsenosides.

plasma triacylglycerols due to their inhibitory action of pancreatic lipase on dietary fat. The delay in intestinal fat absorption was also behind the anti-obesity effects observed for Korean white ginseng extract in high-fat diet-induced obese mice (Lee et al. 2010). American ginseng (*Panax quinquefolium*) is a native plant from North America. The saponins isolated from stems and leaves of *Panax quinquefolium* may prevent fat storage in adipose tissue and postprandial elevations of plasma triacylglycerols by inhibiting the intestinal absorption of dietary fat through the inhibition of pancreatic lipase activity (Liu et al. 2008).

Platycodi radix, widely used in traditional oriental medicines as a remedy for respiratory disorders, is rich in saponins, which are responsible for a diversity of effects including anti-inflammation, anti-allergy, antitumor and immunostimulation (Zhao et al. 2010). Given its inhibitory action on pancreatic lipase (Xu et al. 2005), with platycodin-D as the most efficient compound (Zhan and Kim 2004), it ameliorated high fat-induced obesity in mice and rats (Zhao et al. 2010). SK1 is an edible saponin-rich compound from *Platycodi radix* that is able to reduce body weight and fat accumulation by increasing faecal lipid outputs in high-fat fed mice (Kim et al. 2009).

The methanolic extract from the pericarps of *Sapindus rarak* (Lerak) shows a pancreatic lipase inhibitory activity that is probably due to diverse saponins and sesquiterpene glycosides (Morikawa et al. 2009). Different triterpenoid saponins isolated from the Mongol and Chinese traditional medicinal herb *Scabiosa tschiliensis* have shown strong inhibition of pancreatic lipase *in vitro* (Zheng et al. 2007). Due to the difficult task of isolating scabiosaponins and the scarceness of this type of saponin in nature, some of them have been successfully synthesized in the laboratory (Guo et al. 2009).

At least three kinds of tea (oolong, green and black) have been used as healthy drinks. Tea saponins suppress the increases in body and parametrial adipose tissue weights and adipocyte diameters induced by a high-fat diet in mice by inhibiting pancreatic lipase and also reduce the elevation in plasma triacylglycerol levels after oral administration of a lipid emulsion. The *Ki* value of tea saponins was determined to be 0.25 mg/mL (Han et al. 2001). Thus, the crude saponin fraction from the flower buds of Chinese tea plant exhibits accelerating effects on gastrointestinal transit in mice and inhibitory effects against porcine pancreatic lipase, and three floratheasaponins (A–C) showed inhibitory effects on serum triglyceride elevation (Lee et al. 2008).

7.5 TRITERPENES AS ANTI-OBESITY AGENTS

Terpenes are the primary constituents of the essential oils of many types of plants and are classified by the number of terpene units in the molecule (diterpenes, triterpenes, among others). The pharmacological relevance of triterpenes has increased during the last two decades demonstrating multi-target properties such as wound healing, anti-inflammatory, antibacterial, antiviral, hepatoprotective and antitumoural effects, combined with low toxicity (Jager et al. 2009). Triterpene extracts are safe and provide a high potential for further pharmaceutical and pharmacological research (Jager et al. 2008), some of them inhibiting pancreatic lipase activity. The bark of birch trees (*Betula alba*) contains pentacyclic triterpenes (Figure 7.10). This triterpene extract is safe and provides a high potential for further pharmaceutical and pharmacological research (Jager et al. 2009), displaying an inhibitory activity on pancreatic lipase (Slanc et al. 2004).

Tremendous health concerns have been raised over a dramatic increase in the prevalence of obesity and related metabolic disorders. Considered as a major life style disorder of developed countries, obesity is prevailing at alarming speed in developing countries is because of industrialization, fast food intake and decrease in physical activity (Cairns 2005). According to the WHO, 65% of the world's population lives in countries where overweight and obesity kills more people than those who are underweight. More than 1.4 million adults (age 20 and older) were overweight in 2008. Among them, over 200 million men and nearly 300 million women were obese (WHO 2014). A vast range of health problems coexists with a weight problem and dysfunction of lipid homeostasis. This interlinked network of metabolic

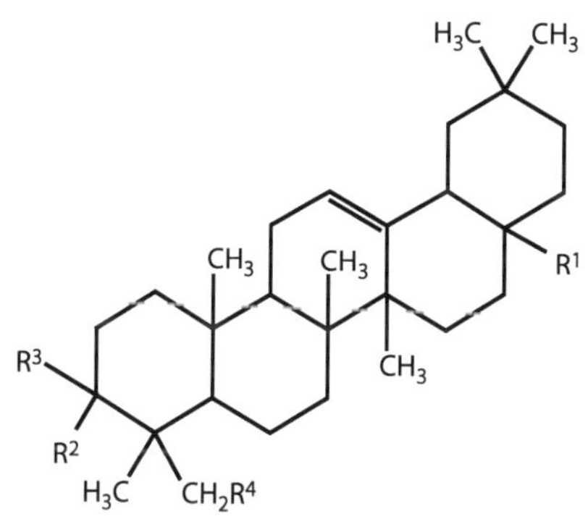

FIGURE 7.10 Structure of pentacyclic triterpenes.

disorders and comorbidities involve serious consequences in cardiovascular anomalies (heart failure, hypertension, pulmonary embolism, etc.), endocrine imbalance (insulin imbalance, glucose intolerance, hypothyroidism, etc.), arthiritis, urinary incontinence and gastrointestinal complications. Apart from obesity and related metabolic disorders disturb lifestyle physically, financially and psychologically. Psychological effects include social discrimination, depression, physical inability, etc., isolating the person from society (Aronne 2002).

A deeper understanding of the process of lipid homeostasis (i.e. absorption, metabolism, storage, deposition and oxidation) has presented a wide verity of enzymatic targets involved (Figure 7.11). Dietary fats are mainly regarded as mixed triglyceride, which undergoes a serious complex of biochemical reactions before absorption in the gastrointestinal tract. Pancreatic, endothelial, hepatic and lipoprotein lipases are members of the human lipase superfamily and possess structural similarity. Other tissues like lung, kidney, skeletal muscles, adipose tissue and placenta also secrete lipase enzymes. Pancreatic acinar cells secrete pancreatic lipase (triacylglycerol acyl hydrolase EC 3.1.1.3), an important enzyme of pancreatic juice responsible for digestion of dietary triglycerides in the small intestine.

Gastric and lingual lipases are responsible for partial hydrolysis of dietary triacylglycerol into free acids and diacylglycerol. This partial digestion in the stomach forms large fat molecule which undergoes emulsification with bile salts to form small droplets of fat. A physical property of emulsion influences the efficiency of

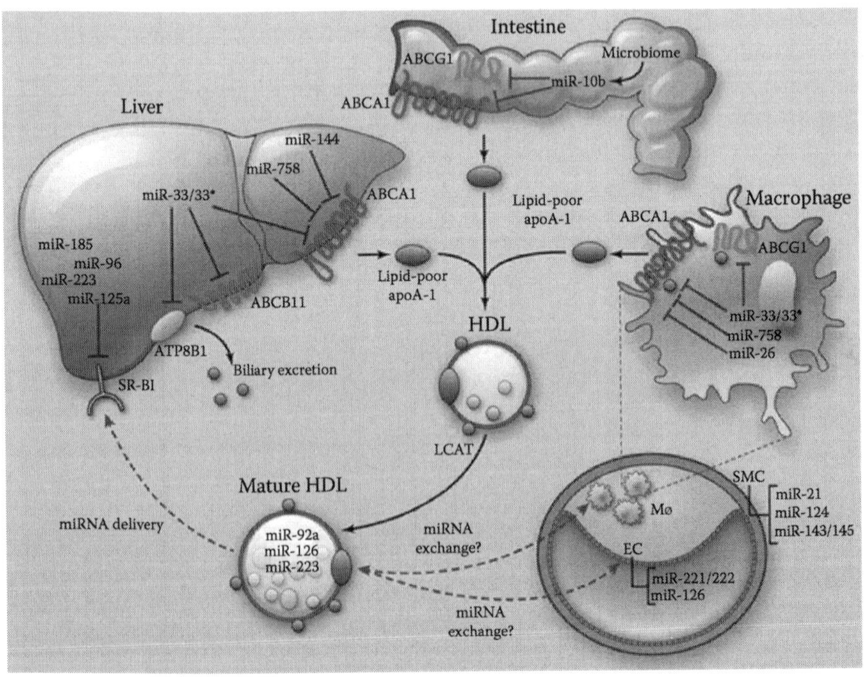

FIGURE 7.11 Metabolism of dietary lipids.

digestion. In the emulsion, dietary triglycerides and diglycerides in the centre of droplet followed by a mixture of polar lipids, phospholipids, cholesterol and free fatty acids; which are later coated with oligosaccharides, denatured proteins and bile salts and form a very complex structure. The pancreatic lipase interacts with emulsion droplets, which continuously change its physical properties as its products are formed, leaving the surface during the process of hydrolysis. Complete hydrolysis process results into free fatty acid, monoacylglycerols and diacylglycerols, which bind with cholesterol, bile salts, fat soluble vitamins and lysophosphatidic acid to form mixed micelles and can be absorbed by enterocytes. Pancreatic lipase uses a pancreatic protein colipase as a cofactor to facilitate lipolytic activity. Phosphatidyl choline inhibits the lipase-substrate complex. Colipase reverses to interact with the scarce surface of the substrate and stabilizes its conformation (Mukherjee et al. 2003; Shi et al. 2004).

7.6 CLINICAL STUDIES ON PANCREATIC LIPASE INHIBITORS

A number of plants and natural products have been screened for their pancreatic lipase inhibitory activity, but just some of them have gone up to clinical studies. In this line, only one product derived from natural compounds (Orlistat) is currently in clinical use, although others are under investigation. Some of them are *Panax ginseng* (Kim and Kang 2005), *Camellia sinensis* (Chantre and Lairon 2002), *Eleutherococcus senticosus* (Lee et al. 2008), *Malus domestica* (Sugiyama et al. 2007) and *Arachis hypogaea* (Moreno et al. 2006).

In one study, Kim and Kang (2005) showed the administration of an extract of *Panax ginseng* in humans for 8 weeks, which decreased circulating cholesterol, triglyceride and low-density lipoprotein levels (LDL). Each subject ingested 2 g of *Panax ginseng* extract three times a day. Lee et al. (2008) reported that healthy postmenopausal women treated for six months with *Eleutherococcus senticosus* supplementation showed significant decreases in serum LDL levels and LDL/HDL ratios. In other study, Sugiyama et al. (2007) assessed six healthy male volunteers that followed a high-fat diet with 40 g of fat with 10 control or 10 apple polyphenol (*Malus domestica*) capsules (600 or 1500 mg). In this study, they demonstrated that apple polyphenols may prevent obesity in humans by a pancreatic lipase inhibitory mechanism.

Green tea (*Camellia sinensis*) has been extensively studied in relation to obesity and other metabolic disorders. Thus, Chantre and Lairon (2002) showed that green tea consumption may be useful to treat obesity by both increasing thermogenesis and inhibiting pancreatic lipase. Thus, a green tea extract showed a direct *in vitro* inhibition of gastric and pancreatic lipases (Chantre and Lairon 2002). In moderately obese patients, green tea lowered body weight by stimulating thermogenesis and increasing energy expenditure when each subject received green tea extract two times/day (two capsules in the morning, two capsules midday). Ingestion of four capsules containing AR25 (Exolise) provided a daily total intake of 375 mg catechins, of which 270 mg was epigallocatechin gallate. Also, He et al. (2009) administered daily 8 g of oolong tea for six weeks to 102 obese subjects. As a result, 70% of the obese subjects decreased more than 1 kg in body weight. *In vitro* studies suggested that the effect of oolong tea on body weight could be partially attributed to the inhibition of pancreatic lipase (Ono et al. 2006).

According to these data, a number of common herbal products that are being studied in animal and human models for obesity treatment contain different metabolites that act on lipid digestion and absorption. However, it is very difficult to establish *in vivo* studies whether these anti-obesity effects are only or mainly due to pancreatic lipase activity inhibition. The clinical implications of this therapeutic approach have yet to be determined.

Pancreatic lipase inhibition is the most widely studied mechanism for the identification of potential anti-obesity agents. Only one blockbuster drug, Orlistat, was approved by the FDA and available for the obesity treatment apart from the centrally acting anti-obesity drugs, is acting through the pancreatic lipase inhibition. Discovery of orlistat was done from the naturally occurring molecule lipstatin. The success of naturally occurring compounds for the treatment of obesity has influenced the research for the identification of newer pancreatic lipase inhibitors that lack unpleasant side effects. Till now, many plant extracts and isolated compounds were identified for the pancreatic lipase inhibition. Other than that, many microbial products and isolated compounds, such as basic protamines, ε-polylysine (Tsujita et al. 1996), polysaccharides like chitosan (Sumiyoshi and Kimura 2006), dietary fibres from wheat bran and cholestyramine (Lairon et al. 1985), soya proteins (Roy and Schneeman 1981) and synthetic compounds etc. have been studied for their inhibitory potential against pancreatic lipase.

7.7 CONCLUSION

Knowledge of herbs has been handed down from generation to generation for thousands of years. Herbal drugs constitute a major part in all traditional systems of medicines. Herbal medicine is a triumph of popular therapeutic diversity. Plants, above all other agents, have been used for medicine from time immemorial because they have fitted the immediate personal need, are easily accessible and inexpensive. In the recent past, there has been a tremendous increase in the use of plant-based health products in developing as well as developed countries resulting in an exponential growth of herbal products globally. In view of this, several attempts have been made to evaluate the anti-obesity potential of Indian traditional medicinal plants and from the experimental results it is noted that there are many plant drugs are effective in controlling and treating obesity. So, some of the potential plant drugs could be selected and new formulations could be developed to eliminate the emerging health issues on obesity.

REFERENCES

Akase, T., Shimada, T., Harasawa, Y., Akase, T., Ikeya, Y., and Nagai, E. 2009. Preventive effects of *Salacia reticulata* on obesity and metabolic disorders in TSOD mice. *Evidence-Based Complementary and Alternative Medicine* 8, 233–239.

Armand, M. 2007. Lipases and lipolysis the human digestive tract: Where do we stand? *Current Opinion in Clinical Nutrition and Metabolic Care* 10, 156–164.

Aronne, L. J. 2002. Classification of obesity and assessment of obesity-related health risks. *Obesity* 10, 105S–115S.

Astrup, A. 2001. The role of dietary fat in the prevention and treatment of obesity. Efficacy and safey of low-fat diets. *International Journal of Obesity and Related Metabolic Disorders* 25, 46–50.

Birari, R. and Bhutani, K. 2007. Pancreatic lipase inhibitors from natural sources: Unexplored potential. *Drug Discovery Today* 12, 879–889.

Bose, M., Lambert, J., Ju, J., Reuhl, K., Shapses, S., and Yang, C. 2008. The major green tea polyphenol, (–)-epigallocatechin-3-gallate, inhibits obesity, metabolic syndrome, and fatty liver disease in high-fat-fed mice. *Journal of Nutrition* 138, 1677–1683.

Bray, G. and Ryan, D. 2007. Drug treatment of the overweight patient. *Gastroenterology* 132, 2239–2252.

Brug, J. and Crawford, D. 2009. The obesity pandemic. Is it bad or worse? *European Journal of Public Health* 19, 570–571.

Cairns, E. 2005. Obesity: The fat lady sings? *Drug Discovery Today* 10, 305–307.

Chantre, P. and Lairon, D. 2002. Recent findings of green tea extract AR25 (Exolise) and its activity for the treatment of obesity. *Phytomedicine* 3, 8–9.

Drew, B., Dixon, A., and Dixon, J. 2007. Obesity management: Update on orlistat. *Vascular Health and Risk Management* 3, 817–821.

García-Lafuente, A., Guillamón, E., Villares, A., Rostagno, M., and Martínez, J. 2009. Flavonoids as anti-inflammatory agents: Implications in cancer and cardiovascular disease. *Inflammtion Research* 58, 537–552.

Gargouri, Y., Julien, R., Pieroni, G., Verger, R., and Sarda, L. 1984. Studies on the inhibition of pancreatic and microbial lipases by soybean proteins. *Journal of Lipid Research* 25, 1214–1221.

Guo, T., Liu, Q., Wang, P., Zhang, L., Zhang, W., and Li, Y. 2009. Facile synthesis of three bidesmosidic oleanolic acid saponins with strong inhibitory activity on pancreatic lipase. *Carbohydrate Research* 344, 1167–1174.

Han, L. K., Kimura, Y., Kawashima, M., Takaku, T., Taniyama, T., and Hayashi, T. 2001. Anti-obesity effects in rodents of dietary teasaponin, a lipase inhibitor. *International Journal of Obesity's* 25, 1459–1464.

Han, L. K., Kimura, Y., and Okuda, H. 1999. Reduction in fat storage during chitin-chitosan treatment in mice fed a high-fat diet. *International Journal of Obesity's* 23, 174–179.

Haralampidis, K., Trojanowska, M., Osbourn, A. 2002. Biosynthesis of triterpenoid saponins in plants. *Advances in Biochemical Engineering/Biotechnology* 75, 31–42.

He, F., Pan, Q., Shi, Y., Duan, C. 2008. Biosynthesis and genetic regulation of proanthocyanidins in plants. *Molecules* 13, 2674–2692.

He, R., Chen, L., Lin, B., Matsui, Y., Yao, X., and Kurihara, H. 2009. Beneficial effects of Oolong tea consumption on diet-induced overweight and obese subjects. *Chinese Journal of Integrative Medicine* 15, 34–41.

Hermsdorff, H. H. M., Volp, A. C. P., and Bressan, J. 2007. Macronutrient profile affects diet-induced thermogenesis and energy intake. *Archives of Latin Nutrition* 57, 33–42.

Heymsfield, S. B., Segal, K. R., Hauptman, J., Lucas, C. P., Boldrin, M. N., and Rissanen, A. 2000. Effects of weight loss with orlistat on glucose tolerance and progression to type 2 diabetes in obese adults. *Archives of Internal Medicine* 160, 1321–1326.

Hosoyama, H., Sugimoto, A., Suzuki, Y., Sakane, I., and Kakuda, T. 2003. Isolation and quantitative analysis of the α-amylase inhibitor in *Lagerstroemia speciosa* (L.) Pers. (Banaba). *Yakugaku Zasshi* 123, 599–605.

Ikeda, I., Tsuda, K., Suzuki, Y., Kobayashi, M., Unno, T., and Tomoyori, H. 2005. Tea catechins with a galloyl moiety suppress postprandial hypertriacylglycerolemia by delaying lymphatic transport of dietary fat in rats. *Journal of Nutrition* 135, 155–159.

Jager, S., Laszczyk, M., and Scheffler, A. 2008. A preliminary pharmacokinetic study of betulin, the main pentacyclic triterpene from extract of outer bark of birch (*Betulae alba* cortex). *Molecules* 13, 3224–3235.

Jager, S., Trojan, H., Kopp, T., Laszczyk, M., and Scheffler, A. 2009. Pentacyclic triterpene distribution in various plants – Rich sources for a new group of multi-potent plant extracts. *Molecules* 14, 2016–2031.

Jiang, W., Li, W., Han, L., Liu, L., Zhang, Q., and Zhang, S. 2006. Biologically active triterpenoid saponins from *Acanthopanax senticosus. Journal of Natural Products* 69, 1577–1581.

Karu, N., Reifen, R., and Kerem, Z. 2007. Weight gain reduction in mice-fed *Panax ginseng* saponin, a pancreatic lipase inhibitor. *Journal of Agricultural and Food Chemistry* 55, 2824–2828.

Kawaguchi, K., Mizuno, T., Aida, K., Uchino, K. 1997. Hesperidin as an inhibitor of lipases from porcine pancreas and *Pseudomonas. Bioscience, Biotechnology, and Biochemistry* 61, 102–104.

Kim, H. and Kang, M. 2005. Screening of Korean medicinal plants for lipase inhibitory activity. *Phytotherphy Research* 19, 359–361.

Kim, J., Moon, K., Seo, K., Park, K., Choi, M., and Do, G. 2009. Supplementation of SK1 from Platycodi radix ameliorates obesity and glucose intolerance in mice fed a high-fat diet. *Journal of Medicinal Food* 12, 629–636.

Kimura, H., Ogawa, S., Jisaka, M., Kimura, Y., Katsube, T., and Yokota, K. 2006. Identification of novel saponins from edible seeds of Japanese horse chestnut (*Aesculus turbinata* Blume) after treatment with wooden ashes and their nutraceutical activity. *Journal of Pharmaceutical and Biomedical Analysis* 41, 1657–1665.

Kimura, H., Ogawa, S., Katsube, T., Jisaka, M., Yokota, K. 2008. Antiobese effects of novel saponins from edible seeds of Japanese horse chestnut (*Aesculus turbinata* BLUME) after treatment with wood ashes. *Journal of Agricultural and Food Chemistry* 56, 4783–4788.

Knuckles, B. 1988. Effect of phytate and other myo-inositol phosphate esters on lipase activity. *Journal of Food Science* 53, 250–252.

Kobayashi, M., Ichitani, M., Suzuki, Y., Unno, T., Sugawara, T., Yamahira, T. 2009. Black-tea polyphenols suppress postprandial hypertriacylglycerolemia by su pressing lymphatic transport of dietary fat in rats. *Journal of Agricultural and Food Chemistry* 5657, 7131–7136.

Kopelman, P., Bryson, A., Hickling, R., Rissanen, A., Rossner, S., and Toubro, S. 2007. Cetilistat (ATL-962), a novel lipase inhibitor: A 12-week randomized, placebo-controlled study of weight reduction in obese patients. *International Journal of Obesity's* 31, 494–499.

Kurihara, H., Shibata, H., Fukui, Y., Kiso, Y., Xu, J., and Yao, X. 2006. Evaluation of the hypolipemic property of *Camellia sinensis*var. *ptilophylla* on postprandial hypertriglyceridemia. *Journal of Agricultural and Food Chemistry* 54, 4977–4981.

Kusano, R., Andou, H., Fujieda, M., Tanaka, T., Matsuo, Y., and Kouno, I. 2008. Polymer-like polyphenols of black tea and their lipase and amylase inhibitory activities. *Chemical and Pharmaceutical Bulletin* 56, 266–272.

Kwon, C., Sohn, H., Kim, S., Kim, J., Son, K., and Lee, J. 2003. Anti-obesity effect of *Dioscorea nipponica* Makino with lipase-inhibitory activity in rodents. *Bioscience, Biotechnology, and Biochemistry* 67, 1451–1456.

Lairon, D., Lafont, H., Vigne, J. L., Nalbone, G., Lonardi, J., and Hauton, J. C. 1985. Effects of dietary fibers and cholestyramine on the activity of pancreatic lipase *in vitro. American Journal of Clinical Nutrition* 42, 629–638.

Lee, Y., Cha, B., Yamaguchi, K., Choi, S., Yonezawa, T., and Teruya, T. 2010. Effects of Korean white ginseng extracts on obesity in high-fat diet-induced obese mice. *Cytotechnology* 62, 367–376.

Lee, Y., Chung, H., Kwak, H., and Yoon, S. 2008. The effects of *A. senticosus* supplementation on serum lipid profiles, biomarkers of oxidative stress, and lymphocyte DNA damage in post-menopausal women. *Biochemical and Biophysical Research Communications* 375, 44–48.

Lee, I., Lee, J., Baek, N., and Kim, D. 2005. Antihyperlipidemic effect of crocin isolated from the fructus of *Gardenia jasminoides* and its metabolite Crocetin. *Biological and Pharmaceutical Bulletin* 28, 2106–2110.

Lei, F., Zhang, X. N., Wang, W., Xing, D. M., Xie, W. D., and Su, H. 2007. Evidence of anti-obesity effects of the pomegranate leaf extract in high-fat diet induced obese mice. *International Journal of Obesity's* 31, 1023–1029.

Li, C. C., Wu, S. J., Chang, C. H., and Ng, L. T. 2007. Antioxidant activity of *Cinnamomum cassia. Phytotherphy Research* 17, 726–730.

Little, T., Horowitz, M., and Feinle-Bisset, C. 2007. Modulation by high-fat diets of gastrointestinal function and hormones associated with the regulation of energy intake: Implications for the pathophysiology of obesity. *American Journal of Clinical Nutrition* 86, 531–541.

Liu, W., Zheng, Y., Han, L., Wang, H., Saito, M., and Ling, M. 2008. Saponins (Ginsenosides) from stems and leaves of *Panax quinquefolium* prevented high-fat diet-induced obesity in mice. *Phytomedicine* 15, 1140–1145.

Lomba, A., Martinez, J. A., Garcia-Daz, D., Paternain, L., Marti, A., Campion, J., and Milagro, F. 2010. Weight gain induced by an isocaloric pair-fed high fat diet: A nutri-epigenetic study on FASN and NDUFB6 gene promoters. *Molecular Genetics and Metabolism* 101, 273–278.

Lowe, M. 2002. The triglyceride lipases of the pancreas. *Journal of Lipid Research* 43: 2007–2016.

Martins, F., Noso, T., Porto, V., Curiel, A., Gambero, A., and Bastos, D. H. M. 2010. Mate tea inhibits *in vitro* pancreatic lipase activity and has hypolipidemic effect on high-fat diet-induced obese mice. *Obesity* 18, 42–47.

McClendon, K., Richo, D., Uwaifo, G. 2007. Orlistat: Current status in clinical therapeutics. *Expert Opinion Drug Safety* 8, 727–744.

McDougall, G. J., Kulkarni, N. N., and Stewart, D. 2009. Berry polyphenols inhibit pancreatic lipase activity *in vitro. Journal of Food Chemistry* 115, 193–199

Mella, A. T., Koss Twardy, S. G., and Zhi, J. 1996. The effect of orlistat, an inhibitor of dietary fat absorption, on the absorption of vitamins A and E in healthy volunteers. *Journal of Clinical Pharmacology* 36, 647–653.

Mobbs, C., Mastaitis, J., Yen, K., Schwartz, J., Mohan, V., and Poplawski, M. 2010. Low-carbohydrate diets cause obesity, low-carbohydrate diets reverse obesity: A metabolic mechanism resolving the paradox. *Appetite* 48, 135–138.

Moreno, D., Ilic, N., Poulev, A., Brasaemle, D., Fried, S., and Raskin, I. 2003. Inhibitory effects of grape seed extract on lipases. *Nutrition* 19, 876–879.

Moreno, D., Ilic, N., Poulev, A., and Raskin, I. 2006. Effects of *Arachis hypogaea* nutshell extract on lipid metabolic enzymes and obesity parameters. *Life Science* 78, 2797–2803.

Morikawa, T., Xie, Y., Asao, Y., Okamoto, M., Yamashita, C., and Muraoka, O. 2009. Oleanane-type triterpene oligoglycosides with pancreatic lipase inhibitory activity from the pericarps of *Sapindus rarak. Phytochemistry* 70, 1166–1172.

Mukherjee, M. 2003. Human digestive and metabolic lipases-a brief review. *Journal of Molecular Catalysis B: Enzymatic* 22, 369–376.

Nakai, M., Fukui, Y., Asami, S., Toyoda-Ono, Y., Iwashita, T., and Shibata, H. 2005. Inhibitory effects of oolong tea polyphenols on pancreatic lipase *in vitro*. *Journal of Agricultural and Food Chemistry* 53, 4593–4598.

Neovius, M., Johansson, K., and Rssner, S. 2008. Head-to-head studies evaluating efficacy of pharmacotherapy for obesity: A systematic review and meta-analysis. *Obesity Reviews* 9, 420–427.

Ninomiya, K., Matsuda, H., Shimoda, H., Nishida, N., Kasajima, N., and Yoshino, T. 2004. Carnosic acid, a new class of lipid absorption inhibitor from sage. *Bioorganic and Medicinal Chemistry Letters* 14, 1943–1946.

Ono, Y., Hattori, E., Fukaya, Y., Imai, S., and Ohizumi, Y. 2006. Anti-obesity effect of *Nelumbo nucifera* leaves extract in mice and rats. *Journal of Ethnopharmacology* 106, 238–244.

Quesada, H., Del Bas, J. M., Pajuelo, D., Daz, S., Fernandez-Larrea, J., and Pinent, M. 2009. Grape seed proanthocyanidins correct dyslipidemia associated with a high-fat diet in rats and repress genes controlling lipogenesis and VLDL assembling in liver. *International Journal of Obesity's* 33, 1007–1012.

Raghavendra, M. P. and Prakash, V. 2002. Phenylboronic acid – A potent inhibitor of lipase from *Oryza sativa. Journal of Agricultural and Food Chemistry* 50, 6037–6041.

Roy, D. M. and Schneeman, B.O. 1981. Effect of soy protein, casein and trypsin inhibitor on cholesterol, bile acids and pancreatic enzymes in mice. *Journal of Nutrition* 111, 878–885.

Rubio, M., Gargallo, M., Millán, A., and Moreno, B. 2007. Drugs in the treatment of obesity: Sibutramine, orlistat and rimonabant. *Public Health Nutrition* 10, 1200–1205.

Schrauwen, P. and Westerterp, K. R. 2000. The role of high-fat diets and physical activity in the regulation of body weight. *British Journal of Nutrition* 84, 417–427.

Sheng, L., Qian, Z., Zheng, S., and Xi, L. 2006. Mechanism of hypolipidemic effect of crocin in rats: Crocin inhibits pancreatic lipase. *European Journal of Pharmacology* 543, 116–122.

Shi, Y. and Burn, P. 2004. Lipid metabolic enzymes: Emerging drug targets for the treatment of obesity. *Nature Reviews Drug Discovery* 3, 695–710.

Slanc, P., Doljak, B., Kreft, S., Lunder, M., Janes, D., and Strukelj, B. 2009. Screening of selected food and medicinal plant extracts for pancreatic lipase inhibition. *Phytotherphy Research* 23, 874–877.

Slanc, S., Shobana, D., and Akhilender Naidu, K. 2004. Antioxidant activity of selected Indian spices. *Prostaglandins, Leukotrienes, Essential Fatty Acids* 62, 107–110.

Song, M., Lv, N., Kim, E., Kwon, K., Yoo, Y., and Kim, J. 2009. Anti-obesity activity of aqueous extracts of *Rhizoma dioscoreaetokoronis* on high-fat diet-induced obesity in mice. *Journal of Medicinal Food* 12, 304–309.

Sparg, S. G., Light, M. E., and van Staden, J. 2004. Biological activities and distribution of plant saponins. *Journal of Ethnopharmacology* 94, 219–243.

Stevenson, E., Astbury, N., Simpson, E., Taylor, M., and Macdonald, I. 2009. Fat oxidation during exercise and satiety during recovery are increased following a low-glycemic index breakfast in sedentary women. *Journal of Nutrition* 139, 890–897.

Sugimoto, S., Nakamura, S., Yamamoto, S., Yamashita, C., Oda, Y., Matsuda, H. 2009. Brazilian natural medicines. III. Structures of triterpene oligoglycosides and lipase inhibitors from mate, leaves of *Ilex paraguariensis. Chemical and Pharmaceutical Bulletin* 57, 257–261.

Sugiyama, H., Akazome, Y., Shoji, T., Yamaguchi, A., Yasue, M., and Kanda, T. 2007. Oligomeric procyanidins in apple polyphenol are main active components for inhibition of pancreatic lipase and triglyceride absorption. *Journal of Agricultural and Food Chemistry* 55, 4604–4609.

Sumiyoshi, M. and Kimura, Y. 2006. Low molecular weight chitosan inhibits obesity induced by feeding a high-fat diet long-term in mice. *Journal of Pharmacy and Pharmacology* 58, 201–207.

Tanaka, K., Tamaru, S., Nishizono, S., Miyata, Y., Tamaya, K., and Matsui, T. 2010. Hypotriacylglycerolemic and antiobesity propertie of a new fermented tea product obtained by tea-rolling processing of third-crop green tea (*Camellia sinensis*) leaves and Loquat (*Eriobotrya japonica*) leaves. *Bioscience, Biotechnology, and Biochemistry* 74, 1606–1612.

Terra, X., Montagut, G., Bustos, M., Llopiz, N., Ardvol, A., and Blad, C. 2009. Grape-seed procyanidins prevent low-grade inflammation by modulating cytokine expression in rats fed a high-fat diet. *Journal of Nutritional Biochemistry* 20, 210–218.

Tormo, M.A., Gil-Exojo, I., De Tejada, A. R., and Campillo, J. E. 2006. White bean amylase inhibitor administered orally reduces glycaemia in type 2 diabetic rats. *British Journal of Nutrition* 96, 539–544.

Toyoda-Ono, Y., Yoshimura, M., Nakai, M., Fukui, Y., Asami, S., and Shibata, H. 2007. Suppression of postprandial hypertriglyceridemia in rats and mice by oolong tea polymerized polyphenols. *Bioscience, Biotechnology, and Biochemistry* 71, 971–976.

Tsujita, T., Matsuura, Y., and Okuda, H. 1996. Studies on the inhibition of pancreatic and carboxylester lipases by protamine. *Journal of Lipid Research* 37, 1481–1487.

Tsujita, T., Takaku, T., and Suzuki, T. 2008. Chestnut astringent skin extract, an alpha-amylase inhibitor, retards carbohydrate absorption in rats and humans. *Journal of Nutritional Science and Vitaminology* 54, 82–88.

Tsujita, T., Sumiyoshi, M., Takaku, T., Momsen, W., Lowe, M., and Brockman, H. 2003. Inhibition of lipases by epsilon-polylysine. *Journal of Lipid Research* 4, 2278–2286.

Uchiyama, S., Taniguchi, Y., Saka, A., Yoshida, A., and Yajima, H. 2011. Prevention of diet-induced obesity by dietary black tea polyphenols extract *in vitro* and *in vivo*. *Nutrition* 27, 287–292.

Vincken, J., Heng, L., de Groot, A., and Gruppen, H. 2007. Saponins, classification and occurrence in the plant kingdom. *Phytochemistry* 68, 275–297.

Viner, R. M., Hsia, Y., Tomsic, T., and Wong, I. C. K. 2010. Efficacy and safety of anti-obesity drugs in children and adolescents: Systematic review and meta-analysis. *Obesity Review* 11, 593–602.

Weibel, E. K., Hadvary, P., Hochuli, E., Kupfer, E., and Lengsfeld, H. 1987. Lipstatin, an inhibitor of pancreatic lipase, produced by *Streptomyces toxytricini*. I. Producing organism, fermentation, isolation and biological activity. *Journal of Antibiotics* 40, 1081–1085.

WHO. 2014. Obesity and overweight. Geneva, Fact sheet No 311 http://www.who.int/mediacentre/factsheets/fs311/en/1341.

Won, S., Kim, S., Kim, Y., Lee, P., Ryu, J., and Kim, J. 2007. Licochalcone A: A lipase inhibitor from the roots of *Glycyrrhiza uralensis*. *Food Research International* 40, 1046–1050.

Xu, B., Han, L., Zheng, Y., Lee, J., and Sung, C. 2005. *In vitro* inhibitory effect of triterpenoidal saponins from *Platycodi radix* on pancreatic lipase. *Archives of Pharmaceutical Research* 28, 180–185.

Yamagishi, S., Matsui, T., Ueda, S., Fukami, K., and Okuda, S. 2009. Clinical utility of acarbose, an alpha-glucosidase inhibitor in cardiometabolic disorders. *Current Drug Metabolism* 10, 159–163.

Yamamoto, J., Schaab, A., Kreuter, M., and Drewe, S. 2000. *Piper cubeba* demonstrates anti-estrogenic and anti-inflammatory properties. *Planta Medica* 74, 142–146.

Yang, S., Ye, Y., and Li, B. 2003. 1'S-1'-Acetoxychavicol acetate isolated from Alpinia galanga inhibits human immunodeficiency virus type 1 replication by blocking Rev Transport. *Journal of General Virology* 87, 2047.

Yoshikawa, M., Shimoda, H., Nishida, N., Takada, M., and Matsuda, H. 2002. *Salacia reticulata* and its polyphenolic constituents with lipase inhibitory and lipolytic activities have mild antiobesity effects in rats. *Journal of Nutrition* 132, 1819–1824.

Yoshizumi, K., Hirano, K., Ando, H., Hirai, Y., Ida, Y., Tsuji, T. 2006. Lupane-type saponins from leaves of *Acanthopanax sessiliflorus* and their inhibitory activity on pancreatic lipase. *Journal of Agricultural and Food Chemistry* 54, 335–341.

Zhan, S. and Kim, T. 2004 The ability of an ethanol extract of *Cinnamomum cassia* to inhibit Src and spleen tyrosine kinase activity contributes to its anti-inflammatory action. *Journal of Ethnopharmacology* 139, 566–573.

Zhang, J., Kang, M., Kim, M., Kim, M., Song, J., and Lee, Y. 2008. Pancreatic lipase inhibitory activity of *Taraxacum officinale in vitro* and *in vivo. Nutrition Research and Practice* 2, 200–203.

Zhao, H. and Kim, Y. 2004. Determination of the kinetic properties of platycodin D for the inhibition of pancreatic lipase using a 1,2-diglyceride-based colorimetric assay. *Archives of Pharmacology* 27, 1048–1052.

Zhao, J., Pang, Y., and Dixon, R. 2010. The mysteries of proanthocyanidin transport and polymerization. *Plant Physiology* 153, 1–18.

Zheng, Q., Li, W., Han, L., and Koike, K. 2007. Pancreatic lipase-inhibiting triterpenoid saponins from *Gypsophila oldhamiana. Chemical and Pharmaceutical Bulletin* 55, 646–650.

8 Promise of Medicinal Plants for Neurodegenerative Disease

Alzheimer's, A Review

Kalidass Subramanaiam

CONTENTS

8.1 Introduction 133
8.2 Alzheimer's Disease (AD) 134
8.3 Conclusion 154
References 154

8.1 INTRODUCTION

Ayurvedic literature describes the nervous system and its disorders. In Sanskrit, 'Vata Vyadhi' is termed as a nervous system disorder which is thought to be brought on by imbalances of Vata, the biological air humor, a kind of energy which moves through the brain and the nerves (ativism considers the wind or air as the nerve impulses which travel throughout the body) controlling both voluntary and involuntary functions. Hence, any disturbances which cause an imbalance of Vata are said to be cause for some weakness, disturbance or hypersensitivity of the nervous system.

To modify the advancement and symptoms of mental disorders, herbal medicine offers several recourses. Preparation and marketing of medicinal plant-based drugs has been in current vogue, and the scientific and commercial magnitude appears to be gathering momentum in health-relevant areas. The products derived from these plants are standardized carefully, and their efficacy and safety for their specific implementation has been demonstrated (Perry et al. 1999; Howes et al. 2003; Abascal and Yarnell 2004; Kumar 2006; Kennedy and Wightman 2011). This review explains the use of several medicinal plants, their qualities and their activity towards Alzheimer's disease(AD).

8.2 ALZHEIMER'S DISEASE (AD)

AD is an age-related, chronic, irreversible and progressive neurodegenerative disease characterized by memory loss and cognitive dementia function (Essa and Reshmi 2012). This brain disorder was first discovered in 1906 by German Physician Alois Alzheimer (Anil et al. 2014). It is estimated that, along with the aging population, the number of people affected by AD is expected to reach 106.8 million by 2050 (Essa and Reshmi 2012).

Early-onset AD shows the loss of short-term memory, an incapability of learning new things and forgetting names and addresses (Rammohan et al. 2012; Nahid and Zahra 2014). In severe cases, patients' memories become completely lost, they lack self-restraint and even forget their way home. They become totally dependent on others, so comprehensive care is necessary. Symptoms usually appear after age 60, but some early-onset forms show symptoms due to specific genetic defects (Rammohan et al. 2012). The neocortex and hippocampus regions of the brain are associated with cognitive functions and are mostly affected by the pathology of AD (Nahid and Zahra 2014).

The extracellular deposition of amyloid Beta (Aβ) plaques, intraneuronal neurofibrillary tangles (NFT) consisting of abnormally phosphorylated tau protein, inflammatory processes and disturbance of neurotransmitters are the pathological features identified in the central nervous system (CNS) in AD (Singh et al. 2010; Carmo and Cuello 2013). So far, there is no effective treatment available to treat or cure AD and the drugs available for treatment only address the symptoms with limited effectiveness (Rammohan et al. 2012). Pharmacological treatment mechanisms of AD include three classifications of drugs: those based on disease-modifying therapies like vitamin E; those based on the rectification of neurotransmitters like cholinesterase inhibitors; and psychotherapy to manage symptoms (Ashley 2003; Nahid and Zahra 2014).

The most accepted treatment of AD at present are cholinesterase inhibitors that can inhibit the acetylcholinesterase enzyme (AchE) to increase the acetylcholine levels in the brain. Rivastigmine, tacrine, donepezil, galantamine and recently-prescribed methyl-D-aspartate receptor antagonists (memantine) are the acetylcholinesterase inhibitors in use so far, however, there is no complete cure for AD (Nahid and Zahra 2014; Orhan and Aslan 2009). It is believed that therapeutic interventions to postpone the onset or advancement of AD would reduce the number of cases dramatically in the next 50 years (Alzheimer's Association 2010; Rammohan et al. 2012).

Herbal medicines are considered to be effective medicines with more therapeutic benefits and fewer side effects. They also offer several options to modify the progress and symptoms of AD (Rammohan et al. 2012; Nahid and Zahra 2014). The present review is focused onto the several medicinal plants used for AD. Table 8.1 summarizes the report on medicinal plants and its action against Alzheimer's disease.

TABLE 8.1
Report on Medicinal Plants and Their Action against Alzheimer's Disease

S. No.	Plant	Classification	Results	References
1	*Acorus calamus*	Kingdom: Plantae Division: Magnoliophyta Class: Liliopsida Order: Lamiales Family: Scrophulariaceae Genus: *Acorus* Species: *calamus*	Memory-enhancing properties for memory deterioration, learning performance and behaviour modification. Inhibits Acetylcholinesterase enzyme (AchE). Contains majority of α- and β-asarone. Used for the treatment of memory loss and related symptoms.	Lannert and Hoyer 1998; Anil et al., 2014
2	*Angelica archangelica*	Kingdom: Plantae Division: Magnoliophyta Class: Magnoliopsida Order: Apiales Family: Apiaceae Genus: *Angelica* Species: *archangelica*	*A. archangelica* increases blood flow in the brain and also inhibits AChE *in vitro*. Contains similar substances used in AD drugs with fewer side effects.	Anil et al. 2014
3	*Bacopa monniera*	Kingdom: Plantae Division: Angiospermae Class: Dicotyledonae Order: Lamiales Family: Scrophulariaceae Genus: *Bacopa* Species: *monniera*	Contains a compound bacoside with nootropic activity and also reverses the depletion of Acetylcholine (Ach), reduces choline acetyltransferase ChAT activity and decreases muscarine receptor binding in the frontal cortex and hippocampus. Improves memory and cognitive improvement.	Vohora et al. 1990; Bhattacharya 2000; Stough et al. 2001 Russo and Borrelli 2005; Kumar 2006; Narendra et al. 2011

(*Continued*)

TABLE 8.1 (CONTINUED)
Report on Medicinal Plants and Their Action against Alzheimer's Disease

S. No.	Plant	Classification	Results	References
4	*Bertholettia excelsa*	Kingdom: Plantae Division: Magnoliophyta Class: Magnoliopsida Order: Ericales Family: Lecythidaceae Genus: *Bertholetta* Species: *excelsa*	Contains a high concentration of lecithin, which contains choline, a building block for acetylcholine, which enhances the concentration of acetylcholine in AD patients.	Anil et al. 2014
5	*Biota orientalis*	Kingdom: Plantae Division: Tracheophyta Class: Pinopsida Order: Pinales Family: Cupressaceae Genus: *Biota* Species: *orientalis*	Improves memory dysfunction.	Keyvan et al. 2007
6	*Catharanthus roseus*	Kingdom: Plantae Division: Magnoliophyta Class: Magnoliopsida Order: Gentianales Family: Apocynaceae Genus: *Catharanthus* Species: *roseus*	Vinpocetine is a chemical derived from vincamine, a constituent found in the leaves of *C. roseus*, and is used as a treatment for memory loss and mental impairments. Has the potential to enhance cerebral blood flow and neuroprotective effects.	Lorincz et al. 1976; Dezsi et al. 2002; Szilagyi et al. 2005; Narendra et al. 2011

(*Continued*)

TABLE 8.1 (CONTINUED)
Report on Medicinal Plants and Their Action against Alzheimer's Disease

S. No.	Plant	Classification	Results	References
7	*Celastrus paniculatus*	Kingdom: Plantae Division: Magnoliophyta Class: Magnoliopsida Order: Celastrales Family: Celastraceae Genus: *Celastrus* Species: *paniculatus*	Possesses antioxidant and anti-inflammatory activity, memory improvement potential and cognitive-enhancing activity.	Nalini et al. 1995; Gattu et al. 1997; Kumar and Gupta 2002a; Russo and Borrelli 2005
8	*Centella asiatica*	Kingdom: Plantae Division: Angiospermae Class: Dicotyledonae Order: Umbelliferae Family: Apiaceae Genus: *Centella* Species: *asiatica*	Prevents dementia, possesses tranquilizing and potentially cholinomimetic activities *in vivo*, which may be due to the presence of the triterpenoid brahminoside. It modulates dopaminergic, serotonic and adrenergic systems *in vivo* and improves learning and memory. Enhances cognitive function.	Sakina and Dandiya 1990; Kumar and Gupta 2002b; Narendra et al. 2011; Anil et al. 2014
9	*Clitoria ternatea*	Kingdom: Plantae Division: Magnoliophyta Class: Magnoliopsida Order: Fabales Family: Fabaceae Genus: *Clitoria* Species: *ternatea*	Memory-enhancing effects. These effects are associated with increased levels of ChAT and Ach *in vivo*.	Taranalli and Cheeramkuzhy 2000; Rai et al. 2002

(Continued)

TABLE 8.1 (CONTINUED)
Report on Medicinal Plants and Their Action against Alzheimer's Disease

S. No.	Plant	Classification	Results	References
10	*Collinsonia canadensis*	Kingdom: Plantae Division: Magnoliophyta Class: Magnoliopsida Order: Lamiales Family: Lamiaceae Genus: *Collinsonia* Species: *canadensis*	Prevents the breakdown of acetylcholine. Contains chief chemical constituents carvacol and thymol, which are used for AD.	Anil et al. 2014
11	*Commiphora wightii*	Kingdom: Plantae Division: Tracheophyta Class: Magnoliopsida Order: Sapindales Family: Burseraceae Genus: *Commiphora* Species: *wightii*	Contains the major constituent of guggulipids, a potential cognitive enhancer for improvement of memory in scopolamine-induced memory deficits. *C. whighitti* acts on impairment in learning and memory and decreases choline actyltransferase levels in hippocampus. Shows maximum effects on memory functions and dementia disorder.	Lannert and Hoyer 1998; Anil et al. 2014
12	*Convolvulus pluricaulis*	Kingdom: Plantae Division: Tracheophyta Class: Magnoliopsida Order: Solanales Family: Convolvulaceae Genus: *Convolvulus* Species: *pluricaulis*	Hippocampal regions that associate with the learning and memory functions show a dose-dependent increase in acetylcholine esterase activity in the CA1 and CA3 area with *C. pluricaulis* treatment.	Sharma et al. 2010

(*Continued*)

TABLE 8.1 (CONTINUED)
Report on Medicinal Plants and Their Action against Alzheimer's Disease

S. No.	Plant	Classification	Results	References
13	*Coptis chinensis*	Kingdom: Plantae Division: Magnoliophyta Class: Magnoliopsida Order: Ranunculales Family: Ranunculaceae Genus: *Coptis* Species: *chinensis*	Possesses AChE inhibition, nerve growth factor (NGF) enhancement and monoamine oxidase (MAO) inhibition *in vivo*. They also possess *in vitro* antioxidant activity and anti-inflammatory activity.	Hsieh et al. 2000; Mani et al. 2010
14	*Coriandrum sativum*	Kingdom: Plantae Division: Magnoliophyta Class: Magnoliopsida Order: Apiales Family: Apiaceae Genus: *Coriandrum* Species: *sativum*	Well-known producers for anti-anxiety, analgaesic, anticonvulsant, anti-inflammatory and nerve soothing properties. Improves memory and also demonstrates AChE inhibitory activity.	Mani et al. 2010
15	*Crocus sativus*	Kingdom: Plantae Division: Magnoliophyta Class: Liliopsida Order: Liliales Family: Iridaceae Genus: *Crocus* Species: *sativus*	Improves ethanol-induced impaired learning and behaviour in mice by inhibiting the impairment of hippocampal synaptic plasticity. Inhibits Aβ fibrillogenesis and exerted antioxidant effects *in vivo*.	Abe and Saito 2000

(Continued)

TABLE 8.1 (CONTINUED)
Report on Medicinal Plants and Their Action against Alzheimer's Disease

S. No.	Plant	Classification	Results	References
16	*Curcuma longa*	Kingdom: Plantae Division: Magnoliophyta Class: Liliopsida Order: Zingiberales Family: Zingiberaceae Genus: *Curcuma* Species: *longa*	Possesses antidepressant activity in mice following oral administration, which was associated with the inhibition of brain MAO type A.	Yu et al. 2002
17	*Cyperus rotundus*	Kingdom: Plantae Division: Magnoliophyta Class: Liliopsida Order: Cyperales Family: Cyperaceae Genus: *Cyperus* Species: *rotundus*	Pinene, a smaller concentration of cineole, terpenes and isociprol (a new alcohol) are seen in the rhizome and are rich in essential oils. Some of the chemical compounds possess anti-AChE activity. Studies on *C. rotundus* show improvement in special learning and memory as well as static avoidance of learning.	Rabiei et al. 2013; Sharma and Gupta 2007
18	*Euphorbia royleana*	Kingdom: Plantae Division: Magnoliophyta Class: Magnoliopsida Order: Euphorbiales Family: Euphorbiaceae Genus: *Euphorbia* Species: *royleana*	Potential treatment for AD. Possess adaptogenic or antistress and immunomodulatory activity. *E. royleana* affects some events in cortical and basal forebrain cholinergic signal transduction cascade in rat brains.	Narendra et al. 2011, Schliebs et al. 1997

(*Continued*)

TABLE 8.1 (CONTINUED)
Report on Medicinal Plants and Their Action against Alzheimer's Disease

S. No.	Plant	Classification	Results	References
19	*Evodia rutaecarpa*	Kingdom: Plantae Division: Magnoliopsida Class: Dicotyledons Order: Rutales Family: Rutaceae Genus: *Evodia* Species: *rutaecarpa*	Strongly inhibits AChE *in vitro* and reverses scopolamine-induced memory impairment in rats.	Park et al. 1996
20	*Galanthus caucasicus*	Kingdom: Plantae Division: Magnoliophyta Class: Liliopsida Order: Liliales Family: Liliaceae Genus: *Galanthus* Species: *caucasicus*	Galantamine, a drug isolated from *G. caucasicus*, has shown to modulate allosterically nicotinic Ach receptors on cholinergic neurons to increase acetylcholine release.	Narendra et al., 2011
21	*Galanthus nivalis*	Kingdom: Plantae Division: Magnoliophyta Class: Liliopsida Order: Liliales Family: Liliaceae Genus: *Galanthus* Species: *nivalis*	The chief chemical constituent of the *G. nivalis* is Galanthamine. Acetylcholinesterase (AChE) inhibitors, which are also called 'anticholinesterase drugs', are approved to be a promising treatment approach for AD.	Bores et al. 1996; Narendra et al. 2011; Anil et al. 2014

(*Continued*)

TABLE 8.1 (CONTINUED)
Report on Medicinal Plants and Their Action against Alzheimer's Disease

S. No.	Plant	Classification	Results	References
22	*Galanthus woronowii*	Kingdom: Plantae Division: Tracheophyta Class: Liliopsida Order: Liliales Family: Amaryllidaceae Genus: *Galanthus* Species: *woronowii*	Possess the potent AChE inhibitor alkaloid galanthamine, a selective, reversible and competitive AChE inhibitor. Galathamine improves memory and enhances cognitive performance in AD patients.	Marco and Carreiras 2006
23	*Ginkgo biloba*	Kingdom: Plantae Division: Ginkgophyta Class: Ginkgoopsida Order: Ginkgoales Family: Ginkgoaceae Genus: *Ginkgo* Species: *biloba*	Contains a neuroprotective effect against Aβ and nitric oxide (NO)–induced toxicity in the neuronal cell culture and could reduce apoptosis both *in vitro* and *in vivo*. The improved blood supply to the brain thereby ensures its efficient functioning and enhanced cognitive performance.	Barth et al. 1991; Maurer et al. 1997; Narendra et al. 2011; Loffler et al. 2001; Anil et al. 2014
24	*Glycyrrhiza glabra*	Kingdom: Plantae Division: Magnoliophyta Class: Asterids Order: Fabales Family: Asteraceae Genus: *Withania* Species: *somnifera*	Decreases Amyloid β production and β Amyloid plaque formation and improves memory impairment.	Zhiyuan et al. 2010; Narendra et al. 2011; Anil et al. 2014

(Continued)

TABLE 8.1 (CONTINUED)
Report on Medicinal Plants and Their Action against Alzheimer's Disease

S. No.	Plant	Classification	Results	References
25	*Hedyotis diffusa*	Kingdom: Plantae Division: Trecheophyta Class: Magnoliopsida Order: Gentianales Family: Rubiaceae Genus: *Hedyotis* Species: *diffusa*	Possesses neuroprotective effects.	Schindowski et al. 2001
26	*Huperzia serrata*	Kingdom: Plantae Division: Lycopodiophyta Class: Lycopodiopsida Order: Lycopodiales Family: Lycopodiaceae Genus: *Huperzia* Species: *serrata*	Drug isolated from *H. serrata* possesses antioxidant and neuroprotective properties, suggesting its potential in the treatment of AD. They also inhibit acetylcholinesterase (such as tacrine and donepezil) and improve memory and mental functioning in patients with AD and other severe conditions.	Pang et al. 1994; Raves et al. 1997; Kozikowski et al. 1999; Narendra et al. 2011; Anil et al. 2014
27	*Hypericum perforatum*	Kingdom: Plantae Division: Magnoliophyta Class: Magnoliopsida Order: Theales Family: Clusiaceae Genus: *Hypericum* Species: *perforatum*	Possess nootropic activity *in vivo*, which may be due to adrenergic (α- and β-receptor) and serotonergic (5HT1A) antagonistic activity.	Kumar et al. 2000; Khalifa 2001; Kumar et al. 2002c

(Continued)

TABLE 8.1 (CONTINUED)
Report on Medicinal Plants and Their Action against Alzheimer's Disease

S. No.	Plant	Classification	Results	References
28	*Lavandula officinalis*	Kingdom: Plantae Division: Magnoliophyta Class: Magnoliopsida Order: Lamiales Family: Lamiaceae Genus: *Lavandula* Species: *officianalis*	Shows calming and soothing effects through gamma-aminobutyric acid (GABA) in CNS area, which may due to the presence of geraniol, linalool, linalyl acetate, cineol, borneol, aflapin, camphor, butyric acid, valerianic acid, ursolic acid and luteolin flavonoids. In rat models, it showed a neuroprotective effect with a reduction in the neurologic loss, stroke volume, carbonyl and reactive oxygen species. It improves special learning and memory, motor coordination and static avoidance learning.	Hosseinzadeh and Nassin 2003; Wang et al. 2012; Rabiei et al. 2014b
29	*Lipidium meyenii*	Kingdom: Plantae Division: Magnoliophyta Class: Magnoliopsida Order: Capparales Family: Brassicaceae Genus: *Lipidium* Species: *meyenii*	Improves memory, learning and experimental memory impairment induced by ovariectomy, due in part, to its antioxidant and AChE inhibitory activities.	Fu and Li 2009; Anil et al. 2014

(*Continued*)

TABLE 8.1 (CONTINUED)
Report on Medicinal Plants and Their Action against Alzheimer's Disease

S. No.	Plant	Classification	Results	References
30	*Lycopodium serratum*	Kingdom: Plantae Division: Tracheophyta Class: Lycopodiopsida Order: Lycopodiales Family: Lycopodiaceae Genus: *Lycopodium* Species: *serratum*	Produces huperzine A, a potential therapeutic agent for the treatment of AD. Used to treat fever, inflammation, blood disorders and schizophrenia. Highly scrupulous, reversible and potent AChE inhibitor. Its activity is superior or similar to that of physostigmine, galanthamine, donepezil and tacrine. Gives protection against oxidative injury induced by Aβ and neuronal apoptosis, regulates the nerve growth factor and reduces inflammation, apoptosis, mitochondrial dysfunction and glutamate-induced toxicity.	Liu et al. 1986; Wang and Tang 1998; Zangara 2003; Zhang 2008
31	*Magnolia offinalis*	Kingdom: Plantae Division: Tracheophyta Class: Magnoliopsida Order: Magnoliales Family: Magnoliaceae Genus: *Magnolia* Species: *offinalis*	Inhibits the memory impairment induced by scopolamine through the inhibition of AChE.	Lannert and Hoyer 1998; Anil et al. 2014

(*Continued*)

TABLE 8.1 (CONTINUED)
Report on Medicinal Plants and Their Action against Alzheimer's Disease

S. No.	Plant	Classification	Results	References
32	*Matricaria recutita*	Kingdom: Plantae Division: Magnoliophyta Class: Magnoliopsida Order: Asterales Family: Asteraceae Genus: *Matricaria* Species: *recutita*	Stimulates the brain, dispels weariness, calms the nerves, counteracts insomnia, aids in digestion, breaks up mucus in the throat and lungs, aids the immune system, relieves anxiety and, in higher doses, leads to drowsiness.	Anil et al. 2014
33	*Melissa officinalis*	Kingdom: Plantae Division: Magnoliophyta Class: Magnoliopsida Order: Lamiales Family: Lamiaceae Genus: *Melissa* Species: *officinalis*	Possesses *in vitro* AChE inhibitory and antioxidant activity.	Kennedy et al. 2002; Akhondzadeh et al. 2003a; Anil et al. 2014
34	*Moringa oleifera*	Kingdom: Plantae Division: Magnoliophyta Class: Magnoliopsida Order: Capparales Family: Moringaceae Genus: *Moringa* Species: *oleifera*	Improves memory and learning. Treatment with *M. oleifera* leaves showed a significant increase in antioxidant enzymes such as superoxide dismutase (SOD) and catalase and a significant reduction in lipid peroxidase (LPO) levels. The antioxidant activity of the plant is believed to be the reason behind the improved cognitive functions.	Ganguly et al. 2005

(*Continued*)

TABLE 8.1 (CONTINUED)
Report on Medicinal Plants and Their Action against Alzheimer's Disease

S. No.	Plant	Classification	Results	References
35	*Nardostachys jatamansi*	Kingdom: Plantae Division: Tracheophyta Class: Magnoliopsida Order: Dipsacales Family: Valerianaceae Genus: *Nardostachys* Species: *jatamansi*	*N. jatamansi* eases all of the symptoms of chronic fatigue syndrome (CFS) in rats. Reverses the CFS triggered increase in lipid peroxidation, nitrite, superoxide dismutase levels and low catalase levels. Improves learning, memory and also reverses amnesia induced by diazepam and scopolamine. Aging-induced amnesia due to the natural aging of mice was also reversed, suggesting that the compounds in this plant may prove to be useful in restoring memory in older individuals, as well as in patients with age-associated dementia.	Joshi and Parle 2006; Lyle et al. 2009
36	*Paeonia suffruticosa*	Kingdom: Plantae Division: Magnoliophyta Class: Magnoliopsida Order: Dilleniales Family: Paeoniaceae Genus: *Paeonia* Species: *suffruticosa*	Possesses anti-inflammatory and antipyretic activity. The major polyphenol 1,2,3,4,6-Penta-O-galloyl-b-D-glucopyranose (PGG) present in *Paeonia suffruticosa*, inhibits Aβ1-42.	Lyle et al. 2009; Anil et al. 2014

(*Continued*)

TABLE 8.1 (CONTINUED)
Report on Medicinal Plants and Their Action against Alzheimer's Disease

S. No.	Plant	Classification	Results	References
37	*Panax ginseng*	Kingdom: Plantae Division: Eudicots Class: Asterids Order: Apiales Family: Araliaceae Genus: *Panax* Species: *ginseng*	Improves neurological performance and memory ability of vascular dementia (VD) in rats through mechanisms related to antiapoptosis.	Narendra et al. 2011; Anil et al. 2014
38	*Phyllanthus emblica*	Kingdom:Plantae Division: Magnoliophyta Class: Magnoliopsida Order: Euphorbiales Family: Euphorbiaceae Genus: *Phyllanthus* Species: *emblica*	Highly nutritious, contains vitamins, minerals and aminoacids. Experiments in rats induced with scopolamine and sodium nitrite to deficit memory were improved by treatment of methanol extract of *P. emblica*. The fruit possesses anti-oxidative, anti-inflammatory, and adaptogenic activity. The presence of ascorbic acid and other polyphenolic compounds present in the fruit may be the reason behind the neuroprotective effect in the central nervous system (CNS).	Ashwlayan et al. 2011
39	*Piper methysticum*	Kingdom: Plantae Division: Magnoliophyta Class: Magnoliopsida Order: Piperales Family: Piperaceae Genus: *Piper* Species: *methysticum*	Mood and cognition performance enhancement.	Thompson et al. 2004

(Continued)

TABLE 8.1 (CONTINUED)
Report on Medicinal Plants and Their Action against Alzheimer's Disease

S. No.	Plant	Classification	Results	References
40	*Polygala tenuifolia*	Kingdom: Plantae Division: Magnoliophyta Class: Magnoliopsida Order: Fabales Family: Polygalaceae Genus: *Polygala* Species: *tenuifolia*	Upregulated ChAT activity increased NGF secretion *in vitro*. *P. tenuifolia*, and also demonstrated *in vitro* AChE inhibitory activity.	Park et al. 2002
41	*Poncirus trifoliata*	Kingdom: Plantae Division: Magnoliophyta Class: Magnolipsida Order: Sapindales Family: Rutaceae Genus: *Poncirus* Species: *trifoliata*	Contains naturally occurring AChE inhibitors which increase the acetylcholine level and thereby boost cholinergic neuronal transmission. It helps in cognitive performance and signal transduction in nerve cells.	Kim et al. 2009
42	*Prunella vulgaris*	Kingdom: Plantae Division: Magnoliophyta Class: Magnolipsida Order: Lamiales Family: Lamiaceae Genus: *Prunella* Species: *vulgaris*	Used to treat inflammation, eye pain, headaches and dizziness. Contains several compounds like oleic acid, ursolic acid, butyric acid, flavonoids and rosmarinic acid, and also possesses anti-allergic, anti-inflammatory, antioxidant, antimicrobial and antiviral activity. Helps in memory and learning by increasing cholinergic neurotransmitters and methyl-D-aspartate receptor signaling.	Lamaison et al. 1991; Psotová et al. 2009; Park et al. 2010

(*Continued*)

TABLE 8.1 (CONTINUED)
Report on Medicinal Plants and Their Action against Alzheimer's Disease

S. No.	Plant	Classification	Results	References
43	*Rosmarinus officinalis*	Kingdom: Plantae Division: Magnoliophyta Class: Magnoliopsida Order: Lamiales Family: Lamiaceae Genus: *Rosmarinus* Species: *officinalis*	The essencial oil from *R. officinalis* produces a significant decline in the performance of working memory, and impaired reaction times for both memory and attention-based tasks. The findings of the study indicate that the olfactory properties of this essential oil can produce objective effects on cognitive performance, as well as subjective effects on mood. They are active in inhibition of AChE or ß-amyloid deposits, inhibition *in vitro* and also may have anti-BuChE activity.	Ozarowski et al. 2009; Narendra et al. 2011; Anil et al. 2014
44	*Salvia lavandulaefolia*	Kingdom: Plantae Division: Magnoliophyta Class: Magnoliopsida Order: Lamiales Family: Lamiaceae Genus: *Salvia* Species: *lavandulaefolia*	Strong AChE inhibitory activity. This is due to the presence of cyclic monoterpenes 1,8-cineole and α-pinene.	Tildesley et al. 2009
45	*Salvia militiorrhiza*	Kingdom: Plantae Division: Magnoliophyta Class: Magnoliopsida Order: Lamiales Family: Lamiaceae Genus: *Salvia* Species: *miliorrhiza*	Demonstrated *in vitro* antioxidant activity.	Hsieh et al. 2000

(*Continued*)

TABLE 8.1 (CONTINUED)
Report on Medicinal Plants and Their Action against Alzheimer's Disease

S. No.	Plant	Classification	Results	References
46	*Salvia officinalis*	Kingdom: Plantae Division: Magnoliophyta Class: Magnoliopsida Order: Lamiales Family: Lamiaceae Genus: *Salvia* Species: *officinalis*	Exhibited *in vitro* AChE and BuChE inhibitory activity.	Baricevic et al. 2001; Akhondzadeh et al. 2003b; Narendra et al. 2011; Anil et al. 2014
47	*Scrophularia buergeriana*	Kingdom: Plantae Division: Magnoliophyta Class: Magnoliopsida Order: Lamiales Family: Scrophulariaceae Genus: *Scrophularia* Species: *buergeriana*	Contains neuroprotective activity due to the presence to the harpagide and aglycone of harpagide type iridoids.	Kim et al. 2002
48	*Tinospora cordifolia*	Kingdom: Plantae Division: Magnoliophyta Class: Magnoliopsida Order: Ranunculales Family: Menispermaceae Genus: *Tinospora* Species: *cordifolia*	In normal and memory-deficient animals, it possesses memory-enhancing properties for learning. The mechanism of *T. Cordifolia*'s for cognitive enhancement is by immunostimulation and synthesis of acetylcholine, this supplementation of choline enhances the cognitive function.	Lannert and Hoyer 1998; Anil et al. 2014

(*Continued*)

TABLE 8.1 (CONTINUED)
Report on Medicinal Plants and Their Action against Alzheimer's Disease

S. No.	Plant	Classification	Results	References
49	*Uncaria rhynchophylla*	Kingdom: Plantae Division: Tracheophyta Class: Magnoliopsida Order: Gentianales Family: Rubiaceae Genus: *Uncaria* Species: *rhynchophylla*	Possesses anti-amyloidogenic activity. Use of the thioflavin T method showed a significant reduction in Aβ fibrillation. The presence of oxindole and indole alkaloids *U. rhynchophylla* extract are believed to be behind the neuroprotective effect. Studies suggested that *U. rhynchophylla* extract cannot only the reason behind prevention of Aβ fibril formation, but also disassemble preformed Aβ fibrils.	Fujiwara et al. 2006
50	*Urtica dioica*	Kingdom: Plantae Division: Magnoliophyta Class: Magnoliopsida Order: Urticales Family: Urticaceae Genus: *Urtica* Species: *dioica*	Contains the mineral boron, which is reported to enhance the levels of estrogen which can be good for short-term memory.	Keyvan et al. 2007; Anil et al. 2014

(*Continued*)

TABLE 8.1 (CONTINUED)
Report on Medicinal Plants and Their Action against Alzheimer's Disease

S. No.	Plant	Classification	Results	References
51	*Withania somnifera*	Kingdom: Plantae Division: Magnoliophyta Class: Magnoliopsida Order: Solanales Family: Solanaceae Genus: *Withania* Species: *somnifera*	Improves cognitive dysfunction and neuroprotective effect *in vivo*.	Dhuley 2001; Rasool 2007; Narendra et al. 2011; Anil et al. 2014
52	*Zizyphus jujuba*	Kingdom: Plantae Division: Magnoliophyta Class: Magnoliopsida Order: Rosales Family: Rhamnaceae Genus: *Ziziphus* Species: *jujuba*	Possesses soothing properties. Used as a drug for reducing anxiety and helps to strengthens stomach, spleen, and gastrointestinal system. Contains terpenoids, flavonoids and alkaloid and phenol glycosidic compound. May contribute to improve AD symptoms and motor loss, due to its increased acetylcholine level in the cholinergic terminals. Showed curative effects on rats with learning, memory, motor coordination and behavioural disorders caused by Meynert nucleus lesion.	Oda 1999; Rabiei et al. 2014a; Rabiei 2014

8.3 CONCLUSION

The above-mentioned studies on medicinal plants suggest various options to cure or slow the progress of neurodegenerative diseases like AD. The FDA (the U.S. Food and Drug Administration) approved drugs which are not offering a satisfactory result for the complete cure and treatment of AD. Also, they cause various side effects like headaches, nausea, vomiting, confusion. Furthermore, these drugs can only treat the symptoms and not the underlying cause of the disease. This reason challenged researchers to find an alternative solution for the treatment and management of neurodegenerative diseases with fewer side effects. Medicinal plants, which are richly available throughout the world, proved to be effective therapeutics for various diseases. The use of plants as medicine have been identified throughout the history of humankind. So, it is not a surprise that the plants found in folk remedies can treat memory disorders like Alzheimer's disease. Pharmaceutical companies are facing difficulties in the process of drug discovery for neurodegenerative diseases because of their high expense, risk, inefficiency and side effects. The prime benefit of plants is that they have a lower toxicity compared to pharmaceutical agents. Any disease can be cured if the treatment is started early, so the sooner the treatment is started, the better will be the outcome.

The medicinal plants listed above highlight their effectiveness in treating Alzheimer's or memory-related disorders. These plants have proven to have antioxidant and anti-inflammatory activity, improve memory, cognitive-enhancing activity, memory-enhancing properties for memory deterioration, improve learning performance and improve memory dysfunction. In order to use medicinal plants as a treatment, they should be compared with the available sources of drugs or pharmaceutical treatments currently available in order to improve their validation of clinical trials. Moreover, in order to resolve the potency of these substances in the cognitive disorders, large-scale production and multicenter studies are crucial. Until then, this review compiles the little information on the benefit of some medicinal plants used for the treatment of Alzheimer's disease.

REFERENCES

Abascal, K. and Yarnell, E. 2004. Alzheimer's disease: Part 2 – A botanical treatment plan. *Alternative and Complementary Therapies* 10: 67–72.

Abe, K. and Saito, H. 2000. Effects of saffron extract and its constituent crocin on learning behaviour and long term potentiation. *Phytotherapy Research* 14: 149–152.

Akhondzadeh, S., Noroozian, M., Mohammadi, M., Ohadinia, S., Jamshidi, A. H., and Khani, M. 2003a. *Melissa officinalis* extract in the treatment of patients with mild to moderate Alzheimer's disease: A double blind randomized, placebo controlled trial. *Journal of Neurology, Neurosurgery and Psychiatry* 74: 863–866.

Akhondzadeh, S., Noroozian, M., Mohammadi, M., Ohadinia, S., Jamshidi, A. H., and Khani, M. 2003b. *Salvia officinalis* extract in the treatment of patients with mild to moderate Alzheimer's disease: A double blind, randomized and placebo-controlled trial. *Journal of Clinical Pharmacy and Therapeutics* 28: 53–59.

Alzheimer's Association. 2010. Alzheimer's disease facts and figures. Alzheimer's Dementia. 158–194.

Anil, K. S., Vijay, N., and Om, P. B. 2014. Medicinal plants with a potential to treat Alzheimer's and associated symptoms. *International Journal of Nutrition, Pharmacology, Neurological Diseases* 14: 84–91.

Ashley, I. B. 2003. The metallobiology of Alzheimer's disease. *Trends in Neuroscience* 26(4): 207–214.

Ashwlayan, V. D. and Singh, R. 2011. Reversal effect of *Phyllanthus Emblica* (Euphorbiaceae) Rasayana on memory deficits in mice. *International Journal for Applied Pharmaceutics* 3: 10–15.

Baricevic, D., Sossa, S., and Della, L. R. 2001. Topical and anti-inflammtory activity of *Salvia officinalis* L. leaves: The relevances of ursoloic acid. *Journal of Ethnopharmacology* 75: 125–132.

Barth, S. A., Inselmann, G., Engemann, R., and Heidemann, H. T. 1991. Influences of *Ginkgo biloba* on cyclosporin A induced liver peroxidation in human liver microsomes in comparison to vitamine E, gluthatione and N-acetylcysteine. *Biochemical Pharmacology* 41: 1521–1526.

Bhattacharya, S. K., Bhattacharya, A., Kumar, A., and Ghosal, S. 2000. Anti-oxidant activity of *Bacopa monneira* in rat frontal cortex, striatum and hippocampus. *Phytotherapy Research* 14: 174–179

Bores, G. M., Huger, F. P., Petko, W., Mutlib, A. E., Camacho, F., and Rush, D. K. 1996. Pharmacological evaluation of novel Alzheimer's disease therapeutics: Acetylcholinesterase inhibitors related to galanthamine. *Journal of Pharmacology and Experimental Therapeutics* 277: 728–38.

Carmo, S. D., Cuello, A. C. 2013. Modelling of Alzheimer's disease in transgenic rats. *Molecular degeneration* 8.

Cuellar, M. J., Giner, R. M., Recio, M. C., Maez, S., and Rios, J. L. 2001. Topical anti-inflammatory activity of some Asian medicinal plants used in dermatological disorders. *Fitoterapia* 72: 221–229.

Dezsi, L., Kis-Varga, I., Nagy, J., Komlodi, Z., and Kurpati, E. 2002. Neuroprotective effects of vinpocetine *in vivo* and *in vitro*. Apovincaminic acid derivatives as potential therapeutic tools in ischemic stroke. *Acta Pharmaceutica Hungarica* 72(2): 84–91.

Dhuley, J. N. 2001. Nootropic-like effect of Ashwagandha (*Withania somnifera* L.) in mice. *Phytotherapy Research* 60: 173–178.

Essa, M. M. and Reshmi, K. V. 2012. Neuroprotective Effect of Natural Products against Alzheimer's Disease. *Neurochemical research* 37(9): 1829–42.

Fu, L. M. and Li, J. T. 2009. A systematic review of single Chinese herbs for Alzheimer's disease treatment. *Evidence Based Complementary and Alternative Medicine.*

Fujiwara, H., Iwasaki, K., Furukawa, K., Seki, T., He, M., Maruyama, M., Tomita, N., Kudo, Y., Higuchi, M., Saido, and T. C. 2006. *Uncaria rhynchophylla*, a Chinese medicinal herb, has potent antiaggregation effects on Alzheimer's beta-amyloid proteins. *Journal of Neuroscience Research* 84: 427–433.

Fujiwara, H., Tabuchi, M., Yamaguchi, T., Iwasaki, K., Furukawa, K., Sekiguchi, K., Ikarashi, Y., Kudo, Y., Higuchi, M., and Saido, T. C. 2009. Uncaria rhynchophylla, a Chinease medicinal herb, Paeonia suffruticosa and its active constituent 1,2,3,4,6-penta-O-galloyl-beta-d-glucopyrunose have a potent anti-aggregation effects on Alzheimer's amyloid beta proteins *in vitro* and *in vivo*. *Journal of Neurochemistry* 109: 1648–1657.

Ganguly, R., Hazra, R., Ray, K., and Guha, D. 2005. Effect of *Moringa Oleifera* in experimental model of Alzheimer's disease: role of antioxidant. *Annual review Neuroscience* 12.

Gattu, M., Boss, K. L., Terry, A. V., and Buccafusco, J. J. 1997. Reversal of scopolamine induced deficits in navigational memory performance by the seed oil of *Celastrus paniculatus*. *Pharmacology Biochemistry and Behaviour* 57: 793–799.

Hosseinzadeh, H. and Nassin, M. A. 2003. Anticonvulsant, sedative and muscle relaxant effects of carbenoxolone in mice. *BMC Pharmacology* 3: 3.

Howes, M. J., Perry, N. S., and Houghton, P. J. 2003. Plants with traditional uses and activities, relevant to the management of Alzheimer's disease and other cognitive disorders. *Phytotherapy Research* 17: 1–18.

Hsieh, M. T., Peng, W. H., Wu, C. R., and Wang, W. H. 2000. The ameliorating effects of the cognitive-enhancing Chinese herbs on scopolamine induced amnesia in rats. *Phytotherapy Research* 14: 375–377.

Joshi, H. and Parle, M. 2006. *Nardostachys j. atamansi* improves learning and memory in mice. *Journal of Medicnal Food* 9: 113–118.

Kennedy, D. O., Scholey, A. B., Tildesley, N. T. J., Perry, E. K., and Wesnes, K. A. 2002. Modulation of mood and cognitive performance following acute administration of *Melissa officinalis* (lemon balm). *Pharmacology, Biochemistry and Behaviour* 72: 953964.

Kennedy, D. O. and Wightman, E. L. 2011. Herbal extracts and phytochemicals: Plant secondary metabolites and the enhancement of human brain function. *Advances in Nutrition* 2: 32–50.

Keyvan, D., Damien, D. H. J., Heikki, V., and Raimo, H. 2007. Plants as Potential Sources for Drug Development against Alzheimer's Disease. *International Journal of Biomedical and Pharmaceutical Science* 1: 83–104.

Khalifa, A. E. 2001. *Hypericum perforatum* as a nootropic drug: Enhancement of retrieval memory of a passive avoidance conditioning paradigm in mice. *Journal of Ethnopharmacology* 76: 49–57.

Kim, J. K., Bae, H., Kim, M. J., Choi, S. J., Cho, H. Y., Hwang, H. J., Kim, Y. J., Lim, S. T., Kim, E. K., and Kim, H. K. 2009. Inhibitory effect of *Poncirus Trifoliate* on acetylcholinesterase and attenuating activity against trimethyltin-induced learning and memory impairment. *Bioscience Biotechnology Biochemistry* 73: 1105–1112.

Kim, S. R., Lee, K. Y., Koo, K. A., Sung, S. H., Lee, N. G., Kim, J., and Kim, Y. C. 2002. Four new neuroprotective iridoid glycosides from *Scrophularia buergeriana* roots. *Journal of Natural Products* 65: 1696–1699.

Kozikowski, Alan P. and Tueckmantel, Werner. 1999. Chemistry, Pharmacology, and Clinical Efficacy of the Chinese Nootropic Agent Huperzine A. *Accounts of Chemical Research* 32(8): 641–650.

Kumar, V. 2006. Potential medicinal plants for CNS disorders: an overview. *Phytotherapy Research* 20: 1023–1035.

Kumar, M. H. V. and Gupta, Y. K. 2002a Antioxidant property of *Celastrus paniculatus* Wild.: A possible mechanism in enhancing cognition. *Phytomedicine* 9: 302–311.

Kumar, M. H. V. and Gupta, Y. K. 2002b. Effect of different extracts of *Centella asiatica* on cognition and markers of oxidative stress in rats. *Journal of Ethnopharmacology* 79: 253–260.

Kumar, V., Khanna, V. K., Seth, P. K., Singh, P. N., and Bhattacharya, S. K. 2002c. Brain neurotransmittor receptor binding and nootropic studies on Indian *Hypericum perforatum* Linn. *Phytotherapy Research* 16: 210–216.

Kumar, V., Singh, P. N., Muruganandam, A. V., and Bhattacharya, S. K. 2000. Effect of Indian *Hypericum perforatum* Linn. On animal models of cognitive dysfunction. *Journal of Ethnopharmacology* 72: 119–128.

Lamaison, J. L., Petitjean-Freytet, C., and Carnat, A. 1991. Medicinal Lamiaceae with antioxidant properties, a potential source of rosmarinic acid. *Pharmaceutica Acta Helvetiae* 66(7): 185–188.

Lannert, H. and Hoyer, S. 1998. Intracerebroventricular administration of streptozotocin causes long-term diminutions in learning and memory abilities and in cerebral energy metabolism in adult rats. *Behavioral Neuroscience* 112: 1199–208.

Liu, S., Zhu, Y. L., Yu, C. M., Zhou, Y. Z., Han, Y. Y., and Wu, F. W. 1986. The structures of huperzine A and B, two new alkaloids exhibiting marked anticholinesterase activity. *Canadian Journal of Chemistry* 64(4): 837–839.

Loffler, T., Lee, S. K., Noldner, M., Chatterjee, S. S., Hoyer, S., and Schliebs, R. 2001. Effect of *Ginkgo biloba* extract (EGb 761) on glucose metabolism-related markers in streptozotocin-damaged rat brain. *Journal of Neural Transmission* 108: 1457–1474.

Lorincz, C., Szasz, K., and Kisfaludy, L. 1976. The synthesis of ethyl apovincaminate. *Arzneimittelforschung* 26(10a): 1907.

Lyle, N., Gomes, A., Sur, T., Munshi, S., Paul, S., Chatterjee, S., and Bhattacharyya, D. 2009. The role of antioxidant properties of *Nardostachys jatamansi* in alleviation of the symptoms of the chronic fatigue syndrome. *Behavioural Brain Research* 202: 285–290.

Mani, V., Parle, M., Ramasamy, K., and Abdul Majeed, A. B. 2010. Reversal of memory deficits by *Coriandrum sativum* leaves in mice. *Journal Science Food Agriculture* 91: 186–192

Marco, L. and Md Carreiras, C. 2006. Galanthamine, a natural product for the treatment of Alzheimers disease. *Recent Patents on CNS Drug Discovery* 1: 105–111.

Maurer, K., Ihl, R., Dierks, T., and Frölich, L. 1997. Clinical efficacy of *Ginkgo biloba* extract EGb 761 in dementia of the Alzheimer type. *Journal of Psychiatric Research* 31: 645–655.

Nahid, J. and Zahra, R. 2014. A review study on medicinal plants used in the treatment of learning and memory impairments. *Asian Pacific Journal of Tropical Biomedicine* 4(10): 780–9.

Nalini, K., Karanth, K. S., Rao, A., and Aroot, A. R. 1995. Effects of *Celastrus paniculatus* on passive avoidance performance and biogenic amine turnover in albino rats. *Journal of Ethnopharmacology* 47: 101–108.

Narendra, S., Pandey, B. R., and Pankaj, V. 2011. An Overview of Phytotherapeutic Approach in prevention and Treatment of Alzheimer's Syndrome and Dementia. *International Journal of Pharmaceutical Sciences and Drug Research* 3(3): 162–172.

Oda, Y. 1999. Choline acetyltransferase: the structure, distribution and pathologic changes in the central nervous system. *Pathology International* 49(11): 921–937.

Orhan, I. and Aslan, M. 2009. Appraisal of scopolamine-induced antiamnesic effect in mice and *in vitro* antiacetylcholinesterase and antioxidant activities of some traditionally used Lamiaceae plants. *Journal of Ethnopharmacology* 122(2): 327–332.

Ozarowski, M., Mikolajczak, P. L., Bobkiewlcz-Kozlowska, T., Kujawski, R., and Mrozikiewicz, P. M. 2009. Neuroactive compounds from medicinal plants of the Lamiaceae family showing potentially beneficial activity in treatment of Alzheimer's disease. *Herba Polonica* 55(4): 148–163.

Pang, Y. P., and Kozikowski Alan, P. 1994. Prediction of the binding sites of huperzine A in acetylcholinesterase by docking studies. *Journal of Computer Aided Molecular Design* 8: 669–681.

Pankaj, S. G., Himakshi, S., Georgia, A. C., and Ranjit, K. G. 2014. Development of a novel cellular model Alzheimer's disease utilizing neurosphere cultures derived from B6C3-Tg(APPswe, PSEN1dE9)85Dbo/J embryonic mouse brain. *SpringerPlus* 3(161).

Park, C. H., Choi, S. H., Koo, J. W., Seo, J. H., Kim, H. S., Jeong, S. J., and Suh, Y. H. 2002. Novel cognitive improving and neuroprotective activities of *Polygala tenuifolia* Willdenow extract, BT-11. *Journal of Neuroscience Research* 70: 484–492.

Park, C. H., Kim, S. H., Choi, W., Lee, Y. J., Kim, J. S., Kang, S. S., and Yoo, H. S. 1996. Novel anticholinesterase and antiamnesic activities of dehydroevodiamine, a constituent of *Evodia ruraecarpa. Planta Medica* 62: 405–409.

Park, S. J., Kim, D. H., Lee, I. K., Jung, W. Y., Park, D. H., and Kim, J. M. 2010. The ameliorating effect of the extract of the flower of Prunella vulgaris var. lilacina on drug-induced memory impairments in mice. *Food and Chemical Toxicology* 48(6): 1671–1676.

Perry, E. K., Pickering, A. T., Wang, W. W., Houghton, P. J., and Perry, N. S. 1999. Medicinal plants and Alzheimer's disease: From ethnobotany to phytotherapy. *Journal of Pharmacy and Pharmacology* 51: 527–534.

Psotová Kolář, M., Soušek, J., Švagera, Z., Vičar, J., and Ulrichová, J. 2003. Biological activities of Prunella vulgaris extract. *Phytotherapy Research* 17(9): 1082–1087.

Sharma, R. and Gupta. R. 2007. Cyperus rotundus extract inhibits acetylcholinesterase activity from animal and plants as well as inhibits germination and seedling growth in wheat and tomato. *Life Sciences* 80(24–25) 2389–2392.

Rabiei, Z., Hojjati, M., Rafieian-Kopaeia, M., and Alibabaei Z. 2013. Effect of Cyperus rotundus tubers ethanolic extract on learning and memory in animal model of Alzheimer's. *Biomedicine and Aging Pathology* 3(4): 185–191.

Rabiei, M. and Rafieian, B. 2014. Effects of Zizyphus jujuba extract on motor coordination impairment induced by bilateral electric lesions of the nucleus basalis of Meynert in rat. *Physiolology and Pharmacology* 17 (4): 469–477.

Rabiei, Z., Rafieian-Kopaei, M., Heidarian, E., Saghaei, E., and Mokhtari, S. 2014a. Effects of Zizyphus jujube extract on memory and learning impairment induced by bilateral electric lesions of the nucleus basalis of Meynert in rat. *Neurochemical Research* 39(2): 353–360.

Rabiei, Z., Rafieian-Kopaei, M., Mokhtari, S., Alibabaei, Z., and Shahrani, M. 2014b. The effect of pretreatment with different doses of Lavandula officinalis ethanolic extract on memory, learning and nociception. *Biomedicine and Aging Pathology* 4(1): 71–76.

Rai, K. S., Murthy, K. D., Karanth, K. S., Nalini, K., Rao, M. S., and Srinivasan, K. K. 2002. *Clitoria ternatea* root extract enhances acetylcholine content in rat hippocampus. *Fitoterapia* 73: 685–689.

Rammohan, V. R., Oliver, D., Varghese, J., and Dale, E. B. 2012. Ayurvedic medicinal plants for Alzheimer's disease: A review. *Alzheimer's Research and Therapy* 4(22): 4–9.

Rasool, M. and Varalakshmi, P. 2007. Protective effect of *Withania somnifera* root powder in relation to lipid peroxidation, antioxidant status, glycoproteins and bone collagen on adjuvant-induced arthritis in rats. *Fundamnetal and Clinical Pharmacology* 21: 157–164.

Raves, M. L., Harel, M., Pang, Y. P., Silman, I., Kozikowski, A. P., and Sussman, J. L. 1997. Structure of acetylcholinesterase complexed with the nootropic alkaloid. huperzine A. *Natural Structural and Biology* 4: 57–63.

Rigney, U., Kimber, S., and Hindmarch, I. 1999. The effects of acute doses of standardized *Ginkgo biloba* extract on memory and psychomotor performance in volunteers. *Phytotherapy Research* 13: 408–415.

Rubio, J., Qiong, W., Liu, X., Jiang, Z., Dang, H., and Chen, S. L. 2011. Aqueous Extract of Black Maca (*Lepidium meyenii*) on Memory Impairment Induced by Ovariectomy in Mice. *Evidance Based Complementary and Alternative Medicine.*

Russo, A. and Borrelli, F. 2005. *Baccopa monniera*, a reputed nootropic plant: An overview. *Phytomedicine* 12: 305–317.

Sakina, M. R. and Dandiya, P. C. 1990. A psycho-neuropharmacological profile of *Centella asiatica* extract. *Fitoterapia* 61: 291–296.

Schindowski, K., Leutner, S., Kressmann, S., Eckert, A., and Muller, W. E. 2001. Age-related increase of oxidative stress-induced apoptosis in mice prevention by *Ginkgo biloba* extract (EGb 761). *Journal of Neural Transmission* 108: 969–978.

Schliebs, R., Liebmann, A., Bhattacharya, S. K., Kumar, A., Ghosal, S., and Bigl, V. 1997. Systemic administration of defined extracts from *Withania somnifera* (Indian Ginseng) and Shilajit differentially affects cholinergic but not glutamatergic and GABAergic markers in rat brain. *Neurochemistry International* 30(2): 181–90.

Sharma, K., Bhatnagar, M., and Kulkarni, S. K. 2010. Effect of *Convolvulus pluricaulis* Choisy and *Asparagus racemosus* Wild on learning and memory in young and old mice: A comparative evaluation. *Indian Journal of Experimental Biology* 48: 479–485.

Singh, A. K., Gupta, A., Mishra, A. K., Gupta, V., Bansal, P., and Kumar, S. 2010. Medicinal plant for curing Alzheimer's Disease. *International Journal for of Pharmaceutical and Biological Archives* 1(2): 108–114.

Stough, C., Lloyd, J., Clarke, J., Downey, L.A., Hutchison, C. W., Rodgers, T., and Nathan, P. J. 2001. The chronic effects of an extract of *Bacopa monniera* (Brahmi) on cognitive function in healthy human subjects. *Psychopharmacology* 156: 481–484

Szilagyi, G., Nagy, Z., Balkay, L., Boros, I., Emri, M., Lehel, S., Marian, T., Molnar, T., Szakall, S., Tron, L., Bereczki, D., Csiba, L., Fekete, I., Kerenyi, L., Galuska, L., Varga, J., Bonoczk, P., Vas, A., and Gulyas, B. 2005. Effects of vinpocetine on the redistribution of cerebral blood flow and glucose metabolism in chronic ischemic stroke patients: A PET study. *Journal of Neurological Sciences* 229–230: 275–84.

Taranalli, A. D. and Cheeramkuzhy, T. C. 2000. Influence of *Clitoria ternatea* extracts on memory and central cholinergic activity in rats. *Pharmaceutical Biology* 38: 51–56.

Thompson, R., Ruch, W., and Hasenöhrl, U. 2004. Enhanced cognitive performance and cheerful mood by standardised extract of *Piper methysticum* (Kava-kava). *Human Psychopharmacology: Cinical and Experimental* 19: 243–250.

Tildesley, N. T. J., Kennedy, D. O., Perry, E. K., Ballard, C. G., Wesnes, K. A., and Scholey, A. B. Positive modulation of mood and cognitive performance following administration of acute doses of *Salvia lavandulaefolia* essential oil to healthy young volunteers. *Physiology and Behaviour* 83: 699–709.

Vohora, S. B., Shah, S. A., and Dandiya, P. C. 1990. Central nervous system studies on an ethanol extract of *Acorus calamus* rhizomes. *Journal of Ethnopharmacology* 8: 53–62.

Wang, H. and Tang, X. C. 1998. Anticholinesterase effects of huperzine A, E2020, and tacrine in rats. *Zhongguo Yao Li Xue Bao* 19(1): 27–30.

Wang, D., Yuan, X., Liu, T., Liu, L., Hu, Y., and Wang, Z. 2012. Neuroprotective activity of lavender oil on transient focal cerebral ischemia in mice. *Molecules* 17: 9803–9817.

Yu, Z. F., Kong, L. D., and Chen, Y. 2002. Antidepressant activity of aqueous extracts of *Curcuma longa* in mice. *Journal of Ethnopharmacology* 83:161–165.

Zangara, A. 2003. The psychopharmacology of huperzine A: An alkaloid with cognitive-enhancing and neuroprotective properties of interest in the treatment of Alzheimer's disease. *Pharmacology Biochemistry and Behavior* 75(3): 675–686.

Zhang, H. Y., Zheng, C. Y., Yan, H., Wang, Z. F., Tang, L. L., and Gao, X. 2008. Potential therapeutic targets of huperzine A for Alzheimer's disease and vascular dementia. *Chemico Biological Interactions* 175(1–3): 396–402.

Zhiyuan, Z., Chenjing, L., Wang, X., Zhengyi, Y., Jing, C., Lihong, H., Hualiang, J., and Shen X. 2010. 2, 2',4'-Trihydroxychalcone from *Glycyrrhiza glabra* as a new specific BACE1 inhibitor efficiently ameliorates memory impairment in mice. *Journal of Neurochemistry* 114: 374–385.

9 Plant Sources as Potential Therapeutics for Alzheimer's Disease

Azhwar Raghunath, Kiruthika Sundarraj, Vishnu Vignesh Kanagaraj and Ekambaram Perumal

CONTENTS

9.1 Introduction 162
9.2 Therapeutic Strategies for AD 163
9.2.1 Aβ and Senile Plaque 163
9.2.2 Secretases 163
9.2.3 Tau Protein and Neurofibrillary Tangles 164
9.2.4 Oxidative Stress 164
9.2.5 Inflammation 164
9.2.6 Cholinergic System 164
9.2.7 Cholinesterase 165
9.3 Plants as a Source for AD Therapeutics 165
9.3.1 *Catharanthus roseus* 166
9.3.2 *Celastrus paniculatus* 166
9.3.3 *Centella asiatica* 178
9.3.4 *Curcuma longa* 178
9.3.5 *Euphorbia royleana* Boiss 178
9.3.6 *Galanthus caucasicus* 179
9.3.7 *Ginkgo biloba* 179
9.3.8 *Glycyrrhiza glabra* 180
9.3.9 *Huperzia serrata* 180
9.3.10 *Melissa officinalis* 181
9.3.11 *Panax ginseng* 181
9.3.12 *Salvia officinalis* 181
9.3.13 *Withania somnifera* 182
9.4 Conclusion 182
Acknowledgements 183
References 183

9.1 INTRODUCTION

Alzheimer's disease (AD) is a slow, progressive and a chronic neurodegenerative disease that results in loss of memory, cognition, language skills and drastic behavioural changes which cause 60–70% of dementia. The word 'Alzheimer's' is derived from the German psychiatrist Dr. Alois Alzheimer (1901), who identified the first case of AD in his 50-year-old woman patient, August Deter. He carefully followed her case until she died in 1906 and he then publicly announced the results of his study. Emil Kraepelin described this disease as one which has its own pathological features and also named Alzheimer's disease as presenile dementia, a subtype of senile dementia (Berrios 1990). In 1977 at a conference held for scientists who study AD, he concluded that presenile dementia and senile dementia had the same pathological condition, but the causes for each disease are different. Since then, the term 'Alzheimer's disease' was made official in the medical nomenclature (Amaducci et al. 1986). Epidemiological data indicates a potentially considerable increase in the prevalence of the disease over the next two decades. AD affects up to 5% of people over 65 years, rising to 20% of those over 80 years. AD was estimated to double every 20 years to 66 million and 115 million by 2030 and 2050, respectively (World Alzheimer's Report 2015).

In AD, the neurons in the brain, which produce biochemicals called neurotransmitters, and acetylcholine (ACh) will break their connection with their neighboring neurons and ultimately die. The two main hallmarks by which AD progresses are amyloid plaque, a protein fragment which fails to degrade and accumulates around the neurons, hindering their function and neurofibrillary tangles that are insoluble, twisted tangles of a protein called tau protein that eventually builds inside each affected nerve cell, killing it from within.

The causes of AD include aging, neurofibillary tangles, senile plaque, genetics and others. Each of the above-mentioned causes has its own pathological mechanism by which it progresses to AD (Munoz and Feldman 2000). Aging has always been a risk factor for many diseases, including AD. During the process of cellular respiration, reactive oxygen species (ROS) or free radicals, are produced, which play a crucial role in the development of age-related AD (Smith et al. 1995). Upon upregulation of the activity, an antioxidant enzyme follows with the oxidative damage done to proteins and membrane lipids (Sayre et al. 1997; Smith et al. 1997). A study conducted by researchers at the University of British Columbia indicated an abnormally increased level of iron-handling protein called melanotransferrin, which was observed in the serum of AD patients, thus indicating the mishandling of iron in the neuronal system, and hence it may use as a potential diagnostic tool for the initial study of age-related AD (Kennard et al. 1996).

Inflammation also serves as a mean for AD. The brains of AD patients show evidence of mild active inflammation, including microglial with complementing activation and the presence of inflammatory cytokines (McGeer and McGeer 1995). Significant research has shown that people who take anti-inflammatory agents on a longer basis have a decreased prevalence of AD. On the other hand, women who consumed estrogen were less likely to be diagnosed with AD, and there are supporting studies that showed signs of improvement of AD in women patients (Birge 1997;

Kawas et al. 1997). A cholinesterase inhibitor tacrine combined with estrogen may have a greater therapeutic effect (Schneider 1995).

Abnormal phosphorylated fibrillary proteins, which aggregate in the neuronal cytoplasm, are called neurofibrillary tangles. Neurons, which are very complex with their branching structures, have to maintain a healthy cytoskeleton system, but the failure of this causes the tangle formation (Munoz and Feldman 2000). Neurofibrillary tangles are highly accumulated in the hippocampus, which is primarily involved in the storage of permanent memories. The basal forebrain that provides the cholinergic innervations to the cortex is also equally affected, resulting in severe neurodegeneration (Munoz and Feldman 2000).

Senile plaques are complex and made up of a sugar polymer component called glycosaminoglycans, which aids in the assembly of the beta-amyloid (Aβ) of the transmembrane protein. Senile plaque deposition will start slow, but it eventually builds up through an orderly, sequential manner which finally develops into AD (Mackenzie 1994). The amount of senile plaque will increase with age. Butyrylcholinesterase may also play an essential role in the maturation process of the plaques (Guillozet et al. 1997).

Familial AD, which is a kind of autosomal dominant disease caused by point mutation of the gene coding for Aβ precursor protein, is located on chromosome 21. The mutation causes varied effects on the formation of the protein Aβ, sometimes as a long amino acid chain or vice-versa. Other genes include the genes which code for presenilin 1 and presenilin 2 proteins, which also accounts for familial AD (Gomez-Isla et al. 1997). Both Aβ precursor proteins and presenilin 1 and 2 proteins are helpful in organising vesicular trafficking (Beyreuther and Master 1997). Other major causative factors for AD includes accidents and severe head trauma (Schofield et al. 1997). Patients who suffer from severe epilepsy have a very high risk of developing senile plaques at a younger age than the normal population (Mackenzie and Miller 1994).

9.2 THERAPEUTIC STRATEGIES FOR AD

AD pathology displays multifactorial and complex pathways (Small and Mayeux 2005). Pharmacological interventions on all these factors and pathways were investigated, and more are being discovered as the years unfold. These possible mechanisms could be intervened with plants and their derivatives to prevent and treat AD.

9.2.1 Aβ and Senile Plaque

Secretases play an important role in the formation of senile plaques as a result of proteolytic cleavage of the amyloid precursor protein (APP) into Aβ – a primary component of senile plaques. Aβ aggregation and deposition leads to neurodegeneration (Castro et al. 2002). Plant derivatives are capable of intervening in the formation and deposition of Aβ fibrils, and the clearance of senile plaques are possible therapeutic approaches in AD.

9.2.2 Secretases

APP processing involves two pathways – amyloidogenic and non-amyloidogenic. β-secretase and γ-secretase of the amyloidogenic pathway cleave the extracellular and transmembrane domains of the APP respectively, and results in abundant Aβ(1-40) and fewer Aβ(1-42) fragments (Scorer 2001). Amyloid fibrils are immediately formed by the aggregation of pathogenic Aβ(1-42) fragments. The membrane-bound α-secretase also cleaves APP, but no Aβ fragment is generated, as it falls under a non-amyloidogenic pathway (Blennow 2006). Plants and their compounds may offer the activation of α-secretase in the non-amyloidogenic pathway and inhibition of β-secretase and γ-secretase in the amyloidogneic pathway.

9.2.3 Tau Protein and Neurofibrillary Tangles

The hyperphosphorylation of tau proteins leading to the formation of intercellular neurofibrillary tangles is one of the pathologies in AD. The assembly and stability of microtubules are maintained by tau axonal protein, hyperphosphorylation of tau hinders, microtubule dynamics and impaired trafficking of transport within axons, which ultimately alter neuronal function. The impediment of tau hyperphosphorylation and aggregation with the help of plant bioactive constituents offers a promise in the treatment and prevention of AD.

9.2.4 Oxidative Stress

A vast array of plant products exhibit antioxidant properties and can be utilized as AD therapeutics. The impaired metal homeostasis (Sayre et al. 2001), mitochondrial dysfunction (Law et al. 2001), Aβ fragments and microglial activation (Varadarajan et al. 2000) lead to ROS generation. All these triggers' degeneration pathways eventually lead into the progression of AD. Enzymatic and non-enzymatic antioxidants prevent damage induced by ROS under normal physiological conditions (Valko et al. 2007). The imbalance created due to excessive ROS leads to the pathogenesis of AD. A handful of plant compounds displayed the ability to scavenge these free radicals, preventing oxidative damage and their effects were tested in the CNS, which could pave the way for AD treatment.

9.2.5 Inflammation

Inflammatory markers, the activation of microglial and pro-inflammatory cytokines, were identified in the brains of patients with AD. Anti-inflammatory agents are being tested to mitigate the production of inflammatory markers. A wide variety of plant constituents have displayed anti-inflammatory potentials, which could pave the way for anti-inflammatory targets in the treatment of AD.

9.2.6 Cholinergic System

The degeneration of cholinergic neurons results in the deficiency of neurotransmitters and this contributes to cognitive deficits. Aβ formation, senile plaque formation, ROS and astrocyte phagocytic activity are the causes of cholinergic neuron degeneration and thus disrupt the transmission pathway (Small and Mayeux 2005). Hence restoration of the cholinergic system enhances cognitive impairment.

Plants and their bioactive principles have displayed their ability to restore the cholinergic systems.

9.2.7 Cholinesterase

Cholinergic deficit is the main hallmark of AD. Acetylcholinesterase (AChE) hydrolyzes ACh into choline and acetate in the brain. Hence, cholinesterase inhibition serves as the most effective treatment approach to alleviate symptoms of AD. The extracts from different plant sources are proven to inhibit cholinesterase, thereby enhancing memory and cognition. Increased ACh levels trigger non-amyloidogenic processing, thus lowering Aβ formation and toxicity.

9.3 PLANTS AS A SOURCE FOR AD THERAPEUTICS

Nature is a rich source of biological and chemical diversity. The unique and complex structures of natural products cannot be obtained easily by chemical synthesis. A number of plants in the world have been used in traditional medicinal remedies. Plant species and their derivatives have been used by humans for over 5000 years as medicines (Goldman 2001; Liu and Wang 2008). The bioactive phytochemicals from the extracts of plant species are classified into a number of different chemical groups: alkaloids, phenolics and terpenoids (Samuelsson 2004). A single entity or mixture of components in these groups offer the most fecund potential for the treatment of AD. Scientific research has opened new directions for the use of plants and their derivatives for AD therapeutics from ethnobotanical evidence through cultural, empirical and complementary medical uses of plants. Research in the past decades has revealed that plants are a welcome source of therapeutics for AD, due to their low toxic potential when compared to pharmaceuticals. In humans, plant derivatives as food supplements have been proved to possess inherent abilities to enhance cognitive function and prevent neurodegeneration (Spencer 2007).

One of the primary occurrences of AD is the deficits in the cholinergic nervous system which can be evidenced from the behavioural and cognitive symptoms (Bartus et al. 1982). The use of AChE inhibitors served as an effective therapeutic strategy for many years. AChE breaks down ACh in the synaptic cleft and AChE inhibitors prevent this breakdown, thus providing symptomatic treatment of AD (Cummings 2000; Nordberg and Svensson 1998). Although the pathogenesis of AD is not completely explored, the existence of Aβ plaques, loss of nerve cells and neurofibrillary tangles still occur within the brain. Aβ is the principal component in amyloid plaques generated with β-secretase and γ-secretase by proteolytic cleavage of APP into 39-42 amino acid Aβ peptide (Hardy and Allsop 1991). Aβ peptide, once formed, aggregates and is deposited in the cell membranes and the dendrites of the neurons, and thus triggers dementia leading to AD. The suppression of Aβ production could serve as a therapeutic target in AD.

All the drugs currently available for AD are symptomatic and have their own limitations; they hardly have influence over the advancement of AD and cause side effects in patients. Hence, the hunt for more cogent and competent drugs is always needed. Plants with a potential for AD therapeutics have drawn the heed of the scientific community. Plant species and their derivatives have offered plenty of choices to alter the progression and symptoms of AD. A variety of plants produce alkaloids, flavonoids, lignans, polyphenols, sterols, tannins, and triterpenes and these phytochemicals manifest antioxidant, antiamyloidogenic, anticholinesterase, anti-inflammatory and hypolipidemic properties (Howes et al. 2003; Kumar 2006; Kelley and Knopman 2008; Kennedy and Wightman 2011). In humans, plant derivatives as food supplements have been proven to possess an inherent ability to enhance cognitive function and prevent neurodegeneration (Spencer 2007). Chapter 9 discusses the plants that offer potential therapeutics for AD and are identified in Table 9.1.

B. monnieri is a nerve tonic which improves memory and intelligence and is used in the treatment of mental and neurological disorders in aging (Morgan and Stevens 2010). *B. monnieri* has the potential to interact with the monoaminergic neurotransmitter system of the central nervous system (Sheikh et al. 2007). *Bacopa* recuperated the lost memory during the AD treatment by reducing the AChE activity and enhanced learning and memory processes (Joshi and Parle 2006). Along with Alzheimer's and dementia, their potential to treat neuritis was proved in a clinical study on humans (Bhalla et al. 2008). In a rat AD model, *B. monnieri* inhibited degeneration of cholinergic system and enhanced cognitive effect (Uabundit et al. 2010) whereas the Aβ levels diminished in AD mouse models (Holcomb et al. 2006). The extracts of *B. monnieri* prevented neuronal cell death induced by Aβ and suppressed AChE activity (Limpeanchob et al. 2008). The antioxidant activity of *B. monnieri* extract was well-evidenced from decreased ROS in neurons (Bhattacharya et al. 2000). *B. monnieri* was observed to increase the efficacy of memory and attention disorders and was established in many clinical studies (Pravina et al. 2007).

9.3.1 *Catharanthus roseus*

C. roseus has an important alkaloid constituent: vinpocetine from its leaves. Vinpocetine has a tendency to improve cerebral blood flow and thereby offer neuroprotective effects (Dézsi et al. 2002). Due to these inherent properties, vinpocetine is utilized in memory impairments (Szatmari and Whitehouse 2003). The wide use of vinpocetine in the treatment of cognitive decline dates back to 1970 in Japan and Europe. Vinpocetine enhances glucose metabolism and cerebral blood flow in different lobes of the brain (Szilágyi et al. 2005). Vinpocetine inhibited Aβ triggered cytokine production and inflammation (Liu et al. 2014). Several clinical studies have shown the beneficial effects of vinpocetine in patients with mild to moderate dementia (Peruzza et al. 1986). The leaf extracts have shown to possess AChE inhibitory potential and low cholinergic receptor affinity (Pereira et al. 2010). The components of *C. roseus* (ajmalicine, cathranthine, serpentine, tabersonine and vindoline) possessed anti-inflammatory properties, since inflammation is the primary feature in AD (Manigandan et al. 2014).

9.3.2 *Celastrus paniculatus*

Besides its many pharmacological properties, *C. paniculatus* is used for the treatment of brain-related diseases. The seeds produce Jyothismati oil, which promotes

TABLE 9.1
Potential plants for AD therapeutics

Plant Species	Country or Region of Origin	Bioactive Constituents	Mode of Action	References
Andrographis paniculata **Acanthaceae**	India and Sri Lanka	Diterpenoids, flavonoids and polyphenols	Inhibition of AChE	Mukherjee et al. 2007
Salsola oppositifolia **Amaranthaceae**	Morocco	Alkaloids	Inhibition of AChE and BuChE	Tundis et al. 2009
Salsola soda **Amaranthaceae**	Italy	Alkaloids	Inhibition of AChE and BuChE	Tundis et al. 2009
Salsola tragus **Amaranthaceae**	Georgia	Alkaloids	Inhibition of AChE and BuChE	Tundis et al. 2009
Crinum jagus **Amaryllidaceae**	Africa	Caryophyllene, eicosane, tetratriacontane, tetratetracontane and cis-decahydronaphathalene	Inhibition of AChE	Elufioye et al. 2010
Crinum moorei **Amaryllidaceae**	South Africa	Amaryllidaceae alkaloids	Inhibition of AChE	Fawole et al. 2010
Galanthus nivalis L. **Amaryllidaceae**	Netherlands	Galanthamine	Cholinesterase inhibition	Bores et al. 1996
Nerine undulate **Amaryllidaceae**	Eastern Cape province, South Africa	Amaryllidaceae alkaloids	Inhibition of AChE	Cahlikova et al. 2011a
Scadoxus multiflorus **Amaryllidaceae**	South Africa	Amaryllidaceae alkaloids	Inhibition of AChE	Cahlikova et al. 2011a
Sprekelia formaosissima **Amaryllidaceae**	Mexico and Guatemala	Amaryllidaceae alkaloids	Inhibition of AChE	Cahlikova et al. 2011a
Zephyranthes grandiflora **Amaryllidaceae**	Central America	Amaryllidaceae alkaloids	Inhibition of AChE	Cahlikova et al. 2011b

(Continued)

TABLE 9.1 (CONTINUED)
Potential Plants for AD Therapeutics

Plant Species	Country or Region of Origin	Bioactive Constituents	Mode of Action	References
Harpephyllum caffrum **Anacardiaceae**	South Africa	Proanthocyanidins, gallotannins and flavonoids	Inhibition of AChE	Moyo et al. 2010
Sclerocarya birrea **Anacardiaceae**	Southern Africa	Proanthocyanidins, gallotannins and flavonoids	Inhibition of AChE	Moyo et al. 2010
Spondias mombin **Anacardiaceae**	Americas, Africa, India, Bangladesh and Sri Lanka	Phenolics and carotenoids	Inhibition of AChE	Elufioye et al. 2010
Geissospermum Vellosii **Apocynaceae**	South America	Alkaloids	Cholinesterase inhibition	Lima et al. 2009
Acorus calamus L. **Araceae**	Czech Republic	A- and β-asarone	Inhibition of AChE	Lannert and Hoyer 1998
Colocasia antiquorum **Araceae**	Southern China, India, Africa and the West Indies	Amaryllidaceae alkaloids	Inhibition of AChE	Fawole et al. 2010
Pinellia ternate **Araceae**	China, Japan and Korea	Alkaloids	Inhibition of AChE	Yang et al. 2010
Panax Ginseng **Araliaceae**	Manchuria and China	Saponins, protopanaxadiol, protopantriol and oleanolic acid	Memory-enhancing action for learning impairment	Lannert and Hoyer 1998
Phoenix dactylifera **Arecaceae**	Iraq	Flavonoid	Cholinesterase inhibiton	Sekeroglu et al. 2012
Leopoldia comosa **Asparagaceae**	Southeastern Europe	Quercetin	Cholinesterase inhibition	Loizzo et al. 2011
Aloe ferox **Asphodelaceae**	Southern Africa	Amaryllidaceae alkaloids	Inhibition of AChE	Fawole et al. 2010
Achyrocline tomentosa **Asteraceae**	Northern and Central Argentina	Polyphenolic compounds	Cholinesterase inhibition	Carpinella et al. 2009

(Continued)

TABLE 9.1 (CONTINUED)
Potential Plants for AD Therapeutics

Plant Species	Country or Region of Origin	Bioactive Constituents	Mode of Action	References
Arnica chamissonis ssp. *Foliosa* **Asteraceae**	North America	Flavonoids, flavonols, flavones and helenelin	Inhibition of AChE and BuChE	Carpinella et al. 2010
Chromolaena tequendamensis **Asteraceae**	Colombia	Polyphenolic compounds	*In vitro* inhibition of AChE	Nino et al. 2006
Eupatorium viscidum **Asteraceae**	Argentina	Polyphenolic compounds	Cholinesterase inhibition	Carpinella et al. 2009
Pulicaria stephanocarpa **Asteraceae**	Yemen	Cadinol, spathulenol, caryophyllene, cadience and muurolene	Cholinesterase inhibition	Nasser et al. 2012
Schistocarpha sinforosi **Asteraceae**	Colombia	Polyphenolic compounds	*In vitro* inhibition of AChE	Nino et al. 2006
Trichocline reptans **Asteraceae**	South America	Polyphenolic compounds	Cholinesterase inhibition	Carpinella et al. 2009
Berberis darwinii **Berberidaceae**	Southern Chile and Argentina	Isoquinoline alkaloids	*In vitro* inhibition of AChE	Habtemariam 2011
Buddleja salviifolia **Buddlejaceae**	Southern and Eastern Africa	Polyphenolic compounds	Inhibition of AChE	Adewusi et al. 2011
Boswellia socotranao **Burseraceae**	North Africa and Arabian Peninsula	Mono-, di- and triterpenes	Inhibition of AChE	Bakthir et al. 2011
Commiphora whighitti **Burseraceae**	India	Guggulsterone	Decreased choline actyltransferase levels in hippocampus	Rubio et al. 2011
Lipidium Meyenii Walp **Brassicaceae**	South America	Macamides	Cholinesterase inhibition	Fu and Li 2009

(Continued)

TABLE 9.1 (CONTINUED)
Potential Plants for AD Therapeutics

Plant Species	Country or Region of Origin	Bioactive Constituents	Mode of Action	References
Cistus laurifolius **Cistaceae**	Spain and Southern France	3-O-methlquercetin and 3,7-O-dimethylquercetin	Cholinesterase inhibition	Akkol et al. 2012
Urtica dioica L. **Clusiaceae**	Europe, Asia, Northern Africa and Western North America	Kaempferol, lection, lecithin, lignin, linoleic acids, quercetin and terpenes	Enhanced levels of estrogen, which can be beneficial for short-term memory	Keyvan et al. 2007
Terminalia bellirica **Combretaceae**	Southeast Asia	Gallic acid and ellagic acid	Inhibition of AChE	Nag and De, 2011
Evolvulus alsinoides **Convolulaceae**	Burkina Faso	Phenols, flavonoids and tannins	Inhibition of AChE	Mukherjee et al. 2007
Ipomoea asarifolia **Convolulaceae**	Southern India	Alphatocopherol, ascorbic acid and phenolics	Inhibition of AChE	Feitosa et al. 2011
Kalanchoe brasiliensis **Crassulaceae**	South America	Alphatocopherol, ascorbic acid and phenolics	Inhibition of AChE	Feitosa et al. 2011
Eureiandra balfourii **Cucurbitaceae**	Yemen	3,7-dimethylocatn-2-one, butyl 2-methylbutyrate, linalool, linalyl acetate, zingiberence and γ-cadinene	Inhibition of AChE	Bakthir et al. 2011
Juniperus phoenicea **Cupressaceae**	Morocco, Portugal and Italy	Phenolic compounds	Inhibition of AChE	Tavares et al. 2012
Juniperus turbinate **Cupressaceae**	Southern to tropical Africa	Phenolic compounds	Inhibition of AChE	Tavares et al. 2012
Rhododendron yedoense var. *poukhanense* **Ericaceae**	Republic of Korea	Gallic acid and quercetin	Inhibition of AChE	Lee et al. 2011

(*Continued*)

TABLE 9.1 (CONTINUED)
Potential Plants for AD Therapeutics

Plant Species	Country or Region of Origin	Bioactive Constituents	Mode of Action	References
Eucommia ulmoides **Eucommiaceae**	China	Epicatechin, catechin, n-octacosanoid acid, rutin and chlorogenic acid	Inhibition of AChE in a dose-dependent manner	Kwon et al. 2011
Alchornia laxiflora **Euphorbiaceae**	Central African Republic	Quercetin and rutin	Inhibition of AChE	Elufioye et al. 2010
Cephalocroton socotranus **Euphorbiaceae**	Yemen	Alkaloids	Inhibition of AChE	Bakthir et al. 2011
Jatropha curcas **Euphorbiaceae**	South and Central America	Alphatocopherol, ascorbic acid and phenolics	Inhibition of AChE	Feitosa et al. 2011
Jatropha gossypiifolia **Euphorbiaceae**	Mexico and tropical South America	Alphatocopherol, ascorbic acid and phenolics	Inhibition of AChE	Feitosa et al. 2011
Acacia nilotica **Fabaceae**	Africa	Terpenes, tannins, flavonoids and condensed tannins	Inhibition of AChE	Crowch and Okello 2009
Cassia obtusifolia **Fabaceae**	Argentina	Flavonoids	Inhibition of AChE	Carpinella et al. 2009
Chamaecrista mimosides **Fabaceae**	Ethiopia	Polyphenolic compounds	Inhibition of AChE	Adewusi et al. 2011
Genista tenera **Fabaceae**	Madeira Islands	Flavones and flavonoids	Inhibition of AChE	Rauter et al. 2009
Glycyrrhiza glabra **Fabaceae**	China and India	2', 4', 4-trihydroxychalcone	Prevents brain cell death	Bilge and Ilkay 2005
Peltophorum pterocarpum **Fabaceae**	Southeastern Asia	Phytol and sitosterol	Inhibition of AChE	Elufioye et al. 2010
Schotia brachypetala **Fabaceae**	South Africa	Polyphenolic compounds	Inhibition of AChE	Adewusi et al. 2011

(*Continued*)

TABLE 9.1 (CONTINUED)
Potential Plants for AD Therapeutics

Plant Species	Country or Region of Origin	Bioactive Constituents	Mode of Action	References
Senna alata **Fabaceae**	Tropical Africa	Alphatocopherol, ascorbic acid and phenolics	Inhibition of AChE	Feitosa et al. 2011
Spatholobus suberectus **Fabaceae**	China	Liquiritigenin	Inhibition of AChE in a dose-dependent manner	Lin et al. 2008
Trigonella foenum-graecum **Fabaceae**	Indian subcontinent	Alkaloids	Inhibition of AChE	Satheeskumar et al. 2010
Ginkgo biloba L. **Ginkgoaceae**	India	Terpenoids and flavonoid glycosides	Antiamyloid aggregation	Luo et al. 2002
Callophyllum inophyllurn **Guttiferaceae**	Vanuatu and South Pacific	–	Inhibition of AChE	Elufioye et al. 2010
Hypericum perforatum **Hypericaceae**	Europe and Asia	Flavonoids, flavonols, flavones, helenelin	Inhibition of AChE and BuChE	Carpinella et al. 2010
Illicium verum **Illiciaceae**	China and Vietnam	Anethole	Cholinesterase inhibition	Bhadra et al. 2011
Collinsonia canadensis **Lamiaceae**	North America	Carvacol and thymol	Acetylcholine breakdown prevention	Singhal et al. 2012
Cyclotrichium niveum **Lamiaceae**	Iran	Flavonoids	Inhibition of AChE and BuChE	Orhan et al. 2009
Hyssopus officinials **Lamiaceae**	Croatia, Hungary and France	Flavonoids, flavonols, flavones, helenelin	Inhibition of AChE and BuChE	Carpinella et al. 2010
Lavandula viridis **Lamiaceae**	Northern and Eastern Africa	Rosmarinic acid and luteolin-7-O-glucoside	*In vitro* and *in vivo* inhibition of AChE	Costa et al. 2011
Marrubium vulgare **Lamiaceae**	Europe, Northern Africa and Southwestern Asia	Diterpenes, phenylethanoids and flavonoidderivatives	Inhibition of AChE and BuChE	Orhan et al. 2010

(Continued)

TABLE 9.1 (CONTINUED)
Potential Plants for AD Therapeutics

Plant Species	Country or Region of Origin	Bioactive Constituents	Mode of Action	References
Melissa officinalis L. **Lamiaceae**	Western Asia	Terpenoids, geranial, neral and citronella	Inhibition of AChE	Ji et al. 2010
Origanum ehrenbergii **Lamiaceae**	Western and Southwestern Eurasia	Thymol and carvacrol	Cholinesterase inhibition	Loizzo et al. 2009
Origanum majorana **Lamiaceae**	Middle Eastern countries	Sabinene, 4-terpinel	Inhibition of AChE	Mossa and Nawwar 2011
Origanum syriacum **Lamiaceae**	West Asia	Thymol and carvacrol	Cholinesterase inhibition	Loizzo et al. 2009
Pycnostachys reticulata **Lamiaceae**	Central and South America	Amaryllidaceae alkaloids	Inhibition of AChE	Fawole et al. 2010
Rosmarinus officinalis **Lamiaceae**	England	Diterpenes quinones, apigenin, carcacrol, eugenol, oleanolic acid and ursolic acid	COX-2 inhibitors	Habtemariam 2016
Salvia chionantha **Lamiaceae**	Turkey	Germacrene D, spathulenol	Inhibition of AChE	Tel et al., 2010
Salvia fruticose **Lamiaceae**	Southern Italy and Canary Islands	Flavonoids	Inhibition of AChE	Senol et al. 2010
Salvia leriifolia **Lamiaceae**	Central and South America	Camphor, 1,8-cineole, camphene	Cholinesterase inhibition	Loizzo et al. 2009
Salvia officinalis **Lamiaceae**	Western and Southern Balkans	Carnosic and rosmarinic acid	Cholinesterase inhibition	Obulesu and Rao 2011
Bertholletia excels **Lecythidaceae**	Brazil	Lecithin	Enhanced concentration of acetylcholine	Chunhieng et al. 2008

(*Continued*)

TABLE 9.1 (CONTINUED)
Potential Plants for AD Therapeutics

Plant Species	Country or Region of Origin	Bioactive Constituents	Mode of Action	References
Huperzia serrata **Lycopodiaceae**	Hubei and China	Lycopodium alkaloids	Cholinesterase inhibition	Marston et al., 2002
Teucrium royleanum **Lamiaceae**	India	β-santalene and cis-α-bisabolene	Cholinesterase inhibition	Ahmad et al., 2007
Magnolia officinalis **Magnoliaceae**	South Korea	Magnolol	Inhibition of AChE	Lannert and Hoyer 1998
Tinospora cordifolia **Menispermaceae**	India	Magnoflorine, tinocordiside and syringin	Memory-enhancing property by immunostimulation and synthesis of acetylcholine	Lannert and Hoyer 1998
Stephania pierrei **Menispermaceae**	Cambodia	Alkaloids and phenolics	Cholinesterase inhibition	Tappayuthpijarn et al. 2011
Dorstenia gigas **Moraceae**	Indonesia and Southern Asia	Gancaonin Q and stipulin	Inhibition of AChE	Bakthir et al. 2011
Ficus religiosa **Moraceae**	India	Flavonoids and phenols	*In vitro* inhibition of AChE	Vinutha et al. 2007
Embelia ribes **Myristicaceae**	India	Vilangin	*In vitro* inhibition of AChE	Vinutha et al. 2007
Orchis mascula **Orchidaceae**	India	β-ocimene	*In vitro* inhibition of AChE	Ashraf et al. 2011
Paeonia lactiflora **Paeoniaceae**	Japan	Albiflorin and oxypaeoniflorin	Inhibition of AChE in a dose-dependent manner	Lin et al. 2008
Paeonia veitchii **Paeoniaceae**	Central China	Galloylpaeoniflorin, benzoic acid and palmitic acid	Inhibition of AChE in a dose-dependent manner	Lin et al. 2008

(*Continued*)

TABLE 9.1 (CONTINUED)
Potential Plants for AD Therapeutics

Plant Species	Country or Region of Origin	Bioactive Constituents	Mode of Action	References
Corydalis intermedia **Papaveraceae**	Northern Europe and Asia	1,8-cineole, α-terpineol and terpinen-4-ol	Inhibition of AChE in a dose-dependent manner	Adsersen et al. 2006
Corydalis solida ssp. Slivenensis **Papaveraceae**	North Europe and Asia	A-pinene and eugenol	Inhibition of AChE in a dose-dependent manner	Adsersen et al. 2006
Emblica officinalis **Phyllantaceae**	India	Gallic acid and ellagic acid	Inhibition of AChE in a dose-dependent manner	Nag and De 2011
Pinus halepensis **Pinaceae**	Mediterranean region	A and β pinene	Inhibition of AChE	Ustun et al. 2012
Pinus heldreichii subsp. *Leucodermis* **Pinaceae**	Balkans and Southern Italy	Terpinolene, linaly acetate and terpinen-4-ol	Inhibition of AChE	Bonesi et al. 2010
Pinus nigra subsp. *Nigra* **Pinaceae**	Southern Mediterranean Europe from Spain	Terpinolene, linaly acetate and terpinen-4-ol	Inhibition of AChE	Bonesi et al. 2010
Pinus nigra subsp. *Calabrica* **Pinaceae**	Southern Mediterranean Europe from Spain	Terpinolene, linaly acetate and terpinen-4-ol	Inhibition of AChE	Bonesi et al. 2010
Pinus pinaster **Pinaceae**	Mediterranean region	A and β pinene	Inhibition of AChE	Ustun et al. 2012
Piper nigrum **Piperaceae**	Southern India	Phenolic amides	Cholinesterase inhibition	Tappayuthpijarn et al. 2011
Cymbopogon jawarancusa **Poaceae**	Asia, Africa and Australia	Active principles of methanolic extracts	Inhibition of AChE	Ashraf et al. 2011
Cymbopogon schoenanthus **Poaceae**	Eastern and Western India	Limonene, β-phellandrene and α-terpineol	Inhibition of AChE	Khadri et al. 2008

(Continued)

TABLE 9.1 (CONTINUED)
Potential Plants for AD Therapeutics

Plant Species	Country or Region of Origin	Bioactive Constituents	Mode of Action	References
Fallopia multiflora **Polygonaceae**	China	Emodin and physcion	Inhibition of AChE in a dose-dependent manner	Lin et al. 2008
Rheum palmatum **Polygonaceae**	China	Flavan-3-ols, dianthrones and gallic acid	Inhibition of AChE in a dose-dependent manner	Lin et al. 2008
Ruprechita apetala **Polygonaceae**	Argentina	Polyphenolic compounds	Cholinesterase inhibition	Carpinella et al. 2009
Portulaca oleracea **Portulacaceae**	Northern Africa	Alkaloids	Inhibition of AChE	Yang et al. 2012
Rhamnus prinoides **Rhamnaceae**	Africa	Physcion, emodinanthrone and prinoidin	Inhibition of AChE	Crowch and Okello 2009
Leucosidea sericea **Rosaceae**	Southern Africa	Alkaloids, iridoids and phenolic compounds	Inhibition of AChE	Aremu et al. 2011
Galium odaoratum **Rubiaceae**	Europe, Spain and Ireland	Flavonoids, flavonols, flavones and helenelin	Inhibition of AChE and BuChE	Carpinella et al. 2010
Morinda lucida **Rubiaceae**	Borneo, New Guinea and Northern Australia	Sabinene and β-pinene	Inhibition of AChE	Elufioye et al. 2010
Ruta graveolens **Rutaceae**	Balkan Peninsula	Flavonoids, flavonols, flavones and helenelin	Inhibition of AChE and BuChE	Carpinella et al. 2010
Zanthoxylum coco **Rutaceae**	Central region of Argentina	Sabinene and β-pinene	Cholinesterase inhibition	Carpinella et al. 2009
Bacopa monnieri **Scrophulariaceace**	Wetlands of Southern India, Australia, Europe, Africa, Asia and North and South America.	Bacosides, brahmine, herpestine, apigenin, hersaponin, monnierasides cucurbutaciss and plantainoside B	Cholinesterase inhibition	Goswami et al. 2011

(Continued)

TABLE 9.1 (CONTINUED)
Potential Plants for AD Therapeutics

Plant Species	Country or Region of Origin	Bioactive Constituents	Mode of Action	References
Withania somnifera **Solanaceae**	India	Withanamides, withaferin-A and sitoindosides VII-X	Differential inhibition of AChE and enhanced M1-muscarinic receptor binding in brain	Schliebs et al. 1997
Witheringia coccolobodies **Solanaceae**	Neotropicales	Polyphenolic compounds	*In vitro* inhibition of AChE	Nino et al. 2006
Angelica archangelica L. **Umbelliferae**	Europe	Methoxsalen	*In vitro* inhibition of AChE	Carpinella et al. 2009
Centella asiatica L. **Umbelliferae**	India	Asiaticoside, oxyasiaticoside, centelloside, brahmoside, brahminoside, thankunoside, isothankunoside and sapogenins	Recuperates Aβ pathology, modulates adrenergic, dopaminergic and serotonergic systems	Kumar and Gupta 2003
Nardostachys jatamansi **Valerianaceae**	India and China	Oleanolic acid and nardosinone	*In vitro* inhibition of AChE	Vinutha et al. 2007
Curcuma longa L. **Zingiberaceae**	India	Curcumin, demethoxycurcumin and bisdemethoxycurcumin	Anti-amyloid properties	Mishra and Palanivelu 2008
Kaempfera parviflora **Zingiberaceae**	Thailand	5,7-dimethoxyflavone and 5-hydroxy-3,7,4'-trimethoxyflavone	Cholinesterase inhibition	Tappayuthpijarn et al. 2011

memory. By modulating the monoamines norepinephrine, dopamine and serotonin, *C. paniculatus* oil improves the learning and memory processes (Nalini et al. 1995). In addition, *C. paniculatus* seed oil elevates the cholinergic activity in rat brains by decreasing AChE activity and thereby improving memory performance (Da Rocha et al. 2011). Cognition-enhancement properties were revealed from the recuperation of spatial memory impairment through the release of muscarinic receptor blockages (George et al. 2010). The methanolic extracts from the inflorescences of *C. paniculatus* exhibited anti-inflammatory properties and can be utilized in AD treatment (Ahmad et al. 1994). *C. paniculatus* has been shown to posess antioxidant potential, cognition enhancement, concentration improvement and the ability sharpen memory (Bhanumathy et al. 2010).

9.3.3 *Centella asiatica*

C. asiatica has the potential to prevent cognitive deficits and used in the treatment of AD (Kumar and Gupta 2003). The extract from *C. asiatica* was proven to change Aβ pathology in the brains of mice (Dhanasekaran et al. 2009). The extract alleviates the oxidative stress response generated during neurodegeneration in AD (Subathra et al. 2005). This herb has the potential to increase memory, intelligence and longevity (Cervenka and Jahodar 2006). In the brain of PSAPP (APP/Sw x PS1M 146L) mice, the extracts of *C. asiatica* recuperated the Aβ pathology, and hence could play a potential role in the treatment and prevention of AD (Xu et al. 2008; Da Rocha et al. 2011). *C. asiatica* improved memory and learning by modulating adrenergic, dopaminergic and serotoneegic systems *in vivo* (Nalini et al. 1992).

9.3.4 *Curcuma longa*

Curcumin obtained from *C. longa* is an excellent anti-inflammatory agent, and chronic inflammation is one of the factors in the pathogenesis in AD. Thus, curcumin serves as an agent in the treatment of AD. Aβ levels were decreased in mice with AD when treated with curcumin. Curcumin has antioxidant and non-steroidal anti-inflammatory properties and oxidative stress and inflammation are primary pathogenesis factors in AD, hence curcumin eases AD symptoms. Aging increases the pathogenesis of AD, as there is loss of superoxide dismutase and sodium-potassium ATPase activities. Curcumin revealed to increase these levels, even with aging (Bala et al. 2006). Curcumin reduced the senile plaques and effectively cleared the Aβ by increasing the phagocytosis of Aβ (Yang et al. 2005). Besides reducing plaque deposition, curcumin reversed the amyloid pathology in mice with AD (Yang et al. 2005). The rhizome aqueous extract was shown to possess antidepressant activity, which is essential for the management of AD (Yu et al. 2002). All these properties of curcumin have been considered for AD clinical trials (Kelley and Knopman 2008).

9.3.5 *Euphorbia royleana* Boiss

Shilajit is a substance found in the latex of *E. royleana* Boiss, which has immunomodulatory activities (Pandey and Tewari 1975). Shilajit interacts with the cholinergic signal pathways and therefore can act as a potential therapeutic agent in AD treatment. Shilajit is adaptogenic, immunomodulatory and reduces stress (Winston et al. 2007). These unique properties make it an ideal candidate for the treatment of AD. Shilajit in combination with *Withania somnifera* influences cholinergic signal transduction cascade in the cortical and basal forebrain (Schliebs et al. 1997). Tau protein buildup is a hallmark of AD. Fulvic acid, the primary constituent of shilajit, strongly inhibited tau protein aggregation and is one of the few polyphenolic compounds that exhibits this property (Guzmán-Martinez et al. 2013). Fulvic acid is not only capable of inhibiting tau filament formation, but it also breaks apart and untangles the tau filaments (Cornejo et al. 2011). In addition, exposure to fulvic acid resulted in the increased neurite outgrowth in the neural cell cultures. This special feature revealed from *E. royleana* Boiss may open up new possibilities towards AD treatment.

9.3.6 *Galanthus caucasicus*

G. caucasicus produces an alkaloid named galantamine in its bulbs and flowers, which acts as a competitive AChE inhibitor and mitigates the symptoms of AD (Theodorou 2006). Patients in the early stages of AD benefitted from galantamine treatment and exhibited improved cognition, as revealed from the randomized trials (Raskind et al. 2000). Galantamine possesses long-acting, selective, competitive and reversible AChE inhibitory activity for mild to moderate AD (Heinrich and Teoh 2004). The inhibition of AChE increases ACh levels in the synaptic cleft. In addition, it has been shown to slow down the process of neurological degeneration in AD. Galantamine offers neuroprotection (Egea et al. 2012), inhibits Aβ aggregation and therefore cytotoxicity (Matharu et al. 2009), scavenges ROS in neurons protecting them from oxidative damage (Tsvetkova et al. 2013) and promotes neurogenesis (Kita et al. 2014). Cholinergic and glutamatergic dysfunction are the symptoms of AD where galantamine, in combination with memantine, alleviates the impairments of both systems. Thus *G. caucasicus* may serve as a therapeutic agent in AD through neurogenesis and neuroprotection.

9.3.7 *Ginkgo biloba*

G. biloba has the potential to diminish memory loss, improve brain activity and retard the degeneration of neurons in AD (DeKosky et al. 2008). The extract of this plant contains ginkgolides, which possess antioxidants, cholinergic properties and neuroprotective properties and are used to alleviate the detrimental effects of AD (Perry et al. 1999). EGB761 is an extract from *G. biloba* that prevented Aβ toxicity in neurons and proved to be efficacious in AD treatment (Kanowski and Hoerr 2003). In both AD and control patients, the extracts from *G. biloba* enhanced cognitive functions (Rigney et al. 1999). In addition, the extract prevented Aβ and nitric oxide–induced toxicity and reduced apoptosis in neuronal cell culture (Bastianetto et al. 2000; Schindowski et al. 2001). In addition to its antioxidant activities, the extract facilitated a blood supply to the brain, thus enriching cognitive performance (Loffler et al. 2001). In aging and dementia, the efficient recuperation of cognitive deficits with *G. biloba* extracts was evidenced through many studies. Neurotoxicity in AD was mediated by Ca^{2+} dyshomeostasis (Berrocal et al. 2009). Ginkgolide B was observed to prevent Aβ-triggered Ca^{2+} influx through N-methyl-D-aspartic acid (NMDA) receptors. The hyperphosphorylation of tau protein is a major event in the AD pathogenesis, where EGb761 was found to dephosphorylate the hyperphosphorylated tau protein in the hippocampus and cortex of mice (Watanabe et al. 2001). The nerve growth factor (NGF) level was modulated in AD. Bilobalide, and a terpenoid constituent of EGb761 upregulated NGF, was reported along with the glial-derived neurotrophic factor and vascular endothelial growth factor in rat cortical astrocytes (Zheng et al. 2002).

9.3.8 *Glycyrrhiza glabra*

G. glabra roots and rhizomes exhibit antimicrobial, anti-inflammatory, antiulcer, anxiolytic and expectorant activities, hence they are used widely as medicine. A natural product from *G. glabra*, known as 2, 2', 4-trichydroxychalcone, prevents the generation of Aβ and its plaque formation, thus improving memory deficits (Dhingra et al. 2004) by inhibiting β-site amyloid precursor protein cleaving enzyme 1 (Zhu et al. 2010). Flavonoids, such as licorice, chalcone and licorice isoflavones from *G. glabra*, exhibited strong antioxidant activity scavenging free radicals, which may help in the prevention of AD. Glyderinine, glabridin and lichochalocone A from *G. glabra* exhibited *in vivo* anti-inflammatory effects (Azimov et al. 1998; Furuhashi et al. 2005). The improvement in learning and memory was evident in mice upon the administration of licorice extract (Parle et al. 2004). The AChE inhibitory activity of *G. glabra* might assist the improvement of memory through cholinergic pathways (Dhingra et al. 2006). The development of inflammatory lesions is part of the pathology of AD. *G. glabra* has anti-inflammatory effects that aid in the prevention of such lesions.

9.3.9 *Huperzia serrata*

H. serrata is a Chinese herb which contains an alkaloid Huperzine A, which is a natural cholinesterase inhibitor (Ashani et al. 1992). This alkaloid is commercially available as a food supplement and as a drug for memory loss and mental disabilities. Huperzine A demonstrates AChE inhibitor activity, antioxidant properties, memory improvement and neuroprotection (Kozikowski and Tuckmantel 1999). With the exhibition of these properties, Huperzine A significantly improved the symptoms of AD in a double-blind clinical trial of over 450 patients in China (Zhang et al. 2002). The neuroprotective effect of *H. serrata* was evidenced from protection against glutamate toxicity through antagonism at NMDA receptors (Ved et al. 1997; Wang et al. 1999). Huperzine A prevents the deleterious effect of Aβ and recuperates cholinergic and monoaminergic dysfunction in rats (Liang et al. 2008). The production of amyloidogenic fragments was suppressed by increased α-secretase activity with huperzine A treatment (Peng et al. 2006). This might diminish Aβ toxicity and accumulation of fibrillary amyloids in the brains of AD patients. *H. serrata* is a high potential AChE inhibitor with strong memory improvement capacity, which modulates Aβ metabolism and neuroprotective activity, and receives much attention as a promising candidate for AD treatment.

9.3.10 *Melissa officinalis*

M. officinalis exhibits ACh receptor activity in the CNS with both nicotinic and muscarinic interacting abilities (Perry et al. 1999). Reports showed its inherent capability to modulate mood, improve cognition and minimize agitation in AD patients (Kennedy et al. 2002). It has been widely used as a memory-improving medicine in the European system of traditional medicine (Howes et al. 2003). Ethanolic leaf extracts were reported to bind muscarinic receptors (Wake et al. 2000), whereas methanolic leaf extracts exhibited mood and attention-improving properties (Kennedy et al. 2002). Hydroalcoholic extracts are effective in cognition deficits in mild to moderate AD patients (Akhondzadeh et al. 2003a). In addition, essential oils and extracts had AChE inhibition effects and antioxidant properties, respectively (Ferreira et al. 2003). The cognition-enhancing potential of the extract is attributable to cholinergic binding properties. Aromatherapy using essential oil from *M. officinalis* reported to decrease agitation in severe dementia patients (Ballard et al. 2002).

9.3.11 *Panax ginseng*

The neuroprotective potential of ginseng was validated in AD experimental models. *P. ginseng* has shown to be effective in the improvement of cognitive performance in AD patients (Lee et al. 2008). Ginseng was found to regenerate axons and synapses (Tohda et al. 2005) and increase hippocampal synaptic densities (Mook-Jung et al. 2001). Ginsenosides, which are steroid-like compounds, are the active components in ginseng (Chen et al. 2006). Ginsenosides have the potential to minimize Aβ, reduce the Aβ inhibition of hippocampal cholinergic transmission and hinder Aβ–induced memory loss (Wang et al. 2006). Ginseng was shown to improve the cognitive and psychomotor performance by increasing brain cholinergic function and restore the damaged networks. In healthy humans, ginseng enhanced cognitive performance (Kennedy et al. 2001). Furthermore, *P. ginseng* increased cerebral blood flow and scavenged ROS due to its antioxidant potential (Kitts et al. 2000; Kim et al. 2002). The nonsaponin fraction enhanced memory and learning in aged rats, whereas polyacetylenic alcohols promoted neuritogenesis (Yamazaki et al. 2001).

9.3.12 *Salvia officinalis*

S. officinalis offered benefits in cognition and exhibited cholinesterase inhibitor effects in patients with AD (Akhondzadeh et al. 2003b). Both AChE and butyrylcholinesterase inhibitory activities with reduced anxiety and increased alertness were exhibited by ethanolic leaf extracts (Perry et al. 1996; Kennedy et al. 2006). In addition, leaf alcoholic extracts offered protection against Aβ–induced neurotoxicity (Iuvone et al. 2006). Memory performance was improved in an elderly cohort study with ethanolic extract of *S. officinalis* (Scholey et al. 2008). *S. officinalis* significantly improved dementia rating scores along with the AD clinical assessment cognitive subscale in a study with 30 AD patients at 16 weeks (Akhondzadeh et al. 2003b). Rosmarinic acid inhibited Aβ fibril formation and destabilized the fibrils. In addition, it prevented Aβ-induced tau protein hyperphosphorylation (Iuvone et al. 2006).

9.3.13 *Withania somnifera*

W. somnifera is known as a nerve tonic and attenuates neuritic atrophy, which leads to neurodegenerative disorders. Ashwagandha helps the body to alleviate stress and is used immensely as an adaptogen. Adaptogens from other sources are stimulating, but ashwagandha invokes a calming effect not only in AD patients, but also in mammals (Auddy et al. 2008). Of all the steroid compounds in ashwagandha, withanamides have a great potential to scavenge ROS that are produced at the initiation and progression of AD, and to prevent amyloid plaques that trigger cell death in neurons (Jayaprakasam et al. 2010). Cognition and memory improvement was evidenced from the increased cholinergic activities in rats, thus increasing ACh content and eventually decreasing AChE activity (Schliebs et al. 1997). The semipurified root extract of ashwagandha recuperated behavioral defects, lessened plaque loads and prevented accumulation of Aβ in mice with AD (Sehgal et al. 2012). The roots are the source for glycowithanolides withaferin-A and sitoindosides VII-X. These compounds recuperate the cognitive defects in the AD model and attenuate the dendritic and synoptic losses in mice (Kuboyama et al. 2006). In addition, these compounds enhance the outgrowth of axons and dendrites and induce synaptogenesis (Tohda et al. 2000). The oral administration of these metabolites from *W. somnifera* reduces the symptoms of AD. *W. somnifera* is an excellent anti-AD plant, and further clinical trials will make it an effective therapeutic against AD.

9.4 CONCLUSION

The benefits offered by plants as AD therapeutics are immense: they are inexpensive, are taken as food supplements and contain fewer side effects. The dietary supplementation of medicinal plants' bioactive components in daily life offers great promise in the prevention of AD. Such bioactive components possess antioxidant properties, memory enhancement, toxicity prevention and cognitive function improvement. These bioactive principles from medicinal plants reduce neurodegeneration through interaction with neuronal signaling pathways. This routine supplementation prevents the onset of many neurodegenerative diseases in addition to cognitive improvement. Multiple factors are responsible for the pathogenesis of AD; therefore, one therapeutic approach will end up incompetent. The plants that offer multicomponents with a wide range of therapeutic targets is appealing and promising, and should receive greater attention for AD treatment. Future research should head in the direction of including two or more plant constituents that could serve as multipotent agents in AD models and patients. The efficacy of these plants and their active principles necessitate a rigorous assessment before being utilized for AD therapeutics.

With more investigation, the plants and their active principles will offer an effective therapeutic regimen for AD in the near future.

ACKNOWLEDGEMENTS

Mr. Azhwar Raghunath acknowledges the UGC-BSR fellowship (UGC-BSR No. F7-25/2007), funded by UGC-BSR, New Delhi, India. The authors thank the Department of Science and Technology, Science and Engineering Research Board (EMR/2014/000600) and Empowerment and Equity Opportunities for Excellence in Science (SB/EMEQ-246/2014).

REFERENCES

Adersen, A., Gauguin, B., and Gudiksen, L. et al. (2006). Screening of plants used in Danish folk medicine to treat memory dysfunction for acetylcholinesterase inhibitory activity. *J. Ethnopharmacol.* 104, 418–22.

Adewusi, E. A., Moodley, N., and Steenkamp, V. (2011). Antioxidant and acetylcholinesterase inhibitory activity of selected South African medicinal plants. S *Afr. J. Bot.* 77, 638–44.

Ahmad, F., Khan, R. A., and Rasheed, S. (1994). Preliminary screening of methanolic extracts of *Celastrus paniculatus* and *Tecomella undulata* for analgesics and anti-inflammatory activities. *J. Ethnopharmcol.* 42, 193–8.

Ahmad, B., Mukarram Shah, S. M., Khan, H. et al. (2007). Enzyme inhibition activities of *Teucrium royleanum. J. Enzym. Inhib. Med.* 22, 730–2.

Akhondzadeh, S., Noroozian, M., Mohammadi, M. et al. (2003a). Melissa officinalis extract in the treatment of patients with mild to moderate Alzheimer's disease: A double blind, randomized, placebo-controlled trial. *J. Neurol. Neurosurg. Psychiatry.* 74, 863–66.

Akhondzadeh, S., Noroozian, M., Mohammadi, M. et al. (2003b). Salvia officinalis extract in the treatment of patients with mild to moderate Alzheimer's disease: A double-blind, randomized, and placebo-controlled trial. *J. Clin. Pharm. Ther.* 28, 53–9.

Akkol, E. K., Orhan, I. E., and Yesilada, E. (2012). Acetylcholinesterase and antioxidant effects of the ethanol extract, ethanol fractions and isolated flavonoids from *Cistus laurifolius*. L. leaves. *Food Chem.* 131, 626–31.

Amaducci, L. A., Rocca, W. A., and Schoenberg, B. S. (1986). Origin of the Distinction between Alzheimer's Disease and Senile Dementia: How History Can Clarify Nosology. *Neurol.* 36, 1497–9.

Aremu, A. O., Amoo, S. O., Ndhlala, A. R. et al. (2011). Antioxidant activity, acetylcholinesterase inhibiton, iridoid content and mutagenic evalution of *Leucosidea sericea. Food Chem. Toxicol.* 49, 1122–8.

Ashani, Y., Peggins III, J. O., Doctor, B. P. (1992). Mechanism of inhibition of cholinesterases by huperzine A. *Biochem. Biophys. Res. Commun.* 184, 719–26.

Ashraf, M., Ahmad, K., Ahmad, I. et al. (2011). Acetylcholinesterase and NADH oxidase inhibitory activity of some medicinal plants. *J. Med. Plant. Res.* 5, 2086–9.

Auddy, B., Hazra, J., Mitra, A. A. (2008) Standardized *Withania somnifera* extract significantly reduces stress-related parameters in chronically stressed humans: A double-blind, randomized, placebo-controlled study. *J. Am. Nutra. Assoc.* 11, 50–6.

Azimov, M. M., Zakirov, U. B., Radzhapova, S. H. D. et al. (1998). Pharmacological study of the anti-inflammatory agent glyderinine. *Farmakol Toksikol.* 51, 90–3.

Bakthir, H., Ali, N. A. A., Arnold, N. et al. (2011). Acetylcholenesterase activity of endemic plant extracts from Soqotra. *Afr. J. Tradit. Complemnt. Altern. Med.* 8, 296–9.

Bala, K., Tripathy, B. C., and Sharma, D. (2006). Neuroprotective and anti-aging effects of curcumin in aged rat brain regions. *Biogerontology.* 7, 81–9.

Ballard, C. G., O'Brien, J. T., Reichelt, K. et al. (2002). Aromatherapy as a safe and effective treatment for the management of agitation in severe dementia: The results of a double-blind, placebo-controlled trial with Melissa. *J. Clin. Psychiatry* 63, 553–8.

Bartus, R. T, Dean III, R. L., Beer, B. et al. (1982). The cholinergic hypothesis of geriatric memory dysfunction. *Science* 217, 408–14.

Bastianetto, S., Ramassamy, C., Dore, S. et al. (2000). The *Ginkgo biloba* extract (EGb 761) protects hippocampal neurons against cell death induced by β-amyloid. *Eur. J. Neurosci.* 12, 1882–90.

Berrios, G. E. (1990). Alzheimer's Disease: A Conceptual History. *Int. J. Ger. Psychiatry* 5, 355–65.

Berrocal, M., Marcos, D., Sepúlveda, M. R. et al. (2009). Altered Ca2+ dependence of synaptosomal plasma membrane Ca2+-ATPase in human brain affected by Alzheimer's disease. *FASEB J.* 23, 1826–34.

Beyreuther, K. and Masters, C. L. (1997). Alzheimer's disease. The ins and outs of amyloid-beta. *Nature* 389, 677–8.

Bhadra, S., Mukherjee, P. K., Kumar, N. S. (2011). Anticholinesterase activity of standardized extract of *Illicium verum* Hook f fruits. *Fitoterapia* 82, 342–46.

Bhalla, M., Raghuvanshi, P., Sharma, S. C. et al. (2008). Management of Neuritis and Parkinsonism by *Bacopa monniera*. Proceedings: Relevance of Modern Methods of Pharmacological Studies to Traditional Medicine. 77.

Bhanumathy, M., Harish, M. S., Shivaprasad, H. N. et al. (2010). Nootropic activity of *Celastrus paniculatus* seed. *Pharm. Biol.* 48, 324–37.

Bhattacharya, S. K., Bhattacharya, A., Kumar, A. et al. (2000). Antioxidant activity of *Bacopa monniera* in rat frontal cortex, striatum and hippocampus. *Phytother. Res.* 14, 174–9.

Bilge, S. and Ilkay, O. (2005). Discovery of drug candidates from some Turkish plants and conservation of biodiversity. *Pure Appl. Chem.* 77, 53–64.

Birge, S. J. (1997). The role of estrogen in the treatment of Alzheimer's disease. *Neurol.* 48, S36–41.

Blennow, K., de Leon, M. J., and Zetterberg, H. (2006). Alzheimer's disease. *Lacet.* 368, 387–403.

Bonesi, M., Menichini, F., Tundis, R. et al. (2010). Acetylcholinesterase and butyrylcholinesterase inhibitory activity of *Pinus* species essential oils and their constituents. *J. Enzym. Inhib. Med. Ch.* 25, 622–8.

Bores, G. M., Huger, F. P., Petko, W. et al. (1996). Pharmacological evaluation of novel Alzheimer's disease therapeutics: Acetylcholinesterase inhibitors related to galanthamine. *J. Pharmacol. Exp. Ther.* 277, 728–38.

Cahlikova, L., Benesova, N., Macakova, K. et al. (2011a). GC/MS analysis of three amaryllidaceae species and their cholinesterase activity. *Nat. Prod. Commun.* 6, 1255–8.

Cahlikova, L., Valterova, I., Macakova, K. et al. (2011b). Analysis of Amaryllidaceaee alkaloids from *Zephyranthes grandiflora* by GC/S and their cholinesterase activity. *Rev. Bras. Farmacogn.* 21, 575–80.

Carpinella, M. C., Andrione, D. G., Ruiz, G. et al. (2009). Screening for Acetylcholinesterase inhibitory activity in plant extracts from Argentina. *Phytother. Res.* 24, 259–63.

Carpinella, M. C., Andrione, D. G., Ruiz, G. et al. (2010). Screening of traditional European herbal medicines for acetylcholinesterase and butyrylcholinesterase inhibitory activity. *Phytother. Res.* 24, 259–63.

Castro, A., Conde, S., Rodriquez-Franco, M. I. et al. (2002). Non-cholinergic pharmacotherapy approaches to the future treatment of Alzheimer's disease. *Mini Rev. Med. Chem.* 2, 37–50.

Cervenka, F. and Jahodar, L. (2006). Plant metabolites as nootropics and cognitives. *Ceska Slov. Farm.* 55, 219–29.

Chen, F., Eckman, E. A., and Eckman, C. B. (2006). Reductions in levels of the Alzheimer's amyloid beta peptide after oral administration of ginsenosides. *FASEB J.* 20, 1269–71.

Chunhieng, T., Hafidi, A., and Pioch, D. et al. (2008). Detailed study of Brazil nut (*Bertholletia excels*) oil micro-compounds: Phospholipids, tocopherols and sterols. *J. Braz. Chem. Soc.* 19, 1374–80.

Cornejo, A., Jiménez, J. M., Caballero, L. et al. (2011). Fulvic acid inhibits aggregation and promotes disassembly of tau fibrils associated with Alzheimer's disease. *J. Alzheimer's Dis.* 27, 143–53.

Costa, P., Goncalves, S., Andrade, P. B. et al. (2011). Inhibitory effect of *Lavandula viridis* on Fe^{2+} induced lipid peroxidation, antioxidant and anti-cholinesterase properties. *Food Chem.* 126, 1779–86.

Crowch, M. C. and Okello, J. E. (2009). Kinetics of Acetylcholinesterase inhibitory activities by aquoes extratcs of *Acacia nilotica* and *Rhamnus prinoides*. *Afr. J. Pharm. Pharmacol.* 3, 469–75.

Cummings, J. L. (2000). Cholinesterase inhibitors: A new class of psychotropic compounds. *Am. J. Psychiatr.* 157, 4–15.

Da Rocha, M. D., Viegas, F. P., Campos, H. C. et al. (2011). The role of natural products in the discovery of new drug candidates for the treatment of neurodegenerative disorders II: Alzheimer's disease. *CNS Neurol. Disord. Drug Targets* 10, 251–70.

DeKosky, S. T., Williamson, J. D., Fitzpatrick, A. L. et al. (2008). *Ginkgo biloba* for prevention of dementia: A randomized controlled trial. *JAMA* 300, 2253–62.

Dézsi, L., Kis-Varga, I., Nagy, J. et al. (2002) Neuroprotective effects of vinpocetine *in vivo* and *in vitro*. Apovincaminic acid derivatives as potential therapeutic tools in ischemic stroke. Acta Pharma Hung. 72, 84–91.

Dhanasekaran, M., Holcomb Leigh A., Hitt Angie, R. et al. (2009). *Centella asiatica* Extract Selectively Decreases, Amyloid β Levels in Hippocampus of Alzheimer's disease Animal Model. *Phytother. Res.* 23, 14–19.

Dhingra, D., Parle, M., and Kulkarni, S. K. (2004). Memory enhancing activity of *Glycyrrhiza glabra* in mice. *J. Ethnopharmcol.* 91, 361–65.

Dhingra, D., Parle, M., Kulkarni, S. K. (2006). Comparative brain cholinesterase inhibiting activity of *Glycyrrhiza glabra*, *Myristica fragrans*, ascorbic acid and metrifonate in mice. *J. Med. Food* 9, 281–3.

Egea, J., Martin-de-Saavedra, M. D., Parada, E. et al. (2012). Galantamine elicits neuroprotection by inhibiting iNOS, NADPH oxidase and ROS in hippocampal slices stressed with anoxal/reoxygenation. *Neuropharmacol.* 62, 1082–90.

Elufioye, T. O., Obuotor, E. M., and Sennuga, A. T. (2010). Acetylcholinesterase and butyrylcholinesterase inhibitry activity of some selected Nigerian medicinal plants. *Rev. Bras. Farmacogn.* 20, 472–7.

Fawole, O. A., Amoo, S. O., Ndhala, A. R. et al. (2010). Anti-inflammatory, anticholinesterase, antioxidant and phytochemical properties of medicinal plants used for pain related ailments in South Africa. *J. Ethnopharmacol.* 127, 235–41.

Feitosa, C. M., Freitas, R. M., and Luz, N. N. N. et al. (2011). Acetylcholinesterase inhibition by some promising Brazilian medicinal plants. *Braz. J. Biol.* 71, 783–9.

Ferreira, A. R., Furstenau, L., Blanco, C. et al. (2003). Role of hippocampal M1 and M4 muscarinic receptor subtypes in memory consolidation in the rat. *Pharmacol. Biochem. Behav.* 74, 411–5.

Fu, L. M. and Li, J. T. (2009). A systematic review of single Chinese herbs for Alzheimer's disease treatment. *Evid. Based Complement Alternat. Med.*, 640284.

Furuhashi, I., Iwata, S., Shibata, S. et al. (2005). Inhibition by licochalcone A, a novel flavonoid isolated from liquorice root, of IL-1beta-induced PGE2 production in human skin fibroblasts. *J. Pharm. Pharmacol.* 57, 1661–6.

George, L., Kumar, B. P., Rao, S. N. et al. (2010). Cognitive enhancement and Neuroprotective effect of *Celastrus paniculatus* Wild. seed oil (Jyothismati oil) on male Wistar rats. *J. Pharmaceut. Sci. Tech.* 2, 130–8.

Goldman, P. (2001). Herbal medicines today and the roots of modern pharmacology. *Ann. Intern Med.* 135, 594–600.

Gomez-Isla, T., Hollister, R., West, H. et al. (1997). Neuronal loss correlates with but exceeds neurofibrillary tangles in Alzheimer's disease. *Ann. Neurol.* 41, 17–24.

Goswami, S., Saoji, A., Kumar, N. et al. (2011). Effect of *Bacopa monnieri* on cognitive functions in Alzheimer's disease patients. *Int. J. Collab. Ras. Intern. Med. Public Health* 3, 285–293.

Guillozet, A. L., Smiley, J. F., Mash, D. C. et al. (1997). Butyrylcholinesterase in the life cycle of amyloid plaques. *Ann. Neurol.* 42, 909–18.

Guzmán-Martinez, L., Farías, G. A., and Maccioni, R. B. (2013). Tau oligomers as potential targets for Alzheimer's diagnosis and novel drugs. *Front. Neurol.* 4, 167.

Habtemariam, S. (2011). The therapeutic potential of *Berberis darwinii* stem bark. quantification of berberine and *in vitro* evidence for Alzheimer's disease therapy. *Nat. Prod. Commun.* 6, 1089–90.

Habtemarim, S. (2016). The therapeutic potential of Rosemary (*Rosmarinus officinalis*) diterpenes for Alzheimer's disease. *Evid. Based Complement Alternat. Med.*, 2680409.

Hardy, J. and Allsop, D. (1991). Amyloid deposition as the central events in the aetiology of Alzheimer's disease. *Trends Pharmacol. Sci.* 12, 383–88.

Heinrich, M. and Lee, Teoh H. (2004). Galanthamine from snowdrop – The development of a modern drug against Alzheimer's disease from local Caucasian knowledge. *J. Ethnopharmacol.* 92, 147–62.

Holcomb, L. A., Dhanasekaran, M., Hitt, A. R. et al. (2006). *Bacopa monniera* extract reduces amyloid levels in PSAPP mice. *J. Alzheimer's Dis.* 9, 243–51.

Howes, M. J., Perry, N. S., and Houghton, P. J. (2003). Plants with traditional uses and activities, relevant to the management of Alzheimer's disease and other cognitive disorders. *Phytother. Res.* 17, 1–18.

Iuvone, T., Filippis, D. D., Esposito, G. et al. (2006). The spice sage and its active ingredient rosmarinic acid protect PC12 cells from amyloid-B peptide-induced neurotoxicity. *J. Pharmacol. Exp. Ther.* 317, 1143–9.

Jayaprakasam, B., Padmanabhan, K., and Nair, M. G. (2010). Withanamides in *Withania somnifera* fruit protect PC-12 cells from beta-amyloid responsible for Alzheimer's disease. *Phytother. Res.* 24, 859–63.

Ji, Y. A., Suna, K., Sung, E. J. et al. (2010). Effect of licorice (Glycyrrhiza urelensis fisch) on amyloid-beta-induced neurotoxicity in PC12 cells. Food Sci. Biotechnol., 19, 1391–5.

Joshi, H. and Parle, M. (2006) Brahmi rasayana improves learning and memory in mice. *Evid. Based Complement Alternat. Med.* 3, 79–85.

Kanowski, S. and Hoerr, R. (2003). Ginkgo biloba extract EGb 761 in dementia: Intent-to-treat analyses of a 24-week, multi-center, double-blind, placebo-controlled, randomized trial. *Pharmacopsychiatry* 36, 297–303.

Kawas, C., Resnick, S., Morrison, A. et al. (1997). A prospective study of estrogen replacement therapy and the risk of developing Alzheimer's disease: The Baltimore Longitudinal Study of Aging. *Neurol.* 48, 1517–21.

Kelley, B. J. and Knopman, D. S. (2008). Alternative medicine and Alzheimer's disease. *Neurologist* 14, 299–306.

Kennard, M. L., Feldman, H., Yamada, T. et al. (1996). Serum levels of the iron binding protein p^{97} are elevated in Alzheimer's disease. *Nature Med.* 2, 1230–5.

Kennedy, D. O., Pace, S., Haskell, C. et al. (2006). Effects of cholinesterase inhibiting Sage (*Salvia officinalis*) on mood, anxiety and performance on a psychological stress or battery. *Neuropsychopharmacology* 31, 845–52.

Kennedy, D. O., Scholey, A. B., Tildesley, N. T. et al. (2002). Modulation of mood and cognitive performance following acute administration of *Melissa officinalis* (lemon balm). *Pharmacol. Biochem. Behav.* 72, 953–64.

Kennedy, D. O. and Wightman, E. L. (2011). Herbal extracts and phytochemicals: Plant secondary metabolites and the enhancement of human brain function. *Adv. Nutr.* 2, 32–50.

Kennedy, D. O., Scholeya, A. B., and Wesnes, K. A. (2001). Dose-dependent changes in cognitive performance and mood following acute administration of ginseng to healthy young volunteers. *Nutr. Neurosci.* 4, 295–310.

Keyvan, D., Damien, D. H. J., Heikki, V. et al. (2007). Plants as potential sources foe drug development against Alzheimer's disease. *Int. J. BioMed. Pharm. Sci.* 1, 83–104.

Khafri, A., Serralheiro, M. L. M., Nogueira, J. M. F. et al. (2008). Antioxidant and antiacetylcholinesterase activites of essential olis from *Cymbopogon schoenanthus* L. Spreng (2008) Determination of chemical composition by GC-mass spectrometry and 13C NMR. *Food Chem.* 109, 630–7.

Kim, C. S., Park, J. B., Kim, K. J. et al. (2002). Effect of Korean red ginseng on cerebral blood flow and superoxide production. *Acta. Pharmacol. Sin.* 23, 1152–6.

Kita, Y., Ago, Y., Higashino, K. et al. (2014). Galantamine promotes adult hippocampal neurogenesis via M1 muscarinic and α7 nicotinic receptors in mice. *Int. J. Neuropsychopharmacol.* 17, 1957–68.

Kitts, D. D., Wijewickreme, A. N., and Hu, C. (2000). Antioxidant properties of a North American ginseng extract. *Mol. Cell. Biochem.* 203, 1–10.

Kozikowski, A. P. and Tuckmantel, W. (1999). Chemistry, Pharmacology, and Clinical Efficacy of the Chinese Nootropic Agent Huperzine A. *Acc. Chem Res.* 32, 641–50.

Kuboyama, A. T., Tohda, C., Komatsu, K. et al. (2006). Withanoside IV and its active metabolite, sominone, attenuate Abeta (25-35)-induced neurodegeneration. *Eur. J. Neurosci.* 23, 1417–26.

Kumar, M. H. V. and Gupta, Y. K. (2003). Effect of *Centella asiatica* on cognition and oxidative stress in an intracerebroventricular streptozotocin model of Alzheimer's disease in rats. *Clin. Exp. Pharmacol. Physiol.* 30, 336–42.

Kumar, V. (2006). Potential medicinal plants for CNS disorders: An overview. *Phytother. Res.* 20, 1023–35.

Kwon, S. H., Lee, H. K., Kim, J. A. et al. (2011). Neuroprotective effects of *Eucommia ulmoides* oliv bark on amyloid beta (25-35)–induced learning and memory impairments in mice. *Neuroscience Letters* 487, 123–7.

Lannert, H. and Hoyer, S. (1998). Intrecerobroventricular administration of steptozotocin causes long-term diminutions in learning and memory abilities and in cerebral energy metabolism in adult rats. *Behav. Neurosci.* 112, 1199–208.

Law, A., Gauthier, D., and Quirion, R. (2001). Say NO to Alzheimer's disease: The putative links between nitric oxide and dementia of the Alzheimer's type. *Brain Res. Brain Res. Rev.* 35, 73–96.

Lee, S. H., Sanchti, S. A., Bafna, M. R. et al. (2011). Acetylcholinesterase inhibitory and anti oxidant properties of *Rhododendron yedoense* var. *poukhanense* bark. *J. Med. Plants Res.* 5, 248–54.

Lee, S. T., Chu, K., and Sim, J. Y. (2008). Panax ginseng enhances cognitive performance in Alzheimer's disease. *Alzheimer's Dis. Assoc. Disord.* 22, 222–6.

Liang, Y. Q., Huang, X. T., Tang, X. C. (2008). Huperzine A reverses cholinergic and monoaminergic dysfunction induced by bilateral nucleus basalis magnocellularis injection of beta-amyloid peptide (1-40) in rats. *Cell. Mol. Neurobiol.* 28, 87–101.

Lima, J. A., Costa, R. S., Epifanio, R. A. et al. (2009). Geissospermum vellosi stembark. Anticholinesterase activity and improvement of scopolamine-induced memory deficits. *Pharmocol. Biochem. Be.* 92, 508–13.

Limpeanchob, N., Jaipan, S., Rattana Karuna, S. et al. (2008). Neuroprotective effect of *Bacopa monnieri* on beta-amyloid-induced cell death in primary cortical culture. *J. Ethnopharmacol.* 120, 112–7.

Lin, H. Q., Ho, M. T., Lau, L. S. et al. (2008). Anti-acetylcholinesterase activities of traditional Chinese medicine for treating Alzheimer's disease. *Chem. Biol. Interact.* 175, 352–54.

Liu, R. T., Wang, A., To, E. et al. (2014). Vinpocetine inhibits amyloid-beta-induced activation of NF-B, NLRP3 inflammasome and cytokine production in retinal pigment epithelial cells. *Exp. Eye Res.* 127, 49–58.

Liu, Y. X. and Wang, M. W. (2008). Botanical drugs: Challenges and opportunities: Contribution to Linnaeus memorial symposium 2007. *Life Sci.* 82, 445–9.

Loffler, T., Lee, S. K., Noldner, M. et al. (2001). Effect of *Ginkgo biloba* extract (EGb 761) on glucose metabolism-related markers in streptozotocin-damaged rat brain. *J. Neural Transm.* 108, 1457–74.

Loizzo, M. R., Menichini, F., Conforti, F. et al. (2009). Chemical analysis, antioxidants, anti-inflammatory and anticholinesterase activities of *Origanum ehrenbergii* Boiss and *Origanum syriacum* L. essential oils. *Food Chem.* 117, 174–80.

Loizzo, M. R., Tundis, R., Menichini, F. et al. (2011). Radical scavenging activity and cholinesterase inhibitory activity of Leopodia comosa bulbs. *Progr. Nutr.* 13, 300–3.

Luo, Y., Smith, J. V., Paramasivam, V. et al. (2002). Inhibition of amyloid-beta aggregation and caspase-3 activation by the *Ginkgo biloba* extracts EGb761. *Proc. Natl. Acad. Sci. USA* 99, 12197–202.

Mackenzie, I. R. (1994). Senile plaques do not progressively accumulate with normal aging. *Acta. Neuropathol. (Berl).* 87, 520–5.

Mackenzie, I. R. and Miller, L. A. (1994). Senile plaques in temporal lobe epilepsy. *Acta. Neuropathol. (Berl).* 87, 504–10.

Manigandan, V., Gurudeeban, S., Satyavani, F. et al. (2014). Molecular Docking Studies of *Rhizophora mucronata* Alkaloids Against Neuroinflammatory Marker Cyclooxygenase 2. *Int. J. Biol. Chem.* 8, 91–9.

Marston, A., Kissling, J., and Hostettmann, K. (2002). A rapid TLC bioautographic method for the detection of acetylcholinesterase and butyrylcholinesterase inhibitors in plants. *Phytochem. Anal.* 13, 51–4.

Matharu, B., Gibson, G., Parsons, R. et al. (2009). Galantamine inhibits beta-amyloid aggregation and cytotoxicity. *J. Neuro. Sci.* 280, 49–58.

McGeer, P. L. and McGeer, E. G. (1995). The inflammatory response system of brain: Implication for therapy of Alzheimer's and other neurodegenerative diseases. *Brain Res. Brain Res. Rev.* 21, 195–218.

Mishra, S. and Palanivelu, K. (2008). The effect of curcumin (tumeric) on Alzheimer's disease: An overview. *Ann. Indian Acad. Neurol.* 11, 13–9.

Mook-Jung, I., Hong, H. S., Boo, J. H. et al. (2001). Ginsenoside Rb1 and Rg1 improve spatial learning and increase hippocampal synaptophysin level in mice. *J. Neurosci. Res.* 63, 509–15.

Morgan, A. and Stevens J. (2010). Does Bacopa monnieri improve memory performance in older persons? Results of a randomized, placebo-controlled, double-blind trial. *J. Altern. Complement. Med.* 16, 753–9.

Mossa, A. T. and Nawwar, G. A. (2011). Free radical scavenging and antiacetylcholinesterase activities of *Origanum majorana* L. essential oil. *Hum. Exp. Toxicol.* 30, 1501–13.

Moyo, M., Ndhlala, A. R., Finnie, J. F. et al. (2010). Phenolic composition antioxidant and acetylcholinesterase inhibitory activities of *Sclerocarya birrea* and *Harpephyllum caffrum* extracts. *Food Chem.* 123, 69–76.

Mukherjee, P. K., Kumar, V., and Houghton, P. J. (2007). Screening of Indian medicinal plants for acetylcholinesterase inhibitory activity. *Phytother. Res.* 21, 1142–5.

Munoz, D. G. and Feldman, H. (2000). Causes of Alzheimer's disease David G. CMAJ. 162, 65–72.

Nag, G. and De, B. (2011). Acetylcholinesterase inhibitory activity of *Terminalia chebula*, *Terminala bellerica*, and *Emblica officinalis* and some phenolic compounds. *Int. J. Pharm. Pharm. Sci.* 3, 121–4.

Nalini, K., Karanth, K. S., Rao, A. et al. (1995). Effects of *Celatrus paniculatus* on passive avoidance performance and biogenic amine turnover in albino rats. *J. Ethnopharm.* 47, 101–8.

Nasser, A., Awadh, Ali, and Rebecca, A. (2012). Chemical composition, Antimicrobial, Antiradical, and Anticholineterase activity of the essential oil of *Pulicaria stephanocarpa* from soqotra. *Nat. Prod. Commun.* 7, 113–6.

Nino, J., Hernandez, J. A., Correa, Y. M. et al. (2006). *In vitro* inhibition of acetylcholinesterase by crude plant extracts from Colombian flora. *Mem. Inst. Oawaldo Cruz.* 101, 783–5.

Nordberg, A. and Svensson, A. L. (1998). Cholinesterase inhibitors in the treatment of Alzheimer's disease: A comparison of tolerability and pharmacology. *Drug Sar.* 19, 465–80.

Obulesu, M. and Rao, D. M. (2011). Effect of plant extracts on Alzheimer's disease insight into therapeutic avenues. *J. Neurosci. Rural Pract.* 2, 56–61.

Orhan, I., Senol, F. S., and Gulpinar, A. R. (2009). Acetylcholinesterase inhibitory and antioxidant properties of *Cyclotrichium niveum*, *Thymus praecox* subsp. *caucasicus* var. *caucasicus*, *Echinacea purpurea* and *E. pallida*. *Food Chem. Toxicol.* 47, 1304–10.

Orhan, I. E., Belhattab, R., Senol, F. S. et al. (2010). Profiling of cholinesterase inhibitory and antioxidant activities of *Artemisia absinthium*, *A. herba alba*, *A. fragrans*. *Marrubium vulgare*, *M. astranicum*, *Origanum vulgare* subsp. *glandulossum* and essential oil analysis of two Artemisia species. *Ind. Crop Prod.* 32, 566–71.

Pandey, H. C., Tewari, L. C. et al. (1975). Latex of *Euphorbia royleana* Bioss. The Source of Gomutra Silajit (Silajatu) – An Ancient Miraculous Drug of India. *Quart J. Crude Drug Res.* 13, 135–42.

Parle, M., Dhingra, D., and Kulkarni, S. K. (2004). Memory-strengthening activity of *Glycyrrhiza glabra* in exteroceptive and interoceptive behavioral models. *J. Med. Food.* 7, 462–6.

Peng, Y., Jiang, L., Lee, D. Y. et al. (2006) Effects of huperzine A on amyloid precursor protein processing and beta-amyloid generation in human embryonic kidney 293 APP Swedish mutant cells. *J. Neurosci. Res.* 84, 903–11.

Pereira, D. M., Ferreres, F., Oliveira, J. M. et al. (2010). Pharmacological effects of *Catharanthus roseus* root alkaloids in acetylcholinesterase inhibition and cholinergic neurotransmission. *Phytomedicine* 17, 646–52.

Perry, E. K., Pickering, A. T., Wang, W. W. et al. (1999). Medicinal plants and Alzheimer's disease: From ethnobotany to phytotherapy. Medical Research Council, Newcastle General Hospital, Newcastle upon Tyne. *J. Pharm. Pharmacol.* 51, 527–34.

Perry, N., Court, G., Bidet, N. et al. (1996). European herbs with cholinergic activities: Potential in dementia therapy. *Int. J. Geriatr. Psychiatry* 11, 1063–9.

Peruzza, M. and DeJacobis, M. (1986). A double-blind placebo controlled evaluation of the efficacy and safety of vinpocetine in the treatment of patients with chronic vascular or degenerative senile cerebral dysfunction. *Adv. Ther.* 3, 201–9.

Pravina, K., Ravindra, K. R., Goudar, K. S. et al. (2007). Safety evaluation of BacoMind in healthy volunteers: A phase I study. *Phytomed.* 14, 301–8.

Raskind, M. A., Peskind, E. R., and Wessel, T. (2000). Galantamine in AD: A six-month, randomized, placebo-controlled trial with a six-month extension. *Neurol.* 54, 2261–8.

Rauter, A. P., Martins, A.; Lopes, R. et al. (2009). Bioactivity studies and chemical profile of the antidiabetic plant *Genista tenera*. *J. Ethnopharmacol.* 122, 384–93.

Rigney, U., Kimber, S., and Hindmarch, I. (1999). The effects of acute doses of standardized *Ginkgo biloba* extract on memory and psychomotor performance in volunteers. *Phytotherayp. Research* 13, 408–15.

Rubio, J., Qiong, W., Liu, X., Jiang, Z., Dang, H., Chen, S. L. et al. (2011). Aqueous extract of black maca (*Lepidium meyenii*) on memory impairment induced by ovariectomy in mice. *Evid. Based Complement Alternat. Med.*, 253958.

Samuelsson, G. (2004). *Drugs of Natural Origin, A Textbook of Pharmacognosy*. 5th Edition. Swedish Pharmaceutical Press, Stockholm.

Satheeshkumar, N., Mukherjee, P. K., and Bhadra, S. (2010). Acetylcholinesterase enzyme inhibitory potential of standardized extracts of *Trigonella foenum-graecum* L. and its constituents. *Phytomed.* 17, 292–5.

Sayre, L. M., Smith, M. A., and Perry, G. (2001). Chemistry and biochemistry of oxidative stress in neurodegenerative disease. *Curr. Med. Chem.* 8, 721–38.

Sayre, L. M., Zelasko, D. A., Harris, P. L. et al. (1997). 4-Hydroxynonenal-derived advanced lipid peroxidation end products are increased in Alzheimer's disease. *J. Neurochem.* 68, 2092–7.

Schindowski, K., Leutner, S., Kressmann, S. et al. (2001). Age-related increase of oxidative stress-induced apoptosis in mice prevention by *Ginkgo biloba* extract (EGb 761). *J. Neural Transm.* 108, 969–78.

Schliebs, R., Liebmann, A., Bhattacharya, S. K. et al. (1997). Systemic administration of defined extracts from *Withania somnifera* (Indian Ginseng) and Shilajit differentially affects cholinergic but not glutamatergic and GABAergic markers in rat brain. Neurochem. Int. 30, 181–90.

Schneider, L. S. (1995). Estrogen replacement therapy may enhance response to tacrine in women with Alzheimer's disease. *Neurology* 45, A228.

Schofield, P. W., Tang, M., Marder, K. et al. (1997). Alzheimer's disease after remote head injury: An incidence study. *J. Neurol. Neurosurg. Psychiatry* 62, 119–24.

Scholey, A. B., Tildesley, N. T., Ballard, C. G. et al. (2008). An extract of *Salvia* (sage) with anticholinesterase properties improves memory and attention in healthy older volunteers. *Psychopharmacology (Berl).* 198, 127–39.

Scorer, C. A. (2001). Preclinical and clinical challenges in the development of disease modifying therapies for Alzheimer's disease. *Drug Discov. Today* 6, 1207–19.

Sehgal Gupta, A., Valli, R. K. et al. (2012). *Withania somnifera* reverses Alzheimer's disease pathology by enhancing low-density lipoprotein receptor-related protein in liver. *Proc. Natl. Acad. Sci. USA* 109, 3510–5.

Sekeroglu, N., Senol, F. S., Orhan, I. E. et al. (2012). *In vitro* prospective effects of various traditional herbal coffes consumed in anatolia linked to neurodegenaration. *Food Res. Int.* 45, 197–203.

Senol, F. S., Orhan, I., Celep, F. et al. (2010). Survey of 55 Turkish *Salvia* taxa for their acetylcholinesterase inhibitory and antioxidnat activities. *Food Chem.* 120, 34–43.

Sheikh, N., Ahmad, A., Siripurapu, K. B. et al. (2007). Effect of Bacopa monniera on stress-induced changes in plasma corticosterone and brain monoamines in rats. *J. Ethnopharmacol.* 111, 671–6.

Singhal AK, Naithani V, Bangar OP (2012) Medicinal plants with a potential to treat Alzheimer and associated symptoms. Int J Nutr Pharmacol Neurol Dis. 2, 84-91.

Small, S. A. and Mayeux, R. (2005). Alzheimer's disease and related dementias. Ed. by L. P. Rowland. *Merritt's Neurology*. Lippincott Williams & Wilkins, Philadelphia, USA 107, 771–6.

Smith, M. A., Richey, H. P., Sayre, L. M. et al. (1997). Widespread peroxynitrite-mediated damage in Alzheimer's disease. *J. Neurosci.* 17, 2653–7.

Smith, M. A., Sayre, L. M., Monnier, V. M. et al. (1995). Radical aging in Alzheimer's disease. *Trends Neurosci.* 18, 172–6.

Spencer, J. P. E. (2007). The interaction of flavonoids within neuronal signalling pathways. *Genes Nutr.* 2, 257–73.

Subathra, M., Shila, S., Devi, M. A. et al. (2005). Emerging role of Centella asiatica in improving age-related neurological antioxidant status. *Exp. Gerontol.* 40, 707–15.

Szatmari, S. Z. and Whitehouse, P. J. (2003). Vinpocetine for cognitive impairment and dementia. *Cochrane Database Syst. Rev.* 1, CD003119.

Szilágyi, G., Nagy, Z., Balkay, L. et al. (2005). Effects of vinpocetine on the redistribution of cerebral blood flow and glucose metabolism in chronic ischemic stroke patients: A PET study. *J. Neurol. Sci.* 229–230, 275–84.

Tappayuthpijam, P., Itharat, A., Makchuchit, S. et al. (2011). Acetylcholinesterase inhibitory activity of Thai traditional nootropic remedy and its herbal ingredients. *J. Med. Assoc. Thai.* 94, S183–9.

Tavares, L., McDougall, G. J., Fortalezas, S. et al. (2012). The neuroprotective potential of phenolic-enriched fractions from four *Juniperus* species found in portugal. *Food Chem.* 135, 562–70.

Tel, G., Ozturk, M., Duru, M. E. et al. (2010). Chemical composition of the essential oil and hexane extract of *Salvia chionantha* and their antioxidant and anticholinesterase activities. *Food Chem. Toxicol.* 48, 3189–93.

The Canadian Study of Health and Aging. (1994). The Canadian Study of Health and Aging: Risk factors for Alzheimer's disease. *Neurology* 44, 2073–80.

Theodorou, M. (2006). NNFCC Project Factsheet: Sustainable Production of the Natural Product Galanthamine (Defra), NF0612. R & D Programme "Non-Food Uses of Crops – Supply Chain Assessment and Development."

Tohda, C., Kuboyama, T., and Komatsu, K. (2005). Search for natural products related to regeneration of the neuronal network. *Neurosignals* 14, 34–5.

Tsvetkova, D., Obreshkova, D., Zheleva-Dimitrova, D. et al. (2013). Antioxidant activity of galantamine and some of its derivatives. *Curr. Med. Chem.* 20, 4505–608.

Tundis, R., Menichini, F., Conforti, F. et al. (2009). A potential role of alkaloid extracts from *Salsola* species in the treatment of Alzheimer's disease. *J. Enzyme Inhib. Med. Chem.* 24, 818–24.

Uabundit, N., Wattanathorn, J., Muci Mapura, S. et al. (2010). Cognitive enhancement and neuroprotective effects of *Bacopa monnieri* in Alzheimer's disease model. *J. Ethnopharmacol.* 127, 26–31.

Ustun, O., Senol, F., Kurkcuoghu, M. et al. (2012). Investigation on chemical composition, anticholinesterase and antioxidant activities of extracts and essential oils of Turkish *Pinus* species and pycnogenol. *Ind. Crop Prod.* 38, 115–23.

Valko, M., Leibfritz, D., Monco, J. et al. (2007). Free radicals and antioxidants in normal physiological functions and human disease. *Int. J. Biochem. Cell Biol.* 39, 44–84.

Varadarajan, S., Yatin, S. M., Aksenova, M. et al. (2000). Alzheimer's amyloid β-peptide-associated free radical oxidative stress and neurotoxicity. *J. Struct. Biol.* 130, 184–208.

Ved, H. S., Koenig, M. L., Dave, J. R. et al. (1997). Huperzine A, a potential therapeutic agent for dementia, reduces neuronal cell death caused by glutamate. *Neuroreport* 8, 963–8.

Vinutha, B., Prashanth, D., Salma, K. et al. (2007). Screening of selected Indian medicinal plants for acetylcholinesterase inhibitory activity. *J. Ethnopharmacol.* 109, 359–63.

Wake, G., Court, J., Pickering, A. et al. (2000). CNS acetylcholine receptor activity in European medicinal plants tranditionally used to improve failing memory. *J. Ethnopharmacol.* 69, 105–14.

Wang, L. C., Wang, B., Ng, S. Y. et al. (2006). Effects of ginseng saponins on beta-amyloid-induced amnesia in rats. *J. Ethnopharmacol.* 103, 103–8.

Wang, X. D., Zhang, J. M., Yang, H. H. et al. (1999). Modulation of NMDA receptor by huperzine A in rat cerebral cortex. *Zhongguo Yao Li Xue Bao* 20, 31–5.

Watanabe, C. M., Wolffram, S., Ader, P. et al. (2001). The *in vivo* neuromodulatory effects of the herbal medicine *Ginkgo biloba*. *Proc. Natl. Acad. Sci. USA* 98, 6577–80.

Winston, D. and Maimes, S. (2007). *Adaptogens: Herbs for Strength, Stamina and Stress Relief, Healing.* Arts Press, VT.

World Alzheimer's Report. 2015. The global impact of dementia. An anlaysis of prevanlence, incidence, cost and trends.

Xu, Y., Cao, Z., Khan, I. et al. (2008). *Centella asiatica* extract enhances phosphorylation of cyclic AMP response element binding protein in neuroblastoma cells expressing amyloid beta peptide. *J. Alzheimer's Dis.* 13, 341–49.

Yalla Reddy, K., Mohana Lakshmi, S., and Saravana Kumar, A. (2010). Review on effect of natural memory enhancing drugs on dementhia. *Int. J. Phytopharmacol.* 1, 1–7.

Yamazaki, M., Hirakura, K., Miyaichi, Y. et al. (2001). Effect of polyacetylenes on the neurite outgrowth of neuronal culture cells and scopolamine-induced memory impairment in mice. *Biol. Pharm. Bull.* 24, 1434–6.

Yang, F., Lim, G. P., Begum, A. N. et al. (2005). Curcumin inhibits formation of amyloid beta oligomers and fibrils, binds plaques, and reduces amyloid *in vivo*. *J. Biol. Chem.* 280, 5892–901.

Yang, Z., Zhang, D., Ren, J. et al. (2010). Acetylcholinesterase inhibitory activity of the alkaloids from traditional Chinese herbal medicine for treating Alzheimer's disease. *Med. Chem. Res.* 21, 734–38.

Yang, Z., Zhang, D., Ran, J. et al. (2012). Acetylcholinesterase inhibitory activity of the total alkaloid from traditional Chinese herbal medicine for treating Alzheimer's disease. *Med. Chem. Res.* 21, 734–8.

Yu, Z. F., Kong, L. D., and Chen, Y. (2002). Antidepressant activity of aqueous extracts of *Curcuma longa* in mice. *J. Ethnopharmacol.* 83, 161–5.

Zhang, Z., Wang, X., and Chen, Q. (2002). Clinical efficacy and safety of huperzine alpha in treatment of mild to moderate Alzheimer's disease: A placebo-controlled, double-blind, randomized trial. *Zhonghua Yi Xue Za Zhi* 82, 941–4.

Zheng, S. X., Zhou, L. J., Chen, Z. L. et al. (2002). Bilobalide promotes expression of glial cell line-derived neurotrophic factor and vascular endothelial growth factor in rat astrocytes. *Acta. Pharmacol. Sin.* 21, 151–5.

Zhu, Z., Li, C., Xu, W. et al. (2010). 2,2',4'-Trihydroxychalcone from *Glycyrrhiza glabra* as a new specific BACE1 inhibitor efficiently ameliorates memory impairment in mice. *J. Neurochem.* 114, 374–85.

10 *Schefflera* Genus (Araliaceae)

A Review of Its Botanical Description, Ethnobotanical Uses, Phytochemistry, Pharmacology and Toxicology

Puthanpura Sasidharan Sreeja, Karuppusamy Arunachalam and Parimelazhagan Thangaraj

CONTENTS

10.1 Introduction 194
10.2 Botanical Description and Taxonomy of *Schefflera* 195
10.3 Ethnomedicinal Uses of *Schefflera* 195
10.4 Pharmacological Studies and Phytochemical Screening 198
10.4.1 Pharmacological Studies 198
10.4.1.1 Antimicrobial Activity 198
10.4.1.2 Antiplasmodial Activity 198
10.4.1.3 Anti-Inflammatory Activity 198
10.4.1.4 Anti-Tumor Activity 199
10.4.1.5 Anti-Allergic Activity 199
10.4.1.6 Antioxidant Activity 199
10.4.1.7 Toxicity Studies 200
10.4.1.8 Miscellaneous 200
10.4.2 Phytochemistry 201
10.4.2.1 Phytochemical Screening 201
10.4.2.2 Compound Isolation 202
10.5 Conclusion 216
References 217

10.1 INTRODUCTION

The angiosperm family Araliaceae presently comprises roughly 1600 species, over half of which can be traced to a single genus, *Schefflera*, whose circumscription has been broadened substantially over the past three decades (Plunkett et al. 2005). In recent decades, *Schefflera* J. R. Forst. and G. Forst. have been re-circumscribed to incorporate several formerly separate genera, making it the largest and most widely-ranging genus of Araliaceae family, with over 1100 formally accepted species and several hundred more awaiting description (Wang et al. 2013).

Schefflera is a large genus with over 650 species of polygamous or dioecious trees, tall shrubs or climbers distributed in tropical and subtropical regions (Deepa and Nalini 2014). The recognition of these clades is based on molecular data, but they match the informal sub-generic groupings proposed by Frodin (Plunkett et al. 2005) more closely on the basis of morphology and geography. *Schefflera* ranges nearly throughout the more humid tropical and subtropical parts of the world (save for the Mascarenes or Pacific Islands east and north of Samoa), with some taxa occurring in warm–temperate areas, but at lower elevations in the warmest regions, the genus is poorly represented or absent. The plants range from terrestrial to epiphytic dwarf shrubs, to tall trees and large woody climbers and are generally characterised by their unarmed woody habit; digitate, compound leaves with sheathing petiolar bases often arranged in palmoid rosettes; terminal (or pseudo lateral) paniculate or compound umbellate inflorescences bearing umbellules, heads, racemules or spicules; and more or less fleshy fruits topped by sometimes broadly lobed or toothed calyx-rim, a disk and (1-)2–32(-100) free or united styles or sessile stigmas. Most species are more or less homoblastic; with the exception of some such as *S. heptaphylla* and *S. morototoni*; marked heteroblasty is limited to a few insular groups or island representatives of otherwise continental groups, such as the Sri Lankan *S. heterobotrya* (Frodin et al. 2010).

Since the 1980s, this genus has attracted more attention for phytochemical studies. Guo et al. (2005) reviewed the chemical composition of *Schefflera* and the emergence of a large number of chemical constituents and pharmacological activity were reported. Recently, from the genus-isolated 199 compounds (most of which are pentacyclic triterpenoids and glycosides, in addition to some long-chain compounds, steroids and sesquiterpenes) organic acids have been reported (Wang et al. 2013). The genus has relatively few reports of pharmacological activity, mainly in the analgesic anti-inflammatory, antitumor, antibacterial, antiviral, receptor binding activity, anti-allergic, anti-malarial and other aspects. Currently, the possible toxicity of synthetic drugs has been criticized. Thus, the interest in natural drugs, especially of plant origin, has greatly increased in recent years (Jayaprakash and Rao 2000). Hence, searching for alternative and effective medicines from plants against different diseases has become an important concern all over the world. Synthetic drugs cause side effects and are also costly for the poor communities of the developing world. Furthermore, the drug may be associated with adverse effects, including hypersensitivity and immune suppression (Bridge and Zhao 2012).

Therefore, this review was designed with the aim to compile the available fragmented literature on *Schefflera* species and suggest measures on newer and safer herbal drugs for many diseases. Furthermore, this review will provide knowledge on ethnomedicines and phytochemistry of those *Schefflera* species that have pharmacological potential. Above all, this review will provide baseline information for chemists, pharmacists and pharmacologists to carry out in-depth *in vitro* and *in vivo* activities for the development of novel drugs from *Schefflera* species, with lower cost and fewer side effects.

10.2 BOTANICAL DESCRIPTION AND TAXONOMY OF *SCHEFFLERA*

Schefflera J. R. Forst. and G. Forst. belongs to the family Araliaceae. *Schefflera* plants are a genus of plants named in honor of Jacob Christian Scheffler, an 18th-century German botanist. A number of them are popular as houseplants, including *S. arboricola*, also known as the 'dwarf umbrella tree' (Elizabeth 1969). The circumscription of the genus has varied greatly. The plants are trees, shrubs or lianas, growing 1–30 m (3 ft, 3 in–98 ft, 5 in) tall, with woody stems. They rarely have palmately compound, digitate or unifoliolate leaves. Petioles are long; the stipules usually connate within the petiole. The flowers produced by some species are small and yellow-green in color and the petals grow in long clusters of around two dozen, making a spear formation. The natural habitat of *Schleffera* plants is subtropical and they grow naturally in Southeast Australia, Latin America, Africa, Asia, Papua New Guinea and Taiwan.

In an ideal environment, they are capable of growing to heights of 40 feet. As an evergreen shrub, *Schefflera* is available all year round and the most common species are *S. actinophylla* (Umbrella plant) and *S. arboricola*. Flowers are in panicles of umbels or compound racemes, usually terminal. Fruit are sub-globose, five to six angles dry drupe. Plants in the *Schefflera* genus are green and leafy, although some species will also bear flowers. When planted outdoors, they can survive but require particular conditions replicating their original habitat. (Gamble 1919; Reyger 1766). The scientific names and the distribution of the species were confirmed with the databases of the Plant list and Missouri Botanical Garden (MOBOT) (available at http://www.theplantlist.org; http://www.tropicos.org).

A - Inflorcence; B - Leaf morphology; C - Flowers; D - Fruits; E - Habit. (http://keyserver.lucidcentral.org; www.theplantlist.org)

10.3 ETHNOMEDICINAL USES OF *SCHEFFLERA*

The ethnomedicinal uses of *Schefflera* (Table 10.1) include treatment for asthma, liver diseases, rheumatism, arthritis, sprains, fracture, stomach pain, antipyretic, anti-inflammatory, analgesic, migraines and use as a general tonic (Ragasa and Lin 2005). *S. octophylla* (Lour.) Harms, a traditional Chinese herb mainly distributed in Southeast Asia, possesses antinociceptive, anti-inflammatory, anti–rheumatoid arthritis activity (Yanfen Chena et al. 2015) and can be prepared as a medicinal bath for eczema and dermatitis (Li et al. 2006).

Xi-long Zheng and Fu-wu Xing (2009) reported that alcohol decoction of the roots, stem and leaves of *S. arboricola* and *S. heptaphylla* is used for rheumatoid

TABLE 10.1
Ethnomedicinal Uses of Schefflera Plants

Scientific Name	Common Name(s)	Parts Used	Traditional Uses	Preparation and Mode of Administration	References
S. octophylla (Lour.) Harms	Chân chim bảy lá	Leaves and bark	Itching	Leaves and bark are crushed, then applied on the affected part	Hoang et al. 2008
S. abyssinica Harms.		Leaves	Itching	Mixed with butter	Zerabruk and Yirga 2012
		Shoot tip	Toothache	Mixed with milk	Zerabruk and Yirga 2012
S. bengalensis Gamb.	Au Cha Pee	Leaves and stem	Postpartum depression, bone and joint pain, antispasmodic	Decoction, oral, bathing	Kannika et al. 2011
S. arboricola (Hayata) Merr.	hào yìng	Leaves	Sore swellings	Mashed material, external	Zheng et al. 2009
		Leaves and stem	Postpartum care	Boil and apply as a wash or hot compress	Homervergel and Young 2014
S. barteri Harms.	Guetsa		Pregnancy and childbirth–related issues		Yemele et al. 2015
S. elliptica (Blume) Harms	Dandoleseh	Leaves	Bone fractures	Paste/topical	Punnam et al. 2015
		Stem	Postpartum care	Boil and apply as a wash or hot compress	Homervergel and Young 2014

(Continued)

TABLE 10.1 (CONTINUED)
Ethnomedicinal Uses of Schefflera Plants

Scientific Name	Common Name(s)	Parts Used	Traditional Uses	Preparation and Mode of Administration	References
S. longipedicellata (Lecomte) Bernardi	Membolo vatsila	Leaves	Epilepsy, common cold, gonorrhoea		Mendrika et al. 2013
Schefflera sp.	jankang	Leaves	Spleen fever	Decoction	Danna et al. 1995
S. venulosa (Wight & Arn.) Harms	Lyagro, Mashing pal, Simal	Leaves and flowers	Armpit odour	Paste or juice are mixed with yak butter	Ghimire et al. 2001
S. brevipedicellata Harms	wo mia	Shrub	Abortion	Decoction, mashed, orally, dressing	Abdolbaset et al. 2011
S. chinensis (Dunn) H. L. Li	lo ben	Tree	Rheumatism, bone fractures	Mashed and heated by fire, plaster, dressing	Abdolbaset et al. 2011
S. heptaphylla (L.) Frodin		Young leaves	Edible	Young leaves, fresh, boiled or roasted	Hugo et al. 2012
		Roots	Nephrolithiasis	Root, collected in the morning, macerate, decoct, drink; Alt. Stem, collect in the afternoon or evening, from the West side of tree, while holding breath, macerate, decoct, drink	
		Bark	Neonatal rash, Itch	Boil, wash	
		Wood	Stomachache	Roast, decoct, drink	
		Leaves	Fever	Roast, poultice on body, massage	

arthritis and injury. As per the reports of Kalpana and Rishi (2009) *S. venulosa* possesses anti rheumatic activity and its stem and twigs are crushed together with black pepper seed and water for menstruation disorders. The roots of *S. chinensis* and leaves of *S. rubriflora* are used for curing rheumatism and bone fractures (Pei 1985; Abdolbaset 2011) and *S. capitate* is traditionally used to recover bone dislocation by wrapping its leaf paste in a cloth and externally applied along with sesame oil on the affected place (Maria et al. 2014). As per the report of Witthawaskul et al. (2003), *S. leucantha* is used as a single herbal drug for the treatment of chronic asthma and for the prevention of asthma attacks. *S. brevipedicellata* Harms and *S. elliptica* (Blume) Harms are used for curing headaches when applied as a poultice (Abdolbaset et al. 2011; Homervergel and Young 2014).

10.4 PHARMACOLOGICAL STUDIES AND PHYTOCHEMICAL SCREENING

Several *Schefflera* species possessed a wide range of biological activities and contained various phytochemicals.

10.4.1 Pharmacological Studies

The former reports on the pharmacological properties of *Schefflera* plants are discussed in the following sections.

10.4.1.1 Antimicrobial Activity

The *S. racemosa* plant extract shows potential antibacterial activity against *Escherichia coli* (Malar et al. 2015). Potduang et al. (2007) reported that the ethanol extract of *S. leucantha* has a broad spectrum of antimicrobial effect against various respiratory tract infection–causing microorganisms. The studies on southern China's 21 kinds of medicinal plants found *S. octophylla* has some anti-RSV (respiratory syncytial virus) (IC_{50} 12.5 mg L-1). Also the 3,5-O-caffeoyl acid II and 3,4-O-caffeoyl acid II from *S. heptaphylla* has anti-RSV virus activity (IC_{50} 1.2, 1.16 μg/mL) (Li et al. 2004, 2005; Xu et al. 2006). Caffeoylquinic acid derivatives (3, 4-Di-O-caffeoylquinic acid, 3, 5-Di-O-caffeoylquinic acid) isolated from *S. heptapylla* possess anti-RSV activity (Yaolan et al. 2005).

10.4.1.2 Antiplasmodial Activity

S. umbellifera's ethanolic and ethyl acetate extracts and isolated compound betulin exhibited good antiplasmodial activity against the *Plasmodium falciparum*, a malaria-causing agent, in an *in vitro* assay (Tetyana et al. 2002; Mthembu et al. 2010).

10.4.1.3 Anti-Inflammatory Activity

S. arboricola has been used as most important plant for the preparation of Shuang-Qi gout capsule in China. It can cure rheumatoid arthritis, rheumatism, joint pain, fever, sore throat, etc. and is also found to be effective clinically for inflammation, pain, fever and gout arthritis (Kodithuwakku et al. 2013). Chen et al. (2015) found that the ethanol extract of *S. octophylla* showed significant dose-dependent, anti-inflammatory, antinociceptive and anti–rheumatoid arthritis activities. It was found that the levels of TNF-α, IL-1β and IL-6 in ethanol extract (600 mg/kg) and $CHCl_3$ fraction (300 mg/kg) groups that were significantly lower. Ethyl acetate extract of *S. arboricola* shows good anti–rheumatoid arthritis activity (Liu et al. 2012).

10.4.1.4 Anti-Tumor Activity

The two lupane type triterpenoid saponins, 3β-O-(α-L-rhamnopyranosyl-(1→2)-α-L-arabinopyranosyl)-lup-20(29)-ene-28-O-β-D-glucopyranosyl ester and 3β-O-α-L-arabinopyranosyl-lup-20(29)-ene-28-O-β-D-glucopyranosyl ester, were isolated from *S. rotundifolia* to inhibit proliferation activity for J774.A1, HEK-293 and WEHI-164 cells (Braca et al. 2004). But the oleanane type triterpenoid saponins 3β-O-[β-D-glucopyranosyl-(1→2)-β-D-glucopyranosyl-(1→3)-β-D-xylopyranosyl)-16α-hydroxyolean-12-en-28-O-(β-D-galactopyranosyl) ester-30-oic acid and 3β-O-[β-D-glucopyranosyl-(1→3)-β-D-xylopyranosyl)-16α-hydroxyolean-12-en-28 O-(β-D-galactopyranosyl) ester-30-oic acid isolated from *S. faguetigueti* possesses significantly more inhibitory proliferative activity for these three cell lines (Cioffi et al. 2003). Falcarindiol; C18-acetylenverbindung; 3α-hydroxyoleane-12; 28-dioic acid methyl ester (methyl ester 22); 3α-hydroxylup-20(29)-ene-23; 28-dioic acid methyl ester; a mixture of γ-tocopherol, docosyl and tetracosyl; 4-hydroxy-trans-cinnamate of docosyl and tetracosyl; and a mixture of hexacosyl 4-hydroxy-cis-cinnamate were obtained from *S. taiwaniana* and shows inhibitory activity towards the HUGC, HONE1 cell to a certain extent and t-murrolol possess significant cytotoxic activity against A-549 (EC_{50} 3.2 μg/mL), MCF (EC_{50} 0.6 μg/mL) and HT-29 cells (EC_{50} 1.8 μg/mL) (Kuo et al. 2002). Liu et al. (2005) found that the Chrysophanol, 2, 6-dimethoxy-p-benzoquinone separated from the dense pulse *S. venulosa*, and could inhibit K562 cells, thus showing a value-added activity, and separated ten palmitic acid also inhibits the proliferation activity on FT210 cells. Yao et al. (2009) found that the essential oil of *S. heptaphylla* possessed significant antiproliferative activity against the human cancer cell lines MCF-7, A375 and HepG2 cells.

10.4.1.5 Anti-Allergic Activity

Matsui et al. (2010) studied Langerhans cells and mast cells for the effects of *S. leucantha* on allergy mediators, and found *S. leucantha* ethanol extract can inhibit the production of allergy mediators to some extent. Falcarinol, heptadeca-1,9(Z)-diene-4,6-diyne-3-ol isolated from the plant *S. arboricola* has been reported to cause allergic contact dermatitis (Hansen et al. 1986).

10.4.1.6 Antioxidant Activity

Deepa and Nalini (2014) on a comparative study between *S. racemosa* and *S. stellata*, found that leaf aqueous extracts exhibited potent antioxidant activity. The aqueous leaf and bark extracts of *S. stellata* showed potent antioxidant activity when compared to the standard antioxidant, ascorbic acid. The maximum TPTZ-Fe (III) complex reduces the ability of TPTZ-Fe (II), and the maximum reducing ability power was seen in the aqueous leaf extracts of *S. stellata* and *S. racemosa*. The aqueous extracts of *S. venulosa* flower and *S. wallichiana* leaf showed potent antioxidant activity (Deepa and Nalimi 2013).

The *S. racemosa* plant extract was found to exhibit excellent antioxidant property for the DPPH, ABTS and FRAP assays (Malar Selvi et al. 2015). The DPPH radical scavenging ability of ethanol leaf extract of *S. leucantha* suggested that the leaves as a source of antioxidants might be effective in diseases caused by overproduction of radicals (Potduang et al. 2007). The leaf extract of *S. odorata* experimentally proved that it could quench hydroxyl free radicals suggesting its antioxidant activity (Gloria and Cristina 2001).

10.4.1.7 Toxicity Studies

Witthawaskul et al. (2003) studied the acute toxicity of water extracts and saponin mixtures, and the subacute toxicity of the saponin mixture of *S. leucantha*. Acute toxicity experiments show that a single oral dose of 5000 mg kg-1 of extract in rats did not produce any mortality and abnormal behaviour. In subacute toxicity experiments, rats were orally given a saponin mixture of 1000 mg/kg-1 for 14 d and no rats died or showed other symptoms of poisoning. After weighing the organs, researchers found the saponin mixture in rats increased liver weight and reduced testicular weight, suggesting that the plant saponin mixture may affect liver and kidney function.

The acute toxicity study on *S. barteri* showed no mortality at a dose limit of 16000 mg/kg b.w. by oral administration. Subacute treatment significantly ($p < 0.05$) increased the level of serum transaminase, proteins and HDL cholesterol. Therefore, the methylene chloride/methanol mixture stem bark extract of *S. barteri* is considered relatively harmless. On the other hand, the extract significantly ($p < 0.05$) reduced the level of leucocytes as well as neutrophils, basophils and monocytes in females. No significant variation of serum creatinine, LDL cholesterol, serum triglycerides, as well as liver, spleen, testicle and ovary proteins was noted. Hence subacute administration is associated with side effects on the central nervous system, immune system, liver and testis. The acute administration of the stem bark extract of *S. barteri* is associated with signs of toxicity; administration over a long duration provokes hepatotoxicity and testis and lung toxicity (Atsafack et al. 2015).

10.4.1.8 Miscellaneous

Zhu et al. (1999) and Chen et al. (2002) found that the ethanol extracts of the leaves and roots of *S. bodinieri* show a strong binding affinity to α1- and α2-adrenergic, sulphonylureas, GABAA and GABB receptors; and also three separate compounds, namely bodinone, bodinone glycoside; and D-sorbitol, which showed selective binding affinity to muscarine receptors, a trisaccharide bound to Ca2+ channel and 5HT-2 receptors, stigmasterol 3-O-glucoside bound to 5HT-2 receptors and bodirin A bound to dopamine-2 receptors. Matsui et al. (2010) reported that *S. leucantha* ethanol extract significantly inhibited the production of the eosinophil chemo attractant CCL5 and the type 2 T helper (TH_2) associated chemokine CCL17 from PEG stimulated from *Staphylococcus aureus* and histamine release from mast cells and significantly decreased the release of histamine from mast cells by IgE-mediated degranulation.

Saponins from the leaf extract of *S. odorata* can serve as modulators of the cell signalling pathway. The leaf extract acts as an extra cellular signal to switch off extracellular enzymes that are required for the proliferation of ASM (airway smooth muscle cells) cells. The signal can then be transmitted into the nucleus, which in turn promotes DNA fragmentation of A549 lung cancer cell lines, suggesting that *S. odorata* can induce A549 cells to undergo programmed cell death. Recent experiments have also revealed the potential immune modulating property of the leaf extract (Gloria and Cristina 2001).

10.4.2 Phytochemistry

10.4.2.1 Phytochemical Screening

Phytochemical studies on plants in the *Schefflera* genus have revealed the presence of triterpenes, triterpenoid glycosides and saponins (Sabulal et al. 2008). Phytochemical analysis of the stem bark extract of *S. barteri* revealed the presence of anthocyanins, anthraquinons and saponins (Table 10.2) (Atsafack et al. 2015).

TABLE 10.2
Phytochemicals Reported from *Schefflera* Genus

Plant Species	Parts Used	Phytochemical Name	Source
S. octophylla	Bark	Oleanolic acid, asiatic acid	Schmidt et al. 1984; Sung et al. 1992
S. impressa C. B. Clarke	Bark	Oleanolic acid, β-amyrin, hederagenin	Schmidt et al. 1984; Srivastava and Jain 1989; Srivastava 1992
S. arboricola	Roots	Succinic acid, fumaric acid, tartaric acid	Po et al. 2000
		Oleanolic acid, 3-acctyloleanolic acid, mesembryanthemoidigenic acid, quinatic acid, betulinic acid	Guo et al. 2005
S. venulosa	Leaves	β-amyrin, betulinic acid, betulinic acid glucoside	Purohit and Rawat 1993
S. kwangsiensis Merr. ex Li	Aerial part	Ursolic acid	Xu et al. 2006
S. umbellifera	Leaves	Betulin	Mthembu et al. 2010
S. heptaphylla	Leaf stalks	Oleanolic acid	Wu et al. 2011
S. odorata Blanco	Leaves	Oleanolic acid, lutein	Consolacion and Kathleen 2005
S. abyssinica (Hochst. Ex A. Rich.) Harms	Leaves	Oleanolic acid, hederagenin	Tapondjou et al. 2006
S. delavayi		β-Amyrin, oleanonic acid	Jian and Xian 1990
S. capitata		Capitogenic acid	Jain and Khanna 1982

The preliminary screening for phytochemicals on *S. racemosa* leaf extracts showed positive results for saponins, tannins, flavonoids, cardiac glycosides and steroids, reducing sugars and terpenoids. Whereas *S. stellata* leaf and bark extracts showed the presence of saponins, tannins, flavonoids, terpenoids, cardiacglycosides and reducing sugars mostly in polar and semipolar solvents (Deepa et al. 2014). The phytochemical investigation revealed the presence of saponins, tannins, flavonoids, alkaloids, cardiac glycosides and reducing sugars in *S. venulosa* and *S. wallichiana* (Deepa and Nalimi 2013) (Table 10.2).

10.4.2.2 Compound Isolation

Four novel triterpene glycosides named bodinitins derivatives have been identified based on 1H-1H COSY and 13C-1H COSY NMR spectroscopic data as 28-O-[αL-rhamopyranosyl-(1→4)-O-β-D-glucopyranosyl-(1→6)]-β-D-glucopyranoside of demethyl isoaleuritolic acid, 28-O-[α-L-rhamopyranosyl(1→4)-O-β-D-glucopyranosyl(1→6)]-β-D-glucopyranoside of isoaleuritolic acid, 28-O-[α-L-rhamnopyranosyl(1→4)-O-β-D-glucopyranosyl(1→6)]-β-Dglucopyranoside of 3-oxo-8-demethyliso aleuritolic acid and 28-O-[α-L rhamopyranosyl (1→4)-O-β-D-glucopyranosyl(1→6)]-β-D-glucopyranoside of 3-oxoiso aleuritolic acid (Min et al. 1996).

By the end of 20th century, many authors have started isolating several compounds from *Schefflera* plants. Compounds like 3cY-hydroxy-lup-20(29)-ene-23, 28dioic acid and 3α,11-α-hydroxylup-20(29)-ene-23, 28-dioic acid isolated from leaves and 3a-hydroxy-lup-20(29)-ene-23,28-dioic acid, Scheffursoside C, D, E, F and Scheffoleoside D, F, A, B from bark of *Scheflera octophylla* (Adam et al. 1982; Chizuko et al. 1994; Sung and Adam 1991). Melek et al. (2003) isolated 3-O-[α-L-rhamnopyranosyl-(1→4)-β-D-glucuronopyranosyl] oleanolic acid, 3-O-[α-L-rhamnopyranosyl-(1→4)-β-D-glucuronopyranosyl] 16α-O-Hechinocystic acid and 3-O-[α-L-rhamnopyranosyl-(1→4)-[α-L-arabinopyranosyl-(1→2)]-β-D-glucuronopyranosyl] oleanolic acid and Schefflerin A, B, C, D, F, G (Zhao et al. 2010) and 3-0-[α-L-rhamnopyranosyl-(1→2)-β-D-glucuronopyranosyl]-oleanolic acid 28-0- β-D-glucopyranoside (Abdel 2001) from leaves and stem of *Schefflera arboricola*. Leaves of *S. umbellifera* undergone isolation of Chloroquine, 7-hydroxy-6-methoxycoumarin and ent-kaur-16-en-19-oic acid by Mthembu et al. (2010). 3α,13-dihydroxy-urs-11-en-23,28-dioicacid-13,28-lactone, 3α-hydroxylup-20(29)-ene-23,28-dioicacid, Scheffleoside A, D, 3α-hydroxylup-20(29)-ene-23, 28-dioicacid28-0-[α-L-rhamnopyranosyl-(1-4)-0-β-D-glucopyranosyl-(1-6)-β-D-glucopyranoside isolated from the leaf stalks of *S. heptaphylla* (Wu et al. 2011). The entire isolated compound list showed in Table 10.3 and 10.4.

Recently, Yan et al. (2015) isolated Schekwangsienside Ia, Ib, IIa, IIb, III, IV, V, VI, VIIa, VIIb and VIII and Scheffarboside C, D, 3β-O-β-D-xylopyranosyl-(1-3)-α-L-rhamnopyranosyl-(1-2)-α-L-arabinopyranopyranosyl-hederagenin-28-*O*-α-L-rhamnopyranosyl-(1-4)-)-β-D-glucopyranosyl-(1-6)-β-D-6'-acetyl-glucopyranosyl ester (Bridge et al. 2012) and Scheffleside A, B, C, D, E, F, G and H (Ying et al. 2014) from the aerial part of the plant *S. kwangsiensis*.

TABLE 10.3
The List of Compounds Isolated from *Schefflera* Genus

Plant Species	Parts Used	Compound Name	Sources
S. octophylla (Lour.) Harms	Bark	3α-hydroxy-lup-20(29)-ene-23,28-dioic acid	Schmidt et al. 1984
		3α-hydroxy-urs-12-ene-23,28-dioic acid	Sung et al. 1992
		3-epi-betulinic acid	Sung et al. 1991
		3-epi-betulinic acid 3-0-sulfate	Kitajima et al. 1990
		Betulinic acid 3-0-sulfate	Kitajima et al. 1990
		3α-hydroxylup-20(29)-ene-23, 28-dioic acid	Lischewski et al. 1984
		3α-hydroxylup-20(29)-ene-23, 28-dioic acid fatty acid ester	Schmidt et al. 1984
		3α-hydroxylup-20(29)-ene-23, 28-dioic acid 28-0-[α-L-rhamnopyranosyl-(1-4)-0-β-D-glucopyranosyl-(1-6)- β-D-glucopyranoside	Sung et al. 1991
		3-epi-betulinic acid 3-0- β-D-glucopyranoside	Kitajima et al. 1990
		3α,11α-dihydroxylup-20(29)-ene-23,28-dioic acid 28-0-[α-L-rhamnopyranosyl-(1-4)-0- β-D-glucopyranosyl-(1-6)]- β-D-glucopyranoside	Sung et al. 1991
		3-epi-betulinic acid 3-0-[α-L-rhamnopyranosyl-(1→4)-0- β-D-glucopyranosyl-(1→6)- β-D-glucopyranoside	
		3-epi-betulinic acid 3-0-sulfate 28-0-[α-L-rhamnopyranosyl-3α-OSO_3I (1→4)-0- β-D-gluconopyranosyl-(1→6)- β-D-gluconopyranoside	
		3-epi-betulinic acid 3-0- β-D-glucopyranoside 28-0-[α-L-rhamnopyranosyl-(1-4)-0- β-D-glucopyranosyl-(1→6)]- β-D-glucopyranoside	

(*Continued*)

TABLE 10.3 (CONTINUED)
The List of Compounds Isolated from *Schefflera* Genus

Plant Species	Parts Used	Compound Name	Sources
		3-epi-betulinic acid 3-0-β-D-6' acetylglucopyranoside	Sung et al. 1992
		28-0-[α-L-rhamnopyranosyl-(1→4)-0- β-D-gluconopyranosyl-(1→6)]- β-D-glucopyranoside	
	Leaves	28-O-[a-L-rhamnopyranosyl(1→4)-o-β-D-glucopyranosyl (l-+6)-]-β-D-glucopyranosides of 3α-hydroxy-lup-20(29)-ene-23,28-dioic acid	Sung et al. 1991
		3a,l la-dihydroxy-lup-20(29)-ene-23,28-dioic acid and 3-epi-betulinic acid	
		3a,l la-dihydroxylup-20(29)-ene-23,28-dioic acid	Lischewski et al. 1984
S. impressa (C. B. Clarke), Harms	Stem and bark	3/3,23-dihydroxy-urs- 12-en-28-oic acid-3-0-p-D-glucuronopyranoside 6'-O-methyl ester and 4-epihederagenin-3-O-B-D-glucuronopyranoside-6'-O-methyl ester along with oleanolic acid; hederagenin; 23-hydroxy	Srivastava and Jain 1989
		Ursolic acid and hederagenin-3-0-P-D-glucuronopyranoside-6'-O-methyl ester	
		Hederagenin-3-O-β-D-glucuronopyranoside-6'-O-methyl ester	
		4-epihederagenin-3-O- β-D-glucuronopyranoside-6'-O-methyl ester	
		Hederagenin-3-O-β-D-6'acetylglucuronopyranoside	Srivastava and Jain 1989
		α–(α-amyrin)	Srivastava 1992
		3β, 23-dihydroxyl-urs-12-en-28-oic acid	Srivastava and Jain 1989
		3β, 23-dihydroxyl-urs-12-en-28-oic acid 3-0-β-D-glucuronopyranoside 6'-0-methyl ester	
		3α, 11 α -hydroxylup-20(29)-ene-28-oic acid	Srivastava 1992
		3α, 11α-dihydroxylup-20(29)-ene-28-oic acid	
		28-0- β-D-glucopyranosyl-(1-6)- β-D-glucopyranoside	

(Continued)

TABLE 10.3 (CONTINUED)
The List of Compounds Isolated from *Schefflera* Genus

Plant Species	Parts Used	Compound Name	Sources
S. bodinieri (H. Lév.) Rehder	Roots	28-O-[a-L-rhamopyranosyl-(1→ 4)-O-fl-o-glucopyranosyl-(1→ 6)-]-fl-D-glucopyranoside of 3fl-hydroxy-isopolygalic-13(14)-ene-28-acid, 28-O-[a-L-rhamopyranosyl(1 → 4)-O-fl-O-glucopyranosyl(1 → 6)-]-fl-D-glucopyranoside of 3-oxo-isopolygalic-13(14)-ene-28-acid, 28-O-fl-o-glucopyranosyl(1 → 6)-fl-D-glucopyranoside cf 3fl-hydroxy-isopolygalic- 13(14)-ene-28-acid and 28-0 [a-L-rhamopyranosyl-(1 → 4)-O-fl-D-glucopyranosyl-(1 → 6)-]-D-glucopyranoside of 3fl-hydroxy-18-methyl-polygalic-13(14)-ene-28-acid	Min et al. 1996
	Leaves	3α-hydroxyl-20-demethylisoaleuritolic-14(15)-ene-28, 29-dioic acid	Zhu et al. 1996
		3-oxo-20-demethylisoaleuritolic-14(15)-ene-28,29-dioic acid	Zhu et al. 1996
		28-0-[α-L-rhamnopyranosyl-(1→4)-0-β-D-glucopyranosyl-(1→6)]-0-β-D-glucopyranoside (bodinone glycoside)	Zhu and Li 1999
		3β-hydroxyl-isopolygalic-13(14)-ene-28-oic acid 28-0-[α-L-rhamnopyranosyl-(1→4)-0-β-D-glucopyranosyl-(1→6)]-β-D-glucopyranoside (bodirin A)	
		3-oxo-isopolygalic-13(14)-ene-28-oic acid 28-0-[α-L-rhamnopyranosyl-(1→4)-0-β-D-glucopyranosyl-(1→6)]-β-D-glucopyranoside	
		3β-hydroxyl-isopolygalic-13(14)-ene-28-oic acid 28-0-β-D-glucopyranosyl-(1→5)-β-D-glucopyranoside	
		3β-hydroxl-18-methyl-polygalic-13(14)-ene-28-oic acid 28-0-[α-L-rhamnopyranosyl-(1→4)-0-β-D-glucopyranosyl-(1→6)]-β-D-glucopyranoside	

(*Continued*)

TABLE 10.3 (CONTINUED)
The List of Compounds Isolated from *Schefflera* Genus

Plant Species	Parts Used	Compound Name	Sources
	Leaves and stem	3-O-α-L-rhamnopyranosyl-(1→4)-[β-D-galactopyranosyl-(1→2)]-β-D-glucuronopyranosyl] oleanolic acid	Melek et al. 2003
		3-O-β-D-apiofuranosyl-(1→4)-[α-L-arabinopyranosyl-(1→2)]-β-D-glucuronopyranosyl] oleanolic acid	
		3-0-[β-D-apiofurnosyl-(1→4)-β-D-glucuronopyranosy]-oleanolic acid 28-0-β-D-glucopyranoside	
		3-0-α-L-rhamnopyranosyl-(1→4)-[α-L-arabinopyranosyl-(1→2)]-β-D-glucuronopyranosyl-oleanolic acid 28-0- β-D-glucopyranoside	
		3-0-α-L-rhamnopyranosyl-(1→4)-[β-D-galactopyranosyl-(1→2)]- β-D-glucuronopyranosyl-oleanolic acid 28-0- β-D-glucopyranoside	
		3-0- β-D-apiofuranosyl-(1→4)-[α-L-arabinopyranosyl-(1→2)]- β-D-glucuronopyranosyl-oleanolic acid 28-0- β-D-glucopyranoside	
		3-0-[α-L-arabinopyranosyl-(1→2)- β-D-glucuronopyranosyl]-oleanolic acid 28-0- β-D-glucopyranoside	Abdel 2001
		Falcarinol, heptadeca-l,9(Z)-dien4,6-diyn-3-o1. The isolation of (Ebfi-famesene, phytol and	Lene and Per 1986
		24B-ethylcholesta-522(E)-diene-3gol (poriferasterol)	
S. venulosa (Wight & Arn.) Harms	Leaves	Lup-20(29)-en-28-oic-3-O-β-D-glucopyranosyl (2-1)-O-β-D-glucopyranoside	Purohit et al. 1991
		3-*O*-β-D-glucopyranosyl-(1→2)-β-D-glucopyranosyl-27-oxo-betulinic acid-28-*O*-β-D-glucopyranoside	Peng et al. 2012
		3-*O*-β-D-glucuronopyranosyl-27-oxo-betulinic acid-28-*O*-β-D-glucopyranoside	
		3β-0-β-D-glucopyranosyl-(1-2)-β-D-glucopyranosyl-27-oxo-betulinic acid-28-0-β-D-glucopyranoside	
		3β-0-β-D-glucuronopyranosyl-27-oxo-betulinic acid-28-0-β-D-glucopyranoside	
		Lup-20(29)-en-28-oic-3-0-β-D-glucopyranosyl-(2-1)-0-β-D-glucopyranoside	Purohit et al. 1991

(*Continued*)

TABLE 10.3 (CONTINUED)
The List of Compounds Isolated from *Schefflera* Genus

Plant Species	Parts Used	Compound Name	Sources
S. kwangsiensis Merr. ex H.L.Li	Aerial part	Hederagenin-3-O-β-D-glucuronopyranosyl-(1-2)-α-L-arabinopy-ranoside	Xu et al. 2006
		23,28-dihydroxyolean-12 (13)-en-3-O-[β-D-glucuronopyranosyl-(1-3)- α-L-rhamnopyranosyl-(1-2)]- α-L-arabinopyranosyl-(1-4)- α-L-arabinopyranoside	
		G (kalopanaxsaponin G)	Bridge et al. 2012
		Tauroside St-H_1	
		Saponins HCS-B	
		B(kalopanaxsaponin B)	
		3β-O-β-D-xylopyranosyl-(1-3)-α-L-rhamnopyranosyl-(1-2)-α-L-arabinopyranosyl-hederaagenin-28-O-α-L-rhamnopyranosyl-(1-4)-O-β-D-glucopyranosyl-(106)-β-D-glucopyranosyl ester	
		3-*O*-[β-*D*-glucopyranosyl-(1-4)- β-*D*-xylopyranosyl-(1-3)-α-L-rhamnopyranopyranosyl-(1-2)-α-L-arabinopyranosyl]-oleanolic acid 28-*0*-α-*L*-rhamnopyranosyl-(1-4)-*0*-β-D-glucopyranosyl-(1-6) β-D-glucopyranosyl ester	
		Oleanolic acid 3-O-b-D-glucopyranosyl (1 → 2) [a-L-arabinopyranosyl (1 → 4)]-b-D-(6-O-methyl) glucuronopyranoside (1), 22a-hydroxyoleanolic acid 3-O-a-L-arabinopyranosyl (1 → 4)-b-D-glucuronopyranoside (2), hedera- genin 3-O-a-L-arabinopyranosyl (1 → 4)-b-D-glucuronopyranoside (3) and oleanolic acid 28-O-b-D-glucopyranosyl (1 → 2)-b-D-glucuronopyranosyl ester	Cun et al. 2014

(*Continued*)

TABLE 10.3 (CONTINUED)
The List of Compounds Isolated from *Schefflera* Genus

Plant Species	Parts Used	Compound Name	Sources
S. heptaphylla (L.) Frodin	Stem and bark	3-oxo-urs-20-en-23,28-dioic acid 28-O-α-L-rhamnopyranosyl-(1 → 4)-β-D-glucopyranosyl-(1 → 6)-β-D-glucopyranoside	Chun et al. 2014
		3α-hydroxy-urs-20-en-23,28-dioic acid 28-O-β-Dglucopyranosyl-(1 → 6)-β-D-glucopyranoside, 3α-hydroxy-urs-20-en-23	
		28-dioic acid 23-O-β-Dglucopyranosyl, 28-O-α-L-rhamnopyranosyl-(1 → 4)-β-Dglucopyranosyl- (1 → 6)-β-D-glucopyranoside	
		3-oxo-urs-12-en-24-nor-oic acid 28-O-α-L-rhamnopyranosyl-(1 → 4)-β-D-glucopyranosyl-(1 → 6)-β-D-glucopyranoside	
S. heptaphylla (L.) Frodin		3α-hydroxy-20β-hydroxyursan-23,28-dioic acid δ-lactone	
		23-O-β-D-glucopyranoside	
	Leafstalks	3,4-di-*O*-caffeoylquinic acid, 3,5-di-*O*-caffeoylquinic acid, and 3-*O*-caffeoylquinic acid	Yaolan et al. 2005
		3-oxo-urs-12-en-28-oic acid	Wu et al. 2011
		Scheffuroside D	
		Scheffuroside F	
		3-O-a-L-arabinopyranosylhederagenin (cauloside A)	Tapondjou et al. 2006
		3-O-b-D-glucuronopyranosyl-hederagenin (copteroside B)	
		3-O-b-D-glucopyranosyl-(1/2)-a-L-arabinopyranosyl-oleanolic acid (Fatsiaside C1)	
		3-O-b-D-glucopyranosyl-(1/3)-a-L-arabinopyranosyl-oleanolic acid (Guaianin N)	
		3-O-b-D-glucopyranosyl-(1/3)-a-L-arabinopyranosyl-hederagenin (collinsonidin)	
		3-O-a-L-arabinopyranosyl-oleanolic acid-28-O-a-L-rhamnopyranosyl-(1/4)-b-D-glucopyranosyl-(1/6)-b-D-glucopyranosyl ester (Ciwujianoside C3)	

(Continued)

TABLE 10.3 (CONTINUED)
The List of Compounds Isolated from *Schefflera* Genus

Plant Species	Parts Used	Compound Name	Sources
S. heptaphylla (L.) Frodin		3-O-a-L-arabinopyranosyl-hederagenin-28-O-a-L-rhamnopyranosyl-(1/4)-b-D-glucopyranosyl-(1/6)-b-D-glucopyranosyl ester (9) (cauloside D)	
		3-O-b-D-glucopyranosyl-(1/2)-b-D-glucuronopyranosyl-oleanolic acid-28-O-b-D-glucopyranosyl ester (chikusetsusaponin V)	
		3-O-b-D-glucopyranosyl-(1/2)-a-L-arabinopyranosyl-oleanolic acid-28-O-a-L-rhamnopyranosyl-(1/4)-b-D-glucopyranosyl-(1/6)-b-D-glucopyranosyl ester (Ciwujianoside A1)	
		3-O-b-D-glucopyranosyl-(1/3)-a-L-arabinopyranosyl-hederagenin-28-O-a-Lrhamnopyranosyl-(1/4)-b-D-glucopyranosyl-(1/6)-b-D-glucopyranosyl ester	
		3-O-b-D-glucopyranosyl-(1/2)-a-L-arabinopyranosyl-hederagenin-28-O-a-L rhamnopyranosyl-(1/4)-b-D-glucopyranosyl-(1/6)-b-D-glucopyranosyl ester (cauloside G)	
S. sessiliflora (Marchal) Frodin & Govaerts		Scheffleraside A (1), scheffleraside B (2); together with two known saponins, chikusetsusaponin IVa (3), 3-O-[a-L-rhamnopyranosyl-(1 →3)]-b-D-glucuronopyranosyl hederagenin	Nguyen et al. 2015
S. delavayi (Franch.) Harms		Scheffleraside I	Jian and Xian 1990
		Oleanolic acid 3-O-α-L-arabinopyranoside	
		Schefferaside II	

(Continued)

TABLE 10.3 (CONTINUED)
The List of Compounds Isolated from *Schefflera* Genus

Plant Species	Parts Used	Compound Name	Sources
S. divaricata (Blume) Koord.		3 β-O-β-D-galactopyranosyl-(1-3)-[β-D-glucopyranosyl-(1-4)]-β-D-glucopyranosylolean-12-en-28-oic acid	Tommasi and Pizza 1997
		3β-O-β-D-galactopyranosyl-(1-3)-[β-D-glucuronopyranosyl-(1-4)]- β-D-glucopyranosylolean-12-en-28-oic acid	
		3β-O-α-L-arabinopyranosyl-(1-3)-[β-D-xylopyranosyl-(1-2)]- β-D-glucuronopyranosyl-16 α-hydroxyolean-12-ene-28,30-dioic acid	
		3β-O-α-L-arabinopyranosyl-(1-3)-[β-D-xylopyranosyl-(1-2)]- β-D-glucuronopyranosyl-16 α-O-(3-hydroxy-3-methy-butanoyl) olean-12-ene-28,30-dioic acid	
		3β-0-β-D-xylopyranosyl-(1-2)-β-D- β-D-glucuronopyranosyllup-20 (29)-en-28-oic acid	
		3β-0-β-D-glucopyranosyl-(1-3)-[β-D-xylopyranosyl-(1-2)]- β-D-glucuronopyranosyl-23-hydroxylup-20(29)-en-28-oic acid	
		3β-0-β-D-xylopyranosyl-(1-2)-β-D-glucuronopyranosyl-23-hydroxylup-20(29)-en-28-oic acid	
		3β-0-β-D-glucuronopyranosyl-23-hydroxylup-20(29)-en-28-oic acid	
		3β-D-0-β-D-glucopyranosyl-(1-3)-β-D-glucuronopyranosyl-23-oxo-lup-20(29)-en-28-oic acid	
		3β-0-β-D-xylopyranosyl-(1-2)-β-D-glucuronopyranosyl-23-oxo-lup-20(29)-en-28-oic acid	
		3β-0-β-D-glucopyranosyl-(1-2)-β-D-glucuronopyranosyl-23-oxolup-20(29)-en-28-oic acid	
		3β-0-β-D-glucuronopyranosyl-23-oxolup-20(29)-en-28-oic acid	

(Continued)

TABLE 10.3 (CONTINUED)
The List of Compounds Isolated from *Schefflera* Genus

Plant Species	Parts Used	Compound Name	Sources
S. capitata M. J. Cannon & Cannon		Oleanolic acid 3-O-α-L-arabinopyranoside	Jain and Khanna 1982
		Scheffleroside	Jain et al. 1977
		Hederagenin-3-o- α-L-arabinopyranoside	Jain and Khanna 1982
S. leucantha R. Vig.	Leaves	3-O-[a-L-rhamnopyranosyl (1 +2)-,!I-D-glucopyranosyl (1+2)-&D-glucuronopyranosyl J betulinic acid, 3-O-[a-t_-rhamnopyranosyl (I -2)~/3-D-xylopyranosyl (1+2)$-D-glucuronopyranosyl]betulinic acid and 3-O-[a-L-rhamnopyranosyl(I+2)-/l-D-glucopyranosyl (1 +2)-P-D-glucuronopyranosyl]oleanolic acid	Orasa et al. 1994
		3-O-[α-L-rhamnopyranosyl-(1-2)-β-D-glucopyranosyl-(1-2)-β-D-glucuronopyranosyl] oleanolic acid	Pancharoen et al. 1994
		3-0-[α-L-rhamnopyranosyl-(1-2) -β-D-glucopyranosyl-(1-2)-β-D-glucuronopyranosyl] betulinic acid	
		3-0-[α-L-rhamnopyranosyl-(1-2)-β-D-xylopyranosyl-(1-2)-β-D-glucuronopyranosyl] betulinic acid	
S. taiwaniana (Nakai) Kaneh.		3α, 23-dihydroxyolean-12-en-28-oic acid	Kuo et al. 2002
		3α-dihydroxyolean-12-en-23,28-dioic acid	
		3α-hydroxylup-20(29)-ene-23, 28-dioic acid	
		3α, 11 α -hydroxylup-20(29)-ene-28-oic acid	

(Continued)

TABLE 10.3 (CONTINUED)
The List of Compounds Isolated from *Schefflera* Genus

Plant Species	Parts Used	Compound Name	Sources
S. rotundifolia (Ten.) Frodin		3β-O-(β-D-glucopyranosyl-(1-3)-α-L-rhamnopyranosyl-(1-2)-α-L-arabinopyranosyl-hederaagenin-28-O-α-L-rhamnopyranosyl-(1-4)-O-β-D-glucopyranosyl-(106)-β--D-glucopyranosyl) ester	Braca et al. 2004
		3β-O-(β-D-glucopyranosyl-(1-3)-α-L-rhamnopyranosyl-(1-2)-α-L-arabinopyranosyl-hederaagenin-28-O-β-D-glucopyranosyl ester	
		3β-O-(α-L-rhamnopyranosyl-(1-2)-α-L-arabinopyranosyl-(1-2) hederaagenin-28-O-(β-D-glucopyranosyl-(1-4)- β-D-glucopyranosyl ester	
		3β-*O*-[α-L-rhamnopyranosyl-(1-3)-α-L-rhamnopyranosyl-(1-2)-α-L-arabinopyranosyl)-olean-12-ene-28-*O*-(1-6)-β-D-glucopyranosyl-(1-4)-β-D-glucopyranosyl) ester	
		3β-*O*-[α-L-rhamnopyranosyl-(1-2)-α-L-rhamnopyranosyl-olean-12-ene-28-*O*-(1-6)-β-D-glucopyranosyl-(1-4)-β-D-glucopyranosyl) ester	
		3β-0-β-D-xylopyranosyl-olean-12-ene-28-0-(α-L-rhamnopyranosyl-(1-4)-β-D-glucopyranosyl-(1-4)-β-D-glucopyranosyl) ester	
		3β-0-(α-L-rhamnopyranosyl-(1-2)-α-L-arabinopyranosyl)-lup-20(29)-ene-28-0- β-D-glucopyranosyl ester	
		3β-0-α-L-arabinopyranosyl-lup-20 (29)-ene-28-0- β-D-glucopyranosyl ester	

(*Continued*)

TABLE 10.3 (CONTINUED)
The List of Compounds Isolated from *Schefflera* Genus

Plant Species	Parts Used	Compound Name	Sources
S. fagueti Baill.		3β-0-[β-D-glucopyranosy-(1-2)- β-D-glucuronopyranosyl]-oleanolic acid 28-0- β-D-glucopyranoside	Cioffi et al. 2003
		3β-0-[β-D-glucopyranosy-(1-3)- β-D-xylopyranosyl)-16α-hydroxyolean-12-en-28-0-(β-D-galactopyranosyl)ester-30-oic acid	
		3β-0-(β-D-glucopyranosyl-(1-3)-α-L-rhamnopyranosyl-(1-2)-α-L-arabinopyranosyl-(1-6)-β-D-galactopyranosyl) ester	
		3β-0-(β-D-glucopyranosyl-(1-3)-α-L-rhamnopyranosyl-(1-2)-α-L-arabinopyranosyl)-23-hydroxylup-12-en-28-0-(α-L-rhamnopyranosyl-(1-4)-β-D-glucopyranosyl-(1-6)-β-D-galactopyranosyl) ester	
		3β-0-(α-L-rhamnopyranosyl-(1-2)-α-L-arabinopyranosyl)-23-hydroxylup-12-en-28-0-(α-L-rhamnopyranosyl-(1-4)-β-D-glucopyranosyl-(1-6)-β-D-galactopyranosyl) ester	
		3β-0-(α-L-rhamnopyranosyl-(1-2)-α-L-arabinopyranosyl)-lup-12-en-28-0-(α-L-rhamnopyranosyl-(1-4)-β-D-glucopyranosyl-(1-6)-β-D-galactopyranosyl) ester	

TABLE 10.4
Some of the Common Compounds Isolated from *Schefflera* Genus

Compound	Molecular Formula	Structure	Molecular Weight (g/mol)	Structure ID
Oleanolic acid	$C_{30}H_{48}O_3$	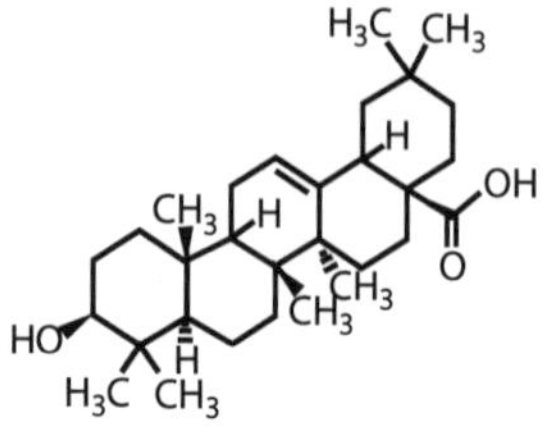	456.70032	Pubchem CID: 10494
Asiatic acid	$C_{30}H_{48}O_5$	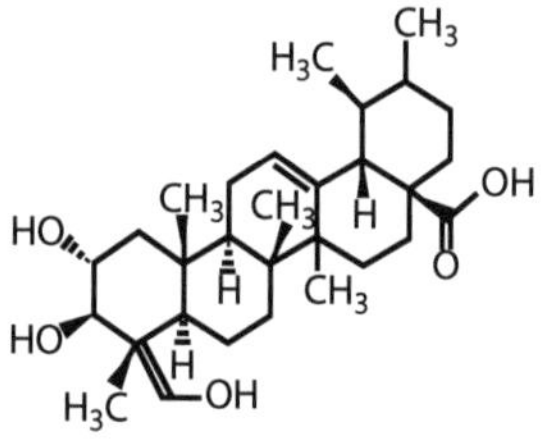	488.69912	Pubchem CID: 24825670
β-Amyrin	$C_{30}H_{50}O$	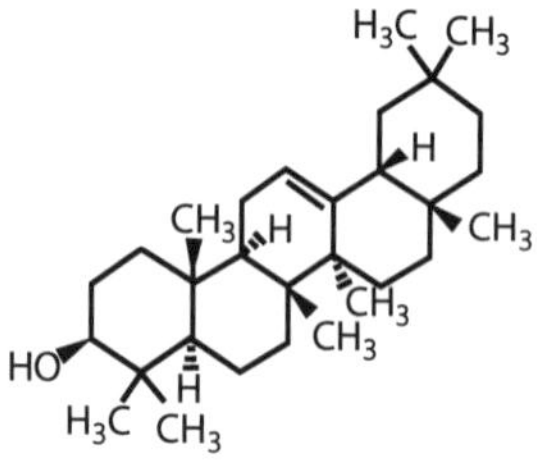	426.7174	Pubchem CID: 73145
Hederagenin	$C_{30}H_{48}O_4$	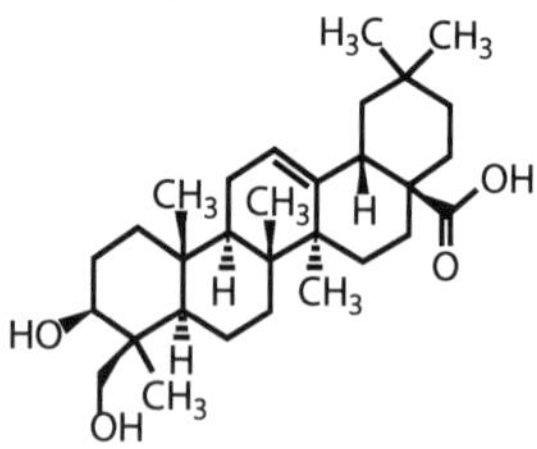	472.69972	Pubchem CID: 73299
Succinic acid	$C_4H_6O_4$	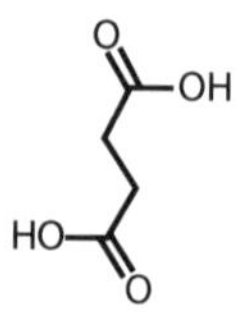	118.08804	Pubchem CID: 1110

(*Continued*)

TABLE 10.4 (CONTINUED)
Some of the Common Compounds Isolated from *Schefflera* Genus

Compound	Molecular Formula	Structure	Molecular Weight (g/mol)	Structure ID
Fumaric acid	$C_4H_4O_4$	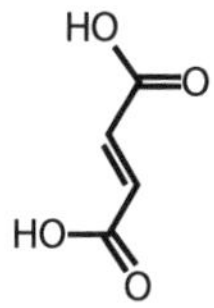	116.07216	Pubchem CID: 444972
Tartaric acid	$C_4H_6O_6$	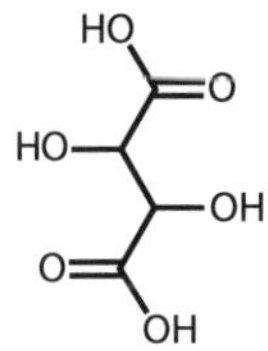	150.08684	Pubchem CID: 444305
3-acctyloleanolic acid	$C_{32}H_{50}O_4$	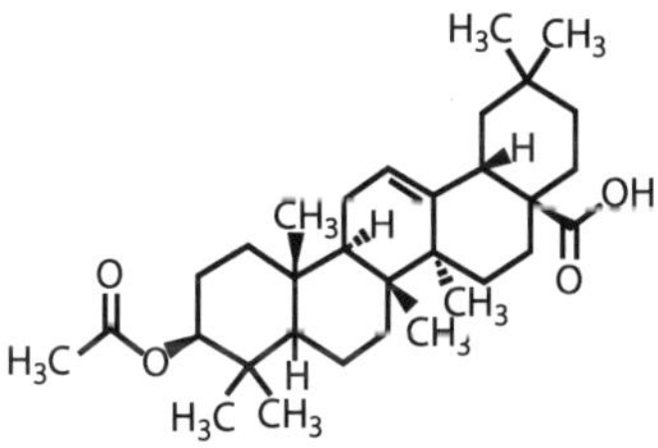	498.737	Pubchem CID: 151202
Mesembryanthe-moidigenic acid	$C_{30}H_{48}O_4$	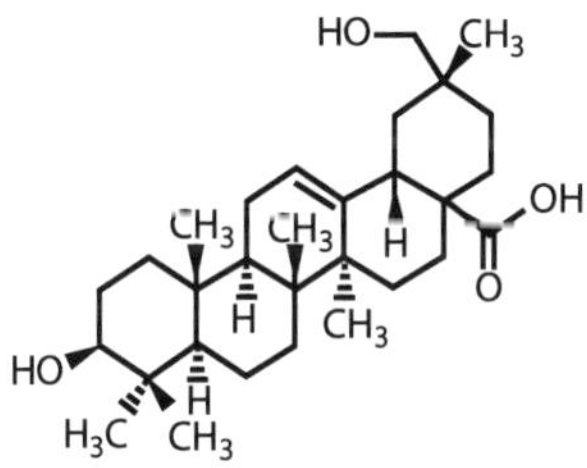	472.69972	Pubchem CID: 6452118
Quinatic acid	$C_{29}H_{44}O_4$	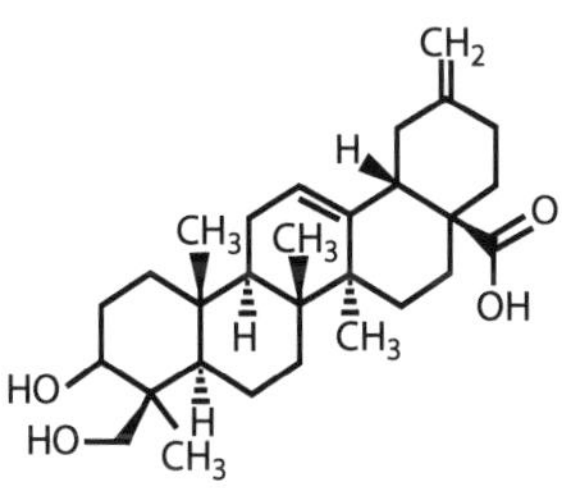	456.65726	Pubchem CID: 14191225

(*Continued*)

TABLE 10.4 (CONTINUED)
Some of the Common Compounds Isolated from *Schefflera* Genus

Compound	Molecular Formula	Structure	Molecular Weight (g/mol)	Structure ID
Betulinic acid	$C_{30}H_{48}O_3$	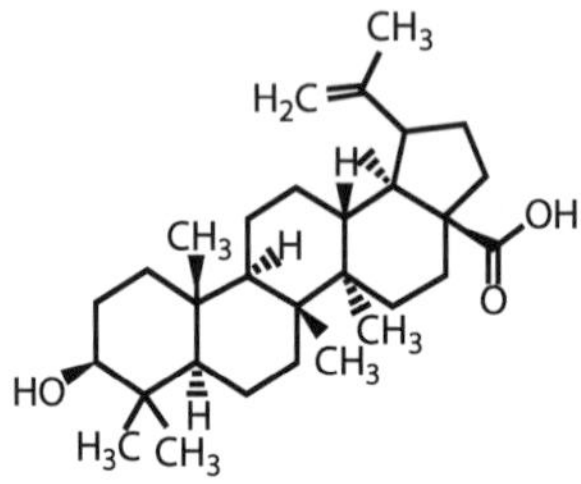	456.70032	Pubchem CID: 64971
Ursolic acid	$C_{30}H_{48}O_3$	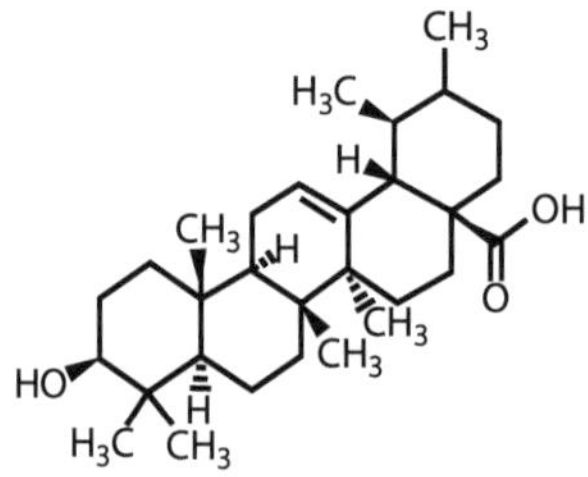	456.70032	Pubchem CID: 64945
Lutein	$C_{40}H_{56}O_2$	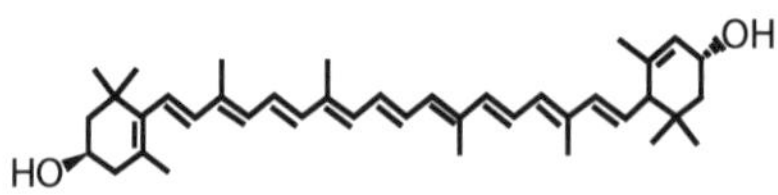	568.871	ChemSpider ID: 4444655

10.5 CONCLUSION

Schefflera is a worldwide-distributed species traditionally used in local remedies for various diseases. It is distributed mainly in tropical areas, rarely in temperate areas, it is mostly used by tribal peoples. Various reports about the pharmacological properties of *Schefflera* species and the compounds separated from it have been circulated. Therefore, in this literature we took an attempt to comprise all the reports and make sure all the information is correct and in one place. Traditionally many *Schefflera* species reported for various diseases (rheumatoid, inflammation, itches, allergy, among others) and some of the pharmacological properties have been evaluated in experimental animal models. The extracts of some species found effective against cancer cell lines, althought there is still a need to investigate other properties *in vivo* and study their mechanisms of action. Many effective compounds have been isolated from these plants for years. Even though there were many reports about these plants' pharmacological properties, there has not been a confirmation through clinical trial. This may be because of the lack of information available, so this review may help to inspire the serious study of *Schefflera* plants to develop therapeutics.

REFERENCES

Abdel, K. S. M. 2001. Two Triterpenoid saponins from *Schefflera arboricola* family Araliaceae. *Al-Azhar Journal of Pharmaceutical Sciences* 28, 39–47.

Abdolbaset, G., Gerhard, L., Liu, F., and Joachim, S. 2011. Ethnobotanical study of medicinal plants utilised by Hani ethnicity in Naban River Watershed National Nature Reserve, Yunnan, China. *Journal of Ethnopharmcology* 134, 651–667.

Adam, G., Lischewski, M., Phiet, H. V., Preiss, A., Schmidt, J., and Sungt, T. V. 1982. 3α hydroxy-lup-20(29)-ene-23, 28-dioic acid from *Schefflera octophylla. Phytochemisty* 21, 1385–1387.

Atsafack, S. S., Kuiate, J. R., Mouokeu, R. S., Koanga, M. M. L., Tchinda, A. T., Dieu, T. J., Magnifouet, N. H., Ebelle, E. R. M., Biyiti, L., and Ngono, N. R. A. 2015. Toxicological studies of stem bark extract from *Schefflera barteri* Harms (Araliaceae). *BMC Complementary and Alternative Medicine* 15, 44.

Braca, A., Autore, G., and Desimone, F. 2004. Cytotoxic saponins from. *Schefflera rotundifolia. Planta Medica* 70, 960–966.

Bridge, S. J. and Zhao, Y. W. 2012 Chinese peach glycosides study. *Chinese Herbal Medicines* 43, 2141–2145.

Chen, L., Zhu, J., and Lin, S. 2002. Chemical composition and physiological activity of the genus *Schefflera* profile. *Chinese Herbal Medicines* 25, 360–363.

Chen, Y., Tao, S., Zeng, F., Xie, L., and Shen, Z. 2015. Antinociceptive and anti-inflammatory activities of *Schefflera octophylla* extracts. *Journal of Ethnopharmacology* 2, 42–50.

Chizuko, M., Kazuhiro, O., Ryoji, K., Kazuo, Y., Nguyen, M. D., Nguyen, T. N., and Nguyen, K. Q. C. 1994. Oleanane and ursane glycosides from *Schefflera octophylla. Phytochemistry* 37, 1131–1137.

Chun, W., Ying, H. D., Wei, T., Man, M.L., Xia, W., Guo, C. W., Wen, C. Y., Guang, X. Z., and Yao, L. L. 2014. New ursane-type triterpenoid saponins from the stem bark of *Schefflera heptaphylla. Fitoter* 92, 127–132.

Cioffi, G., Braca, A., Autore, G., Morelli, I., Pinto, A., Venturella, F., and De Tommasi, N. 2003. Cytotoxic saponins from. *Schefflera faqueti. Planta Medica* 69, 1–7.

Cláudio, R. N. and Lucia, M. X. L. 2009. Antiplasmodial Natural Products. *Molecules* 16, 2146–2190.

Consolacion, Y. R. and Kathleen, L. 2005. Secondary Metabolites from *Schefflera odorata* Blanco. Philip. *Journal of Science* 134, 63–67.

Cun, Q. W., Ying, W., Wen, J. W., Lei, W., and Wen, C. Y. 2014. New oleanane saponins from *Schefflera kwangsiensis. Phytochemistry Letters* 10, 268–271.

Danna, J. L., Arnason, T. J., Razali, Y., Harini, S. R., Herwasono, S., Cindy, K. A., and John, M. P. 1995. Malaria remedies of the Kenyah of the Apo Kayan, East Kalimantan, Indonesian Borneo: A quantitative assessment of local consensus as an indicator of biological efficacy. *Journal of Ethnopharmacology* 49, 1–16.

Deepa, H. R. and Nalini, M. S. 2013. Phytochemical screening, total phenolic content and *in vitro* antioxidant studies of leaf, bark and flower extracts of *Schefflera* spp. (Araliaceae). *Journal of Applied Pharmaceutical Science* 3, 094–098.

Deepa, H. R. and Nalini, M. S. 2014. Evaluation of phytochemicals, total phenolics and antioxidant activities of *Schefflera* spp. (Araliaceae) from southern India. *Journal of Pharmacognosy and Phytochemistry* 2, 10–14.

Elizabeth, E. 1969. Preface to "Characteres Generum Plantarum" by J. R. and G. Forster, 1776; a translation, *New Zealand Journal of Botany* 7, 311–315.

Frodin, D. G., Porter, P. L., and Gregory, M. P. 2010. Schefflera (Araliaceae): Taxonomic history, overview and progress. *Plant Diversity and Evolution* 128, 561–595.

Gamble, J. S. 1919. *Flora of the Presidency of Madras.* Vol. 1. Adlard & Son Ltd., London, 474.

Ghimire, S. K., Lama, Y. C., Tripathi, G. R., Schmitt, S., and Aumeeruddy-Thomas, Y. 2001. Conservation of plant resources, community development and training in applied ethnobotany at Shey-Phoksundo national park and its buffer zone, Dolpa. Report Series No. 41, WWF Nepal Program, Kathmandu, Nepal.

Guo, F., Lin, S., and Li, Y. 2005. Isolation and identification of triterpenoids from *Schefflera arboricola. Chinese Journal of Medicinal Chemistry* 15, 294–296.

Hansen, L., Hammershoy, O., and Boll, P. M. 1986. Allergic contact dermatitis from falcarinol isolated from *Schefflera arboricola. Dermatitis* 14, 91–93.

Hoang, V. S., Pieter, B., and Paul, A. J. K. 2008. Traditional medicinal plants in ben en National Park, Vietnam. *Blumed* 53, 569–601.

Homervergel, G. O. and Young, D. K. 2014. Quantitative ethnobotanical study of the medicinal plants used by the Ati Negrito indigenous group in Guimaras Island, Philippines. *Journal of Ethnopharmacology* 157, 228–242.

http://keyserver.lucidcentral.org

http://www.theplantlist.org

http://www.tropicos.org

Hugo, J. B., Vichith, L., and Lars, B. 2012. Comparing medicinal plant knowledge using similarity indices: A case of the Brou, Saek and Kry in Lao PDR. *Journal of Ethnopharmacology* 141, 481–500.

Jayaprakash, G. K. and Rao, L. J. 2000. Phenolic constituent from lichen *Paramontrema stuppeum. Food Control* 56, 1018–1022.

Jain, G. K. and Khanna, N. M. 1982. Capitogenic acid: A new triterpene acid from *Schefflera capitata. Indian Journal of Chemistry* 21, 620–622.

Jain, G. K., Sarin, J. P. S. and Khanna, N. M. 1977. Constitution of scheffleroside: A spermicidal saponin from *Schefflera capitata. Indian Journal of Chemistry* 15B, 1130–1139.

Jian, Q. and Xian, Z., 1990. Studies on the Chemical Components Of *Schettlera delavayi* Haums Ex Diels West China. *Journal of Pharmaceutical Sciences* 5, 69–71.

Kalpana, G. and Rishi, R. B. 2009. Ethnomedicinal knowledge and healthcare practices among the Tharus of Nawalparasi district in central Nepal. *Forest Ecology and Management* 257, 2066–2072.

Kannika, P., Tran, V. O., Panee, S., Pratchaya, S., Sunee, C. K., and Surapol, N. 2011. Medicinal plants of the Mien (Yao) in Northern Thailand and their potential value in the primary healthcare of postpartum women. *Journal of Ethnopharmacology* 135, 226–237.

Kitajima, J., Shindo, M., and Tanaka, Y. 1990. Two new triterpenoids sulfates from the Leaves of *Schefflera octophylla. Chemical and Pharmaceutical Bulletin* 38, 714.

Kodithuwakku, N. D., Min, P., Yi-Lin, Z., Yanyan, Z., Yi-Dong, F., Wei-Rong, F., and Yun-Man, L. 2013. Anti-inflammatory and antinociceptive effects of Chinese medicine SQ gout capsules and its modulation of pro-inflammatory cytokines focusing on gout arthritis. *Journal of Ethnopharmacology* 150, 1071–1079.

Kuo, Y. H., Lo, J. M., and Chan, Y. F. 2002. Cytotoxic components from the leaves of *Schefflera taiwaniana. Journal of the Chinese Chemical Society* 49, 427–431.

Lene, H. and Per, M. B. 1986. The Polyacetylenic Falcarinol as the Major Allergen in *Schefflera arboricola. Phytochemistry* 25, 529–530.

Li, S., Long, C., Liu, F., Lee, S., Guo, Q., Li, R. and Liu, Y. 2006. Herbs for medicinal baths among the traditional Yao communities of China. *Journal of Ethnopharmacology* 108(1), 59–67.

Li, Y. L., But, P. P. H., and Ooi, V. E. C. 2005. Antiviral Activity and mode of Action of caffeoylquinic Acids from *Schefflera heptaphylla* (L.) Frodin. *Antiviral Research* 68, 1–9.

Li, Y. L., Ooi, S. M., and Wang, H. 2004. Antiviral activities of Medicinal herbs traditionally Used in southern mainland China. *Phytotherapy Research* 18, 718–722.

Lischewski, M., Schmidt, J., Preiss, A., Phiett, H. V., and Adam, G. 1984. 1 la-dihydroxylup-20(29)-ene-23,28-dioic acid from *Schefflera octophylla. Phytochemistry* 23, 1695–1697.

Liu, R., Gu, Q., and Cui, C. 2005. Dense pulse *Schefflera* chemical composition and anti-tumor activity. *Herbal* 36, 328.

Liu, X., Liu, D., and Liu, H. 2012. Screening of anti-inflammatory and analgesic extraction from *Schefflera arboricola* Hayata. *Anhui Medical and Pharmaceutical* 123–125.

Malar Selvi, B., Padmavathy, S., Brijitta, J., and Kaliaraj, G. S. 2015. Antibacterial and antioxidative activity of Methanolic Extract of *Schefflera racemosa* (Wight) Harms. *International Journal of Biological & Pharmaceutical Research* 2, 25–29.

Maria Francis, J. B. N., Aron, S., and Mehalingam, P. 2014. An ethnobotanical study of medicinal plants used by the Paliyars aboriginal community in Virudhunagar district, Tamil Nadu, India. *Indian Journal of Traditional Knowledge* 13, 613–618.

Matsui, K., Wirotesangthong, M., Thanakijcharoenpath, W., Mungmee, C., and Nishikawa, A. 2010. Inhibitory Effects of *Schefflera leucantha* Extract on Production of Allergic Mediators by Langerhans Cells and Mast Cells. *The Journal of Investigational Allergology and Clinical Immunology* 20, 463–468.

Melek, F. R., Toshio, M., Abdel Khalik, S. M., and El-Gindi, M. R. 2003. Triterpenoid saponins from *Schefflera arboricola. Phytochemistry* 63, 401–407.

Mendrika, R., Alyse, R. K., Harison, R., Vonjison, R., Charlotte, R., Nivo, R., Tabita, R., Fortunat, R., Reza, L., Armand, R., and Rainer, W. B. 2013. Medicinal plants used by women from Agnalazaha littoral forest (Southeastern Madagascar). *Journal of Ethnobiology and Ethnomedicine* 9, 1–13.

Min, Z., Shiling, Y., David, J. P., Greengrasst, P. M., and Norman, G. B. 1996. Triterpene Glycosides From *Schefflera bodinieri* Roots. *Phytochemistry* 43, 1313–1318.

Mthembu, X. S., van Heerden, F. R., and Fouche, G. 2010. Antimalarial compounds from *Schefflera umbellifera. South African Journal of Botany* 76, 82–85.

Nguyen, T. P., Le, T. V. H., Mai, D. T., Le, T. D., Phan, N. M., and Bui, T. D. 2015. Two new oleanane-type triterpene saponins from the leaves of *Schefflera sessiliflora* De P. V. *Phytochemistry Letters* 11, 102–105.

Orasa, P., Ittayat, U., Alter, C. T., and Kelvin, P. 1994. Triterpenoid glycosides from *Schefflera lucantha. Phytochemistry* 35, 987–592.

Pancharoen, O., Tuntiwachiwuttikul, P., Taylor, W. C., and Picker, K. 1994. Triterpenoid glycosides from *Schefflera lucantha. Phytochemistry* 35 (4), 987.

Pei, S. 1985. Preliminary Study of Ethnobotany in Xishuang Banna, People's Republic of China. *Journal of Ethnopharmacoology* 13, 121–137.

Peng, L. F., Xia, W. J., He, L., and Cui, T. 2012. Two new lupane triterpenoides from *Schefflera venulosa*. Chin. *Journal of Natural Medicines* 10, 0081–0083.

Plunkett, G. M., Porter, P., Lowry, I. I., David, G. F., and Jun, W. 2005. Phylogeny and Geography of Schefflera: Pervasive polyphyly in the largest genus of Araliaceae. *Annals of the Missouri botanical Garden* 92, 202–224.

Po, C., Ying, Z., Wanzhi, W., and Shou, Z. Y. 2000. Determination of divalent carboxylic acids in traditional Chinese herbal and patent medicines using HPLC with double cell quartz crystal detector. *Journal of Pharmaceutical and Biomedical Analysis* 23, 387–393.

Potduang, B., Chongsiriroeg, C., Benmart, Y., Giwanon, R., Supatanakul, W., and Tanpanich, S. 2007. Biological Activities of *Schefflera leucantha. African Journal of Traditional, Complementary, and Alternative medicines* 4, 157–164.

Punnam Chander, M., Kartick, C., and Vijayachari, P. 2015. Ethnomedicinal knowledge among Karens of Andaman & Nicobar Islands, India. *Journal of Ethnopharmacology* 162, 127–133.

Purohit, M. C., Pant, G., and Rawat, M. S. M. 1991. A betulinic acid glycoside from *Schefflera venulosa. Phytochemistry* 30, 2419.

Purohit, M. C. and Rawat, M. S. M. 1993. Spermicidal activity and chemical analysis of *Schefflera venulosa. Fitoterapia* 64, 274.

Ragasa, C. Y. and Lim, K. 2005. Secondary metabolites from *Schefflera odorata* Blanco. *Philippine Journal of Science* 13, 63–67.

Reyger, G. T. 1766. *Florae Gedanensis.* Vol. 2. 176.

Sabulal, B., George, V., Pradeep, N. S., and Dan, M. 2008. Volatile oils from the root, stem and leaves of S. stellata (Gaertn.) Harms (Araliaceae): Chemical characterization and antimicrobial activity. *Journal of Essential Oil Research* 20, 79–82.

Schmidt, J., Nam, V. V., Manfred, L., Hoang Van, P., Christine, K., and Gujnter, A. 1984. Long-chain fatty acid esters of 3a-hydroxy-lup-20(29)-ene-23,28-dioic acid and other triterpenoid constituents from the bark of *schefflera octophylla. Phytochemistry* 23, 2081–2082.

Srivastava, S. K. 1992. A new triterpenoid saponin from *Schefflera impressa. Journal of Natural Products* 55, 810–813.

Srivastava, S. K. and Jain, D. C. 1989. Triterpenoid saponins from plants of araliaceae. *Phytochemistry* 28, 644–647.

Sung, T. V. and Adam, G. A. 1991. Sulphated Triterpenoid Saponin From *Schefflera octophylla. Phytochemistry* 30, 2717–2720.

Sung, T. V., Lavaud, C., and Porzel, A. 1992. Triterpenoids and their glycosides from the bark of *Schefflera octophylla. Phytochemistry* 31, 227–231.

Sung, T. V., Steglich, W., and Adams, G. 1991. Triterpene Glycosides From *Schefflera octophylla. Phytochemistry* 30, 2349–2356.

Tapondjou, L. A., Mitaine, O. A. C., Miyamoto, T., Lerche, H., Mirjolet, J. F., Guilbaud, N., Lacaille D. M. A. 2006. Triterpene saponins from *Schefflera abyssinica. Biochemical Systematics and Ecology* 34, 887–889.

Tetyana, P., Prozesky, E. A., Jäger, A. K., Meyer, J. J. M., and Staden, J. 2002. Some medicinal properties of *Cussonia* and *Schefflera* species used in traditional medicine. *South African Journal of Botany* 68, 51–54.

Tommasi, N. D. and Pizza, C. 1997. Triterpenoid s aponins from *Schefflera divaricata. Journal of Natural Products* 60, 663–668.

Wang, Y., Chen, R., and Yu, D. 2013. Advances of chemical constituents and pharmacological activities of *Schefflera* genus. *Zhongguo Zhong Yao Za Zhi* 38, 2254–2263.

Witthawaskul, P., Ampai, P., Kanjanapothi, D., Taesothikul, T., and Lertprasertsuke, N. 2003. Acute and subacute toxicities of the saponin mixture isolated from *Schefflera leucantha* Viguier. *Journal of Ethnopharmacology* 89, 115–121.

Wu, C., Wang, L., Yang, X. X., Duan, Y. H., Dai, Y., Jiang, R. W., Ye, W. C., and Li, Y. L. 2011. A new ursane-type triterpenoid from *Schefflera heptaphylla* (L.) Frodin. *Journal of Asian Natural Products Research* 13, 434.

Xu, J., Peng, X., and Li, W. 2006. *Schefflera* phytochemical, pharmacological and clinical profile. *Hunan University of Traditional Chinese Medicine* 26 (5), 62.

Yanfen, C., Shuhong, T., Fanlin, Z., Luwei, X., and Zhibin, S. 2015. Antinociceptive and anti-inflammatory activities of *Schefflera octophylla* extracts. *Journal of Ethnopharmacology* 17, 42–50.

Yao, L. L., Chung-Man, Y., Lawrence, C. M. C., Ying-Zhou, C., and Vincent, E. C. O. 2009. Chemical Composition and Antiproliferative Activity of Essential Oil from the Leaves of a Medicinal Herb, *Schefflera heptaphylla. Journal of Phytotherapy* 23, 140–142.

Yaolan, L., Paul, P. H. B, and Vincent, E. C. O. 2005. Antiviral activity and mode of action of caffeoylquinic acids from *Schefflera heptaphylla* (L.) Frodin. *Antiviral Research* 68, 1–9.

Yemele, M. D., Telefo, P. B., Lienou, L. L., Tagne, S. R., Fodouop, C. S. P., Goka, C. S., Lemfack, M. C., and Moundipa, F. P. 2015. Ethnobotanical survey of medicinal plants used for pregnant women's health conditions in Menoua division, West Cameroon. *Journal of Ethnopharmacology* 160, 14–31.

Ying, W., Lei, W., Wen-Jing, W., Xiao-Qi, Z., Hai-Yan, T., Qing-Wen, Z., Yao-Lan, L., and Wen-Cai, Y. 2014. New triterpenoid saponins from the aerial parts of *Schefflera kwangsiensis. Carbohydrate Research* 385, 65–71.

Zerabruk, S. and Yirga, G. 2012. Traditional knowledge of medicinal plants in Gindeberet district, Western Ethiopia. *South African Journal of Botany* 78, 165–169.

Zhao, Z. M., Matsunami, K., and Otsuka, H. 2010. Scheffierins A-G, new triterpene glucosides from *the* leaves of *Schefflera arboricola. Chemical and Pharmaceutical Bulletin* 58, 1343–1348.

Zheng, X. L., and Xing, F. W. 2009. Ethnobotanical study on medicinal plants around Mt. Yinggeling, Hainan Island, China. *Journal of Ethnopharmacology* 124, 197–210.

Zhu, M., David, J. P., Greengrasst, P. M., and Bowery, N. G. 1996. Triterpenoids and a triterpene glycoside from *Schefflera bodinieri* leaves. *Phytochemistry* 43, 1307–1311.

Zhu, M., and Li, R. C. 1999. Receptor binding activities of Schefflera triterpenoids and oligosaccharides. *Planta Medica* 65, 99–103.

11 Selective Chinese Viviparous Ferns, Their Bioactive Principles and Economical Values

A Review

Xavier Ravi Baskaran, Antony Varuvel Geo Vigila, Wenbo Liao and Zhang Shouzhou

CONTENTS

11.1 Introduction 223
11.2 Vivipary Plants 224
11.3 Medicinal, Edible Uses and Bioactive Principles of Viviparous Ferns 224
11.4 Plant Hormones and the Role of ABA in Vivipary 229
11.5 Abscisic Acid in Antioxidant Induction 231
References 231

11.1 INTRODUCTION

Ferns first appear in the fossil record 360 million years ago in the late Devonian period. Genetically similar to seed plants, ferns represent a critical clade for comparative evolutionary studies in land plants (Pryer et al. 2002; Barker and Wolf 2010). Thus, as an early tracheophyte, Pteridophytes are an important outgroup for studying the evolution of wood, seeds, pollen, flowers and fruit among other economically important characteristics found in seed plants, as well as the evolution of development in these complex structures and the expansion of gene families associated with seed plant evolution (e.g. transcription-associated proteins) (Nakazato et al. 2008; Barker 2009; Barker and Wolf 2010). Though, we have not yet understood ferns on the basis of special character on vivipary, in some pteridophytes, fertilization takes place inside female gametophytes, on the sporangium on the mother plant itself. Vivipary is considered as an important step in the evolution of seed habit in vascular plants. Furthermore, megaspores remain within the sporangium, which germinate and develop into the gametophyte and are fertilised. The gametophytes of *Selaginella* show complete dependence of gametophyte upon sporophyte, as in angiosperms. Moreover, the viability period of chlorophyll is frequently very short

(Lloyd and Klekowski 1970; Pandey 2003), as seen in many vivipary cases. However, this character is evolutionarily important and rare in homosporous ferns like *Grammitis medialis* (Irudayaraj et al. 2003). In general, vivipary can be divided into two major different forms based on their mode of formation, namely, true vivipary and pseudovivipary. True vivipary describes that plants produce sexual offspring and pseudovivipary describes that plants that produce apomictic or asexual propagules, such as bulbils or plantlets, in the place of sexual reproductive structure with the mother plant itself. Pseudovivipary plants exclude vegetative budding not associated with reproductive axes (Elmqvist and Cox 1996; Tomlinson and Cox 2000).

11.2 VIVIPARY PLANTS

Naturally, some fern genera and families produce viviparous plantlets by asexual reproduction on the sporophyte fronds of *Adiantum*, *Asplenium*, *Camptosorus*, *Cystopteris*, *Diplazium*, *Tectaria*, *Woodwardia*, *Cyathea arborea*, *Asplenium bulbiferum*, *Dennstaedtia scabra*, Grammitidaceae, Hymenophyllaceae and Vittariaceae (Mickel 1967; Faden 1973; Page 1979; Bir 1995; Rozario et al. 2001; Graham 2011). The germination of fern spores within the sporangium without a dormancy period is called vivipary, and contain very few homosporous pteridophytes (Lloyd and Klekowski 1970; Pandey 2003; Irudayaraj et al. 2003). Likewise, the germination of seeds was reported in gymnosperms plants namely, *Biota orientalis* (Gahalain et al. 2006), *Cupressus torulosa* (Majumder et al. 2010), *Ginkgo biloba* (Favre-Ducharte 1958), *Podocarpus makoyi* (Lloyd 1902), *Podocarpus macrophyllus* (Mahabale 1961), *Ephedra trifurca* (Coulter and Chamberlain 1955), *Pinus wallichiana* (Wali and Tiku 1965), Angiosperms (Elmqvist and Cox 1996), mangroves (Farnsworth and Farrant 1998), *Kalanchoe daigremontiana* (Garces et al. 2014), *Eleusine coracana* and *Setaria viridis* (Li 1950), monocot/poaceae (Lee and Harmer 1980). An asexual reproduction in the form of gemmae cup and adventive branches were reported in Bryophytes (Cavers 1903). Table 11.1 summarizes the viviparous ferns (see Figure 11.1), their altitude, bioactive compounds and their ornamental and pharmacological uses.

11.3 MEDICINAL, EDIBLE USES AND BIOACTIVE PRINCIPLES OF VIVIPAROUS FERNS

Some viviparous ferns are edible, for example, the shoots and fiddleheads of the 'hen and chickens fern' (*Asplenium bulbiferum*) were eaten in the form of greens by the early Maori. Their leaves contain antioxidant flavonoids, such as kaempferol glucosides (Imperato 1984). Secondary metabolites, namely, Apigenin-7-O-rhamnoside, luteolin-7-O-dirhamnoside, genkwanin-4'-O-glucosylrhamnoside, apigenin-7-O-rhamnoside-4'-O-glucosyl-rhamnoside and its derivatives and apigenin 7,4'-di-O-rhamnoside were isolated in *Asplenium normale* (Sadamu et al. 2003). Out of eight flavone glycosides from the fronds of *Asplenium normale*, six of them, namely, apigenin 7-O-dirhamnoside, 7-O-glucosylrhamnoside, luteolin 7-O-dirhamnoside, 7-O-glucosylrhamnoside, genkwanin 4'-O-glucosylrhamnoside and vicenin-2 were isolated. The remaining two glycosides were tentatively

TABLE 11.1
Selective Viviparous Ferns in China and Their Various Uses

S. No	Viviparous Ferns	Altitude	Uses	References
1	*Asplenium normale* D. Don	600–2500 m	Flavonoid compounds	Iwashina 1990; Umikalsom and Harborne 1991; Iwashina and Matsumoto 1994; Iwashina et al. 2010; Sadamu et al. 2003
2	*Asplenium ruprechtii* Sa. Kurata	300–2000 m	–	–
3	*Asplenium finlaysonianum* Wall. Ex Hook	700–1100 m	For dysentery	Vishal 2006
4	*Asplenium pellucidum* Lam.	800 m	Kaempferol polyglycosides	Umikalsom et al. 1994
5	*Asplenium yoshinagae* Makino	1100–3300 m	–	–
6	*Asplenium tripteropus* Nakai	400–100 m	–	–
7	*Cyclosorus proliferus* (Retz.) Tardieu ex Tardieu and C. Chr.	100–950 m	–	–
8	*Cyclosorus tylodes* (Kunze) Panigrahi	800–2300 m	To prepare liquor; endangered in South India; Presence of Quercetin	Manickam 1995; Srivastava and Adi community 2009; Pathania et al. 2012
9	*Athyrium clarkei* Bedd.	1500–2700 m	Ornamental fern in China; Food plants for larvae	Zeng 2008; Wikipedia
10	*Athyrium iseanum* Rosenst.	50–2800 m	Food plants for larvae	Wikipedia
11	*Woodwardia orientalis* Sw.	100–1100 m	Polyphenolics herpes simplex virus type 1 (HSV-1), poliovirus type 1 and measles virus; Herbal medicine	Wada et al. 1992; Xu et al. 1993; Tsai and Hwang 1999

(*Continued*)

TABLE 11.1 (CONTINUED)
Selective Viviparous Ferns in China and Their Various Uses

S. No	Viviparous Ferns	Altitude	Uses	References
12	*Woodwardia unigemmata* (Makino) Nakai	450–3000 m	Herbal medicine; anti-HIV activity; anti-influenza herbal drug; Antidiabetic activity; cakes, noodles and liquor	Tsai and Hwang 1999; Chang and Yeung 1988; Collins et al. 1997; Dai et al. 2003; Peng 2008; Yun et al. 2009
13	*Bolbitis heteroclita* (C. Presl) Ching	50–1500 m	Chinese Pharmacopoeia; ornamental	Ma et al. 2010; Wikipedia
14	*Polystichum anomalum* (Hook. ex Arn.) C. Chr.	250–1850 m	Rare in Sri Lanka	Subhash et al. 2008
15	*Polystichum attenuatum* Tagawa and K. Iwats.	800–2200 m	–	–
16	*Polystichum craspedosorum* (Maxim.) Diels.	500–2100 m	Bioaccumulation fern	Zhao and Guo 2013
17	*Polystichum lentum* (D. Don) T. Moore	1800–2120 m	Ornamental and economical value in India	Srivastava and Uniyal 2014
18	*Monachosorum flagellare* (Maxim.) Hayata	800–1500 m	Monachosorins and methylmonachosorin; traditional Chinese medicine	Satake et al. 1985; Hori et al. 1987; Wikibook
19	*Monachosorum henryi* Christ.	500–1600 m	Monachosorins and methylmonachosorin; Febrifuge from poisoning in Bhutan; traditional Chinese medicine	Satake et al. 1985; Hori et al. 1987; Wangchuk 2014; Wikibook
20	*Tectaria fauriei* Tagawa	540–950 m	–	–

FIGURE 11.1 Viviparous ferns of China: (a) *Asplenium tripteropus*; (b) *Cyclosorus proliferus*; (c) *Woodwardia oricentalis*; (d) *Woodwardia unigemmata*; (e) *Polystichum craspedosorum*; (f) *Tectaria fauriei.*

characterized as genkwanin 4'-O-glycoside and 6,8-di-C-glycosylluteolin (Iwashina 1990). Two new flavone rhamnosides, apigenin 7-O-alpha-L-rhamnopyranosyl-(1>4)-O-alpha-L-rhamnopyranoside and apigenin 7-O-alpha-L-rhamnopyranosyl-(1>4)-O-alpha-L-rhamnopyranoside-4'-O-alpha-L-rhamnopyranoside were isolated with two known C-glycosylflavones, vicenin-2 and lucenin-2 from the fronds of *Asplenium normale* D. Don and established by UV, LC-MS as the characterization of acid hydrolysates and 1H and 13C NMR spectroscopy (Iwashina et al. 2010).

Asplenium normale had shown 7-O-dirhamnosides of apigenin, luteolin, genkwanin 4'-O-glucosylrhamnoside and apigenin 7-O-rhamnoside-4'-O-glucosylrhamnoside (Iwashina and Matsumoto 1994). Moreover, Kaempferol 3 O-methyl ether was also found in *A. normale* (Umikalsom and Harborne 1991). The Kaempferol polyglycosides have been reported in the Malaysian vivipary fern, *Asplenium pellucidum* Lam. (Umikalsom et al. 1994). Caroline et al. (2003) studied the phylogenetic analysis of *Asplenium tripteropus.* The roots of *Asplenium finlaysonianum* are pounded into a paste with a small quantity of water for dysentery (Vishal 2006).

Five species of *Woodwardia* rhizomes are edible in China (Liu et al. 2012). The carbohydrate starch is present in the rhizomes of *Woodwardia unigemmata* and also frequently used by people to prepare cakes, noodles and liquor in China (Yun et al. 2009; Dai et al. 2003). Both *Woodwardia* viviparous ferns (*W. unigemmata* and *W. orientalis*, along with some other ferns) have been used for the preparation of Chinese herbal medicine known as Blechni Rhizoma or otherwise known as *Dryopteris crassirhizomae.* Rhizoma has traditionally been used to treat illnesses such as hepatitis B virus (HBV), hepatitis C virus (HCV), human immunodeficiency virus (HIV), injury, swelling, fever, measles and erysipelas (Tsai and Hwang 1999). The partially-purified aqueous extract of *W. unigemmata* exhibited 83.6% of the highest inhibition against HIV-I gp120 and 35 μg/ml of immobilized CD4 receptors; and also showed α-glucosidase (100%), β-glucosidase (27.1%) and β-glucuronidase (44.5%) inhibitions, respectively (Collins et al. 1997). The kaempferol-3-O-α-L-rhamnopyranoside-7-O-α-L-rhamnopyranoside and kaempferol-3-O-α-L-(4-O-acetyl)-rhamnopyranoside-7-O-α-L-rhamnopyranoside were isolated from an alcoholic extract of *W. unigemmata*, and the anti-influenza herbal drug was also developed by Peng (2008). The boiling water extracts of *W. unigemmata* with 26 medicinal plants showed anti-HIV activity; its extracts inhibit HIV in the H9 cells in particular (Chang and Yeung 1988). A new glucoside, Woodorientalis, from the hot aqueous and methanol extracts of the rhizomes of *Woodwardia orientalis* showed *in vitro* antiviral activity against herpes simplex virus type 1 (HSV-1), poliovirus type 1 and the measles virus by a plaque reduction assay (Xu et al. 1993). Four lignans (blechnic acid, 7-epiblechnic acid, 8-epiblechnic acid and brainic acid) were isolated from the Blechnaceous fern, *W. orientalis* (Wada et al. 1992).

According to Wangchuk (2014), based on Bhutanese traditional medicine (BTM), the nodal lump of *Monachosorum henryi* is used as febrifuge to treat poisoning. *M. flagellare* and *M. henryi* are also used in traditional Chinese medicine (Wikibook). The *Polystichum lentum* (D. Don) Moore is a terrestrial fern with erect rhizomes and extensive use as an indoor plant, which gained immense ornamental and economic value in the nursery trade (Srivastava and Uniyal 2014). *Polystichum craspedosorum*, a bio accumulation fern, is used to absorb Arsenic from Arsenic-contaminated water in China (Zhao and Guo 2013), is also a viviparous fern. *P. anomalum* is very rare in Sri Lanka and Southern India (Subhash et al. 2008). In China, *Athyrium clarkei* used as an ornamental fern (Zeng 2008) and their bulbils is positioned subterminally beneath the lamina. They are not aphlebium-like, aphlebia being a non-reproductive, non-detachable, basal, pinna-like outgrowth of a stipe-base, modified into a finely dissected, veinless and often lamina-less organ (Kholia and Fraser-Jenkins 2011). The *Athyrium* species are used as food plants by the larvae of some *Lepidoptera* species, including small angle shades and *Sthenopis auratus* (Wikipedia). Ma et al. (2010) reported the Chinese Pharmacopoeia (2005 version), which listed 51 species from 24 families as medicinal Pteridophyte species, in which *Bolbitis heteroclita* was included and analysed with five DNA sequence markers. The *Bolbitis heteroclita* is cultivated as an ornamental plant in aquariums and garden ponds (Wikipedia). Dried frond powder of *Cyclosorus tylodes* mixed with pounded rice and water is kept for at least 2 d, intended for fermentation to prepare local liquor, Apong (Srivastava and Adi community 2009).

11.4 PLANT HORMONES AND THE ROLE OF ABA IN VIVIPARY

Julius von Sachs, a German botanist (1832–1897) suggested that plants produce, transport and perceive 'organ-forming substances' which are responsible for the formation and growth of different plant organs. Plant hormones such as abscisic acid (ABA), gibberellins (GA), ethylene, brassinosteroids (BR), auxins, cytokinins and other signalling molecules have profound effects on plant development at low concentrations. Hormones are chemical messengers which are useful to communicate between cells, tissues and organs of higher plants. Plant seeds contain an embryo surrounded by covering layers and have the important function of ensuring the establishment of a new plant generation. Plant hormones are extremely important for the regulation of seed dormancy and germination (Koornneef et al. 2002; Finkelstein 2004). Moreover, plant hormones are a special group of chemical substances which control growth and development in plants. Hormones play a vital role in the plant's life cycle, such as cell division and extension, seed and bud dormancy, seed germination, flowering, fruit set and ripening and cutting rooting (Preece and Read 1993; Unsal 1993; Foskett 1994; Eris 1995; Hartmann et al. 1997).

ABA has great importance in the developmental processes, induction of seed dormancy and protein and lipid synthesis, in which the desiccation and inhibition of an embryo is tolerated for vegetative development. In mature plants, ABA responds to drought through stomata aperture, which also acts on an adaptation to stress conditions like low temperature, salinity, hypoxia and pathogen attacks. In general, ABA is considered a hormone with an inhibitory activity on growth (Nambara and Marion-Poll 2005; Pozo et al. 2005). Thus, those plants deficient in ABA production or with increased sensitivity fail to enter quiescence resulting in precocious germination or vivipary, and are accompanied by a reduction of maturation and associated gene expression (Cao et al. 2007). Conversely, dry conditions increase the ABA levels (Rodriguez-Gacio et al. 2009), a condition not favourable for vivipary, which is linked to low, not high levels of ABA (Farnsworth and Farrant 1998). Different production levels and compartmentalization of phytohormones (ABA) as well as different physiological, genetic and ecological factors, individually or in concert, also play a major role in the rate of precocious germination (Cota-Sanchez et al. 2011). Moreover, ABA is a unique molecule found in organisms across kingdoms from bacteria to animals, which suggested its ubiquitous and versatile role in the physiological functions of various organisms. ABA is widely known to be one of the growth regulators of tracheophytes, and is also known universally for its hormonal involvement in stress processes. ABA showed stress-dependent biosynthesis, which transported the target cells that enable the plant to cope better with stressful conditions. The widespread occurrence across the entire tree of life suggests an ancient origin of ABA, which plays a vital role in modulating cellular responses to environmental signals, e.g. water-deficient stress (Cuming et al. 2007). However, ABA is the major hormone involved in induction and maintenance of dormancy by pre-harvest sprouting/viviparous mutants in rice (Fang and Chu 2008).

Furthermore, the absence of xanthophylls was associated with reduced ABA content, which in turn correlated with vivipary. Even though, the kernels of VP-8 had a reduced ABA content, xanthophylls were also present. Seedlings of carotenoid-deficient mutants were rescued from viviparous kernels and contained less ABA than wild-type seedlings grown in the same way (Neill et al. 1986). Precocious germination can be induced in soybean seeds after a slow drying treatment, which causes a decrease of its endogenous ABA (Ackerson 1984b); whereas, mid-stage soybean embryos cultured in the absence of ABA exhibit precocious germination. The young embryos undergo normal embryogenesis in the presence of ABA (Ackerson 1984a). The suppression of precocious germination by ABA content derived from the studies of mutants defective in ABA metabolism (Black 1991; McCarty 1995). Also, the exogenously supplied ABA prevented the germination of excised, immature cotton embryos and embryos placed on ABA–free medium exhibited precocious germination/vivipary (Ihle and Dure 1970). Furthermore, ABA–deficient or ABA–insensitive *Arabidopsis* mutants show reduced seed maturation and dormancy (Finch-Savage and Leubner-Metzger 2006; Koornneef et al. 2002; Leon-Kloosterziel et al. 1996). Unlike in *Arabidopsis*, cereal plants and maize embryos from ABA–deficient mutants germinate precociously (vivipary) on the ear (White et al. 2000).

The following ferns have been reported for the presence of ABA. All the ferns are non-viviparous, such as *Anemia phyllitidis* (Cheng and Schraudolf 1974); *Lygodium japonicum* (Yamane et al. 1980); *Cibotium glaucum* and *Dicksonia antarctica* (Yamane et al. 1988); *Marsilea drummondii* (Pilate et al. 1989); *Pteridium esculentum* and *D. antarctica* and a lycophyte (*Selaginella kraussiana*) (Brodribb and McAdam 2011; McAdam and Brodribb 2012). The presence of ABA inhibits precocious germination or vivipary in ferns as described by Sakata et al. (2014) and ABA–related genes are summarised in Table 11.2.

TABLE 11.2
ABA–Related Genes in Plants

Function	Genes
ABA metabolism	ABA1/ZEP, ABA4, NCEDs, ABA2, AAO3, ABA3, BG1, CYP707As
ABA transport	ABCG25, ABCG40, AIT
ABA signalling	PYR/PYL/RCARs, Group A PP2C, Subclass III SnR K2, AB13, AB14, AB15, SLAC1
Ca^{2+}-dependent factor	CIPK/SnRK3, CDPK, CBL, CaM/CML

11.5 ABSCISIC ACID IN ANTIOXIDANT INDUCTION

ABA can induce the expression of antioxidant genes encoding Cu/Zn-superoxide dismutase (SOD) (Guan and Scandalios 1998a; Kaminaka et al. 1999), Mn-SOD (Zhu and Scandalios 1994; Bueno et al. 1998; Kaminaka et al. 1999), Fe-SOD (Kaminaka et al. 1999) and catalase (CAT) (Anderson et al. 1994; Guan and Scandalios 1998b; Guan et al. 2000) in plants. Moreover, ABA not only induces the expression of antioxidant defence genes, but also enhances the activities of antioxidant enzymes in plants. ABA increases the activities of total SOD, Cu/Zn-SOD, Mn-SOD, Fe-SOD, CAT, ascorbate peroxidase (APX) and glutathione reductase (GR) in plants (Anderson et al. 1994; Prasad et al. 1994; Bueno et al. 1998; Gong et al. 1998; Guan and Scandalios 1998b; Bellaire et al. 2000).

Accessory pigments in photosynthesis and photoprotectors prevent photooxidative damage, furthermore, carotenoids can also be the precursors to the hormone ABA (Bewley 1997). Earlier studies proved that carotenoids could also act as antioxidants to quench excessive free radicals and ROS generated from photooxidation (Hirayama et al. 1994; Woodall et al. 1997; Niyogi et al. 1998; Niyogi 1999). In addition, carotenoids play an essential role in photoprotection in plants. During photosynthesis, excess absorbed energy can be eliminated as heat by de-exciting 1Chl through the process of non-photochemical quenching of chlorophyll fluorescence (NPQ), minimizing the generation of harmful reactive oxygen species (ROS) (Niyogi 1999; Tracewell et al. 2001; Ma et al. 2003).

Usually, plants produce secondary metabolites due to abiotic stress generating the plant hormone ABA. In contrast, the nutrient content of the viviparous plant (due to absence/less of ABA) *Festuca vivipara* was high in comparison with the closely-related species *F. ovina* (Harmer and Lee 1978). Chinese viviparous ferns possess diverse flavonoids, which plays a vital role in medicinal activities. In previous studies, natural flavonoids from plant sources have already been documented. Flavonoids can act as proton donators and proved free radical scavenging activity, anti-inflammatory, oestrogenic, antimicrobial, anti-allergic, cardiovascular, antidementia and cytotoxic anti-tumor activities (Commenges et al. 2000; Wojdylo et al. 2007; Pereira et al. 2009). Selected flavonoids such as quercetin, kaempferol and myricetin were ascertained to be effective inhibitors of platelet aggregation in dogs and monkeys (Osman et al. 1998). In conclusion, the presence of enormous and important bioactivities of viviparous ferns have proven their economic value, which reveal that ferns are not only for ornamental but also medicinal purposes. Even though, viviparous fern possess fewer or no ABA, viviparous ferns produce the secondary metabolites and important medicinal activities. Furthermore, biosynthesis of secondary metabolites and their mechanisms in viviparous plants should be studied for better understanding.

REFERENCES

Ackerson, R. C. 1984a. Regulation of soybean embryogenesis by abscisic acid. *Journal of Experimental Botany* 35, 403–413.

Ackerson, R. C. 1984b. Abscisic acid and precocious germination in soybeans. *Journal of Experimental Botany* 35, 414–421.

Anderson, M. D., Prasad, T. K., Martin, B. A., and Stewart, C. R. 1994. Differential gene expression in chilling-acclimatedmaize seedlings and evidence for the involvement of abscisic acid in chilling tolerance. *Plant Physiology* 105, 331–339.

Barker, M. S. 2009. Evolutionary genomic analyses of ferns reveal that high chromosome numbers are a product of high retention and fewer rounds of polyploidy relative to angiosperms. *American Fern Journal* 99, 136–137.

Barker, M. S. and Wolf, P. G. 2010. Unfurling fern biology in the genomics age. *Bioscience* 60, 177–185.

Bellaire, B. A., Carmody, J., Braud, J., Gossett, D. R., Banks, S. W., Lucas, M. C., and Fowler, T. E. 2000. Involvement of abscisic acid-dependent and -independent pathways in the upregulation of antioxidant enzyme activity during NaCl stress in cotton callus tissue. *Free Radical Research* 33, 531–545.

Bewley, J. D. 1997. Seed germination and dormancy. *Plant Cell* 9, 1055–1066.

Bir, S. S. 1995. Extant ferns: Their fossil base and diversities of morphology, anatomy, cytology, palynology and gametophytes. *Indian Fern Journal* 12 (1–2): IX–XII.

Black, M. 1991. Involvement of ABA in the physiology of developing and mature seeds. In *Abscisic acid: Physiology and biochemistry.* Ed. W. J. Davies, and H. G. Jones, Oxford, Bios Scientific Publishers. 99–124 pp.

Brodribb, T. J., and McAdam, S. 2011 Passive origins of stomatal control in vascular plants. *Science* 331, 582–585.

Bueno, P., Piqueras, A., Kurepa, J., Savoure, A., Verbruggen, N., van Montagu, M., and Inze, D. 1998. Expression of antioxidant enzymes in response to abscisic acid and high osmoticum in tobacco BY-2 cell cultures. *Plant Science* 138, 27–34.

Cao, X., Costa, L. M., Petit, C. B., Kbhaya, B., Dey, N., Perez, P., McCarty, D. R., Marcos, J. F. G., and Becraft, P. W. 2007. Abscisic acid and stress signals induce viviparous 1 expression in seed and vegetative tissues of maize. *Plant Physiology* 143, 720–731.

Caroline, J. V. D. H., Ronald, L. L. V., and Mark, W. C. 2003. Phylogenetic analysis of Asplenium subgenus Ceterach (Pteridophyta: Aspleniaceae) based on plastid and nuclear ribosomal ITS DNA sequences. *American Journal of Botany* 90(3), 481–495.

Cavers, 1903. On asexual reproduction and regeneration in Hepaticae. *New Phytologist* 2(6), 121–133.

Chang, R. S. and Yeung, H. W. 1988, Inhibition of growth of human immunodeficiency virus in vitro by crude extracts of Chinese medicinal herbs. *Antiviral Research* 8, 163–176.

Cheng, J. Y. and Schraudolf, V. 1974. Nachweis von Abscisinsure in Sporen and jungen Prothallien von *Anemia phyllitidis* L. Sw. *Z. Pflanzenphysiol.*, 71, 366–369.

Collins, R. A., Ng, T. B., Fong, W. P., Wan, C. C., and Yeung, H. W. 1997. A comparison of human immunodeficiency virus type 1 inhibition by partially purified aqueous extracts of Chinese medicinal herbs. *Life Science* 60, 345–351.

Commenges, D., Scotet, V., Renaud, S., Jacqmin-Gadda, H., Barberger-Gateau, P., and Dartigues, J. F. 2000. Intake of flavonoids and risk of dementia. *European Journal of Epidemiology* 16, 357–363.

Cota-Sanchez, J. H., Reyes-Olivas, A., and Abreu, D. D. 2011. Vivipary in the cactus family: A reply to Ortega-Baes' et al. evaluation of 25 species from northwestern Argentina. *Journal of Arid Environments* 75, 878–880.

Coulter, J. M. and Chamberlain, C. J. 1955. *Morphology of Gymnosperms.* Central Book Depot. Allahabad, 301–302.

Cuming, A. C., Cho, S. H., Kamisugi, Y., Graham, H., and Quatrano, R. S. 2007. Microarray analysis of transcriptional responses to abscisic acid and osmotic, salt, and drought stress in the moss, *Physcomitrella patens. New Phytologist* 176, 275–287.

Dai, X. L., Li, X. G., and Wu, S. F. 2003. List of Chinese edible pteridophytes. *Quarterly of Forest By-Product and Speciality in China* 4, 5–6.

Elmqvist, T. and Cox, P. A. 1996. The evolution of vivipary in flowering plants. *Oikos* 77(1), 3–9.

Eris, A. 1995. *The Physiology of Horticulture.* Uludag University Press.

Faden, R. B. 1973. Ome notes on the *gemmiferous* species of *Asplenium* in tropical East Africa. *American Fern Journal* 63, 85–90.

Fang, J. and Chu, C. 2008. Abscisic acid and the pre-harvest sprouting in cereals. *Plant Signaling & Behavior* 3(12), 1046–1048.

Farnsworth, E. J. and Farrant, J. M. 1998. Reduction in Abscisic acid are linked with viviparous reproduction in mangroves. *American Journal of Botany* 85(6), 760–769.

Favre-Ducharte, M. 1958. Ginkgo an oviparus plant. *Phytomorphology* 8S. 377–390.

Finch-Savage, W. E. and Leubner-Metzger, G. 2006. Seed dormancy and the control of germination. *New Phytologist* 171, 501–523.

Finkelstein, R. R. 2004. The role of hormones during seed development and germination, 513–537. *Plant Hormones – Biosynthesis, Signal Transduction, Action.* Ed. P. J. Davies, Dordrecht, Kluwer Academic.

Foskett, D. E. 1994. *Plant growth and Development.* Academic Press, Inc.

Gahalain, S. S., Fotedar, R. L., and Gupta, R. C. 2006. Vivipary in *Biota orientalis* endl. – Rare phenomenon in a gymnosperm. *Current Science* 91(12), 1608–1609.

Garces, H. M. P., Koenig, D., Townsley, B. T., Minsung, K., and Sinha, N. R. 2014. Truncation of leafy cotyledon 1 Protein Is Required for Asexual Reproduction in *Kalanchoe daigremontiana. Journal of Plant Physiology* 165, 196–206.

Gong, M., Li, Y. J., and Chen, S. Z. 1998. Abscisic acid-induced thermotolerance in maize seedlings is mediated by calcium and associated with antioxidant systems. *Journal of Plant Physiology* 153, 488–496.

Graham, A. 2011. The Strange Case of *Asplenium bulbiferum. Pteridologist*, 240–245.

Guan, L., and Scandalios, J. G. 1998a. Two structurally similar maize cytosolic superoxide dismutase genes, *Sod4* and *Sod4A*, respond differentially to abscisic acid and high osmoticum. *Plant Physiology* 117, 217–224.

Guan, L., and Scandalios, J. G. 1998b. Effects of the plant growth regulator abscisic acid and high osmoticum on the developmental expression of the maize catalase genes. *Physiologiue Plantarum* 104, 413–422.

Guan, L., Zhao, J., and Scandalios, J. G. 2000. Cis-elements and transfactors that regulate expression of the maize Cat1 antioxidant gene in response to ABA and osmotic stress: H2O2 is the likely intermediary signaling molecule for the response. *Plant Journal* 22, 87–95.

Harmer, R. and Lee, J. A. 1978. The growth and nutrient content of *Festuca vivipara* (l.) Sm. Plantlets. *New Phytologist* 80, 99–106.

Hartmann, H. T., Kester, D. E., Davies, F. T., and Geneve, R. L. 1997. *Plant Propagation Principles and Practices.* New Jersey, Prentice Hall.

Hirayama, O., Nakamura, K., Hamada, S., and Kobayasi, Y. 1994. Singlet oxygen quenching ability of naturally occurring carotenoids. *Lipids* 29, 149–150.

Hori, K., Satake, T., Yabuuchi, M., Saiki, Y., Murakami, T., and Chen, C. M. 1987. Chemical and chemotaxonomical studies of Fillices. LXXV. Chemical studies on the constituents of *Dennstaedtia distenta* MOORE. *Yakugaku Zasshi* 107, 485.

Ihle, J. N., and Dure, L. 1970. Hormonal regulation of translation inhibition requiring RNA synthesis. *Biochemistry and Biophysics Research Co.* 38, 995–1001.

Imperato, F. 1984. Two new kaemferide 3,7-diglycosides from the fern *Asplenium bulbiferum. Chemistry & Industry*, 186–187.

Irudayaraj, V., Manickam, V. S., and Johnson, M. 2003. Vivipary, a rare and evolutionarily important phenomenon in a rare homosporous fern *Grammitis medialis* from the Western Ghats, South India. *Current Science* 85(12), 1666–1667.

Iwashina, T. and Matsumoto, S. 1994. Flavonoid variation and evolution in *Asplenium normale* and related species (Aspleniaceae). *Journal of Plant Research* 107(3), 275–282.

Iwashina, T., Matsumoto, S., Kitajima, J., Nakamura, T., Kokubugata, G., Suleiman, M., and Said, I. M. 2010. Apigenin di- and trirhamnoside from *Asplenium normale* in Malaysia. *Natural Product Communications* 5(1), 39–42.

Iwashina, T., Matsumoto, S., Ozawa, K., and Akuzawa, K., 1990. Flavone glycosides from *Asplenium normale. Phytochemistry* 29(11), 3543–3546.

Kaminaka, H., Morita, S., Tokumoto, M., Masumura, T., and Tanaka, K. 1999. Differential gene expression of rice superoxide dismutase isoforms to oxidative and environmental stresses. *Free Radical Research* 31, S219–S225.

Kholia, B. S. and Fraser-Jenkins, C. R. 2011. Misidentification makes scientific publications worthless – Save our taxonomy and taxonomists. *Current Science* 100(4), 458–461.

Koornneef, M., Bentsink, L., and Hilhorst, H. 2002. Seed dormancy and germination. *Current Opinion in Plant Biology* 5, 33–36.

Lee, J. A. and Harmer, R. (1980) Vivipary, a Reproductive Strategy in Response to Environmental Stress. *Oikos* 35(2), 254–265.

Leon-Kloosterziel, K. M., van de Bunt, G. A., Zeevaart, J. A., and Koornneef, M. 1996. Arabidopsis mutants with a reduced seed dormancy. *Plant Physiology* 110, 233–240.

Li, L. Y. 1950. Vivipary in Some Chinese Plants. Source, *Botanical Gazette* 111(3), 358–359.

Liu, Y., Wujisguleng, W., and Long, C. 2012. Food uses of ferns in China: A review. *Acta Societatis Botanicorum Poloniae* 81(4), 263–270.

Lloyd, F. E. 1902. Vivipary in podocarpus. *Torreya* 2(8), 113–117.

Lloyd, R. M. and Klekowski, E. J. 1970. Spore germination and viability in Pteridophyta: Evolutionary significance of chlorophyllous sp. *Biotropica* 2, 129–137.

Ma, X. Y., Xie, C. X., Liu, C., Song, J. Y., Yao, H., Luo, K., Zhu, Y. J., Gao, T., Pang, X. H., Qian, J., and Chen, S. L. 2010. Species Identification of Medicinal Pteridophytes by a DNA Barcode Marker, the Chloroplast *psbA-trnH* Intergenic Region. *Biological and Pharmaceutical Bulletin* 33(11), 1919–1924.

Ma, Y. Z., Holt, N. E., Li, X. P., Niyogi, K. K., and Fleming, G. R. 2003. Evidence for direct carotenoid involvement in the regulation of photosynthetic light harvesting. *Proceedings of the National Academy of Sciences USA*, 100, 4377–4382.

Mahabale, T. S. 1961. Vivipary in *Podocarpus macrophyllus.* Prof. S. P. Agharkar Commemoration, Vol. Poona, 82–88.

Majumder, S., D'Rozario, A., and Bera, S. 2010. Vivipary in Indian Cupressaceae and its Ecological Consideration. *International Journal Botany* 6, 59–63.

Manickam, V. S. 1995. Rare and endangered ferns of the Western Ghats of South India. *Fern Gazette* 15, 1–10.

McAdam, S. A. M. and Brodribb, T. J. 2012. Fern and lycophyte guard cells do not respond to endogenous Abscisic acid. *Plant Cell* 24, 1510–1521.

McCarty, D. R. 1995. Genetic control and integration of maturation and germination pathways in seed development. *Annual Review of Plant Physiology and Plant Molecular Biology* 46, 71–93.

Mickel, J. T. 1967. An advanced course in pteridophytes biology in Costa Rica. *American Fern Journal* 57, 145–161.

Nakazato, T., Barker, M. S., Rieseberg, L. H., and Gastony, G. J. 2008. Evolution of the nuclear genome of ferns and lycophytes. *Biology and Evolution of Ferns and Lycophytes.* Ed. T. A. Ranker and C. H. Haufler. Cambridge University Press, 175–198.

Nambara, E., and Marion-Poll, A. 2005. Abscisic acid biosynthesis and catabolism. *Annual Review of Plant Biology* 56, 165–185.

Neill, S. J., Horgan, R., and Parry, A. D. 1986. The carotenoid and abscisic acid content of viviparous kernels and seedlings of *Zea mays* L. *Planta* 169(1), 87–96.

Niyogi, K. K. 1999. Photoprotection revisited: Genetic and molecular approaches. *Annual Review of Plant Biology* 50, 333–359.

Niyogi, K. K., Grossman, A. R., and Bjorkman, O. 1998. Arabidopsis mutants define a central role for the xanthophyll cycle in the regulation of photosynthetic energy conversion. *Plant Cell* 10, 1121–1134.

Osman, H. E., Maalej, N., Shanmuganayagam, D., and Folts, J. D. 1998. Grape juice but not orange or grape fruit juice inhibits platelet activity in dogs and monkeys. *Journal of Nutrion* 128, 2307–2312.

Page, C. N. 1979. The diversity of ferns: An ecological perspective. *The Experimental Biology of Ferns*. Ed. A. F. Dyer. Academic Press, 10–56.

Pandey, B. P. 2003. *College Botany: Pteridophyta, Gymnosperms and Palaeobotany*. Vol II. S. Chand & Company Ltd., New Delhi.

Pathania, S., Kumar, P., Singh, S., Khatoon, S., Rawat, A. K. S., Punetha, N., Jensen, D. J., Lauren, D. R., and Somvanshi, R. 2012. Detection of ptaquiloside and quercetin in certain Indian ferns. *Current Science* 102(12), 1683–1691.

Peng, B. 2008. Studies on the extraction and purification technology and quality of *Woodwardia unigemmata* effective fractions. Master's thesis. Beijing University of Traditional Chinese medicine.

Pereira, D. M., Valentao, P., Pereira, J. A., and Andrade, P. B. 2009. From chemistry to Biology. *Molecules* 14, 2202–2211.

Pilate, G., Sossountzov, L., and Miginiac, E. 1989. Hormone levels and apical dominance in the aquatic fern *Marsilea drummondii* A. Br. *Plant Physiology* 90, 907–912.

Pozo, J. C., Lopez-Matas, M. A., Ramirez-Parra, E., and Gutierrez, C. 2005. Hormonal control of the plant cell cycle. *Plant Physiology* 123, 173–183.

Prasad, T. K., Anderson, M. D., and Stewart, C. R. 1994. Acclimation, hydrogen peroxide, and abscisic acid protect mitochondria against irreversible chilling injury in maize seedlings. *Plant Physiology* 105, 619–627.

Preece, J. E., and Read, P. E. 1993. *The Biology of Horticulture*. New York: Wiley.

Pryer, K. M., Schneider, H., Zimmer, E. A., and Ann, B. J. 2002. Deciding among green plants for whole genome studies. *Trends Plant Science* 7, 550–554.

Rodriguez-Gacio, M. C., Matilla-Vazquez, M. A., and Matilla, A. J. 2009. Seed dormancy and ABA signaling: The breakthrough goes on. *Plant Signaling & Behavior* 4, 1035–1048.

Rozario, A. D., Bera, S., and Mukhopadhyay, R. 2001. Viviparous growth of young sporophytes from aphlebiae in *Dennstaedtia scabra* (Wall. Ex Hook.) Moore from Sikkim. *Current Science* 81(4), 347–348.

Sadamu, M., Iwashina, T., Kitajima, J., and Mitsuta, S. 2003. Evidence by flavonoid markers of four natural hybrids among *Asplenium normale* and related species (Aspleniaceae) in Japan. *Biochemical Systematics and Ecology* 31(1), 51–58.

Sakata, Y., Komatsu, K., and Takezawa, D. 2014. ABA as a universal plant hormone. U. Luttge et al. (Eds.), *Prog. Bot.* 75, 57–96. DOI 10.1007/978-3-642-38797-5_2.

Satake, T., Murakami, T., Yokote, N., Saiki, Y., and Chen, C. M. 1985. Chemical and chemotaxonomical studies on Filices LVIII. Chemical studies on the constituents of *Monachosorum arakii* pteridaceae. *Chemical and Pharmaceutical Bulletin* 4175.

Srivastava, R. and Uniyal, P. L. 2014. Developmental Studies in an Ornamental Fern *Polystichum lentum* (D. Don) Moore. *National Academy Science Letters* 37(3), 281–284.

Srivastava, R. C., and Adi community, 2009. Traditional knowledge of Adi Tribe of Arunachal Pradesh on plants. *Indian Journal of Traditional Knowledge* 8(2), 146–153.

Subhash, C., Fraser-Jenkins, C. R., Kumari, A., and Archana, S. 2008. A Summary of the Status of Threatened Pteridophytes of India. *Taiwania* 53(2), 170–209.

Tomlinson, P. B. and Cox, P. A. 2000. Systematic and functional anatomy of seedlings in mangrove Rhizophoraceae: Vivipary explained. *Biological Journal of the Linnean Society* 134, 215–231.

Tracewell, C. A., Vrettos, J. S., Bautista, J. A., Frank, H. A., and Brudvig, G. W. 2001. Carotenoid photooxidation in photosystem II. *Archives of Biochemistry and Biophysics* 385, 61–69.

Tsai, H. H. and Hwang, S. M. 1999. Compositions of matter useful in the treatment of viral infections derived from plant extracts. United States Patent, Patent Number: 5, 989, 556.

Umikalsom, Y., Grayer-Barkmeijer, R. J., and Harborne, J. B. 1994. A comparison of the flavonoids in Athyriaceae and Aspleniaceae. *Biochemical Systematics and Ecology* 22, 587–594.

Umikalsom, Y. and Harborne, J. B. 1991. Flavonoid Distribution in Asplenioid Ferns. *Pertanika* 14(3), 297–300.

Unsal, N. P. 1993. *Plant Growth Substances.* Istanbul University Press.

Vishal, G. 2006. Plants used in folklore medicine by Bangnis of East Kameng, Arunachal Pradesh. *Natural Product Radiance* 5(1), 52–59.

Wada, H., Kido, T., Tanaka, N., Murakami, T., Saiki, Y., and Chen, C. 1992. Chemical and chemotaxonomical studies of ferns: LXXXI. Characteristic lignans of blechnaceous ferns. *CPB* (Tokyo), 40(8), 2099–2101.

Wali, M. K. and Tiku, S. N. 1965. Vivipary in *Pinus wallichiana* A. B. Jackson. *Current Science* 34, 294–294.

Wangchuk, P. 2014. Phytochemical analysis, Bioassays and the identification of drug Lead compounds seven Bhutanese medicinal plants. PhD Dissertation, University of Wollongong.

White, C. N., Proebsting, W. M., Hedden, P., and Rivin, C. J. 2000. Gibberellins and seed development in maize. I. Evidence that gibberellin/abscisic acid balance governs germination versus maturation pathways. *Plant Physiology* 122, 1081–1088.

Wikibook. https://en.wikibooks.org/wiki/Traditional_Chinese_Medicine/From_Maackia_Amurensis_To_Myrsine_Semiserrata.

Wikipedia. http://en.wikipedia.org/wiki/Athyrium.

Wikipedia, https://en.wikipedia.org/wiki/Bolbitis_heteroclita.

Wojdylo, A., Oszmianski, J., and Czemerys, R. 2007. Antioxidant activity and phenolic compounds in 32 selected herbs. *Food Chemistry* 105(3), 940–949.

Woodall, A. A., Britton, G., and Jackson, M. J. 1997. Carotenoids and protection of phospholipids in solution or in liposomes against oxidation by peroxyl radicals: Relationship between carotenoid structure and protective ability. *Biochimica et Biophysica Acta* 1336, 575–586.

Xu, H. X., Kadota, S., Kurokawa, M., Shiraki, K., Matsumoto, T., and Namba, T. 1993. Isolation and Structure of Woodorien, a New Glucoside Having Antiviral Activity, from *Woodwardia orientalis. Chemical and Pharmaceutical Bulletin* 41(10), 1803–1806.

Yamane, H., Fujioka, S., Spray, C. R., Phinney, B. O., MacMillan, J., Gaskin, P., and Takahashi, N. 1988. Endogenous gibberellins from sporophytes of two tree ferns, *Cibotium glaucum* and *Dicksonia antarctica. Plant Physiology* 86, 857–862.

Yamane, H., Sato, Y., Takahashi, N., Takeno, K., and Furuya, M. 1980. Endogenous inhibitors for spore germination in *Lygodium japonicum* and their inhibitory effects on pollen germinations in *Camellia japonica* and *Camellia sinensis. Agricultural and Biological Chemistry* 44, 1697–1699.

Yun, X. L., Zhao, N. W., Pan, L. T., and Zhao, J. H. 2009. Resource distribution and exploitation of edible ferns in Guizhou (II). *Journal of Anhui Agricultural Sciences* 37(33), 16317–16319.

Zeng, H. Y., Kun, Q., Zheng, Z. Q., and Hong, X. 2008. Phenophase observation and introduction and cultivation of ornamental ferns. *Guangxi Zhiwu/Guihai* 28(1), 86–90.

Zhao, J. Y., and Guo, H. M. 2013. Arsenic Uptake from Arsenic-Contaminated Water Using *Pteris vittata* L. and *Polystichum craspedosorum*. Sustainable energy and environmental engineering; Progress in environmental protection and processing of resource. *Applied Mechanics and Materials* 295/298, 1139–1143.

Zhu, D. and Scandalios, J. G. 1994. Differential accumulation of manganese–superoxide dismutase transcripts in maize in response to abscisic acid and high osmoticum. *Plant Physiology* 106, 173–178.

12 Prospects of Local Flora of Trans-Himalayan Region of Ladakh for Various Medicinal Uses

Gyan P. Mishra, Tsering Stobdan, Parimelazhagan Thangaraj, Tania Seth and Bijendra Singh

CONTENTS

12.1 Introduction 239
12.2 Sea buckthorn (*Hippophae rhamnoides*) 240
 12.2.1 Biochemical Compositions 241
 12.2.2 Medicinal Properties 242
12.3 *Capparis spinosa* L. 245
 12.3.1 Biochemical Compositions 246
 12.3.2 Medicinal Properties 248
12.4 *Dactylorhiza hatagirea* (D. don) 249
 12.4.1 Biochemical Compositions 251
 12.4.2 Medicinal Properties 251
12.5 *Inula racemosa* H.L. 252
 12.5.1 Biochemical Compositions 253
 12.5.2 Medicinal Properties 253
12.6 Conclusion 254
References 254

12.1 INTRODUCTION

Ladakh, being a cold arid high-altitude desert of the Trans-Himalayan region of India, has very harsh climatic conditions and a short agriculture season. It lies between long 32° to 36°N and lat 76° to 79°E. The major river of this region is the Indus, which is joined by its tributaries, the Zanskar, the Shingo and the Shyok (Singh 1995; Mishra and Singh 2010; Mishra 2009a; Singh et al. 2011). The Indian part of the Trans-Himalayas spans over 186,000 km^2 above natural tree-line zone, and it is known for its sparsely distributed vegetation and small species diversity. This zone sustains more than 1000 plant species (Kumar et al. 2011).

Ladakh is quite rich in medicinal flora, and many traditional herbs have been shown to have medicinal value that can be used to prevent, alleviate or cure several human diseases (Kumar et al. 2011). These plants were identified by local residents over the years through trial and error. In due time, many of these plants were gradually added to the repositories of medicinal plants and they also became the part of the Amchi system (Mishra et al. 2009b, Bhoyar et al. 2011a, b). The Amchi system is the traditional medicinal system of the cold desert Ladakh, which is basically based on the Tibetan system of medicine. This system provides treatment and relief to indigenous communities (Ballabh and Chaurasia 2011).

In recent years, consumption of herbal medicines is increasing and approximately 80% of the people in developing countries depend on these types of medicines for their primary healthcare needs (Farnsworth et al. 1985). In terms of the wealth of traditional herbal medicinal knowledge systems, India is one of the leading countries in Asia, employing a large number of plant species in Ayurveda (2000 species), Siddha (1121 species), Unani (751 species) and Tibetan (337 species) systems of medicine (Kala 2002). In Chapter 12, we have covered four indigenous plants of the Ladakh region viz. *Hippophae rhamnoides, Capparis spinosa, Dactylorhiza hatagirea* and *Inula racemosa* in terms of its vast medicinal potential.

12.2 SEA BUCKTHORN (*HIPPOPHAE RHAMNOIDES*)

Sea buckthorn (Figure 12.1) is a unique and valuable crop for cold arid regions, having the potential to play a crucial role in the sustainable development of fragile areas of cold arid deserts. The fruit and seeds are the main sources of its nutritional and medicinal values. These beneficial effects have made sea buckthorn products, especially its oil, desirable for various medicinal and cosmetics purposes. A beverage made from sea buckthorn fruit is quite popular in Indian markets and other products (like tea, jam, sauce, wine, vinegar, soap, shampoo, etc.) are gaining popularity.

FIGURE 12.1 Sea buckthorn (the golden bush) growing wild in Ladakh.

Products made from sea buckthorn that have medicinal value are still rare in Indian markets, which should be increased substantially (Stobdan et al. 2011).

12.2.1 Biochemical Compositions

Sea buckthorn fruits contain 60–80% juice rich in sugar, organic acids, amino acids and vitamins, and are among the most nutritious of all berries. The content of soluble sugars ranges from 9.3° to 22.74° Brix and it contains little higher glucose than fructose and xylose. The concentration of vitamins A, B_2 and C is much higher than other fruits and vegetables such as carrots, tomatoes, oranges, etc. Sea buckthorn berries also contain appreciable levels of vitamin B_1 and K. Vitamin C represents a nutrient of major importance in sea buckthorn because of its presence in large quantities ranging from 200–2500 mg/100g. Presence of these antioxidant vitamins in high quantity indicates its strong antioxidant properties. Considering that fresh orange juice contains 35–56 mg/100 mL and Aonla contains 478.5 mg/100 mL, the value of sea buckthorn as a source of vitamin C is apparent. The high vitamin concentration makes sea buckthorn fruit highly suitable for the production of nutritious soft drinks. During the Seoul Olympic Games in 1988, China designated a sea buckthorn sports drink as the official beverages for its athletes. Russian cosmonauts were also supplied with sea buckthorn beverages to enhance their health and resistance to stress (Stobdan et al. 2011; Mishra et al. 2009c, 2011). There are 24 minerals in sea buckthorn juice, including calcium, magnesium, phosphorus, iron, manganese, sodium, potassium and aluminum, among others. Potassium is the most abundant of all the elements, and sea buckthorn juice is also rich in organic acids (2.1–9.1 g/100 mL). Malic and quinic acids are the major acids together constituting around 90% of all the fruit acids (Ma et al. 1989). Eighteen different kinds of free amino acids have been detected in juice, of which eight are essential (Chen et al. 1991).

The results of the Leh-Ladakh samples, showed the presence of high content of multivitamins, including vitamin C (275 mg/100 g), vitamin A (432.4 IU/100 g), vitamin E (3.54 mg/100 g), riboflavin (1.45 mg/100 g), niacin (68.4 mg/100 g), pantothenic acid (0.85 mg/100 g), vitamin B_6 (1.12 mg/100 g) and vitamin B_2 (5.4 mg/100 g). Similarly, the mineral composition revealed a high amount of minerals including potassium (647.2 mg/L), calcium (176.6 mg/L), iron (30.9 mg/L), magnesium (22.5 mg/L), phosphorous (84.2 mg/L), sodium (414.2 mg/L), zinc (1.4 mg/L) and manganese (1.06 mg/L) (Stobdan et al. 2010). Its juice is known to possess antifreeze properties (Bajpai et al. 2009, 2010).

Sea buckthorn leaves also contain many nutrients and bioactive substances such as carotenoids, free and esterified sterols, triterpenols, isoprenols and approximately 15% of proteins. The flavonoid content in leaves ranges from 312–2100 mg/100 g of air-dried leaves. The oil content of sea buckthorn ranges from 1.5% to 3.5% in fruit pulp and about 9.9% to 19.5% in seeds. Oil extracted from seed and pulp differs considerably in fatty acid composition. Oil from the sea buckthorn juice and pulp is rich in palmitic (16:0) and palmitoleic (16:1) acids, while the seed oil contains the essential fatty acids, which are linoleic (18:2) and linolenic (18:3) acids. One of the many special features of sea buckthorn fruit is the exceptionally high content of tocopherols and tocotrienols. The total content of tocotrienols varied from 1.5–8.1 mg/kg^{-1} in berries and from 43–188 mg/kg^{-1} in berry oil (Kallio et al. 2002). The content of these bioactive compounds are among the crucial criteria for defining the quality of sea buckthorn. Oil derived from its juice contains more vitamin E (216 mg/100 g of fruit) than its seeds (64.4–92.7 mg/100 g of seed). The tocopherol fraction consists of approximately 50% α-tocopherol, 40% β-tocopherol and 10% γ-tocopherol (Mironov 1989).

12.2.2 Medicinal Properties

Sea buckthorn has been used in the traditional Tibetan system of medicine for centuries. The medicinal value of sea buckthorn was recorded as early as the 8th century in the Tibetan medicinal classic *rGyud Bzi* (Four Texts of Fundamental Tibetan Medicine). In the book, 84 different prescriptions for the preparation of sea buckthorn medicines is recorded. During the 13th century, the *rGyud Bzi* was disseminated through Mongolia, and since then, sea buckthorn began to be used in traditional Mongolian medicines. The three major species of sea buckthorn have been established in Tibetan medicine as *Sa-sTar* for *H. tibetana*, *Bar-sTar* for *H. rhamnoides*, and *Nam-sTar* for *H. salicifolia*. The classification is based on plant height. *Sa* means 'ground', *Bar* means 'middle' while *Nam* means 'sky'. Of these, *Bar-sTar* is the most commonly used species in Tibetan medicine (Gurmet 2009). The plant is identified as the Ayurvedic medicinal plant *Amlich* or *Badriphal* (Sharma and Chunekar 1998). China listed sea buckthorn in its pharmacopoeia way back in 1977. In the Ladakh region, even today, *Amchies* (local traditional doctor) often prescribe preparations from sea buckthorn for the treatment of common problems like indigestion, throat-infection, gynecological problems, ulcers, gastritis, bronchitis, acidity, diarrhoea, hypertension, blood disorders, fevers, tumours, gallstones, coughs, colds, food poisoning, etc.

Sea buckthorn oil is a valuable product used in medicine as a nutraceutical supplement and in cosmetics. Lipids from sea buckthorn leaves have been recommended for anti-burn and wound healing properties. The presence of a high content of α-tocopherol has significant healthful effects which act as natural antioxidants in the human body. The carotenoid content of sea buckthorn oil ranges from 314–2139 mg/100g (Zhang et al. 1989). It is indicated that the carotenoids consist of approximately 20% β-carotene, 30% γ-carotene, 30% lycopene and 15% oxygen-containing carotenoids. Phytosterol is also constituents of sea buckthorn oil which are capable of lowering plasma cholesterol on consumption by humans. The major phytosterol is sitosterol (β-sitosterol) and 5-avenasterol. The total quantity of phytoesterol in whole sea buckthorn fruits ranges from 340–520 mg/kg and is 4 to 20 times more than soybean oil. A novel triglyceride, 1,3-dicapryloyl-2-linoleoyl glycerol has been isolated and its structure was elucidated by Swaroop et al. (2005).

The plant has been extensively exploited in recent years for treatment of stomach malfunctioning, sluggish digestion, thrombosis, neoplasia, hepatic injuries and tendon and ligament injuries. Clinical trials on patients with ischemic heart disease have shown that the total number of flavonoids of sea buckthorn can reduce cholesterol levels and thereby improve the cardiac functions. The mechanisms of action may include reduced stress of cardiac muscle tissue by the regulation of inflammatory mediators. It is also established that antioxidant-rich sea buckthorn juice reduces the risk factor for coronary heart diseases in humans. No harmful effects of flavonoids were observed in renal or hepatic function. Sea buckthorn is traditionally used in the treatment of gastric ulcer and laboratory studies to confirm the efficacy of the seed oil for this application (Xing et al. 2002). Its function may be to normalize output of gastric acid and reduce inflammation by controlling pro-inflammatory mediators. Clinical trials have also shown that sea buckthorn extracts helped normalize liver enzymes, serum bile acids and immune system markers involved in liver inflammation and degeneration (Gao et al. 2003).

Sea buckthorn oil is also used as for the treatment of oral mucositis, vaginal mucositis, cervical erosion, duodenal ulcers, gastric cancers and skin ulcers (Li 1999). Bioactive oil obtained from young branches and leaves has been incorporated into an ointment for treating a wide variety of skin damage, including burns, bedsores, eczema and radiation injuries. In 1986, many of the Chernobyl nuclear disaster victims were treated with sea buckthorn oil. The oil absorbs ultraviolet light and promotes healthy skin. This unique property of sea buckthorn oil is being recognized and sought after by the cosmetic industry. This property has the potential to protect the skin in the event of nuclear warfare. It is thought that high level of tocopherol in sea buckthorn oil minimizes lipid oxidation, maintaining tissue integrity and reduces skin toughening and wrinkling.

The seed oil of sea buckthorn is characterized by high level of unsaturated linoleic and linolenic acids. These essential fatty-acids are claimed to relieve chronic eczema, cure dermatitis and maintain healthy skin. Flavonoids extracted from the fruits are used especially in the treatment of cardiovascular problems. Laboratory studies also demonstrated that sea buckthorn oil is effective in cancer therapy (Xu 1994). Investigations have also revealed many beneficial effects of tocotrienols present in sea buckthorn fruit and seed oil on human and animals, among which the hypocholesterolemic, antitumor and skin-protecting effects (Kato et al. 1985).

Radioprotection studies at the Institute of Nuclear Medicine and Allied Sciences (INMAS, Delhi) use an herbal preparation of sea buckthorn, RH-3, against whole-body lethal irradiation in mice, which has shown that a dose of 30 mg/kg body weight of RH-3 rendered 82% survival when compared to non-survival in irradiated control. The RH-3 inhibits the Fenton reaction and radiation-mediated regeneration of hydroxyl radicals *in vitro*, superoxide anion mediated Nitroblue tetrazolium reduction and $FeSO_4$ mediated lipid peroxidation in the liver (Goel et al. 2002). Possible contributions towards radioprotective efficiency of RH-3 is the maintenance of chromatin organization, induction of hypoxia, hydrogen atom donation, free radical scavenging and the blocking of cell at G2-M phase by interfering with topoisomerase I activity (Goel et al. 2003). The RH-3 administration protects spermatogenesis by enhancing the spermatogonial proliferation, enhancing the stem cell survival and reducing sperm abnormalities (Goel et al. 2006). It is also found to reduce chromatin compaction and significantly inhibit radiation-induced DNA break. Polyphenol/ flavonoids present in sea buckthorn might be responsible for mitochondrial and genomic DNA protection from radiation-induced damage (Shukla et al. 2006).

Sea buckthorn has been studied extensively for its antioxidant properties. The leaf extract of sea buckthorn has been evaluated on chromium-induced oxidative stress.

Male albino rats fed with potassium dichromate equivalent to a dose of 30 mg/kg body weight of chromium (VI) for 30 d resulted in a decreased body weight and an increased organ to body weight ratio. The leaf extract of sea buckthorn at a concentration of 100 and 250 mg/kg body weight protected the animals from chromium-induced oxidative injury. Similarly, the cytoprotective and antioxidant properties of the alcoholic leaf extract of sea buckthorn against hypoxia-induced oxidative stress in C-6 glioma cells have been demonstrated. Exposure of cells to hypoxia for 12 h resulted in a significant increase in cytotoxicity and a decrease in mitochondrial transmembrane potential compared to the controls. An appreciable increase in nitric oxide, reactive oxygen species (ROS) production and DNA damage has been observed. The pre-treatment of cells with the alcoholic leaf extract of sea buckthorn (200 μg/mL) significantly inhibited cytotoxicity and ROS production, and maintained antioxidant levels similar to that of control cells. Sea buckthorn extraction procedures also had significant effects on antioxidant activities. Microwave-assisted extracts had significantly higher antioxidant capacity than those obtained by ultrasound-assisted extraction and maceration (Sharma et al. 2007).

Immunomodulatory activity of sea buckthorn leaf extract has been evaluated at DIPAS (New Delhi) in adjuvant-induced arthritis (AIA) rat models. Inflammations induced by injecting a complete Freund's adjuvant (CFA) in rats was reduced significantly by administrating sea buckthorn extract on the same day or 5 d prior to inflammation insult. The result suggests the potential use of sea buckthorn for the treatment of arthritis (Ganju et al., 2005). The effect of sea buckthorn leaf extract on nitric oxide (NO) production induced by lipopolysaccharide (LPS) in murine macrophage cell line RAW 264.7 has been studied. The extract showed significant scavenging of NO radicals released by the NO donor. The treatment of macrophage with the extract also causes significant inhibition of inducible nitric oxide synthase. Results suggested significant anti-inflammatory activity of sea buckthorn extract and the potential for the treatment of inflammatory diseases (Padwad et al. 2006). The alcoholic leaf and fruit extract of sea buckthorn at a concentration of 500 μg/mL had a significant cytoprotective effect against sodium nitroprusside (SNP)-induced oxidative stress. These extracts inhibit SNP-induced cytotoxicity free radical production and maintain antioxidant status identical to that of control cells. The alcoholic fruit extract has been found to have significantly higher antioxidant activity than leaf extract against SNP-induced cytotoxicity in murine macrophages (Geetha et al. 2002).

The topical administration of flavone of sea buckthorn on cutaneous wound healing in rats resulted in enhanced wound healing activity, as indicated by improved rate of wound contraction, decrease time taken for epithelialization and the significant increase in hydroxyproline and hexosamine content. The flavone treatment studied at one of the DRDO labs also resulted in a significant decrease in lipid peroxide levels and thus suggests that sea buckthorn flavone promotes wound healing activity (Geetha et al. 2008). Sea buckthorn ointment is found effective in treatment of infected cutaneous wounds in bovines. Application of 5% povidone iodine or sea buckthorn ointment is found better in wound healing of infected cutaneous wounds in calves than liquid paraffin (Mahajan et al. 2005).

In a recent study, sea buckthorn has been evaluated for hepatoprotective activity on carbon tetrachloride (CCl_4)–induced liver injury in male albino rats. Pretreatment of leaf extract at a concentration of 100 and 200 mg/kg body weight significantly protect the animal from CCl_4–induced increase of glutamate oxaloacetate transferase, glutamate pyruvate transferase, alkaline phosphatase and bilirubin and better maintain protein level in serum. The result suggests use of sea buckthorn leaf extract as neutraceutical or food supplement against liver disease (Geetha et al. 2008).

The possible mechanisms of the adaptogenic activity of sea buckthorn during exposure to cold, hypoxia and restraint (C-H-R) stress induces hypothermia and post-stress recovery in rats has been elucidated. Results suggest that sea buckthorn extract treatment causes a trend for shifting anaerobic metabolism to aerobic during C-H-R exposure and post-stress recovery (Saggu and Kumar 2007). Adoptogenic activity of the extract might be due to its antioxidant activity, maintained body glucose level, better utilization of free fatty acids and improved cell membrane permeability (Saggu and Kumar 2008). Sea buckthorn oil has significant anti-atherogenic and cardioprotective activity. The therapeutic value of sea buckthorn against subchronic arsenic toxicity in mice has shown that aqueous extract of sea buckthorn significantly protects against arsenic-induced oxidative stress, but does not chelate arsenic.

In a joint study carried out by different DRDO Life Sciences labs, ethanolic and water extracts of sea buckthorn have shown protective effects against the toxic effect of mustard gas, a chemical warfare agent (Vijayaraghvan et al. 2006). Sea buckthorn extract has also been used against Dengue virus infection in human blood-derived macrophages. Infected cells treated with sea buckthorn extract are able to maintain cell viability of Dengue infected cells at par with some commercially available antiviral drug (Jain et al. 2008). Sea buckthorn seed extract possesses antibacterial activities against *Bacillus cereus*, *B. coagulans*, *B. subtilis*, *Listeria monocytogenes* and *Yersinia enterocolitica*, suggesting the use of sea buckthorn as food preservative (Negi et al. 2005).

Berry oil has shown good potential towards platelet aggregation and the beneficial effects of blood clotting in humans; further studies on the dose–response effect are needed to assess the practical use of berry oil supplements (Stobdan et al. 2011). In arid and semi-arid regions, sea buckthorn has the outstanding capacity to improve the environment and economic development. We can look forward to a continued revelation of sea buckthorn's many gifts through the increasing interest and research into its abundant and valuable properties. Judicious exploitation and utilization of sea buckthorn resources can bring more benefits to humankind throughout the world.

12.3 *CAPPARIS SPINOSA* L.

Capparis spinosa (Figure 12.2) also called the 'caper' and locally known as *Kabra* is one of the oldest of the known medicinal plants of the Amchi system (local doctors) which is occasionally used by local people of Ladakh as a leafy vegetable and forage. It grows wild in the Ladakh region around 10,000 ft above mean sea level. It can be consumed as a cooked vegetable, salad, pickle and condiment. The species is quite hardy and can tolerate temperatures between −5°C and −10°C (Phillips and Rix 1998; Mishra et al. 2009b). Immature flower buds are pickled in vinegar, sauces or preserved in salt and have a high demand in European countries (Bown 1995).

FIGURE 12.2 *Capparis spinosa* growing wild in Ladakh.

Many workers have worked on this plant for both its nutrients and various bioactive compounds, which are summarized below.

12.3.1 Biochemical Compositions

A sample of 100 g of prepared capers contains energy (23 kcal), carbohydrates (5 g), sugars (0.41 g), dietary fibre (3.2 g), fat (0.9 g), protein (2.36 g), vitamin C (4.3 mg), iron (1.7 mg) and sodium (2964 mg) (USDA 2016). Previous chemical studies on capers have shown the presence of alkaloids, lipids, polyphenols, flavonoids, indole and aliphatic glucosinolates (Bhoyar et al. 2011b; Sharaf et al. 2000). Hydroxy cinnamic acids like caffeic acid, ferulic acid, pcoumaric acid and cinnamic acid were reported. Pmethoxybenzoic acid was isolated from aqueous extract of *C. spinosa* aerial parts (Gadgoli and Mishra 1999). Çalis et al. (1999, 2002) have isolated two glucose-containing 1H-indole-3-acetonitrile compounds, capparilosides A and B and two (6S)-hydroxy-3-oxo-aionol glucosides, corchoionoside C and a phenyl glucoside from mature fruits of *C. spinosa*. A qualitative and quantitative analysis of rutin from the leaves, fruits and flowers of *Capparis* growing wild was achieved (Ramezani et al. 2008).

Flavanoid glycosides, like, rutin, quercetin, quercetin-3- rutinoside, were isolated from the floral buds (Giuffrida et al. 2002). Glucocapperin has been shown to be the main glucosinolate in the floral buds (Matthaus and Ozcan 2002). The presence of both flavonoids and hydroxycinnamic acids has also been demonstrated in capers (Bonina 2002). New compounds like Sitosteryl glucoside-6'-octadecan-o-ate and 3-methyl-2-butenyl-glucoside were isolated (Khanfar et al. 2003). The nutrient profiling of caper buds revealed that it is rich in nutrient most of the parameters were on the higher side than the reported values in the USDA database (Table 12.1).

The *Capparis spinosa* leaves collected from nine different sites from three valleys in the Trans-Himalayan region of Ladakh (India) showed maximum DPPH

TABLE 12.1
***Capparis* Flower Bud Nutrient Profile**

Nutrient	Obtained (value/100 g)	USDA Database[a] (value/100 g)
Energy (kcal)	74.21	23.21
Protein (g)	5.53	2.36
Ash (g)	8.69	8.04
Carbohydrate (g)	12.19	4.89
Fibre, total dietary (g)	3.60	3.20
	Minerals	
Calcium (mg)	99.18	40.00
Iron (mg)	2.30	1.67
Magnesium (mg)	56.32	33.00
Zinc (mg)	0.95	0.32
Manganese (mg)	0.42	0.07
	Vitamins	
Vitamin C (mg)	5.52	4.3
Thiamin (mg)	0.05	0.018
Riboflavin (mg)	0.028	0.139
Vitamin B 6 (mg)	0.121	0.023

[a] USDA Nutrient Database for Standard Reference, Release 13 (1999) NBD No. 02054.

and ABTS radical scavenging activity in the leaves samples collected from Skuru and least from Tirchey site. The FRAP assay revealed that Skuru sites possess the maximum antioxidant content compared to the samples collected from any other location. The highest and lowest phenolic and flavonoid content was recorded in Skuru and Tirchey sites, respectively. The total phenolic and flavonoid content correlated reasonably with the IC_{50} of ABTS ($R^2 = 0.741$ and 0.703, respectively) and FRAP ($R^2 = 0.605$ and 0.649, respectively), but correlated poorly with DPPH IC_{50} ($R^2 = 0.303$ and 0.408, respectively). These results thus confirmed the antioxidant potential of *C. spinosa* leaves collected from different locations of the Trans-Himalayas (Bhoyar et al. 2011b; Mishra et al. 2009b, 2010).

The seed oil yield ranged from 27.3–37.6 g/100 g and contained high vitamin E (134 mg/100 g) and tocopherol (4961.8–10009.1 mg/kg), which act as natural antioxidants. The leaf oil was composed of isothiocyanates, n-alkanes, terpenoids, a phenyl propanoid, an aldehyde and a fatty acid. The main components of this oil were thymol (26.4%), isopropyl isothiocyanate (11%), 2-hexenal (10.2%) and butyl isothiocyanate (6.3%). The volatile oils of the ripe fruit and the root were composed mainly of the methyl, isopropyl and sec-butyl isothiocyanates. A protein exhibiting an N-terminal amino acid sequence with some similarity to imidazoleglycerol phosphate synthase was purified from fresh *Capparis* seeds (Lam et al. 2009). A dimeric 62-kDa lectin exhibiting a novel N-terminal amino acid sequence was purified from caper seeds (Lam et al. 2009). Glucosinolates like sinigrin, glucoiberin, and glucocleomin were isolated from the seeds and leaves of *C. spinosa* (Romeo et al. 2007). Three new alkaloids, capparispine, capparispine 26-O-d-glucoside and cadabicine 26-O-d-glucoside hydrochloride, were isolated from the roots of *C. spinosa* (Fu et al. 2007, 2008).

12.3.2 Medicinal Properties

Capparis spinosa is highly nutritious plant having immense medicinal properties. In Ladakh, various medicinal preparations from *Capparis* are used by the Amchis (traditional doctors) for treatment of various ailments. *C. spinosa* extract was able to counteract the inflammatory process induced *in vitro* by IL-1β in human chondrocyte cultures (Panico et al. 2005). Previous chemical studies have reported the presence of alkaloids, lipids, flavonoids and glucosinolates (which are known as flavour compounds), cancer-preventing agents and biopesticides (Bhoyar et al. 2010; Mikkelsen et al. 2000; Germano et al. 2002). Capers are a hepatic stimulant that has been used for improving the functional efficiency of the liver. Recent experimental work also confirms its protective action on the histological architecture of the liver and its positive effect on liver glycogen and serum proteins (Subhose et al. 2005).

Since ancient times, the floral buds were employed as flavouring in cooking and are also used in traditional medicine for their diuretic and antihypertensive properties, prepared in both poultices and tonics (Çalis et al. 1999). *Capparis* flower buds contain 100.51 mg of rutin equivalent/g antioxidant activity (Germano et al. 2002). The main glucosinolate, glucocapparin, amounted to 90% of total glucosinolates (Schraudolf 1989) in caper buds. If taken before a meal, it will increase the appetite (Genders 1994). Unopened flower buds are laxative in nature. Sterols, an important constituent of oil, are capable of lowering plasma cholesterol (Matthaus and Ozcan 2005). From the floral buds, 3-O-rhamnorutinosin kaempferol (Tomas and Ferreres 1978), kaempferol-3-rutinoside and quercetin-7-oglucorhamnoside were isolated. A methanol extract of *C. spinosa* buds afforded significant *in vivo* protection against UVB light-induced skin erythema in healthy human volunteers (Bonina et al. 2002).

The buds are a rich source of compounds known as aldose-reductase inhibitors, and they are effective in preventing the formation of cataracts. Also, the flower buds and roots are used as renal disinfectants and diuretics to treat arteriosclerosis and coughs, and prepared as tonics and compresses for the eyes (Genders 1994; Bown 1995: Batanouny 1999). Methanolic extract of *C. spinosa* buds showed *in vitro* antioxidant, antiviral and immunomodulatory activity; and *in vivo* photoprotective and antiallergic activity against bronchospasm in guinea pigs (Germano et al. 2002). The inhibition of lipid oxidation has been demonstrated in vitro; the mechanism is attributed to a cooperative interaction between the tocopherol, flavonoid and isothiocyanate chemical constituents (Tesoriere et al. 2007).

From the roots, indole glucosinolates like glucobrassicin, neoglucobrassicin and 4-methoxyglucobrassicin were isolated. The homologous polyphenols cappaprenol-12, cappaprenol-13 and cappaprenol-14 with 12, 13 and 14 isoprene units respectively were also isolated (Germano et al. 2002). They are used as drugs for acute viral hepatitis, and are a major constituent of the herbal formulation Liv 52, which is useful in liver disorders (Mathur et al. 1986).

In Ayurveda, the root-bark is given in splenic, renal and hepatic conditions and also as an analgesic, anthelmintic, deobstuent, diuretic, expectorant and vasoconstrictive (Chiej 1984). Internally it is used in the treatment of gastrointestinal infection, diarrhoea and rheumatism (Bown 1995). Externally, it is used to treat skin conditions, capillary weakness and easy bruising (Genders 1994; Bown 1995). Alkaloids, to the tune of 0.91% from root-bark and 0.86% from seeds, are isolated, and the maximum alkaloid content was found in the roots, of which stachydrine constituted 87.43% of the total alkaloid present (Sadykov et al. 1981). In Unani medicine, the decoction of root bark is prescribed as a deobstruent to the liver and spleen, as an anthelmintic and anti-inflammatory agent (Chopra et al. 1999). Four bacterial strains (viz. *Pseudomonas stutzeri* var. mendocina, *Comamonas sp., Agrobacterium tumefaciens* bivar. 2 and *Sphignobacterium sp.*), isolated from the rhizosphere of capers, were found to be able to fix N_2 (Andrade et al. 1997). The ethanolic extract of *C. spinosa* root bark reported anthelmintic activity. It acts as anti-allergic and antihistaminic agent (Zhan 1978).

Capparis spinosa has been reported to possess hepatoprotective activity (Romeo et al. 2007; Gadgoli and Mishra 1995) and p-methoxy benzoic acid was found to be responsible for this activity (Gadgoli and Mishra 1995). Approximately 600 mg of dried, whole-plant extract per day has been used in a mixed preparation in experiments investigating hepatoprotective effects. Similarly, a clinical trial investigating the efficacy of a mixed preparation containing caper extract combined with other extracts found an improvement in liver function laboratory values (Huseini et al. 2005). The aqueous extract of *Capparis spinosa* shows hypotensive activity in spontaneously hypertensive rats (Ali et al. 2007). The alcoholic extract of *C. spinosa* reported antimicrobial activity (Mahasneh 1996). It is used as a metabolic corrective in newborns (Dhurandhar 1973). Rutin or flavanoids from capers, prevented an increase in serum levels of alanine amino transferase, aspartase aminotransferase and aldolase caused by J-irradiation, thereby reducing radiation sickness. It also acted as a coagulation enhancer and diuretic (Altymyshev 1981).

In normal and diabetic (induced) rats fed aqueous extracts of the powdered caper fruits for a two-week period, a reduction in plasma cholesterol and triglycerides was demonstrated (Eddouks 2005). A 2% aqueous gel has been used for antihistaminic effects (Trombetta et al. 2005). The leaves are used for the treatment of gout, coughs, earaches, expelling stomach worm, rheumatism, paralysis, toothaches, as a diuretic and for diabetes control (Andrade et al. 1997; Sharma 2003). The aqueous extract of the aerial parts were reported to possess anti-inflammatory activity and cappaprenol–13 isolated as an anti-inflammatory principle from capers was found to inhibit carrageenan-induced paw edema in rats by 44%, compared to 67% by standard oxyphenbutazone (Al-Said et al. 1988). The juice of the leaves and fruits of capers are used as an anticystic, a fungicide and a bactericide.

12.4 *DACTYLORHIZA HATAGIREA* (D. DON)

Dactylorhiza hatagirea (D. don) (Figure 12.3) is a high-value medicinal plant species abundantly found in different temperate regions of India including certain parts of the Trans-Himalayan regions of Ladakh. It is commonly known as the 'marsh orchid',

FIGURE 12.3 Inflorescence of *Dactylorhiza hatagirea*.

it is also referred as *Munjataka* (Ayurveda), *Angmo-lakpa* (Ladakh), *Salampanja* (Kashmir) and *Hatajari* (Uttaranchal). The name *Dactylorhiza* is derived from the Greek word *dactylos* (finger) and *rhiza* (root) referring to the palmately two to five lobed tubers. Locally in Ladakh, it is known as *Angmo-Lakpa* which means 'fingers of a girl'. It belongs to the family Orchidaceae and this is the only plant from this family which grows at high altitudes of Ladakh. *Dactylorhiza hatagirea syn. Orchis latifolia var. Indica* (Lindley) has been identified as a critically endangered (CAMP status) and critically rare (IUCN status) species and is listed under Appendix II of the Convention of International Trade in Endangered Species (CITES) (Murkute et al. 2011; Warghat et al. 2009, 2013).

The economic potential of the species could be assessed on the basis of its high market value i.e. Rs 2000–2500 kg^{-1} (dry tubers) along with its high annual demand (5000 tons). Marsh orchids are found in the shrubberies, open slopes and marshes in the Himalayas from Pakistan to Southeast Tibet and Nepal. Among different species *D. viridis* is a widespread species in the New World. In India, *D. hatagirea* is an indigenous species and exclusively found in the Uttarakhand, Sikkim, Himachal Pradesh and Ladakh regions of Jammu and Kashmir (Murkute et al. 2011; Warghat et al. 2013).

An extensive survey revealed that *D. hatagirea* is habituated mainly in the Nubra valley (9000–11,000 feet amsl) of the Ladakh region of the Indian Trans-Himalayas. The Nubra valley comprises an acute-angled bend from the valley of Shyok River, further towards the Indus. It is regarded as a cold desert climate characterized by high wind velocity that continues throughout the year, which causes a great variation in temperature (from −40°C in peak winters to 35°C in peak summers). Further, in Nubra, it is widely distributed in Hunder, Skurru, Skampuk, Turtuk, Bogdang, Tirit, Sumur and Tigger villages near running water source (Murkute et al. 2011; Warghat et al. 2009, 2013).

12.4.1 Biochemical Compositions

The mature tubers contain mucilage (45%), starch, glucoside, loroglossin, albumen, volatile oil, phosphate (2.7%), chloride; and the ash contains potassium and lime. Five new compounds known as dactylorhin A-E and two natural compounds known as dactyloses A-B have been reported from the roots of this plant. Dactyloses A-B get synthesized from L-ascorbic acid and 4-hudroxybenzyl alcohol via 2-c-(4-hudroxybenzyl)-α-L-xylo-3-ketohexulofuranosono-1, 4-lactone. However, dactylorhin A and dactylorhin E in enzymatic hydrolysis using almond emulsion give dactylorhin C. Also, dactylorhin D and dactylorhin B in enzymatic hydrolysis using cellulose give a 7a compound (-2-3-dihydroxy-2-2-methylpropyl) butanedioic acid, which in hydrolysis gives Loroglossin.

12.4.2 Medicinal Properties

Dactylorhiza leaves contain loroglossin compounds, which help in the peristaltic movement of the gastrointestinal tract, and its tuber contains dactylorhin compounds, which act as neuroprotective agents against dementia, Alzheimer's disease, depression, anxiety and irritable bowel syndrome, and significantly improved the memory when treated with scopolamine, cycloheximide or alcohol (Murkute et al. 2011; Warghat et al. 2009, 2013).

Bulbous roots are used traditionally in the local Amchi system of medicine as an aphrodisiac, expectorant and nervine tonic. Studies conducted with lyophilized aqueous extract of roots on the effect of sexual behaviour and spermatogenesis in male rats revealed that it increases testosterone production by 2.5 folds and subsequently increases the attraction towards females. The tuber of *Dactylorhiza* yields salep that is being used as a sizing material in silk industry. It is used in the treatment of dysentery and diarrhoea. Salep is also used to relieve hoarseness, diabetes, paralysis, convalesance, impotence, malnutrition, etc. salep boiled with milk is being used as a rejuvenating tonic in Ladakh. This plant is famous not only in Ayurveda but also in Amchi, Unani, etc. About 2.5 gm powder of tubers is considered as a full day's diet in adverse conditions by local people. A decoction of salep with sugar and flavoured with spices has got a tremendous nutraceutical value. Mucilage jelly is also a nutritious by-product and useful in treating diarrhoea, dysentery, chronic fever, leucorrhoea, etc. (Warghat et al. 2009, 2013).

The enormous medicinal potential of *Dactylorhiza* has not been tapped to its full extent till today. Using different mass-spectroscopy methods, the availability of different chemical compounds with high medicinal values should be explored. The standardization of agro-technologies and the development of high-yielding cultivars through conventional breeding methods is also an area left unattended. The efficient use of different DNA markers for characterization will further help the breeding program for exploitation of important medicinal compounds and other agronomical traits (Murkute et al. 2011; Warghat et al. 2013). Due to the overexploitation of one of the critically endangered plants, there is a need to focus on conservation, upgradation and sustainable utilization of this wonder plant (Murkute et al. 2011).

12.5 *INULA RACEMOSA* H.L.

Inula racemosa H.L. (Asteraceae) (Figure 12.4) is a high-value, unconventional medicinal plant of Ladakh region and is locally known as *Manu*. It grows as a wild plant in temperate areas of India viz. Kashmir, Himachal Pradesh and Uttarakhand. In the Ladakh region, it is mainly cultivated in the Leh valley, Nubra valley, Zanskar valley, Suru valley and Kargil up to 3800 m altitude above mean sea level (AMSL). It is a perennial herb up to 1.5 m tall with fragrant, prominent roots and rootstock. The stems are numerous, ascending from the base of the rootstock. The leaves are leathery, rough above and densely hairy below, 25–50 cm long and 10–12 cm wide, and its shape is elliptic–lanceolate. The flower heads are yellowish in colour, have bisexual florets and occur in terminal racemes. In Ladakh, flowering usually occurs from January to July and fruiting in October to November. Fruits (*achenes*) are slender and about 0.5 cm long. There is urgent need

FIGURE 12.4 *Inula racemosa* at flowering and fruiting.

to bring this plant under mass cultivation in Ladakh to harness its full medicinal potential (Chaurasia et al. 2008).

The conditions of the temperate and subalpine regions are most suited for the cultivation of *Inula* species. In Ladakh, *Inula* is easily propagated by seeds as well as rootstocks. For the seeds of *Inula*, chilling treatment is essential for overcoming dormancy. The seeds are sown in the month of November in polyhouse, and germination begins in March. The fruits (*achenes*) mature in winter and can be collected by the end of cold season (March–April) at high hills. *Inula* roots are ready for harvest after about two to three years and they are usually harvested during October to November, or in March in the Ladakh region (Chaurasia et al. 2008). The rootstock and roots should be separated from the aerial portion, chopped into small pieces and allowed to dry in shade.

12.5.1 Biochemical Compositions

This plant is an excellent source of bioactive compounds that have health-promoting properties like alantolactone, isoalantolactone and several other important bioactive constituents (Chaurasia et al. 2008). The bioactive compounds present in the roots of *Inula racemosa* include Inulin, alantolactone, β-sitosterol, isoalantolactone, sesquiterpene lactones, dihydroalantolactone, dihydroisoalantolactone, sitisterol, daucosterol, inunolide, aplotaxene, phenylacctonitrile, isoinunal and glucosides (Raj et al. 2010; Gairola et al. 2014).

12.5.2 Medicinal Properties

Inula is a resilient plant that has multiple uses in the Amchi system of medicine, which makes it an ideal plant for commercial exploitation. It has been used for the treatment of several diseases in traditional medicine in various countries throughout the world. This plant is constituent of several Indian traditional system of medicine like Ayurveda, Siddha, Unani and Amchi (Chaurasia et al. 2008). *Inula racemosa* has been used as traditional medicine in East Asia and Europe. In China, it has been prescribed for abdominal pain, acute enteritis and bacillary dysentery, while Native Americans used this plant for the treatment of tuberculosis (Gairola et al. 2014). It is also known to be quite effective against acidity, gastrointestinal complaints and rheumatism (Angemo et al. 2012), and it also possesses anthelmintic, aphrodisiac and diuretic properties (Tantray et al. 2009).

It can be prescribed in combination with *Guggul* (*Commiphora mukul*) for curing myocardial ischemia. Roots are also used in chronic bronchitis when mixed with half of the amount with *Kuth* (*Saussuria lappa*). It has been used as anthelmintic for children and also as an antiseptic, expectorant and diuretic. Dried rhizomes and roots are used to cure in gastrointestinal troubles and rheumatism (Chaurasia et al., 2008). It has been used as an anthelmintic, antiseptic, expectorant and diuretic (Kumar et al. 2011). Though the cultivation practices of *Inula* were standardized at several locations, the quality and quantity of its bioactive compounds in different valleys in Ladakh under wild and cultivable land have not yet been reported. Standardization related to quality and quantity of bioactive compounds, like sesquiterpene lactones from cultured plantlets, is another area of organized study due to their high medicinal value. The screening of their ecotypes and diversity for morphological, biochemical and genetic levels will enable the researchers to realize the existing population of *Inula*, which will be useful for their conservation and sustained utilization (Chaurasia et al. 2008).

12.6 CONCLUSION

The plant diversity of the cold arid desert of Ladakh has played a vital role in the primary healthcare and also in the daily life of tribal communities for ages (Ballabh and Chaurasia 2011). Sea buckthorn, capers, *Dactylorhiza* and *Inula* have become a vital source for research and development work due to the presence of abundant bioactive compounds. Various plant parts were used in various combinations by the traditional Amchi system of medicine. There is an urgent need for novel techniques and approaches for integrated processing of various plant parts of these high-value medicinal plants into nutraceutical and therapeutic products. There is no doubt that the future holds great promise for these species and and these plants have much to contribute to this planet and its inhabitants.

Due to lack of proper records and over-exploitation of these wild medicinal plants by local peoples, the natural resources, along with related indigenous knowledge, are depleting constantly. It is worthwhile to note that the maintenance of these precious, medicinal genetic resources could be beneficial for humanity as a whole. The hardy nature of these plant species, along with their nutritional and medicinal properties, makes them ideal plants for commercial exploitation in the cold arid region of Ladakh. Keeping this in view, sincere attempts are required from various government agencies working in those areas for the exploration and documentation of traditional knowledge being practiced in Ladakh for the use of medicinal plants.

REFERENCES

Ali, Z. N., Eddouks, M., Michel, J., Sulpice, T., and Hajji, L. 2007. Cardiovascular effect of *Capparis spinosa* aqueous extract. Part III: Antihypotensive effect in spontaneously hypotensive rats. *American Journal of Pharmacology and Toxicology* 2: 111–115.

Al-Said, M. S., Abdelsattar, E. A., Khalifa, S. I., and El-Feraly, F. S. 1988. Isolation and identification of an anti-inflammatory principle from *Capparis spinosa. Pharmazie* 43: 640–641.

Altymyshev, A. 1981. Comparitive action of rutin and flavanoids from the Common Caper on the course of radiation sickness and indexes of the blood coagulation system in rats. *Izv. Akad. Nauk Kirg* SSR 5: 60–63.

Andrade. G., Esteban, E., Velasco, L., Loride, M. J., and Bedmar, E. J. 1997. Isolation and identification of N_2-fixing microorganisms from the rhizosphere of *Capparis spinosa* L. *Plant and Soil* 197: 19–23.

Angmo, K., Adhikari, B. S., and Rawat, G. S. 2012 Changing aspects of traditional healthcare system in western Ladakh, India. *Journal of Ethnopharmacology* 143: 621–630.

Bajpai, P., Murkute, A. A., Mishra, G. P., and Singh, S. B. 2009. Analysis of antifreeze property in Seabuckthorn (*Hippophae rhamnoides*) berry. Souvenir, National Conference on Seabuckthorn and Environment (Sept 25–27, 2009), 59. Organized by DIHAR, DRDO, Leh, J & K, India.

Bajpai, P. K., Murkute, A. A., Mishra, G. P., Stobdan, T., and Singh, S. B. 2010. Molecular characterization of antifreeze properties in Seabuckthorn (*Hippophae rhamnoides*). Souvenir, National conference on Seabuckthorn: Emerging trends in Production to consumption (Feb, 16–18, 2010), 44. Organized by CSKHPKV, Palampur, HP, India.

Ballabh, B. and Chaurasia O. P. 2011. Herbal formulations from cold desert plants used for gynecological disorders. *Ethnobotany Research and Applications* 9: 059–066.

Batanouny, K. H. 1999. Egypt, International Union for Conservation (IUCN): Switzerland: Academy of Scientific Research and Technology, Wild Medicinal Plants in Egypt, 130–131.

Bhoyar, M., Mishra, G. P., Singh, R., and Singh, S. B. 2010. Effects of various dormancy breaking treatments on the germination of wild caper (*Capparis spinosa* L.) seeds from the cold arid desert of Trans–Himalayas. *Indian Journal of Agricultural Research* 80: 620–624.

Bhoyar, M., Mishra, G. P., Singh, R., and Singh, S. B. 2011a. Ethnobotany of traditional wild edible plants from cold arid desert of Ladakh – Potential source of winter vegetables. *The Indian Forester* 138(8): 1029–1033.

Bhoyar, M. S., Mishra, G. P., Naik, P. K., and Srivastava, R. B. 2011b. Estimation of antioxidant activity and total phenolics among natural populations of *Capparis spinosa* leaves collected from cold arid desert of Trans-Himalayas. *Australian Journal of Crop Science* 5(7): 912–919.

Bonina, F., Puglia, C., Ventura, D., Aquino, R., Tortora, S., Sacchi, A., Saija, A., Tomaino, A., Pellegrino, M. L., and de Caprariis, P. 2002. *In vitro* and *in vivo* photo protective effects of lyophilized extract of *Capparis spinosa.* L. buds. *Journal of Cosmetic Science* 53: 321–335.

Bown, D. 1995. *Encyclopaedia of Herbs and their Uses.* Dorling Kindersley, London, ISBN 0-7513-020-31.

Çalis, I., Kuruüzüm, A., Lorenzetto, P. A., and Ruedi, P. 2002. (6S)- Hydroxy-3-oxo-ionol glucosides from *Capparis spinosa* fruits. *Phytochemistry* 59: 451–457.

Çalis, I., Kuruüzüm, A., and Ruedi, P. 1999. 1H-Indole 3 acetonitrile glycosides from *Capparis spinosa* fruits. *Phytochemistry* 50: 1205–1208.

Chaurasia, O. P., Khatoon, N., and Singh, S. B. 2008. Field guide floral diversity of Ladakh. DIHAR-DRDO, Ladakh.

Chen, T. G., Ni, M. K., Li, R., Ji, F., and Chen, T. 1991. Investigation of the biological properties of Central Asian sea buckthorn growing in the province of Kansu, China. *Chemistry of Natural Compounds* 27: 119–121.

Chiej, R. 1984. *Encyclopaedia of Medicinal Plants.* MacDonald, ISBN 0-356-10541-5.

Chopra, R. N., Nayar, S. L., and Chopra, I. C. 1999. *Glossary of Indian Medicinal Plants.* CSIR, New Delhi, India.

Dhurandhar, J. 1973. Bonnisan – A metabolic corrective in gastrointestinal disorders of the newborn (a study of 100 cases). *Probe* 12: 73–78.

Eddouks, M., Lemhadri, A., and Michel, J. B. 2005. Hypolipidemic activity of aqueous extract of *C. spinosa* L. in normal and diabetic rats. *Journal of Ethnopharmacology* 98: 345–350.

Farnsworth, N. R., Akereele, O., and Bingel, A. S. 1985. Medicinal plants in therapy. *Bull World Health Organ.* 63: 965–981.

Fu, X. P., Aisa, H. A., Abdurahim, M., Yili, A., Aripova, S. F., and Tashkhodzhaev, B. 2007. Chemical composition of *Capparis spinosa* fruit. *Chemistry of Natural Compounds* 43: 181–185.

Fu, X. P., Wu, T., Abdurahim, M., Su, Z., Hou, X. L., Aisa, H. A., and Wu, H. 2008. New spermidine alkaloids from *Capparis spinosa* roots. *Phytochemistry Letters* 1: 59–62.

Gadgoli, C. and Mishra, S. H. 1995. Preliminary screening of *Achillea millefolium, Cichorium intybus* and *Capparis spinosa* for antihepatotoxic activity. *Fitoterapia* 66: 319–323.

Gadgoli, C. and Mishra, S. H. 1999. Antihepatotoxic activity of p-methoxy benzoic acid from *Capparis spinosa*. *Journal of Ethnopharmacology* 66: 187–192.

Gairola, S., Sharma, J., and Bedi, Y. S. 2014. A cross-cultural analysis of Jammu, Kashmir and Ladakh (India) medicinal plant use. *Journal of Ethnopharmacology* 55: 925–986.

Ganju, L., Padwad, Y., Singh, R., Karan, D., Chanda, S., Chopra, M. K., Bhatnagar, P., Kashyap, R., and Sawhney, R. C. 2005. Anti-inflammatory activity of Seabuckthorn (*Hippophae rhamnoides*) leaves. *International Journal of Immunopharmacology* 5(12): 1675–1684.

Gao, Z. L., Gu, X., Cheng, F., and Jiang, F. 2003. Effects of sea buckthorn on liver fibrosis: A clinical study. *World Journal of Gastroenterology* 9: 1615–1617.

Geetha, S., Purushothaman, J., Karan, P., Shweta, P., Ratan, K., and Sawhney, R. C. 2008. Hepatoprotective effects of sea buckthorn (*Hippophae rhamnoides* L.) against carbon tetrachloride induced liver injury in rats. *Journal of the Science of Food and Agriculture* 88(9): 1592–1597.

Geetha, S., Sai Ram, M., Singh, V., Ilavazhagan, G., and Sawhney, R. C. 2002. Effect of seabuckthorn on sodium nitroprusside-induced cytotoxicity in murine macrophages. *Biomedicine and Pharmacotherapy* 56(9): 463–467.

Genders, R. 1994. *Scented Flora of the World*. Ed. R. Hay and P. G. Waterman. Bury St. Edmunds: St. Edmundsbury Press.

Germano, M. P., Pasquale, R. D., D'Angelo, V., Catania, S., Silvari, V., and Costa, C. 2002. Evaluation of extracts and isolated fraction from *Capparis spinosa* L. buds as an antioxidant source. *Journal of Agricultural and Food Chemistry* 50: 1168–1171.

Giuffrida, D., Salvo, F., Ziino, M., Toscano, G., and Dugo, G. 2002. Initial investigation on some chemical constituents of Capers from the island of Salina. *Italian Journal of Food Science* 14: 25–33.

Goel, H. C., Kumar, I. P., Samanta, N., and Rana, S. V. S. 2003. Induction of DNA-protein cross-links by Hippophae rhamnoides: Implications in radioprotection and cytotoxicity. *Molecular and Cellular Biochemistry* 245: 57–67.

Goel, H. C., Prasad, J., Singh, S., Sagar, R. K., Kumar, I. P., and Sinha, A. K. 2002. Radioprotection by a herbal preparation of *Hippophae rhamnoides*, RH-3, against whole body lethal irradiation in mice. *Phytomedicine* 9(1): 15–25.

Goel, H. C., Samanta, N., Kannan, K., Kumar, I. P., and Bala, M. 2006. Protection of spermatogenesis in mice against gamma ray induced damage by *Hippophae rhamnoides. Andrologia* 38(6): 199–207.

Gurmet, P. 2009. Seabuckthorn in SOWA-RIGPA (Tibetan Medicine). *Seabuckthorn (Hippophae spp.): The Golden Bush*. Ed. S. K. Dwivedi, T. Parimelazahagan, S. B. Singh, and Z. Ahmed, 105–112. SSPH, Delhi, India.

Huseini, H. F., Alavian, S. M., Heshmat, R., Heydari, M. R., and Abolmaali, K. 2005. The efficacy of Liv-52 on liver cirrhotic patients: A randomized, double-blind, placebo-controlled first approach. *Phytomedicine* 12: 619–624.

Jain, M., Ganju, L., Katiyal, A., Padwad, Y., Mishra, K. P., Chanda, S., Karan, D., Yogendra, K. M. S., and Sawhney, R. C. 2008. Effect of *Hippophae rhamnoides* leaf extract against Dengue virus infection in human blood-derived macrophages. *Phytomedicine* 15(10): 793–799.

Kala, C. P. 2002. *Medicinal Plants of Indian Trans-Himalaya: Focus on Tibetan Use of Medicinal Resources*. Bishen Singh Mahendra Pal Singh, Dehradun.

Kallio, H., Yang, B., and Peippo, P. 2002. Effects of different origins and harvesting time on vitamin C, tocopherols and tocotrienols in seabuckthorn (*Hippophae rhamnoides*) berries. *Journal of Agricultural and Food Chemistry* 50: 6136–6142.

Kato, A., Yamaoka, M., Tanaka, A., Komiyama, K., and Umezawa, I. 1985. Physiological activities of tocotrienols. *Journal of Japan Oil Chemists' Society* 34: 375–376.

Khanfar, M. A., Sabri, S. S., Zarga, M. H. A., and Zeller, K. P. 2003. The chemical constituents of *Capparis spinosa* of Jordanian origin. *Natural Product Research* 17: 9–14.

Kumar, G. P., Kumar, R., Chaurasia, O. P., and Singh, S. B. 2011. Current status and potential prospects of medicinal plant sector in Trans-Himalayan Ladakh. *Journal of Medicinal Plants Research* 5(14): 2929–2940.

Lam, S. K., Han, Q. F., and Ng, T. B. 2009. Isolation and characterization of a lectin with potentially exploitable activities from caper (*Capparis spinosa*) seeds. *Biosci. Rep.* 29: 293–299.

Li, T. S. C. 1999. Sea buckthorn: New crop opportunity. *Prospectives on New Crops and New Uses*. Ed. Alexandria J. Janick, 335–227. ASHS Press, VA.

Ma, Z., Cui, Y., and Feng, G. 1989. Studies on the fruit character and biochemical compositions of some forms with Chinese seabuckthorn (*Hippophae rhamnoides* subsp. *sinensis*). *Proceedings of International Symposium on Seabuckthorn* (*H. rhamnoides* L.), Xian, China.

Mahajan, A., Sharma, S. K., Asrani, R. K., Varshney, A. C., Tyagi, S. P., and Kumar, A. 2005. Treatment of infected cutaneous wounds with seabuckthorn (*Hippophae* sp.) in bovines: A histopathological study. Indian Society for Veterinary Surgery 26(1): 31–33.

Mahasneh, A. M., Abbas, J. A., and El-Oqlah, A. A. 1996. Antimicrobial activity of extracts of herbal plants used in the traditional medicine of Bahrain. *Phytotherapy Reserach* 10: 251–253.

Mathur, S., Prakash, A. O., and Mathur, R. 1986. Protective effect of Liv-52 against beryllium toxicity in rats. *Current Science* 55: 899–901.

Matthaus, B. and Ozcan, M. 2002. Glucosinolate composition of young shoots and flower buds of capers (*Capparis* species) growing wild in Turkey. *Journal of Agricultural and Food Chemistry* 50: 7323–7325.

Matthaus, B. and Ozcan, M. 2005. Glucosinolates and fatty acid, sterol, and tocopherol composition of seed oils from *Capparis spinosa* Var. *spinosa* and *Capparis ovata* Desf. Var. *canescens* (Coss.) Heywood. *Journal of Agricultural and Food Chemistry* 53: 7136–7141.

Mikkelsen, M. D., Hansen, C. H., Wittstock, U., and Halkier, B. A. 2000. Cytochrome P450 CYP79B2 from Arabidopsis catalyzes the conversion of tryptophan to indole-3-acetaldoxime, a precursor of indole glucosinolates and indole-3-acetic acid. *Journal of Biological Chemistry* 275: 33712–33717.

Mironov, V. A. 1989. Chemical composition of *Hippophae rhamnoides* of different populations of the USSR. *Proceedings of International Symposium on Seabuckthorn* (*H. rhamnoides* L.), Xian, China. 67–70.

Mishra, G. P. 2009a. Field Research Laboratory Leh, renamed as Defence Institute of High Altitude Research. *Currrent Science* 96(4): 452.

Mishra, G. P., Kumar, J., Bhoyar, M., Bajpayee, P., Murkute, A. A., and Singh, S. B. 2009c. Seabuckthorn (*Hippophae rhamnoides* l.) genetic diversity characterization and breeding scope. Souvenir, National Conference on Seabuckthorn and Environment: High Altitude Perspectives (Sept 25–27, 2009), 58. Organized by DIHAR-DRDO, Leh, J&K, India.

Mishra, G. P., Murkute, A. A., Kumar, J., Bhoyar, M. S., Kalia, R. K., and Srivastava, R. B. 2011. Seabuckthorn (*Hippophae rhamnoides* l.) genetic diversity characterization and breeding scope. *Deserts: Fauna, Flora and Environment*. Ed Juan, G., Eduardo, G. 1–21. Nova Publications, NY, USA.

Mishra, G. P., Singh, N., Kumar, H., and Singh, S. B. 2010. Protected cultivation for food and nutritional security at Ladakh. *Defence Life Science Journal* 61(2): 219–225.

Mishra, G. P., Singh, R., Bhoyar, M., and Singh, S. B. 2009b. *Capparis spinosa*: unconventional potential food source in cold arid deserts of Ladakh. *Current Science* 96(12): 1563–64.

Mishra, G. P., and Singh, S. B. 2010. Frozen gene bank at frozen mountain. *Current Science* 98(4): 466.

Murkute, A. A., Warghat, A. R., Kumar, G. P., Mishra, G. P., and Jadhav, S. E. 2011. *Dactylorhiza hatagirea* D.don A critically endangered medicinal plant of trans-Himalaya. *Herbal Perspectives: Present and Future*, 38–45. Ed. Parimelazhagan et al. Published by Satish Serial Publishing House, New Delhi.

Negi, P. S., Chauhan, A. S., Sadia, G. A., Rohinishree, Y. S., and Ramteke, R. S. 2005. Antioxidant and antibacterial activities of various seabuckthorn (*Hippophae rhamnoides* L.) seed extracts. *Food Chemistry* 92(1): 119–124.

Padwad, Y., Ganju, L., Jain, M., Chanda, S., Karan, D., Banerjee, P. K., and Sawhney, R. C. 2006. Effect of leaf extract of Seabuckthorn on lipopolysaccharide induced inflammatory response in murine macrophages. *International Journal of Immunopharmacology* 6(1): 46–52.

Panico, A. M., Cardile, V., Garufi, F., Puglia, C., Bonina, F., and Ronsisvalle, G. 2005. Protective effect of *Capparis spinosa* on chondrocytes. *Life Science* 77: 2479–2488.

Phillips, R. and Rix, M. 1998. *Conservatory and Indoor Plants Pan Books*, 1–2. London.

Raj, J., Ballabh, B., Murugan, P. M., Da Silva, J. A. T., Kumar, S., Chaurasia, O. P., and Singh, S. B. 2010. Screening phytochemical constitutents of 21 medicinal plants of trans-Himalayan region. *Medicinal and Aromatic Plant Science and Biotechnology* 4(1): 90–93.

Ramezani, Z., Aghel, N., and Keyghobadi, H. 2008. Rutin from different parts of *Capparis spinosa* growing in Khuzestan/Iran. *Pakistan Journal of Biological Sciences* 11: 768–772.

Romeo, V., Ziino, M., Giuffrida, D., Condurso, C., and Verzera, A. 2007. Flavour profile of capers (*Capparis spinosa* L.) from the Eolian Archipelago by HS-SPME/GC–MS. *Food Chemistry* 101: 1272–1278.

Sadykov, L., Yu, D., and Khodzhimatov, M. 1981. Alkaloids of *Capparis spinosa*. *Dokl Akad Nauktadzh SSR*. 24: 617–620.

Saggu, S. and Kumar, R. 2007. Possible mechanism of adaptogenic activity of seabuckthorn (*Hippophae rhamnoides*) during exposure to cold, hypoxia and restraint (C–H–R) stress induced hypothermia and post stress recovery in rats. *Food and Chemical Toxicology* 45(12): 2426–2433.

Saggu, S. and Kumar, R. 2008. Effect of seabuckthorn leaf extracts on circulating energy fuels, lipid peroxidation and antioxidant parameters in rats during exposure to cold, hypoxia and restraint (C–H–R) stress and post stress recovery. *Phytomedicine* 15(6–7): 437–446.

Schraudolf, H. 1989. Indole glucosinolates of *Capparis spinosa*. *Phytochemistry* 28: 259–260.

Sharaf, M., El-Ansari, M. A., and Saleh, N. A. 2000. Quercetin triglycoside from *Capparis spinosa*. *Fitoterapia* 71: 46–49.

Sharma, R. 2003. *Medicinal Plants of India – An Encyclopedia*, 42–43. Daya Publishing house, New Delhi.

Sharma, S. K., and Chunekar, K. C. 1998. Rashtriya Ayurveda Vidyapeeth, New Delhi.

Sharma, U. K., Sharma, K., Sharma, N., Sharma, A., Singh, H. P., and Sinha, A. K. 2007. Microwave-assisted efficient extraction of different parts of *hippophae rhamnoides* for the comparative evaluation of antioxidant activity and quantification of its phenolic constituents by RP-HPLC. *Journal of Agricultural and Food Chemistry* 56(2): 374–379.

Shukla, S. K., Chaudhary, P., Kumar, I. P., Samanta, N., Afrin, F., Gupta, M. L., Sharma, U. K., Sinha, A. K., Sharma, Y. K., and Sharma, R. K. 2006. Protection from radiation-induced mitochondrial and genomic DNA damage by an extract of *Hippophae rhamnoides*. *Environmental and Molecular Mutagenesis* 47(9): 647–656.

Singh, B. 1995. Vegetable cultivation at Ladakh. DESIDOC, DRDO, New Delhi.

Singh, R., Mishra, G. P., Singh, N., and Singh, S. B. 2011. Protected cultivation and sustainable crop production at high altitudes of Ladakh. Protected cultivation for sustainable horticulture. Ed. Sephia and Sharma, 197–210.Dr YSPUHF, Nauni, HP, India.

Stobdan, T., Chaurasia, O.P., Korekar, G., Mundra, S., Ali, Z., Yadav, A., and Singh, S. B. 2010. Attributes of seabuckthorn (*Hippophae rhamnoides* L.) to meet nutritional requirements in high altitude. *Defence Science Journal* 60(2): 226–230.

Stobdan, T., Yadav, A., Mishra, G. P., Chaurasia, O. P., and Srivastava, R. B. 2011. *Seabuckthorn:* The Super Plant (Production, Characterization, Postharvest and Health applications). 100. DIHAR, DRDO.

Subhose, V., Srinivas P., and Narayana A. 2005. Basic principles of pharmaceutical science in Ayurveda. *Bulletin of the Indian Institute of History of Medicine* 35(2): 83–92.

Swaroop, A., Sinha, A. K., Chawla, R., Arora, R., Sharma, R. K., and Kumar, J. K. 2005. Isolation and Characterization of 1,3-Dicapryloyl-2-linoleoylglycerol: A Novel Triglyceride from Berries of *Hippophae rhamnoides*. *Chemical and Pharmaceutical Bulletin* 53(8): 1021.

Tantray, M. A., Tariq, K. A., Mir, M. M., Bhat, M. A., and Shawl, A. S. 2009. Ethnomedicinal survey of Shopian, Kashmir (J&K), India. *Asian Journal of Traditional Medicines* 4: 1–6.

Tesoriere, L., Butera, D., Gentile, C., and Livrea, M. A. 2007. Bioactive components of caper (*Capparis spinosa* L.) from Sicily and antioxidant effects in a red meat simulated gastric digestion. *Journal of Agricultural and Food Chemistry* 55: 8465–8471.

Tomas, F. and Ferreres, F. 1978. 3-O-Rhamnorutinosyl kaempferol from floral buttons of *Capparis spinosa*. *Revista de Agroquimica y Tecnologia de Alimentos* 18: 232–235.

Trombetta, D., Occhiuto, F., Perri, D., Puglia, C., Santagati, N. A., Pasquale, A. D., Saija, A., and Bonina, F. 2005. Anti-allergic and antihistaminic effect of two extracts of *Capparis spinosa* L. flowering buds. *Phytotheraphy Research* 19: 29–33.

USDA. 2016. Nutrient database. http://www.nal.usda.gov/fnic/foodcomp/search.

Vijayaraghvan, R., Gautam, A., Kumar, O., Pant, S. C., Sharma, M., Singh, S., Kumar, H. T., Singh, A. K., Nivsarkar, M., Kaushik, M. P., Sawhney, R. C., Chaurasia, O. P., and Prasad, G. B. 2006. Protective effect of ethanolic and water extracts of sea buckthorn (*Hippophae rhamnoides* L.) against the toxic effects of mustard gas. *Indian Journal of Experimental Biology* 44(10): 821–831.

Warghat, A. R., Bajpai, P. K., Srivastava, R. B., Chaurasia, O. P., and Sood, H. 2013. Population genetic structure and conservation of small fragmented locations of *Dactylorhiza hatagirea* in Ladakh region of India. *Scientia Horticulturae* 164: 448–454.

Warghat, A. R., Murkute, A. A., Mishra, G. P., Kumar, G. P., and Singh, S. B. 2009. Exploring morphological diversity in *Dactylorhiza hatagirea* (d.don) in the cold arid desert of Ladakh. Souvenir, National Conference on Seabuckthorn and Environment: High Altitude Perspectives (Sept 25–27, 2009), 27. Organized by DIHAR-DRDO, Leh, India.

Xing, J., Yang, B., Dong, Y., Wang, B., Wang, J., and Kallio, H. 2002. Effects of sea buckthorn (*Hippophae rhamnoides* L.) seed and pulp oils on experimental models of gastric ulcer in rats. *Fitoterapia* 73: 644–650.

Xu, M. 1994. Anticancer effects of and direction of research on *Hippophae*. *Hippophae* 7: 41–43.

Zhan, A. 1978. Chemical and Biological characterization of *Capparis spinosa*. *Azerbaidzhanskii Meditsinskii Zhurnal* 55: 70–75.

Zhang, W., Yan, J., Duo, J., Ren, B., and Guo, J. 1989. Preliminary study of biochemical constituents of sea-buckthorn berries growing in Shanxi Province and their changing trend. *Proceedings of International Symposium on Seabuckthorn* (*H. rhamnoides* L.). 96–105. China: Xian.

13 Phytochemistry of *Muntingia calabura* L. Fruits

Kathirvel Preethi

CONTENTS

13.1 Introduction 261
13.2 Botanical Information 261
13.3 Medicinal Value 262
13.4 Phytochemical Investigations 262
13.5 High-Performance Thin-Layer Chromatography Profiling of *M. calabura* Fruits 264
13.6 High Performance Liquid Chromatography Profile of *M. calabura* Fruits 265
13.7 Conclusion 267
References 268

13.1 INTRODUCTION

The use of traditional medicine is widespread, and plants still present a large source of novel, active biological compounds with different activities, including anti-inflammatory, anti-cancer, anti-viral, antibacterial and cardioprotective activities. Antioxidants may play a role in these health-promoting activities (Yan et al. 2002). Plants are potent biochemical factories and have been components of phytomedicine since time immemorial; from them, we are able to obtain from them a wondrous assortment of industrial chemicals. Plant-based natural constituents can be derived from any part of the plant, like the bark, leaves, flowers, roots, fruits, seeds or any part that may contain active components (Makari and Patil 2008).

13.2 BOTANICAL INFORMATION

Muntingia calabura, the sole species in the genus *Muntingia*, is a flowering plant native to southern Mexico, the Caribbean, Central America and western South America, south of Peru and Bolivia. Common names include (English) Jamaican cherry, Panama berry, Singapore cherry and the Strawberry tree. It is a small tree (7–12 m tall) with tiered and slightly drooping branches. It has serrated leaves 2.5–15 cm long and 1–6.5 cm wide. The flowers are small, white and slightly malodorous. It gives rise to 1–1.5 cm light red fruit (Figure 13.1). The fruit is edible, sweet

FIGURE 13.1 *Muntingia calabura* fruits.

and juicy, and contains a large number of tiny (0.5 mm) yellow seeds. This is a fast-growing fruit tree. It is a pioneer species that thrives in poor soil, able to tolerate acidic and alkaline conditions, as well as drought. It is cultivated for its edible fruit, and has become naturalized in some other parts of the tropics, including Southeastern Asia. As a pioneer plant, it could help condition the soil and make it habitable for other plants.

13.3 MEDICINAL VALUE

Traditional medicinal applications of the plant, although not recorded in extensive literature, include its use as an antiseptic to relieve headaches, gastric ulcers and inflammation of the prostate gland (Morton 1987; Zakaria et al. 2007). The extracts of the fruits are used as tranquilizers and tonics (Kaneda et al. 1991). The plant is used to treat measles, mouth pimples and stomachaches (Yasunaka et al. 2005). The fruit is also used to treat incipient cold, as a tranquillizer, as an antispasmodic and as an anti-dyspeptic.

13.4 PHYTOCHEMICAL INVESTIGATIONS

Increasing epidemiological studies have repeatedly revealed the role of consumption of fruits and vegetables in the prevention of degenerative and chronic diseases. The phytochemical properties of *M. calabura* are presented in Table 13.1. These benefits are thought to result from various bioactive components of plant origin, such as vitamins, flavonoids and carotenoids. High intake of fruits and vegetables has shown to

TABLE 13.1
Phytochemicals Present in the *M. calabura* Fruit Extract

S. No	Compound	Methanol	Chloroform	Ethyl Acetate	Petroleum Ether	Butanol
1	Alkaloids	+++[a]	–	–	–	–
2	Flavonoids	+++	+++	+++	+++	+++
3	Saponins	–[b]	++	+	+++	–
4	Steroids	++[c]	+	++	+	+++
5	Sterols	++	+	++	+	+++
6	Tannins	+++	–	+++	–	+
7	Terpenoids	++	+	++	+	+++
8	Gums & Mucilages	–	–	+	–	–
9	Glycosides	+	–	+	–	+

[a] Present in high amount.
[b] Absent.
[c] Moderately present.

be inversely related to incidence of several degenerative diseases, such as coronary heart disease and cancer. Hence the increase in the consumption of fruits may be a practical strategy for prevention. Plant-derived natural products (such as flavonoids, terpenoids, steroids, etc.) have received considerable attention in recent years due to their diverse pharmacological properties, including antioxidant and hepatoprotective activity (Banskota et al. 2000; De Feudis et al. 2003; Takeoka and Dao 2003). Plant secondary metabolites have provided an important source of drugs since ancient times, and now around half of practical drugs used are derived from natural sources (Wang et al. 2008). Phytochemicals are available in vegetables and fruits, which are more appropriately defined as bioactive, non-nutrient plant compounds in citrus fruits, vegetables, grains and other plant foods that have been linked to reduce the major chronic diseases and cancers (Ghodake et al. 2010). Different classes of phytochemicals have been demonstrated to be responsible for various health benefits and disease protection.

This phytochemical screening was done with various solvents for the extraction of phytochemicals. Further, the phytochemicals presented in the *M. calabura* fruit are tabulated in Table 13.2.

High amounts of phenolic compounds are present in *M. calabura* fruits. These phenolics are well-known compounds, owing to the potent antioxidant activities and bioactivities and are also known to diffuse free radicals (Willcox et al. 2004). Flavonoids are considered to be strong scavengers of ROS, which is also present in *M. calabura* fruits. Phytochemicals, such as alkaloids, phenolics, carotenoids and various nitrogenous compounds in fruit and vegetables are reported to account for various bioactivities, like antioxidant, antiproliferation, antifungal, antibacterial and antiviral activities (Dillard and German 2000). *M. calabura* fruits are also rich in dietary proteins

TABLE 13.2
Phytonutrients Present in the *M. calabura* Fruit Extract

S. No	Parameters	mg/100 g of Fresh Weight
1	Total phenols	1491
2	Total flavonoids	300
3	Total anthocyanins	0.3
4	Total ascorbic acid	9861

and essential amino acids. The protein content of the fruit varies, depending on the seasonal influence, age and ripening. The presence of vital phyto nutrients in these fruits paves the way for their utilization as protein supplements to tackle the problem of protein deficiency–associated problems, especially in rural areas.

Anthocyanins are the most common red-purple fruit pigments, and have been extensively studied for their antioxidant anthocyanin contents. Fruits that are not red-purple may contain other potent polyphenolic antioxidants (Einbond et al. 2004). In our study, the ripened fruit of *M. calabura* is a dark red–colored fruit, which may confirm the presence of anthocyanin pigment. Thus, these phytochemicals are also present in *M. calabura* fruits, owing to drug discovery.

Vitamin C is an essential dietary nutrient required as a cofactor for many enzymes, and humans are among the few animals that lack the ability to synthesize this compound from glucose. Hence, it is essential to uptake fruits or vegetables rich in vitamin C. Moreover, epidemiological studies show that individuals with high intakes of vitamin C have a lower risk of numerous chronic diseases, including heart disease, cancer, various eye diseases and neurodegenerative conditions. Vitamin C was found to account for 65–100% of the antioxidant potential of beverages derived from citrus. Although, phenolics appear to be contributors of the antioxidant potential of non-citrus juices. Though *M. calabura* is a non-citrus fruit, it is rich in vitamin C and also a contributor of phenolics.

Section 13.4 presents an overall generalized outline of phytochemistry of *M. calabura*. The enormous therapeutic value of these fruits highlights the need for an in-depth discussion regarding the core phytochemical compounds present in *M. calabura*, as outlined in the preceding sections.

13.5 HIGH-PERFORMANCE THIN-LAYER CHROMATOGRAPHY PROFILING OF *M. CALABURA* FRUITS

Phenolic compounds are further classified into polyphenolics present in excess in *M. calabura* fruits. Many different known and unknown polyphenols are present in *M. calabura* fruits, which were investigated by high-performance thin-layer chromatography (HPTLC), a chromatographic technique, the results of which are presented in Table 13.3.

TABLE 13.3
R_f Values, Height and Area Peaks of Polyphenols and Unknown Compounds

Track	Peak	R_f	Height	Area	Assigned Substance
A	1	0.01	587.0	14041.9	Polyphenol 1
A	2	0.11	129.1	3462.6	Polyphenol 2
A	3	0.17	150.2	5001.3	Polyphenol 3
A	4	0.20	199.2	5959.5	Unknown
A	5	0.40	125.6	7870.0	Polyphenol 4
A	6	0.52	116.2	3753.0	Polyphenol 5
A	7	0.63	31.7	951.1	Unknown
A	8	0.70	30.9	772.0	Unknown
A	9	0.87	19.5	733.3	Unknown

13.6 HIGH-PERFORMANCE LIQUID CHROMATOGRAPHY PROFILE OF *M. CALABURA* FRUITS

Further, in order to separate and detect individual polyphenols in *M. calabura* fruits, analytical high-performance liquid chromatography (HPLC) techniques were used, which revealed the presence of flavonol, a quercetin compound in the fruit.

Quercetin (which contains the highest amount of flavanols in food) and their derivatives (gallic acid and catechol) are reported to possess potent antimutagenesis, antitumor and antimetastasis properties (Vessal et al. 2003). In addition, the combination of quercetin, ellagic acid and some other flavanoids found in *M. calabura* fruits have been credited with health-promoting potential. Further investigations reported phenolic compounds, along with their mass and the compound properties of *M. calabura* fruits. Table 13.4 clearly states the compounds present and their biological properties.

M. calabura fruit is enriched with polyphenols, especially cinnamic acids and their derivatives. The role of hydroxylcinnamic acid compounds as antioxidants and radical scavengers demonstrates that they are a precursor for anthocyanins, which are also present in the fruit. Each of the identified compounds has its correlation with the other, for instance, the total flavonoid content and total phenolic content. All these compounds are directly or indirectly related due to their antioxidant potential. Likewise, flavonols, anthocyanin and flavonoids (such as quercetin) show an indispensable relation with flavonols, owing to their antioxidant activities.

TABLE 13.4
Identified Compounds along with Molecular Mass and Biological Properties

S. No	Compound	Molecular Mass	Biological Properties
		Hydroxy Cinnamic Acids	
1	Pyrocatechin	110.11	Antioxidant
2	Caffeic acid	180.16	Anticancer, Anti-inflammatory
3	Caffeine	194.20	Antioxidant
4	Catechin glycoside	422.40	Antioxidant
5	Diferulic acid	386.56	Antioxidant
		Cinnamic Acid Derivatives	
7	Sinapic acid	224.22	Antioxidant
8	Caffeoyl shikimic acid	336.30	Antioxidant
9	Catechin glycoside	422.40	Antioxidant
10	Ellagic acid	302.20	Antioxidant
11	Cinnamic acid	148.16	Antioxidant
12	Methoxy cinnamaldehyde	162.19	Antioxidant
13	Caffeic acid	180.16	Antioxidant
14	Caffeic acid glucoside	342.31	Antioxidant
		Flavonol	
15	Deoxyquercetin	286.25	Antioxidant, Anti-inflamatory
		Gallic Acid Derivative	
16	Gluco gallic acid	332.27	Antioxidant
		Fatty Acids	
17	Linolenic acid	278.44	Antioxidant
18	Ergosterol	396.66	Antioxidant
19	Palmitolic acid	254.42	Antioxidant
20	Linoleic acid	280.45	Antioxidant, Anticancer
21	Oleic acid	282.47	Antioxidant
22	Arachidic acid	312.54	Antioxidant, Anticancer
23	Stearic acid	284.44	Antioxidant
		Others	
24	Cytisine	190.25	Antioxidant, Bactericide
25	Cucurbic acid	212.29	Antioxidant, Anticancer
26	Pipecolic acid	129.16	Antioxidant, COX-2 Inhibitor
27	Embelic acid	294.40	Antioxidant, Anticancer
28	Rosmarinic acid	360.33	Antioxidant, Anticancer

(*Continued*)

TABLE 13.4 (CONTINUED)
Identified Compounds along with Molecular Mass and Biological Properties

S. No	Compound	Molecular Mass	Biological Properties
29	Isobruceine	522.56	Antidiabetic, Antiseptic
30	Chrysophanol glucoside	416.39	Antidiabetic, Antifungal
31	Aristolochic acid	341.28	Antioxidant
32	Symphytine	381.48	Antioxidant, Antiviral
33	Khellin	260.25	Antioxidant, Antiviral
34	Chlorogenic acid	354.32	Antioxidant
35	Menthyl acetate	198.31	Antioxidant, Immunosupportive
36	Ginsenoside	947.18	Cardiovascular activity
37	Hydroxy methyl benzoic acid	152.15	Antioxidant
38	Cymarose	162.19	Antioxidant, Antibacterial
39	Ajmalicine	352.44	Antioxidant
40	Embelic acid	294.40	Antioxidant, Anticancer
41	Lobelanidine	334.44	Antioxidant, Immune enhancer
42	Rosmarinine	353.42	Antioxidant, Antidiabetic
43	Umbelliferose	504.45	Antioxidant, Antiseptic
44	Hypaconitine	615.73	Antioxidant
45	Delphinine	599.73	Antioxidant

13.7 CONCLUSION

The accumulated research experience, knowledge and practical applications in recent years concerning bioactive compounds, particularly phenolic compounds, have increased a lot. From the results, it seems that *M. calabura* fruit possesses potential antioxidant activity, with high phenolic and flavonoid contents. The recovery of phenols was dependent on the solvent system used; in this study, ethyl acetate and methanol were the best among all the solvents in extracting phenolics efficiently. Results on HPTLC indicated that the *M. calabura* fruit contains polyphenols. The HPLC and LC-MS analysis showed that the major phenolic compounds present in the fruit were cinnamic acid, ferulic acid, sinapic acid, shikimic, catechin and caffeic acid. This high content of cinnamic acid is not frequently encountered in fruits. One of the polyphenolic flavonols, quercetin was identified by analytical HPLC.

Thus, *Muntingia calabura* might be useful in the development of medicinal raw materials from plants. It may therefore be concluded that *M. calabura* fruit may play some contributory role in the prevention of diseases with oxidative stress as one of the etiological factors, in areas where it is widely consumed. Future work should be focused on treatments that have fruit-promoting bioavailability, and also on confirming of the effects of antioxidant compounds from cherries for consumer health.

REFERENCES

Banskota, A. H., Tezuka, Y., Adnyana, I. K., Xiong, Q., Hase, K., Tran, K. Q., and Kadota, S. 2000. Hepatoprotective Effect of *Combretum quadrangulare* and Its Constitutents. *Biological and Pharmaceutical Bulletin* 23(4): 456–460.

De Feudis, F. V., Papadopoulos, V., and Drieu, K. 2003. *Ginkgo biloba* extracts and cancer: A research area in its infancy. *Fundamental & Clinical Pharmacology* 17(4): 405–417.

Dillard, C. J. and German, J. B. 2000. Phytochemicals: Nutraceuticals and human health. *Journal of the Science of Food and Agriculture* 80(12): 1744–1756.

Einbond, L. S., Reynertson, K. A., Luo, X. D., Basile, M. J., and Kennelly, E. J. 2004. Anthocyanin antioxidants from edible fruits. *Food Chemistry* 84(1): 23–28.

Garro, L. C. 1986. Intracultural variation in folk medical knowledge: A comparison between curers and noncurers. *American Anthropologist* 88(2): 351–370.

Ghodake, G. S., Deshpande, N. G., Lee, Y. P., and Jin, E. S. 2010. Pear fruit extract-assisted room-temperature biosynthesis of gold nanoplates. *Colloids and Surfaces B: Biointerfaces* 75(2): 584–589.

Kaneda, N., Pezzuto, J. M., Soejarto, D. D., Kinghorn, A. D., Farnsworth, N. R., Santisuk, T., and Reutrakul, V. 1991. Plant anticancer agents, XLVIII. New cytotoxic flavonoids from *Muntingia calabura* roots. *Journal of Natural Products* 54(1): 196–206.

Makari, H. K. and Patil, R. 2008. *In vitro* antioxidant activity of the hexane and methanolic extracts of *Cordia wallichii* and *Celastrus paniculata. The Internet J. Aesthetic and Antiaging Medicine* 1: 1–10.

Morton, J. F. 1987. Jamaica cherry. *Fruits of Warm Climates.* Ed. J. F. Morton. Miami, Florida 2: 65–69.

Takeoka, G. R. and Dao, L. T. 2003. Antioxidant constituents of almond *Prunus dulcis* (Mill.) DA Webb hulls. *Journal of Agricultural and Food Chemistry* 51(2): 496–501.

Vessal, M., Hemmati, M., and Vasei, M. 2003. Antidiabetic effects of quercetin in streptozocin-induced diabetic rats. *Comparative Biochemistry and Physiology Part C: Toxicology & Pharmacology* 135(3): 357–364.

Wang, S. Y. and Lin, H. S. 2000. Antioxidant activity in fruits and leaves of blackberry, raspberry, and strawberry varies with cultivar and developmental stage. *Journal of Agricultural and Food Chemistry* 48(2): 140–146.

Wang, W., Tai, F., and Chen, S. 2008. Optimizing protein extraction from plant tissues for enhanced proteomics analysis. *Journal of Separation Science* 31(11): 2032–2039.

Willcox, J. K., Ash, S. L., and Catignani, G. L. 2004. Antioxidants and prevention of chronic disease. *Critical Reviews in Food Science and Nutrition* 44(4): 275–295.

Yan, X., Murphy, B. T., Hammond, G. B., Vinson, J. A., and Neto, C. C. 2002. Antioxidant activities and antitumor screening of extracts from cranberry fruit (*Vaccinium macrocarpon*). *Journal of Agricultural and Food Chemistry* 50(21): 5844–5849.

Yasunaka, K., Abe, F., Nagayama, A., Okabe, H., Lozada-Pérez, L., López-Villafranco, E., and Reyes-Chilpa, R. 2005. Antibacterial activity of crude extracts from Mexican medicinal plants and purified coumarins and xanthones. *Journal of Ethnopharmacology* 97(2): 293–299.

Zakaria, Z. A., Kumar, G. H., Zaid, S. N., Ghani, M. A., Hassan, M. H., Hazalin, N. A., and Sulaiman, M. R. 2007. Analgesic and antipyretic actions of *Muntingia calabura* leaves chloroform extract in animal models. *Oriental Pharmacy and Experimental Medicine* 7(1): 34–40.

14 Pharmacognostical and Phytochemical Investigation on *Pterolobium hexapetalum* (Roth.) Sant. & Wagh.

Saikumar Sathyanarayanan, Rahul Chandran, Murugan Rajan and Parimelazhagan Thangaraj

CONTENTS

14.1 Introduction 270
14.2 Materials and Methods 272
14.2.1 Collection and Identification of Plant Material 272
14.2.2 Pharmacognosy Studies 272
14.2.2.1 Anatomy 272
14.2.2.2 Photomicrographs 272
14.2.2.3 Leaf Constants 273
14.2.2.4 Stomatal Number 273
14.2.2.5 Stomatal Index 273
14.2.2.6 Vein-Islet Number 273
14.2.2.7 Fluorescence Analysis 273
14.2.3 Preparation of Plant Extracts 274
14.2.4 Qualitative Phytochemical Screening 274
14.2.4.1 Alkaloids 274
14.2.4.2 Saponins 274
14.2.4.3 Phenolic Compounds 274
14.2.4.4 Tannins 274
14.2.4.5 Flavonoids 275
14.2.4.6 Glycosides 275
14.2.4.7 Flavonol Glycosides 275
14.2.4.8 Cardiac Glycosides 275
14.2.4.9 Phytosterols 275

14.2.4.10 Fixed Oils and Fats 275
14.2.4.11 Gums and Mucilages 275
14.3 Results and Discussion 276
14.3.1 Pharmacognostic Studies of *P. hexapetalum* 276
14.3.1.1 Microscopic Observations of Leaves 276
14.3.1.2 Microscopic Observation of Stems 277
14.3.1.3 Histochemical Observation 278
14.3.1.4 Fluorescence Analysis of Powdered Drugs 278
14.3.2 Qualitative Screening of Phytochemicals 280
14.4 Conclusion 281
References 281

14.1 INTRODUCTION

The use of plants as medicine predates written human history. Knowledge of plant use was widespread in ancient civilizations. Ancient ethnic communities around the world have learnt to utilize their neighbourhood herbal wealth for curative as well as treatment purposes. Among the ancient civilizations, India has been known to be a rich repository of medicinal plants. Until the middle of the 19th century, plants were the main therapeutic agents used by humans, and even today their role in medicine is still relevant. Many rural people hold traditional knowledge of medicinal plants, and the very fact of the existence of such local knowledge is dependent on oral history across generations (Joshi and Joshi 2000; Tabuti et al. 2003). Some of the traditional knowledge systems are well-organized and documented in Indian systems of medicine (Ayurveda, Siddha, and Unani), and the Chinese systems of medicine are about 5000 years old. These traditional systems are reported to make use of about 1000 plant genera and 2500 species. In addition to these traditional systems, there exists a parallel stream of tribal and folk medicine all over the world. The role of indigenous people as protectors of biodiversity and as custodians of ancient knowledge systems, including the medicinal uses of plants, is well recognized (Mukherjee 2009).

Ethnomedicinal investigations are imperative in illuminating significant indigenous plant species, primarily for finding new rudimentary drugs. The documentation of native medicinal knowledge of plant species has contributed to a number of modern drug formulations for basic healthcare (Flaster 1996; Cox 2000). In developing countries and rural societies, the use of medicinal plants is both a valuable resource and a necessity, and furthermore it provides a real alternative for primary healthcare systems. The World Health Organization (WHO 2010) estimated that about 60% of the world's population in developing countries relies on plants for the treatment of various diseases, due to the lack of modern healthcare facilities (Calixto 2005; WHO 2010). In rural communities, medicinal plants have gained attention because of their effectiveness, lack of modern medical alternatives, increasing costs of allopathic medicines and cultural preferences (Heinrich 2000; Tabuti et al. 2003).

In developed countries, many people are turning to herbal remedies especially for minor ailments. Modern scientific medicine still depends on plants (and the knowledge gained from plants) for some essential drugs. In the present movement of 'back to nature' in healthcare, it is very relevant that these valuable plant species should not only be preserved but also cultivated. The international market of medicinal plants is over US $60 billion per year, which is growing at the rate of 7%. China and India are two great producers of medicinal plants, having more than 40% of global biodiversity. There is enormous scope for India also to emerge as a major player in the global herbal market. However, this requires a grand strategic plan, which takes a holistic view of the entire situation to boost the export of Rs. 10,000 crores by 2020 and minimizing the import (Anonymous 2000). A shift from the collection to cultivation of medicinal and aromatic plants will ensure purity, authenticity and a sustainable supply of raw materials required for herbal drugs.

There has been wide increase in the therapeutic importance of herbal medicine and thus, it is essential to obtain a proper quality control profile for various medicinal plants used in traditional systems of medicine. This may be helpful in minimizing the adulteration of these plants which occurs due to improper knowledge. It is also said that correct identification and proper quality assurance of the starting materials is an essential prerequisite to ensure the reproducible quality of herbal medicines (Nagani et al. 2012; Prasad et al. 2012). Standardization will ensure that every packet of medicine that is sold is the correct amount that will induce therapeutic effects. According to the WHO, botanical standards should be proposed as a protocol for the diagnosis of the herbal drug quality control of crude drugs, and its pharmaceuticals can be attempted by different methods of evaluation depending upon morphological and microscopical studies (Kokate et al. 2007). Anatomical studies are helpful in describing a particular drug with a special emphasize on quantitative microscopy (such as sclerides, starch grains, crystals, stomata, trichomes) and qualitative microscopic studies (such as xylem, phloem and other tissues) (Brinda et al. 2000). To check the genuineness of the raw drugs and to detect adulteration of these materials, an authentic pharmacognostic study is needed for each raw drug (Babu et al. 2010)

Usually the drugs are collected by traditional practitioners who have inherited Ayurvedic or other herbal practices. Their identification is mostly based on morphological features or other traditionally known characteristics. In such cases, there is a chance of selecting incorrect raw drugs. Therefore, an extensive anatomical and phytochemical screening is needed for each raw drug used in the formulation to avoid any ambiguity, and such a study will serve also as a reference for further studies (Vaibhav and Kamlesh 2007). Pharmacognostic studies can serve several purposes, inducing batch-to-batch consistency, conformation of correct amount of dosage and positive control to indicate possible loss or degradation (Chaudhury 1992). To avoid the misuse of harmful plant material, it is necessary to scientifically develop pharmacognostical and physicochemical standards of a particular plant material, which may ensure and maintain its quality, efficacy and safety profile (Laloo et al. 2013). Standardization practices continue today because of their biochemical benefits and cultural benefits in many parts of the world, and have made a great contribution towards maintaining human health (Nasreen and Radha 2011).

The importance of medicinal plants lies in their biological active principles, which are the real healers in the process of medication. There are two types of plant chemicals: primary and secondary metabolites. Primary metabolites are universally present in all kinds of plants, whether medicinal or non-medicinal. Secondary metabolites do not have an essential role in plant metabolism, and vary in their distribution from plant to plant. Secondary metabolites are mostly accumulated by plant cells in smaller quantities than primary metabolites. They are synthesized in specialized cells at particular development stages, making their extraction and purification difficult. These secondary metabolites exert a profound physiological effect on the mammalian system and thus are known as the active principles of that plant. The physiological effect of these active principles is used for curing ailments (Padmakumar 2009).

Among these herbals, *Pterolobium hexapetalum* (Caesalpiniaceae) is one of the traditionally used medicinal plants in India. The leaves and bark of this plant are widely used for coughs in children and delivery pains. The plant has a number of therapeutic activity against fever, toothaches, chest pain, dog bites (rabies), vomiting, heat boils, diarrhoea, constipation, piles, bone fractures, jaundice, ulcers, skin infections, wounds and venereal diseases (Ganesan et al. 2004; Duraipandiyan et al. 2006; Pullaiah 2006; Padal et al. 2010; Samuel and Andrews 2010). Based on this ethnobotanical knowledge, the present study was aimed to explore pharmacognostic characters such as microscopy, scanning electron micrographs, fluorescence and preliminary phytochemical investigations of *P. hexapetalum*.

14.2 MATERIALS AND METHODS

14.2.1 Collection and Identification of Plant Material

Fresh leaves and stems of the *P. hexapetalum* were collected during October 2013. The taxonomic identity of the plant was confirmed from Botanical Survey of India (BSI), Southern Regional Centre, Coimbatore, Tamil Nadu. The herbarium specimen was deposited in the Department of Botany at Bharathiar University (Accession No BUBH006241). The fresh leaves and stems were washed under running tap water to remove the surface pollutants and were air-dried under shade. Then they were separately homogenized into a fine powder using a mixer for further studies.

14.2.2 Pharmacognosy Studies

14.2.2.1 Anatomy

The required samples of different parts were cut and removed from the plant and fixed in FAA [formalin (5 mL) + acetic acid (5 mL) + 70% ethyl alcohol (90 mL)]. After 24 h of fixing, the leaf and stem transverse sections were prepared through hand sectioning with a razor blade and the anatomy of the stem was recorded. With the help of the microscope to document the anatomy of the various tissues, the lignified and cellulosic tissues were distinguished using differential staining techniques (Evans 1997).

14.2.2.2 Photomicrographs

Microscopic descriptions of cells are supplemented with micrographs wherever necessary. Photographs of different magnifications were taken with a Nikon Labphoto 2 microscopic unit and scanning electron microscope. For normal observations, bright field was used. For the study of crystals, starch grains and lignified cells, polarized light was employed. Under polarized light, the crystals appear bright against a dark background. The magnifications of the figures are indicated by the scale-bars. Descriptive terms of the anatomical features which are used here in this documentation are given in the standard anatomy books (Evans 1997).

14.2.2.3 Leaf Constants

The important identifying characteristic of leaf constants like stomatal number, stomatal index, vein-islet number and vein termination number, were found out and tabulated (Evans 1997).

14.2.2.4 Stomatal Number

The stomatal number is the average number of stomata per square mm of the epidermis of the leaf. The middle piece of the leaf was cleared by boiling with chloral hydrate solution or alternatively with chlorinated soda. The upper and lower epidermis was peeled separately by means of forceps. It was kept on a slide and mounted in glycerine. A camera lucida and drawing board was arranged for making the drawings to scale. A square of 1 mm was drawn by means of stage micrometre. The slide was placed with the cleared leaf (epidermis) on the stage. The epidermal cell and stomata was traced. The number of stomata present in the area of 1 mm^2 was counted. At least half of the cell's area within the square was included. The result was recorded for each of the ten fields, and the average number of stomata per sq. mm was calculated. (Evans, 1997).

14.2.2.5 Stomatal Index

The stomatal index is the percentage in which the number of stomata forms to the total number of epidermal cells, each stomata being counted as one cell. The stomatal index can be calculated by Equation 14.1 (Evans 1997).

$$\text{Stomatal index S} = \text{S} \times 100 / \text{E} + \text{S} \tag{14.1}$$

where, S = No. of stomata per unit area, E = No. of epidermal cells in the same unit area.

14.2.2.6 Vein-Islet Number

The vein-islet is the small area of green tissue surrounded by the veinlets. The vein-islet number is the average number of vein-islets per square millimetre of a leaf surface. It is determined by counting the number of vein-islets in an area of 4 mm^2 of the central part of the leaf between the midrib and the margin (Evans 1997).

14.2.2.7 Fluorescence Analysis

Fluorescence study is an essential parameter for first-line standardization of crude drugs. The crude powder was subjected to these studies and the fluorescence patterns were noted. The powdered materials were treated separately with different reagents and exposed to visible and ultraviolet light to study their fluorescence behaviour.

The colours obtained by application of different reagents at different wavelengths of radiation were recorded (Ali 2003).

14.2.3 Preparation of Plant Extracts

The powdered leaves and stem materials were packed in small thimbles and separately extracted with organic solvents, such as petroleum ether, ethyl acetate and methanol in the increasing order of their polarity using a Soxhlet apparatus. Before extraction with the next solvent, the thimbles were air-dried each time. Finally, the material was macerated using hot water with constant stirring for 24 h and the water extract was also filtered using Whatman No. 1 filter paper. The different solvent extracts were concentrated by rotary vacuum evaporator and then air-dried.

14.2.4 Qualitative Phytochemical Screening

The leaf and stem extracts of *P. hexapetalum* were analyzed for the presence of major phytochemicals such as alkaloids, saponins, phenolic compounds, tannins, flavonoids, glycosides, flavanol glycosides, cardiac glycosides, phytosterols, fixed oils, fats, gums and mucilages according to standard methods.

14.2.4.1 Alkaloids

Hager's test

About 50 mg of solvent-free extract was stirred with 5 mL of dilute hydrochloric acid and filtered, and 2 mL of Hager's reagent (the saturated aqueous solution of picric acid) was added to the filtrate. A prominent yellow precipitate indicated the test as positive (Wagner et al. 1996).

14.2.4.2 Saponins

Frothing test

An extract of 50 mg was diluted with distilled water and made up to 20 mL. The suspension was shaken in a graduated cylinder for 15 min and a 2 cm layer of foam indicated the presence of saponins (Kokate 1999).

14.2.4.3 Phenolic Compounds

Ferric chloride test

About 50 mg of the extract was dissolved in 5 mL of distilled water. To this, a few drops of 5% of a neutral ferric chloride solution was added. Phenolic compounds were indicated by the presence of a dark green colour (Mace 1963).

14.2.4.4 Tannins

Potassium hydroxide test

The extract (0.5 g) was added into 10 mL of freshly prepared 10% potassium hydroxide (KOH) in a beaker and shaken to dissolve. A dirty precipitate indicated the presence of tannin (Williamson et al. 1996).

14.2.4.5 Flavonoids

Alkaline reagent test

An aqueous solution of the extract was treated with 10% ammonium hydroxide solution. A bulky white precipitate indicated the presence of flavonoids (Raaman and Bruce 2006).

14.2.4.6 Glycosides

Borntrager's test

About 50 mg of extract was hydrolysed with concentrated hydrochloric acid for 2 h on water bath and filtered. To 2 mL of filtered hydrolysate, 3 mL of chloroform was added and shaken. The chloroform layer was separated, and a 10% ammonia solution was added to it. A pink colour indicated the presence of glycosides (Evans 1997).

14.2.4.7 Flavonol Glycosides

Magnesium and hydrochloric acid reduction

An extract of 50 mg was dissolved in 5 mL alcohol and few fragments of magnesium ribbon were added. Concentrated hydrochloric acid was added dropwise into the test tube. The development of a pink or crimson colour indicated the presence of flavonol glycosides (Harborne 1998).

14.2.4.8 Cardiac Glycosides

Keller Killiani test

Total 100 mg of extract was dissolved in 1 mL of glacial acetic acid containing one drop of ferric chloride solution. This was then underlayered with 1 mL of concentrated sulphuric acid. A brown ring obtained at the interface indicated the presence of a deoxy sugar characteristic of cardinolides (Ngbede et al. 2008).

14.2.4.9 Phytosterols

Libermann and Burchard's test

About 50 mg of extract was dissolved in 2 mL of acetic anhydride. To this, one or two drops of concentrated sulphuric acid were added slowly along the sides of the test tube. An array of colour changes showed the presence of phytosterols (Finar 1956).

14.2.4.10 Fixed Oils and Fats

Saponification test

A few drops of 0.5 N alcoholic potassium hydroxide solutions were added to a small quantity of extract along with a drop of phenolphthalein. Then the mixture was heated on boiling water bath for 2 h. The formation of soap or partial neutralization of alkali indicated the presence of fixed oils and fats (Kokate 1999).

14.2.4.11 Gums and Mucilages

Absolute alcohol test

A 100 mg extract was dissolved in 10 mL of distilled water, and to this, 25 mL of absolute alcohol was added with constant stirring. White or cloudy precipitate indicated the presence of gums and mucilages (Whistler and BeMiller 1993).

14.3 RESULTS AND DISCUSSION

For many years, plants have been used as traditional indigenous remedies for a variety of ailments in many parts of the world, especially Asia (Pierson et al. 2012). It is thought that 75–90% of the rural population worldwide still relies on plant medicine (Aju and Ezeibekwe 2010). Among the various medicinal and culinary herbs, some endemic species are of particular interest because they may be used for the production of raw materials or preparations containing phytochemicals with significant antioxidant capacities and health benefits (Exarchou et al. 2002). Researchers have studied the polyphenolic constituents of various legumes and have reported that they contain potential medicinal or nutraceutical properties, including antioxidant activity (Siddhuraju 2006). Therefore, the study of the importance and role of non-nutrient compounds, particularly phenolic acids, flavonoids and high molecular tannins of legumes as natural antioxidants, have greatly increased (Siddhuraju and Becker 2007).

14.3.1 Pharmacognostic Studies of *P. hexapetalum*

14.3.1.1 Microscopic Observations of Leaves

The transverse section of the leaf showed the upper and lower epidermis, mesophyll tissue and prominent vascular bundle (Figure 14.1a and 14.1b). The upper epidermis is formed by the arrangement of rectangular cells and is covered by a

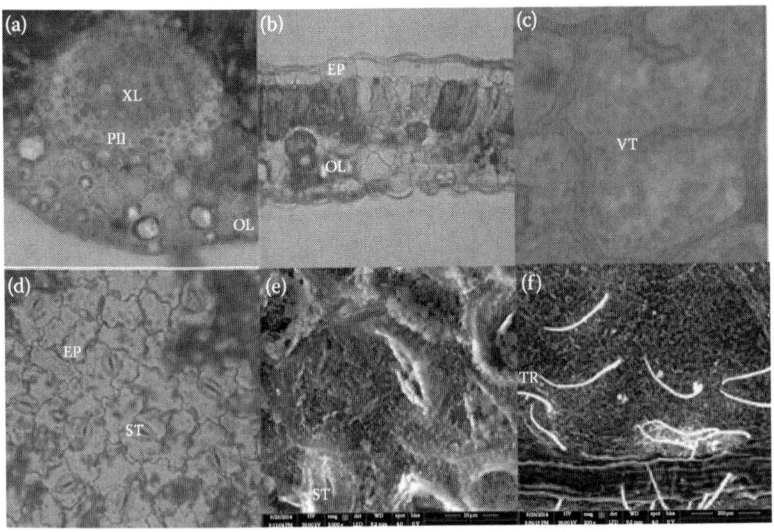

FIGURE 14.1 Microscopic observations of *P. hexapetalum* leaf: (a) T.S. of leaf mid rib (40X); (b) T.S. of leaf lamina region 40x; (c) leaf vein termination 100x; (d) leaf stomata 40x; (e) leaf stomata SEM image; (f) leaf trichome SEM image. PH = phloem, XL = xylem, OL = oil cell, EP = epidermal cell, PA = palisade cell, TR = trichome, ST = stomata, VT = vein termination.

TABLE 14.1
Observation of Leaf Constant of *P. hexapetalum*

S. No	Leaf Constant		Determination/sq mm
1	Stomatal number	Upper epidermis	36.33 ± 1.53
		Lower epidermis	41.33 ± 1.53
2	Stomatal index	Upper epidermis	16.06 ± 0.23
		Lower epidermis	17.06 ± 0.09
3	Vein-islet number	Vein lets number	12.33 ± 1.53
		Vein termination number	7.67 ± 0.58
4	Palisade ratio		1:2

Note: Values are mean of triplicate determination (n=3) ± standard deviation.

thick cuticle. Below the epidermis, a compactly-arranged layer of palisade parenchymatous tissue can be seen followed by three to four layers of loosely arranged spongy parenchymatous tissue consisting of sclerenchymatous patches. The vascular bundle consists of xylem tissue towards upper side and phloem tissue extends towards lower side, and is identified as an endarch vascular bundle. The leaf constant of *P. hexapetalum* is presented in Table 14.1. The upper epidermis of leaf consists 16.06 ± 0.23 stomatal index and 36.33 ± 1.53 stomatal numbers, whereas in the lower side 17.06 ± 0.09 stomatal index and 41.33 ± 1.53 stomatal numbers were observed (Figure 14.1d and 14.1e).

The stoma is surrounded by two prominent guard cells and was covered by two subsidiary cells to show the paracytic type of stomata. Trichomes were observed in the mid-rib and margin of leaves. Scanning electron microscopic examination shows the upper and lower leaf structures and also the clear structure of stomata and trichomes (Figure 14.1e).

14.3.1.2 Microscopic Observation of Stems

The transverse section of the stem shows the single-layered, thick-walled epidermis. Below the epidermis, six to seven layers of parenchymatous hypodermis is present followed by a sclerenchymatous cortex. In between the phloem and xylem, a single-layered cambial ring is present and divides the two tissues as the secondary phloem and xylem. The maceration of stems showed the highly evolute vessels, tracheids and fibre cells. The xylem parenchymas were also observed and the photographs are shown in Figure 14.2. Multicellular curved trichomes are observed in the stem's epidermis. In Figure 14.2, the T.S of stem observed in SEM and many microstructures were clearly noted in SEM image.

The phloem consists of well-developed phloem rays towards the epidermis, starting from the medulla. The protoxylem extends towards medullary region and metaxylem with prominent parenchymatous strands towards the cambium. The central region of the stem showed well-developed parenchymatous medulla (Figures 14.2a, 14.2b, 14.2c and 1.42d).

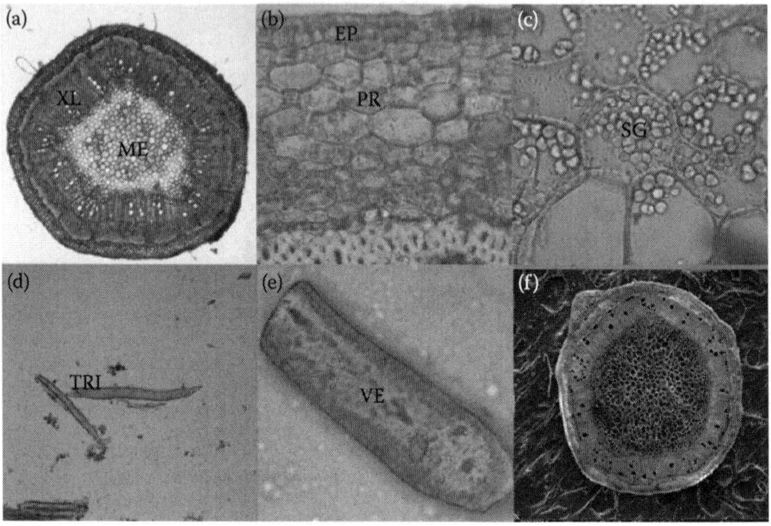

FIGURE 14.2 Microscopic observations of *P. hexapetalum* stem and maceration: (a) T.S. of stem entire view 5X; (b) T.S. of stem 40X; (c) T.S. of stem 100X; (d) xylem fiber tracheid 40X; (e) xylem vessel 40X; (f) T.S. stem entire view SEM image. XL = xylem, ME = medulla, EP = epidermal cell, PR = parenchyma cell, TR = trichome, SG = starch grains, TRI = tracheid, VE = vessel.

14.3.1.3 Histochemical Observation

Stem sections are stained with different type of stains and observed in the microscope. Starch grains and tannin bodies are stained with toluidine blue solution, iodine solution and hydrochloric acid, and were used to observe the crystal and tannin content. Lignified cells are noted as sclerenchymatic cortex regions of stem when stained with iodine solution.

14.3.1.4 Fluorescence Analysis of Powdered Drugs

Observation from fluorescence analysis was tabulated in Table 14.2. The results revealed that under daylight conditions, bark powder as such emitted a light brown colour whereas in UV light, it emitted a dark brown colour where it had excited at 235 nm. Also, most of the combinations emitted black to brown colour. Maximum fluorescence has been observed in stem and bark combinations with powder + 5% NaOH, whereas, minimum emittance was observed by the stem bark in addition to hot water. The highest rate of the colour emission is recorded in leaf combination with powder + 10% HCl followed by powder + 5% NaOH.

Quality control methods play an important role in traditional medicine, which are used as a tool for the identification, authentication and quality control of herbal drugs (Sahoo et al. 2010). The WHO has published *Quality Control Methods for Medicinal Plant Materials*, which describes a recommended test procedure to evaluate the identity, purity and quality of plant-based products. These standardization parameters are essential to publish in the pharmacopoeia. The majority of the information can

TABLE 14.2
Fluorescence Analysis of Powdered Drugs

Reagents with Powder	Leaf		Bark		Pod	
	Visible Light	UV - Light	Visible Light	UV - Light	Visible Light	UV - Light
Normal colour	Green	Green	Light Brown	Brown	Brownish Green	Light Green
Powder + cold water	Light Green	Green	Whitish Brown	Whitish Green	Whitish Brown	Whitish Brown
Powder + hot water	Light Green	Light Green	Whitish Brown	Brown	Whitish Brown	Light Brown
Powder + 70% ethanol	Green	Green	Light Brown	Light Green	Light Green	Green
Powder + 70% methanol	Dark Green	Dark Green	Light Brown	Pale Green	Light Green	Dark Green
Powder + 10% HCl	Brownish Green	Pale Green	Whitish Brown	Light Green	Light Brown	Light Green
Powder + 5% NaOH	Reddish Brown	Brownish Green	Reddish Brown	Dark Green	Reddish Brown	Brownish Green
Powder + conc. H_2SO_4	Dark Brown	Green	Dark Brown	Dark Green	Dark Brown	Dark Brown
Powder + saturated picric acid	Orangish Green	Yellowish Green	Yellowish Brown	Greenish Yellow	Yellowish Brown	Yellowish Green
Powder + acetic acid	Light Green	Green	Dark Brown	Light Green	Orangish Brown	Brownish Green

be obtained from its macroscopy, microscopy, fluorescence parameters and chemical fingerprint of medicinal plant materials. Macroscopic and microscopic methods are the simplest and cheapest methods to establish the correct identity of plant materials (Apraj et al. 2011). The transverse section was prepared with freehand sections of leaves and stems that were stained with safranine to confirm their lignifications. Microscopic observations of the stems during maceration was also carried out, and the specific diagnostic characteristics were recorded, which give a clear idea about the specific histological characteristics of crude drugs, besides the macromorphological and cytomorphological characteristics. While these diagnostic features enable the analyst to know the nature and characteristics of crude drugs, further evaluation of numerical parameters indicate their acceptability by criteria other than the morphological characteristics (Mukherjee 2007).

14.3.2 Qualitative Screening of Phytochemicals

The phytochemical screening on *P. hexapetalum* leaves, bark and pods revealed the presence of secondary metabolites like alkaloids, saponins, phenolic compounds, tannins, flavonoids and glycosides, were found to be variously distributed in all the parts of the plant and are presented in Table 14.3.

The chemical nature of the active constituents present in the plant material can be identified by performing preliminary phytochemical screening of that plant materials. The results from phytochemical analysis revealed the presence of flavonoids, phenols, carbohydrates, alkaloids, steroids, tannins and saponins. The qualitative estimations performed in the study depicted the presence of carbohydrates, flavonoids, phenols and tannins in major quantities, while saponins and alkaloids were

TABLE 14.3
Qualitative Screening of Phytochemicals in *P. hexapetalum*

Chemical Constituents	Leaf	Bark	Pod
Alkaloids	+	+	+
Saponins	+	+	+
Phenolics	+++	+++	+++
Tannins	+++	+++	++
Flavonoids	+++	+++	++
Glycosides	+	+	++
Flavonol glycosides	++	+++	++
Cardiac glycosides	+	++	+
Phytosterols	++	++	+++
Fixed oils and fats	+	+	++
Gums and mucilages	–	–	–

Note: (+): Presence of chemical compound, (–): Absence of chemical compound; (+) < (++) < (+++): Based on the intensity of characteristic colour.

present in quite a considerable amount. Flavonoids have shown potential antioxidant and anti-inflammatory activity which are attributed to an increased capillary permeability, and have been also associated in the treatment of various cardiovascular diseases (Crespy 2002). Tannins have a strong astringent action and are reported to have antibacterial, anti-inflammatory, antiviral and antioxidant activity (Kapu et al. 2001; Schulz et al. 2002). Alkaloids have been reported to possess wide range of therapeutic importance in the fields of cancer, malaria, pain, inflammation, Parkinsonism, hypertension and number of central nervous system disorders (Rhaman and Bruce 2002). Phenolic compounds play beneficial role in active quenching of oxygen-derived free radicals, thus neutralizing them by donating hydrogen atoms or electrons to free radicals. Therefore, they are considered as strong antioxidants with anticarcinogenic, antibacterial and anti-inflammatory activity and are also used in coronary heart disease and some types of tumours (Yildiz et al. 2011; Shukla et al. 2012). The presence of high phenolic compounds in *P. hexapetalum* can act as a good source of natural medicine.

14.4 CONCLUSION

Herbal medicines have been enjoying revitalization all over the world. There are hundreds of medicinal plants that have a long history of curative properties against various diseases and ailments. However, screening of plants for their activity is very crucial and needs imperative attention in order to know the value of the plant. The assessment of the plants for their therapeutic activity is done on the basis of either their chemotaxonomic examination or ethnobotanical information for a particular disease. These obtained data can support the standardization parameters such as microscopy, scanning electron micrographs, fluorescence and preliminary phytochemical investigations, which could be helpful for proper authentication. A set of quality parameters for the standardization of these plants as herbal preparations will ensure the reproducibility of their therapeutic effects. In this manner, the therapeutic potential of these medicinal plants can be best harnessed towards a possible integration into the healthcare system.

REFERENCES

Aju, P. C. and Ezeibekwe, I. O. 2010. Understanding and appreciating the need for biodiversity conservation in Nigeria. *Journal of Medicinal Plants Research* 4(24): 2605–2608.

Ali, M. 2003. *Text Book of Pharmacognosy*, 2nd edition. CBS Publishers and Distributers, New Delhi: 64–283.

Anonymous. 2000. IPGRI, Rome, Italy, Newsletter No. 32 (August), 22–23.

Apraj, V., Thakur, N., Bhagwat, A., Mallya, R., Sawant, L., and Pandita, N. 2011. Pharmacognostic and phytochemical evaluation of *Citrus aurantifolia* (Christm) Swingle peel. *Pharmacognosy Journal* 3(26): 70–76.

Babu, K., Sabesan, G. S., and Rai, S. 2010. Comparative pharmacognostic studies on the barks of four Ficus species. *Turkish Journal of Botany* 34(3): 215–224.

Brinda, P., Saraswathy, A., and Jayaraman, P. 2000. Micromorphological identification of the medicinal bark of *Ventilago madraspatana. Journal of Medicinal and Aromatic Plant Science* 23: 619–622.

Calixto, J. B. 2005. Twenty-five years of research on medicinal plants in Latin America: A personal view. *Journal of Ethnopharmacology* 100(1): 131–134.

Chaudhury, R. R. 1992. *Herbal Medicine for Human Health.* New Delhi: WHO, 1992.

Cox, P. A. 2000. Will tribal knowledge survive the millennium? *Science* 287(5450): 44–45.

Crespy, V., Morand, C., Besson, C., Manach, C., Demigne, C., and Remesy, C. 2002. Quercetin, but not its glycosides, is absorbed from the rat stomach. *Journal of Agricultural and Food Chemistry* 50(3): 618–621.

Duraipandiyan, V., Ayyanar, M., and Ignacimuthu, S. 2006. Antimicrobial activity of some ethnomedicinal plants used by Paliyar tribe from Tamil Nadu, India. *BMC Complementary and Alternative Medicine* 6(1): 35.

Evans, W. C. 1997. Trease and Evans Pharmacognosy, Singapore: Hartcourt Brace and Company Asia Pvt.

Exarchou, V., Nenadis, N., Tsimidou, M., Gerothanassis, I. P., Troganis, A., and Boskou, D. 2002. Antioxidant activities and phenolic composition of extracts from Greek oregano, Greek sage, and summer savory. *Journal of Agricultural and Food Chemistry* 50(19): 5294–5299.

Finar, I. L. 1956. *Organic Chemistry: Stereochemistry and the Chemistry of Natural Products. Volume Two.* Longmans, Green, and Company.

Flaster, T. 1996. Ethnobotanical approaches to the discovery of bioactive compounds. Progress in new crops. *Proceedings of the Third National Symposium.* ASHS Press, Alexandria, 561–565.

Ganesan, S., Suresh, N., and Kesaven, L. 2004. Ethnomedicinal survey of lower Palni hills of Tamil Nadu. *Indian Journal of Traditional Knowledge* 3(3): 299–304.

Harborne, A. J. 1998. *Phytochemical Methods a Guide to Modern Techniques of Plant Analysis.* Springer Science & Business Media, 3rd ed. Chapman and Hall, London, U.K.

Heinrich, M. 2000. Ethnobotany and its role in drug development. *Phytotherapy Research* 14(7): 479–488.

Joshi, A. R. and Joshi, K. 2000. Indigenous knowledge and uses of medicinal plants by local communities of the Kali Gandaki Watershed Area, Nepal. *Journal of Ethnopharmacology* 73(1): 175–183.

Kapu, S. D., Ngwai, Y. B., Kayode, O., Akah, P. A., Wambebe, C., and Gamaniel, K. 2001. Anti-inflammatory, analgesic and anti-lymphocytic activities of the aqueous extract of *Crinum giganteum. Journal of Ethnopharmacology* 78(1): 7–13.

Kokate, C. K. 1999. *Practical Pharmacognosy,* 4th edition. New Delhi, India: Vallabh Prakashan Publication.

Kokate, C. K., Purohit, A. P., and Gokhale, S. B. 2007. *Pharmacognosy.* 39th edition. Published by Nirali Prakashan, Aug 1, 635.

Laloo, D., Kumar, M., Prasad, S. K., and Hemalatha, S. 2013. Quality control standardization of the roots of *Potentilla fulgens* Wall.: A potent medicinal plant of the Western Himalayas and Northeastern India. *Pharmacognosy Journal* 5(3): 97–103.

Liu Willow, J. H. 2011. *Traditional Herbal Medicine Research Methods: Identification, Analysis, Bioassay, and Pharmaceutical and Clinical Studies.* United State/Hoboken, New Jersey: John Wiley & Sons, 223–225.

Mace, M. E. 1963. Histochemical localization of phenols in healthy and diseased banana roots. *Physiologia Plantarum* 16(4): 915–925.

Mukherjee, P. K. 2007. *Quality Control of Herbal Drugs: An Approach to Evaluation of Botanicals.* 2nd ed. New Delhi, India: Business Horizons.

Mukherjee, T. 2009. Medicinal plants: Need for protection. *Pravin Chandra Trivedi Medicinal Plants Utilisation and Conservation.* 2nd revised and enlarged edition, Aavishkar Publishers, Distributors, 391–402.

Nagani, K., Kaneria, M., and Chanda, S. 2012. Pharmacognostic studies on the leaves of *Manilkara zapota* L. (Sapotaceae). *Pharmacognosy Journal* 4(27): 38–41.

Nasreen, S. and Radha, R. 2011. Assessment of Quality of *Withania somnifera* Dunal (Solanaceae) Pharmacognostical and physicochemical profile. *International Journal of Pharmacy and Pharmaceutical Sciences* 3(2): 152–155.

Ngbede, J., Yakubu, R. A., and Nyam, D. A. 2008. Phytochemical screening for active compounds in *Canarium schweinfurthii* (Atile) leaves from Jos North, Plateau State, Nigeria. *Research Journal of Biological Sciences* 3(9): 1076–1078.

Padal, S. B., Murty, P. P., Rao, D. S., and Venkaiah, M. 2010. Ethnomedicinal plants from Paderu division of Visakhapatnam district, AP, India. *Journal of Phytology* 2(8): 70–91.

Padmakumar, T. 2009. Green Healers: A Review. *Pravin Chandra Trivedi Medicinal Plants Utilisation and Conservation.* 2nd revised and enlarged edition, Aavishkar Publishers, Distributors, 95–103.

Pierson, J. T., Dietzgen, R. G., Shaw, P. N., Roberts-Thomson, S. J., Monteith, G. R., and Gidley, M. J. 2012. Major Australian tropical fruits biodiversity: Bioactive compounds and their bioactivities. *Molecular Nutrition & Food Research* 56(3): 357–387.

Prasad, S. K., Sahu, A. N., and Hemalatha, S. 2012. Cytomorphological and physicochemical evaluations of *Cryptocoryne spiralis*. *Journal of Herbs, Spices & Medicinal Plants* 18(4): 304–317.

Pullaiah, T. 2006. *Encyclopedia of World Medicinal Plants*, Vol. 4, Regency Publications, 42.

Raaman, N. 2006. *Phytochemical Techniques.* New India Publishing Agency. Jai Bharat Printing Press. New Delhi, 19–22.

Rathbone, D. A. and Bruce, N. C. 2002. Microbial transformation of alkaloids. *Current Opinion in Microbiology* 5(3): 274–281.

Sahoo, N., Manchikanti, P., and Dey, S. 2010. Herbal drugs: Standards and regulation. *Fitoterapia* 81: 462–471.

Samuel, J. K., and Andrews, B. 2010. Traditional Medicinal Plant wealth of Pachalar and Periyur hamlets Dindigual district, Tamilnadu. *Indian Journal of Traditional Knowledge* 9(2): 264–270.

Schulz, V., Hansel, R., and Tyler, V. E. 2002. Fitoterapia racional: um guia de fitoterapia para as ciências da saúde. *Manole*, 221–227.

Shukla, R., Singh, P., Prakash, B., and Dubey, N. K. 2012. Antifungal, aflatoxin inhibitory and free radical-scavenging activities of some medicinal plants extracts. *Journal of Food Quality* 35(3): 182–189.

Siddhuraju, P. 2006. The antioxidant activity and free radical-scavenging capacity of phenolics of raw and dry heated moth bean (*Vigna aconitifolia*) (Jacq.) Marechal seed extracts. *Food Chemistry* 99(1): 149–157.

Siddhuraju, P. and Becker, K. 2007. The antioxidant and free radical scavenging activities of processed cowpea (*Vigna unguiculata* (L.) Walp.) seed extracts. *Food Chemistry* 101(1): 10–19.

Shinde, V. and Dhalwal, K. 2007. Pharmacognosy: The changing scenario. *Pharmacognosy Reviews* 1(1): 1.

Tabuti, J. R. S., Lye, K. A., and Dhillion, S. S. 2003. Traditional herbal drugs of Bulamogi, Uganda: Plants, use and administration. *Journal of Ethnopharmacology* 88(1): 19–44.

Vaibhav, S. and Kamlesh, D. 2007. Pharmacognosy; the changing scenario. *Pharmacognosy Reviews* 1(1): 1–6.

Wagner, H., Bladt, S., and Zgainski, E. M. 1996. *Plant Drug Analysis.* Springer Verlag: New York, 360.

Whistler, R. L. and BeMiller, J. N. 1993. *Industrial Gums; Polysaccharides and their Derivatives.* Academic Press, London, U.K.

WHO (World Health Organization). 2010. Monographs on Medicinal Plants Commonly Used in the Newly Independent States (NIS).

Williamson, E. M., Okpako, D. T., and Evans, F. J. 1996. *Pharmacological Methods in Phytotherapy Research: Volume 1: Selection, Preparation and Pharmacological Evaluation of Plant Material.* John Wiley & Sons Ltd.

Yildiz, G., Vatan, Ö., Çelikler, S., and Dere, Ş. 2011. Determination of the phenolic compounds and antioxidative capacity in red algae *Gracilaria bursa-pastoris. International Journal of Food Properties* 14(3): 496–502.

15 Antifungal and Anti-Mycotoxigenic Activities of *Pogostemon mollis* Benth. against *Fusarium graminearum*

Kasipandi Muniyandi, Elizabeth George and Parimelazhagan Thangaraj

CONTENTS

15.1 Introduction 285
15.2 Materials and Methods 287
15.2.1 Collection and Identification of Plant Material 287
15.2.2 Chemicals 287
15.2.3 Extraction of Plant Material 287
15.2.4 Antifungal Activity 287
15.2.5 Anti-Mycotoxigenic Activity 287
15.2.5.1 Fungal Culture 288
15.2.5.2 Extraction of Toxin 288
15.2.6 Quantification of Zearalenone Using HPLC 288
15.3 Results and Discussion 288
15.3.1 Antifungal Activity 288
15.3.2 Anti-Mycotoxigenic Activity 290
15.3.3 Identification and Quantification of Zearalenone by HPLC 290
References 294

15.1 INTRODUCTION

Nature is a mine of various herbal medicines. Medicinal and aromatic plants, with their rich source of natural phytochemicals, are the storehouse of these therapeutic entities (Sharma and Bhadange 2013). Medicinal plants are widely used as antimicrobials because of their abundance of active phytochemicals. Recently, many of the pharmaceutical and cosmetic industries have started using these plant-based compounds in their products. Plant-based antimicrobials are eco-friendly and safer to use. Taking into account the vast potential of medicinal plants as antifungal and antibacterial agents, systematic studies should be implemented so they can be further applicable.

Presently, there have been many reported cases about food poisoning due to fungal toxins (Lattanzio et al. 2007; Visconti et al. 2007). Fungal poisoning may lead to vomiting, abdominal pain, pulmonary oedema, convulsions, coma and, in extreme cases, death by cerebral oedema and the fatty involvement of the liver, kidneys and heart. The expression of these diseases may be influenced by age, sex, nutritional status and continuous exposure to the causative agents. Also, there are occurrences of carcinogenicity due to chronic intake of these fungal toxins (International Crops Research Institute for the Semi-Arid Tropics 2000). An extended study on the fungal toxicity will be helpful to understand the level of action of these plant-based extracts. Fungal broth with plant extracts may give diverse responses. The toxins extracted from the fungus can be determined by comparing with standard toxins through HPLC profiling (Zachariazova et al. 2010). Fungal mycelium can also be used to analyse the enzymatic levels and the RNA content and thus gene expression. This enables us to know how the changes occur, and the cause of the difference in growth and toxin levels.

Lamiaceae, or the mint family, is one of the well-known angiospermic families with about 236 genera, and more than 7000 species distributed over the Mediterranean to Central Asia. The most common genera of Lamiaceae are *Salvia, Thymus, Leucas, Ocimum, Plectranthus, Anisomeles, Pogostemon, Rosmarinus, Lavendula, Mentha*, etc. They are mainly used for flavouring, medicinal, culinary and perfumery purposes (Naghibi et al. 2010). The medicinal properties of the plants are widely studied, and many products have also been designed from them. The plants of this family possess anti-inflammatory, analgesic, antipyretic, antifertility, immunoprotective, antimicrobial, antiallergic, antioxidant, cardioprotective, anti-stress and wound healing activities (Joudi et al. 2011; Raja 2012).

Due to the presence of large amounts of bioactive components (George et al. 2016; Muniyandi et al. 2017), Lamiaceae members are widely used for therapeutic preparations. The genus *Pogostemon* consists of more than 85 species. *Pogostemon* is widely celebrated under the banner of patchouli oil. This is an essential oil extracted from *Pogostemon cablin*. The therapeutic effects of *Pogostemon* include antimicrobial, analgesic, anti-inflammatory, aphrodisiac, antidepressant, wound healing and insect repellent properties; and can be used to treat skin infections, diarrhoea, colds, fevers, kidney stones, piles, uterine hemorrhaging, snakebites, etc. (Ashwini et al. 2013). *Pogostemon mollis* Benth. is a medicinal plant of this genus. It is present mainly in the Western Ghats at an altitude above 6000 ft (Gamble 1915). This plant is used by the tribes of India for the treatment of many neurological disorders, for wound healing, etc. (Oudhia, 2012a,b). *P. mollis* is an aromatic plant which contains odoriferous and volatile substances which occur as essential oils in foliage and the plant's aerial parts (Rao et al. 2006). The other species of this genus are widely studied and many herbal products have already been marketed.

15.2 MATERIALS AND METHODS

15.2.1 Collection and Identification of Plant Material

Pogostemon mollis was collected from Kattapettu, Ooty, Tamil Nadu. The taxonomic identity of the plant was confirmed from the Botanical Survey of India, Southern Regional Centre, Coimbatore, Tamil Nadu. An herbarium of the specimen was submitted to the Department of Botany, Bharathiar University, Coimbatore (BUBH-006242). The aerial parts of the plant were collected, washed under running tap water to remove the surface pollutants and air-dried under shade separately. Then they were homogenized into fine powder using a mixer and grinder, which was kept for further studies.

15.2.2 Chemicals

The chemicals were obtained from Himedia Laboratories, Mumbai; Sisco Research Laboratories (SRL), Mumbai; Merck, Bengaluru and Sigma Aldrich, United States. All the chemicals and solvents used in this study were of analytical and HPLC grade.

15.2.3 Extraction of Plant Material

The powdered sample was packed into small thimbles and extracted successively with different organic solvents, ethyl acetate, acetone and methanol in an increasing order of polarity using a Soxhlet apparatus. Each time before extraction with the next solvent, the thimble was dried in hot air oven below 40°C. These different solvent extracts were concentrated by a rotary vacuum evaporator (Equitron, Medica Instruments Mfg. Co., India) and then air-dried. The dried extracts were collected, weighed and stored in deep freezer (–20°C) for further studies.

15.2.4 Antifungal Activity

The strain *Fusarium graminearum* (MTCC-2089) was used for the antifungal activity. Czapek-Dox broth was used to prepare the mother culture of each of the fungal strains. This culture was used to streak the PDA media prepared in petriplates. Later, the wells were made and blocked with agar. The extracts (20 μg) were added to the wells. Standard antibiotic amphotericin b (10 μg) was added to the well in the centre. The plates were kept in 37–40°C for incubation. After 4 to 5 d, fungal growth was monitored and the zone of inhibition was measured.

15.2.5 Anti-Mycotoxigenic Activity

An anti-mycotoxigenic test was carried out to further analyse the effects of the plant extracts on the growth of the fungi.

15.2.5.1 Fungal Culture

Fusarium graminarium (MTCC 2089) was inoculated into 50 mL fungal broth taken in a conical flask. The different plant extracts, namely ethyl acetate, acetone and methanol (10 mg and 25 mg) were added to the broth before inoculation and all the flasks were correctly labeled. Standard antioxidant compounds like quercetin, rutin and BHT were also added in separate flasks. One flask was kept as the control sample without any compound added to it. All the flasks were kept in incubator (40°C) for 15 d.

15.2.5.2 Extraction of Toxin

The determination of zearalenone (ZEA) was carried out as described by Schollenberger et al. (2006). Briefly, extraction was done by mixture of acetonitrile and water followed by liquid/liquid extraction with hexane.

15.2.6 Quantification of Zearalenone Using HPLC

The quantification of ZEA was performed by external standardization. The standard acquired from Sigma Chemicals Co. was diluted in solution. The sample was injected into a LC-6AD (Shimadzhu LC) chromatography system (UFLC) equipped with LC pump system, UV/VIS Detector (SPD-20 A), Luna 5 μ C18 (2)-100A column (250 mm × 4.60 mm) and controlled by LC Solution version 2.1 (Spinco, United States). The wavelengths set in the fluorescence detector were 270 nm and 455 nm, respectively, for excitation and emission of ZEA. The mobile phase was water (A): methanol (B) in gradient mode. The gradient began at 88% A and 12% B during 8 min. From 9–18 min, B concentration raised to 100%. At 19 min, the gradient returned to 88% A and 12% B and remained in this condition until 27 min. The flow rate was 1 mL/min and the injection volume was 20 μL. The accuracy of the method was expressed as the percentage of recovery and evaluated by the coefficient of variation of three repetitions. The quantification was determined by the dilution of standard and fortified samples that generated a detector signal twice in the retention time of the toxin.

15.3 RESULTS AND DISCUSSION

15.3.1 Antifungal Activity

The antifungal activity of the three extracts of *P. mollis* was carried out using the well diffusion method. A different concentration of the extracts ranging from 10–200 μg was checked. It was inferred from the study that antifungal properties were shown by the extracts at a concentration of 200 μg against the tested organism. The results are presented in Figure 15.1. The zone of inhibition was measured and tabulated (Table 15.1). *Fusarium graminearum* showed a good response to the ethyl acetate extract (inhibition zone of 6 mm). Here, Amphotericin b (10 μg) served as the standard drug.

The natural products extracted from aromatic and medicinal plants received particular attention as potential natural agents for food preservation and antimicrobials.

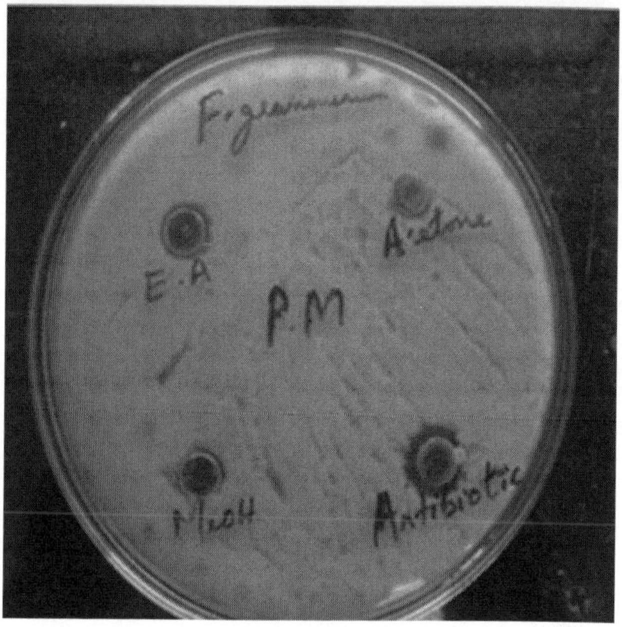

FIGURE 15.1 Antimicrobial activities of *P. mollis*.

TABLE 15.1
Zone of Inhibition (mm)

Samples	*F. graminearum* Zone of Inhibition (mm)
Ethyl acetate	6 ± 0.33^{b}
Methanol	3 ± 0.25^{c}
Acetone	3.5 ± 0.2^{c}
Antibiotics	10 ± 0.4^{a}

Note: Values are mean of triplicate determination (n=3) ± standard deviation; Statistically significant at $p<0.05$ where $^{a} > {}^{b} > {}^{c} > {}^{d}$.

The antimicrobial compounds from plants may inhibit bacterial growth by various mechanisms than which is being done by commercially available synthetic agents (Saravanan and Parimelazhagan 2014). Antimicrobial agents have multiple target sites in microbial cells for the inhibition or destruction of microorganisms, resulting in the microbiostatic and microicidal effect. Cell wall, cell membrane and cytoplasm are the major target sites of these compounds on the microbes (Goel and Sharma 2013).

TABLE 15.2
Reduction of Fungal Dry Weight by Plant Extracts

Sample	Concentration (mg/50 mL)	Dry Weight	% Inhibition
Ethyl acetate	10	11.4784	3.556634766
	25	11.3596	4.554811498
Acetone	10	12.6267	–6.091566751
	25	5.6246	52.74120504
Methanol	10	3.2796	72.44427267
	25	12.2717	–3.108799583
Quercetin	10	12.9467	–8.780258282
BHT	10	12.2617	–3.024777973
Rutin	10	11.8217	0.672172883

15.3.2 Anti-Mycotoxigenic Activity

The anti-mycotoxigenic effect of the *P. mollis* extracts on the *Fusarium graminearum* was analysed and the different parameters of the growth, like mycelium growth and toxin production, was verified. The difference in the mycelial dry weight was calculated and the inhibition percent of the mycelium formation was analysed (Table 15.2). Acetone extract at a lower concentration (10 mg) showed a stimulatory effect on the mycelia growth. But at 25 mg/mL, the growth was inhibited to about 52%. Ethyl acetate extract showed inhibition even at a lower concentration of extract. But the inhibition percentage of mycelia was not increased so much in the higher concentration. But this was not seen in the case of methanol extract.

Different crude extracts of various herbs, spices and aromatic plants rich in polyphenolics are becoming increasingly important in the food industry due to their antifungal, antimycotoxigenic and antioxidant activity. Thus, such plant chemicals can improve the shelf life, quality and nutritional value of various stored commodities (Kumar et al. 2007). The growth of the mycelia and its inhibition was analysed. There is some effect on the fungal growth rate, as seen by the change in the mycelia growth of the treated samples. The effects vary in different extracts, as some have an inhibitory while others a stimulating effect. Here, the toxin extracted from the fungal culture has been analysed. A significant reduction in the toxin level of the extract-treated samples was observed, which is discussed below in detail.

15.3.3 Identification and Quantification of Zearalenone by HPLC

Zearalenone is one of the most dangerous toxins produced by *Fusarium* spp. and causes so many health problems. Quantification of toxin can be regarded as a measure of the extent of fungal growth. Zearalenone was quantified in the samples using the standard graph plotted using the area obtained from the chromatogram of standard toxins in different concentrations. The toxin concentration from different samples calculated using these values showed that acetone extract had the most effect

TABLE 15.3
Quantification of Zearalenone in Fungal Broth by HPLC

Extract	Concentration (mg/50 mL)	Concentration of Zearalenone (mg/mL)
Ethyl acetate	10	0.056
	25	0.289
Acetone	10	0.283
	25	0.445
Methanol	10	0.303
	25	0.329
Quercetin	10	0.129
BHT	10	0.011
Rutin	10	0.033
Control	–	0.502

on the inhibition of toxins when the concentration is compared with the control (Table 15.3). The control sample showed a toxin concentration of 0.502 mg/mL. In the sample where 10 mg of ethyl acetate was added to the fungal broth, toxin production was the least (0.056 mg/mL). The highest toxin level among the treated samples was for acetone at 25 mg (0.445 mg/mL). The chromatograms are presented in Figure 15.2.

The ethyl acetate extract in lower doses gives considerably more inhibition of the toxin than that of the control sample without the addition of any extract. But it seems the stress caused by the excess of the extract may strain the fungal mycelium to produce more toxin than at a lower dose.

The toxin production in the treated samples was much lower when compared to the control. This signifies the effectiveness of the extracts in lowering the toxin production. Even though in many of the cases the toxin production has decreased effectively, it didn't show much consequence in the fungal growth. Similar results were obtained by previous studies in other fungal species (Mossini et al. 2009). The growth pattern is similar to growth of other fungal strains, which is consistent with the findings of other authors where the formation of the toxin was prevented and not the fungal growth (Mossini and Kemmelmeier, 2008). Toxin production generally decreases as mycelium formation decreases. But in the present study, some extracts did not show any reduction in the mycelium growth while toxin level decreased than the control sample. This is analogous to previous research which has shown that the inhibitory potential of toxin production may not always coincide with anti-fungal potentiality against mycelial growth (Mossini et al. 2004). Consequently *P. mollis* can be used as a mycotoxin inhibitory agent and further analysis need not be warranted.

As a whole, the present study revealed that aerial parts of *P. mollis* extract possesses remarkable antioxidant, cytotoxic and antimycotoxigenic activity and thus this plant can be recommended as a therapeutic agent to cure various ailments.

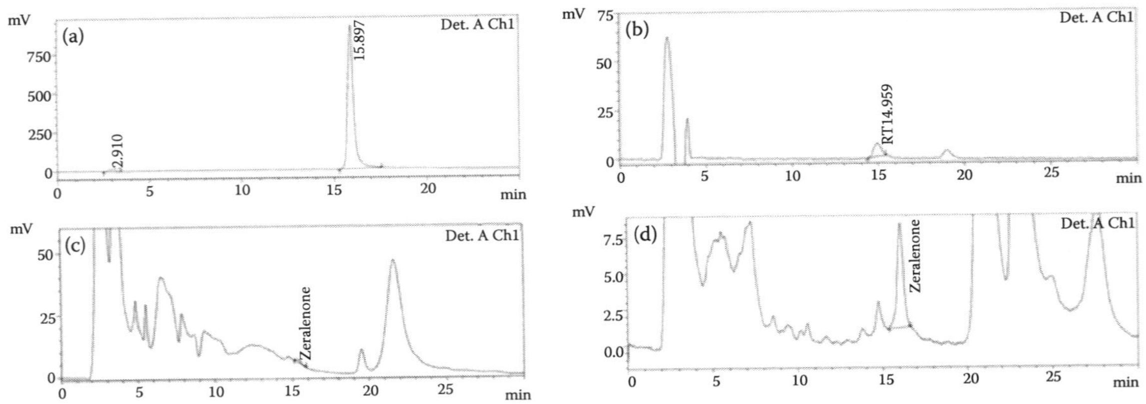

FIGURE 15.2 HPLC analysis of toxin from *F. graminearum*: (a) HPLC analysis of standard toxin (Zearalenone); (b) HPLC analysis of toxin from *F. graminearum*; (c) HPLC analysis of toxin from *F. graminearum* treated with ethyl acetate; (d) HPLC analysis of toxin from *F. graminearum* treated with ethyl acetate extract (25 mg). *(Continued)*

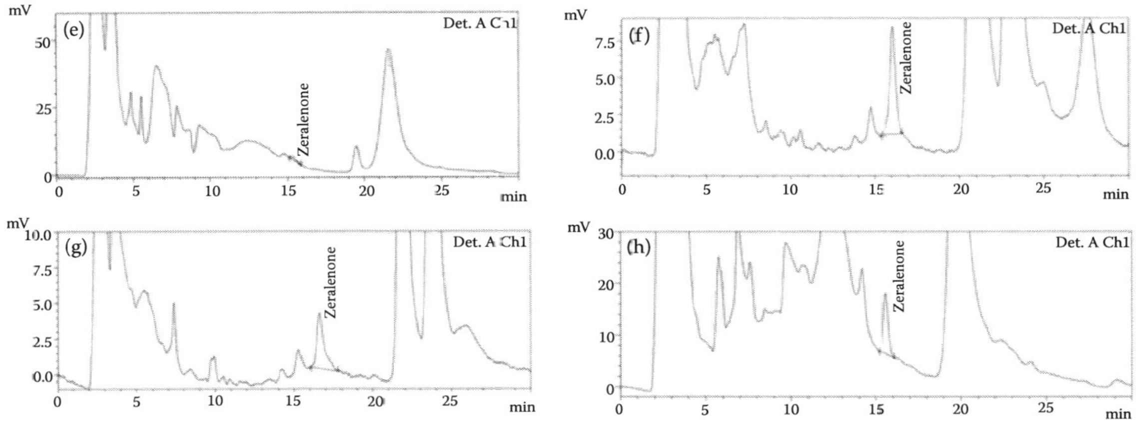

FIGURE 15.2 (CONTINUED) HPLC analysis of toxin from *F. graminearum*: (e) HPLC analysis of toxin from acetone treated culture (10 mg); (f) HPLC analysis of toxin from *F. graminearum* treated with acetone extract (25 mg); (g) HPLC analysis of toxin from *F. graminearum* treated with methanol extract (10 mg); (h) HPLC analysis of toxin from *F. graminearum* treated with methanol extract (25 mg).

REFERENCES

Ashwini, S. V., Khade A. B., Basawaraj, H. C., and Shrishail, G. M. 2013. A comprehensive review on *Pogostemon benghalensis* (Burm. f.) O. Kuntze. *Research and Reviews: Journal of Pharmacognosy and Phytochemistry* 1(1): 10–15.

Gamble, J. S. 1915. *Pogostemon mollis.* Flora of the Presidency of Madras. 2, 1134.

George, E., Kasipandi, M., Vekataramana, M., Kumar, K. N., Allen, J. A., Parimelazhagan, T., and Gopalan, N. 2016. *In vitro* anti-oxidant and cytotoxic analysis of Pogostemon mollis Benth. *Bangladesh Journal of Pharmacology* 11(1): 148–158.

Goel, A., and Sharma, K. 2013. Effect of *Euphorbia pulcherrima* leaf and inflorescence extract on various cytomorphological parameters of *Aspergillus fumigatus. World Academy of Science, Engineering and Technology* 81: 302–305.

International Crops Research Institute for the Semi-Arid Tropics. 2000. Aflatoxins. © International Crops Research Institute for the Semi-Arid Tropics. http://www.icrisat.org/aflatoxin/anamika_Effects_Aflatoxins.asp.

Joudi, L., Bibalani, G. H., and Shadkami, H. 2011. Exploration of medicinal species of Lamiaceae family in Ilkhji and Sharafaldin Regions of Esat Azarbaijan in Iran. *Current Research Journal of Biological Sciences* 3(4): 385–387.

Kumar, R., Mishra, A. K., Dubey, N. K., and Tripathi, Y. B. 2007. Evaluation of *Chenopodium ambrosioides* oil as a potential source of antifungal, antiaflatoxigenic and antioxidant activity. *International Journal of Food Microbiology* 115(2): 159–164.

Lattanzio, V. M. T., Solfrizzo, M., Powers, S., and Visconti, A. 2007. Simultaneous determination of aflatoxins, ochratoxin A and Fusarium toxins in maize by liquid chromatography/tandem mass spectrometry after multitoxin immunoaffinity cleanup. *Rapid Communications in Mass Spectrometry* 21(20): 3253–3261.

Mossini, S. A., Arrotéia, C. C., and Kemmelmeier, C. 2009. Effect of neem leaf extract and Neem oil on Penicillium growth, sporulation, morphology and ochratoxin A production. *Toxins* 1(1): 3–13.

Mossini, S. A. G., de Oliveira, K. P., and Kemmelmeier, C. 2004. Inhibition of patulin production by *Penicillium expansum* cultured with neem (*Azadirachta indica*) leaf extracts. *Journal of Basic Microbiology* 44(2): 106–113.

Mossini, S. A. G., and Kemmelmeier, C. 2008. Inhibition of citrinin production in *Penicillium citrinum* cultures by neem [*Azadirachta indica* A. Juss (Meliaceae)]. *International Journal of Molecular Sciences* 9(9): 1676–1684.

Muniyandi, K., George, E., Mudili, V., Kalagatur, N. K., Anthuvan, A. J., Krishna, K., Thangaraj, P., and Natarajan, G. 2017. Antioxidant and anticancer activities of *Plectranthus stocksii* Hook. f. leaf and stem extracts. *Agriculture and Natural Resources* 51(2): 63–73.

Naghibi, F., Mosaddegh, M., Mohammadi Motamed, M., and Ghorbani, A. 2010. Labiatae family in folk medicine in Iran: From ethnobotany to pharmacology. *Iranian Journal of Pharmaceutical Research* 2: 63–79.

Oudhia, P. 2012a. *Euphorbia hirta.* Compatable herbs in Indian Traditional healing and role of wild Ficus. Research and Media Network. http://researchandmedia.ning.com/profiles/blogs/euphorbia-hirta-compatible-herbs-in-indian-traditional healing.

Oudhia, P. 2012b. https://www.flickr.com/photos/pankajoudhia/8312559685.

Raja, R. R. 2012. Medicinally potential plants of Labiatae (Lamiaceae) family: An overview. *Research Journal of Medicinal Plant* 6(3): 203–213.

Rao, R. R., Sagar, K., and Syamasundar, K. V. 2006. Wild aromatic plant species of western ghats: Diversity, conservation and utilization. *International Seminar on Multidisciplinary Approaches in Angiosperm Systematics*, 358–371.

Saravanan, S. and Parimelazhagan, T. 2014. *In vitro* antioxidant, antimicrobial and anti-diabetic properties of polyphenols of *Passiflora ligularis* Juss. fruit pulp. *Food Science and Human Wellness* 3(2): 56–64.

Schollenberger, M., Müller, H. M., Rüfle, M., Suchy, S., Plank, S., and Drochner, W. 2006. Natural occurrence of 16 Fusarium toxins in grains and feedstuffs of plant origin from Germany. *Mycopathologia* 161(1): 43–52.

Sharma, S. M. and Bhadange, D. G. 2013. Antimicrobial potential of Lamiaceae members. *International Journal of Pharmaceutical Sciences* 3(5): 324–327.

Visconti, A., de Girolamo, A., Lattanzio, V. M., Lippolis, V., Pascale, M., and Solfrizzo, M. 2007. Novel analytical methods for Fusarium toxins in the cereal food chain. *Proceedings of the Conference Advances and Prospects of Research on Fusarium Mycotoxin Contamination of Cereals*, 11–13.

Zachariasova, M., Lacina, O., Malachova, A., Kostelanska, M., Poustka, J., Godula, M., and Hajslova, J. 2010. Novel approaches in analysis of Fusarium mycotoxins in cereals employing ultra-performance liquid chromatography coupled with high resolution mass spectrometry. *Analytica Chimica Acta* 662(1): 51–61.

16 *In Vitro* Calli Induction, Biomass Accumulation and Different Biological Activity of *Leucas aspera* (willd.) Linn.

Masilamani Sri Devi, Krishnamoorthy Vinothini, Blassan P. George, Sudharshan Sekar, Heidi Abrahamse, Bettine van Vuuren and Arjun Pandian

CONTENTS

16.1 Introduction 298
16.1.1 Plant Description 299
16.1.2 Medicinal Uses of *L. aspera* 299
16.1.3 Secondary Metabolites in *L. aspera* 300
16.2 Materials and Methods 300
16.2.1 Chemicals and Reagents 300
16.2.2 Source of Plant Materials 301
16.2.3 Surface Sterilization of Explants 301
16.2.4 Preparation of Plant Tissue Culture Media 301
16.2.5 Explants Inoculation and Callus Induction 301
16.2.6 Organogenic Callus 302
16.2.7 Root Induction 302
16.2.8 Organic Solvent Extraction of Plant Materials 302
16.2.9 Microorganisms 302
16.2.9.1 Antimicrobial Activity 302
16.2.9.2 Antifungal Activity 303
16.2.10 Antiangiogenesis: Chorioallantoic Membrane Assay (CAM) 303
16.2.11 Statistical Analysis 304
16.3 Results and Discussion 304
16.3.1 Explants Inoculation and Callus Formation from Leaf and Stem Explants 304
16.3.2 Root Induction and Formation *In Vitro* Callus 306

16.3.3 Antimicrobial Activity....306
16.3.4 Antiangiogenesis....309
16.4 Conclusion....310
References....312

16.1 INTRODUCTION

Plants have been indispensable sources of medicinal agents for thousands of years and an impressive number of modern drugs have been isolated from natural sources. Studies on natural products are aimed to determine medicinal values of plants by the exploration of existing scientific knowledge, traditional uses and the discovery of potential chemotherapeutic agents (Balunas and Kinghorn 2005). Exploitation and elimination of natural habitats consequently has led to gradual extinction of several medicinal plants, for which micropropagation is an effective approach to conserve such germplasm (Hassan 2008). It also a rapid propagation processes that can lead to the production of virus-free plants (Gonzales et al. 2010). Plant tissue culture is a modern tool available to rapidly propagate plants and there has been an increased interest in recent years in *in vitro* culture techniques (Nagarajan et al. 2009), which offer a viable tool for the mass multiplication of conservation of threatened, rare and endangered medicinal plants (Ajithkumar and Seeni 1998). In addition, plant tissue culture is considered to be the most efficient technology for crop improvement (Hussain et al. 2012). The micropropagation technology has a vast potential to produce plants of superior quality and the isolation of useful variants in well-adapted, high-yielding genotypes with better disease resistance and stress tolerance capacities (Brown and Thorpe 1995).

The preservative effect of many spices and herbs suggests the presence of antioxidative and antimicrobial constituents (Arjun et al. 2012). In recent years, multiple drug resistances in human pathogenic microorganisms have developed due to indiscriminate use of commercial antimicrobial drugs commonly used in the treatment of infectious diseases (Divya et al. 2014). The side effects and the resistance that pathogenic microorganisms build against antibioticsmades causes scientist to pay attention to herbal extracts and biologically active compounds which are good sources of novel antimicrobial agents (Essawi and Sraur 2000). Cancer causes increasing morbidity and mortality every year globally. Tumour growth and systemic metastasis are highly dependent on angiogenesis (Hansen et al. 2000; Siegel et al. 2012).

Angiogenesis is a normal process in growth and development of blood vessels (Zhang et al. 2010) and is a process essential for tumour growth. It is a tightly regulated complex procedure which involves signalling features and extracellular matrixes that induce the migration of endothelial cells in existing blood vessels to target areas which are sources of pro-angiogenic signalling compounds (Ribatti al. 2003). It promotes tumour growth by supplying nutrients and oxygen and removing waste products while facilitating tumour invasion and metastasis (Wu et al. 2012), hence it is a target for cancer chemotherapy. Antiangiogenic therapy aims to prevent the formation of new vessels around tumours and to frustrate the existing abnormal capillary network that supports the tumour (Mahtabifard et al. 2003). Design and synthesis of novel, small molecule antiangiogenic agents are underway in modern medicinal chemistry (Quigley and Armstrong 1998).

Secondary metabolites, such as flavonoids, alkaloids, phenolic acids, lignans, lignins and tannins, are well known free radical scavengers with multiple biological effects, including anticancer and antimicrobial properties (Rachova et al. 2004; Zijuan et al. 2009). Since the secondary metabolites often have a complex stereo structure with many chiral centres which may be essential for the biological activities, many of these cannot be synthesized economically on a commercial basis (Rajendra and D'Souza 2000).

Plant tissue culture techniques are employed to produce large quantities of secondary metabolites, thereby isolating the active components from the callus without exploiting its natural resources (Lee et al. 2011; Arumugam et al. 2011). Based on the medicinal properties and literature survey, there are no reports for plant tissue culture and antiangiogenic properties of *L. aspera*. Therefore, this study was designed to carry out the *in vitro* propagation of *L. aspera* from its leaves, nodes, internodes and seeds.

16.1.1 Plant Description

Plants of genus *Leucas* have been widely employed by traditional healers to cure many disease conditions, which imply that this genus has immense potential for the discovery of new drugs or lead molecules (Hemendra et al. 2011).

Leucas aspera (Willd.) Linn, belongs to Laminaceae family and is a branched herbaceous plant, which is herb erected (ht 15–60 cm), stout and has a hispid acutely quadrangular stem and branches. Its leaves are shortly petiolate or subsessile, linearly lanceolate or linear, obtuse, pubescent (8.0 cm long, 1.25 cm broad) crenate margin; its petioles are 2.5–6 mm long; its flowers sessile small, white, axillary whorls or dense terminal; its bracts are 6 mm long, acute, linear, bristle-tipped, ciliate with long slender hairs; its calyx are variable, tubular (8–13 mm long); curved, contracted above the nutlets, the lower half usually glabrous and membranous, upper half ribbed, hispid; its mouth is small, very oblique, not villous, with small teeth that are triangular, bristle-tipped, ciliate, its upper tooth being the largest. Corolla (1 cm long, tube 5 mm long) are pubescent above, annulate in the middle; its upper lip (3 mm long) is densely white-woolly; lower lip about twice as long, middle lobe rounded, obviate, lateral lobes are subacute and small. Fruits nutlets are 2.5 mm long, brown, oblong, smooth, their inner face angular and outer face rounded (Kirtikar and Basu 1975; Hooker 1984).

16.1.2 Medicinal Uses of *L. aspera*

The plant is used traditionally as an insecticide and antipyretic, the leaves are used for psoriasis, chronic rheumatism, chronic skin eruptions, snakebites, gastrointestinal disorders and respiratory tract disorders (Kirtikar and Basu 1991). The leaf extract has been shown to exhibit antiplasmodial activity against chloroquine-sensitive strains of *Plasmodium falciparum* (Bagavan et al. 2008). Anti-inflammatory activity of various aerial parts has been reported; the flowers can be used as stimulants, diaphoretics, aperients and insecticide (Goudgaon et al. 2003).

The ethanolic root extract showed antinociceptive, cytotoxic and antioxidant activity (Rahman et al. 2007). Extracts of leaf, flowers and seeds exhibited potent larvicidal activity against mosquitos (Kamaraj et al. 2009). Isolated diterpenes showed prostaglandin-induced inhibitory activity (Sadhu et al. 2006), isolated flavonoids and lignins possess antioxidant and prostaglandin inhibitory activity (Sadhu et al. 2003). Whole-plant is traditionally used for analgesic, antipyretic, antirheumatic, anti-inflammatory, antibacterial and antifungal treatment and its paste is applied topically to inflamed areas (Gani 2003; Prajapati et al. 2010).

16.1.3 Secondary Metabolites in *L. aspera*

Whole-plant is reported for the presence of ursolic acid, oleanolic acid and 3-sitosterol (Chaudhury and Ghosh 1969). Aerial parts contains nicotine (Mangathayaru et al. 2006), sterols (Khaleque et al. 1970), alkaloids such as α-sitosterol and β-sitosterol, reducing sugars such as galactose and glucoside (Chatterjee and Majumdar 1969), diterpenes, leucasperones A and B, leucasperols A and B and isopimarane glycosides; together with other compounds like asperphenamate, maslinic acid, (-)-isololiolide, linifolioside (Sadhu et al. 2006), nectandrin B, macelignan, meso-dihydroguaiaretic acid, acacetin, chrysoeriol, apigenin and machilin C, (-)-chicanine, (7R,8R) and (75,85)-licarin A (Sadhu et al. 2003). Among the 25 compounds identified from its leaves, major constituents are u-farnesene (26.4%), x-thujene (12.6%) and menthol (11.3%) (Sadhu et al. 2003). The flowers contain 10 compounds; among them, amyl propionate (15.2%) and isoamyl propionate (14.4%) are dominant (Kalachaveedu et al. 2006). The seed contains linoleic acid (48.11%), oleic acid (42.07%), palmitic acid (6.25%), stearic acid (2.84%) and linolenic acid (0.65%). The unsaponifiables fraction contains 3-sitosterol and ceryl alcohol (Jam et al. 1968; Badami et al. 1975). The shoots contain novel phenolic compounds (4-(24-hydroxy-1-oxo-5-*n*-propyltetracosanyl-phenol) (Misra et al. 1995), aliphatic ketols (28-hydroxypentatriacontan-7-one,7-hydroxydotriacontan-2-one), long chain compounds (1-hydroxytetratriacontan-4-one, 32-methyltetratriacontan-8-ol), nonatriacontane, 5-acetoxytriacontane, β-sitosterol and dotriacontanol (Misra et al. 1992). Leucolactone (I) isolated from the root has been characterized as 3,3,16c-dihydroxyoleanan-28-1,3-oliden (Pradhan et al. 1990).

16.2 MATERIALS AND METHODS

16.2.1 Chemicals and Reagents

$(NH_4)_2NO_3$, KNO_3, $CaCl_2{\cdot}2H_2O$, $MgSO_4{\cdot}7H_2O$, KH_2PO_4, $MnSO_4{\cdot}4H_2O$, $ZnSO_4{\cdot}7H_2O$, H_3BO_3, $Na_2MoO_4{\cdot}4H_2O$, $CuSO_4{\cdot}5H_2O$, $CoCl_2{\cdot}6H_2O$, Na_2EDTA, KI, $FeSO_4{\cdot}7H_2O$, thiamine HCl, pyridoxine HCl, nicotinic acid, meso-inositol, glycine, sucrose, agar, 2,4-dichlorophenoxyacetic acid (2,4-D), indole-3-acetic acid (IAA), indole-3-butyric acid (IBA), naphthalene acetic acid (NAA), benzyl amino purine (BAP), kinetin (KIN), gibberellic acid GA_3), dimethyl sulfoxide (DMSO), phosphate buffer saline (PBS) hexane, ethyl acetate, acetone, chloroform and carbinol were all present. Analytical grade chemicals were supplied by Hi-Media, SRL, Merck and Sigma chemicals.

16.2.2 Source of Plant Materials

Fresh aerial parts (leaves, seeds, axillary buds, nodes and internodes) of the plant *L. aspera* wild grown plant parts were collected from Jeppiaar Engineering College Campus, Chennai, Tamil Nadu, India. Since the explants were collected from the field, they were subjected to the sterilization process.

16.2.3 Surface Sterilization of Explants

Well-grown plant leaves and axillary buds (nodes and internodes) were collected and washed thoroughly under running tap water (25 min) without damage the tissues, to avoid the interaction of microbes such as bacteria and fungi in plant tissue culture. The explants were transferred to a beaker containing sodium hypochlorite (1%) solution with rapid shaking (5–10 min), changing the solution at every 5-min interval. Then, the explants were washed with sterilized distilled water and treated with mercuric chloride (0.5% every 2–3 min). The surface sterilized explants were washed with sterilized distilled water for five times (5 min each). The excess water on the explants was removed by using sterile tissue paper before culture. The explants were cut into small pieces (0.5–1.0 cm) barring the cut ends and transferred to a semi-solid culture medium under aseptic conditions (Arjun 2011).

16.2.4 Preparation of Plant Tissue Culture Media

Murashige and Skoog (MS) medium was used for the tissue culture (Nagarajan et al. 2009). Six individual stock solutions of macro, micro, minor, iron and vitamins were prepared and stored. Iron stock was stored in a black bottle to prevent photolysis of chemicals. All the stock solutions were stored in a refrigerator and used within one year. Meso-inositol, cytokinin and auxin stocks were freshly prepared and used for a month. For preparation of the medium, all the six stock solutions were mixed thoroughly with required amounts of sterile distilled water. Sucrose (3%, 30 g/L), meso-inositol (0.1%, 100 mg/L) and the required amount of plant growth hormones were added to the medium, buffered by 1N HCl or 1N NaOH to adjust the medium pH (5.6–5.8) before the autoclaving medium was solidified by adding agar (0.8%, 8 g/L). Sterile distilled water was used to make the final volume. The medium was poured into culture vials and autoclaved (15 lbs pressure, for 15 min at 121°C); photoperiod light/dark (16/8 h) condition (25 ± 2°C) cool white light (2000 lux) fluorescent tubes were provided (Arjun 2011).

16.2.5 Explants Inoculation and Callus Induction

For callus induction, healthy and disease-free, young green leaves, shoot nodes, internodes and seeds were collected; surface sterilized explants were cut into small pieces (0.5–1.0 cm) and transferred to the MS basal medium. Three different auxins, 2,4-D, IAA, NAA (0.5 mg/L) and cytokinin and BAP (0.5 mg/L) were used for callus induction. Leaf explants cultured in MS medium without any growth regulators were used as a control for callus induction. All treatments were conducted thrice with replicates in each treatment. Surface sterilized explant (0.5–1.0 cm) segments were placed on the solid MS basal medium, supplemented and incubated under dark (25 ± 2°C, 5 d), then transferred into light conditions.

16.2.6 Organogenic Callus

Well grown callus pieces (1.0 g, 25 d old) were transferred to MS medium containing different concentrations of cytokinins; BAP (1.0–2.0 mg/L) in combination with auxins; IAA (1.0–2.0 mg/L) and NAA (1.0–2.0 mg/L) for shoot regeneration. These include IAA+BAP (1+2mg/L), IAA+BAP (2+1mg/L), NAA+BAP (1+2mg/L) and NAA+BAP (2+1mg/L). The organogenic calli were subcultured at 20 d intervals on the same treatment of fresh medium until shoot regeneration commenced. The cultures were incubated in light (25 ± 2°C for four weeks). Remaining calli harvested for the analysis of compounds.

16.2.7 Root Induction

The roots were developed from callus nodes and shoot tips, the callus was transferred to the MS basal medium supplemented with 2,4-D + BAP (3 + 1 mg/L), IBA (0.5 mg/L), IBA (1.5 mg/L), IBA (2.5 mg/L) and GA_3 (1.0 mg/L) for root induction. The cultures were incubated in dark (25 ± 2°C) for two weeks. The cultures were subcultured every 28 d onto the fresh half-strength MS medium supplemented with plant growth regulators. The length of individual roots (in cm) was measured and recorded after six weeks of culture and the mean root length was determined. All treatments were conducted thrice with replicates.

16.2.8 Organic Solvent Extraction of Plant Materials

The seeds and *in vitro* calli were shade-dried and powdered; the powdered samples were cold macerated with different solvents (hexane, ethyl acetate, chloroform, acetone and methanol) for 3 d with occasional stirring. Then all the extracts were filtered through Whatman filter paper and solvent contents were removed at the low temperature (40–45°C) in a rotary evaporator. The evaporated samples were kept in fridge (4°C) for further analysis (Divya et al. 2014).

16.2.9 Microorganisms

Six microorganisms including five bacteria (gram-positive and gram-negative) and one fungal species were used. This includes *Bacillus subtilis, Staphylococcus aureus, Escherichia coli, Vibrio cholera, Salmonella typhi* and *Candida krusie.* All the ATCC cultures were purchased from Pune, India.

16.2.9.1 Antimicrobial Activity

The *in vitro* antimicrobial activity of different solvent extracts from leaves, stems and *in vitro* calli were determined by the well diffusion method described by

Nagarajan et al. (2010). The Muller Hinton Agar (MHA) ingredients are beef infusion (300 g/L), casein acid hydrolysate (17.5 g/L), starch (1.5 g/L) and agar (1.7 g/L) at pH (7.3 ± 0.1) medium was poured onto sterile petriplates. Agar was allowed to set at ambient temperature. Fresh human pathogenic bacteria cultures of two gram-positive bacteria, *B. subtilis* and *S. aureus* and three gram-negative bacteria, *E. coli, S. typhi* and *V. cholera* were spread on the surface of MHA plates using cotton swabs. Wells were cut from the petriplates using a sterile cork (8 mm dia) borer. Different concentrations (25, 50 and 75 μL/mL) of the leaf, stem and *in vitro* calli were loaded into the wells using a sterile micropipette. The inoculated plates were initially incubated for 15 min at room temperature and then they were incubated at 37°C for 24 h. Turbidity was adjusted with sterile broth so as to correspond to 0.5 McFarland standards. Inhibition zones were recorded as the diameter of growth-free zones including the diameter of the well in mm at the end of incubation period.

16.2.9.2 Antifungal Activity

The antifungal activity of hexane, acetone, methanol and aqueous extracts were evaluated (Arjun et al. 2012). The Sabouraud Dextrose Agar (SDA) ingredients are meat peptone (5 g), casein peptone (5 g), dextrose (40 g), agar (15 g) and distilled water (1000 mL), poured onto sterile petriplates. The agar was allowed to set at ambient temperature. Antifungal activity of the extracts was tested against a human pathogenic fungi *C. krusei*. Fresh fungal culture was spread on surface of the SDA plates using a cotton swab. Wells were cut from SDA in the petridishes using a sterile cork (8 mm dia) borer. Different concentrations (25, 50 and 75 μL/mL) of solvent extracts were loaded into the wells using a sterile micropipette. The inoculated plates were initially incubated for 15 min at room temperature and then were incubated at 37°C for 24 h. Then, the plates were examined for any zone of growth inhibition. Inhibition zones were recorded as the diameter of growth free zones including the diameter of well in mm at the end of incubation period. The percentage of inhibition was calculated by the formula:

$$\text{Percentage of inhibition} = \frac{\text{I (Dia. of the inhibition zone in mm)}}{\text{90 (Dia of the petriplate in mm)}} \times 100 \tag{16.1}$$

16.2.10 Antiangiogenesis: Chorioallantoic Membrane Assay (CAM)

Fifty leghorn eggs were purchased from TNUVAS, Chennai and Tamil Nadu. Seven day-old fertilized brown shell eggs were collected from the hatchery and the eggs were cleaned with ethanol (70%). A small window (10 cm^2) was made in the shell of the eggs. Next, air was sucked out from the eggs to bring their membrane down. Different solvent extracts were dissolved in phosphate buffer saline (PBS, 10 mg/mL). Through the window of each egg, a sterile disc of gelatine sponge containing different concentrations and pellets of these solutions (50, 100 and 150 μL/pellet) were added dropwise on gelatine sponges and applied onto the chorioallantonic membrane (CAM). A crude extract and its fraction were implanted inside the egg and at the junction of two blood vessels of the CAM subsequently. The opening was sealed with parafilm and the eggs were re-incubated (37°C) in incubators (72 h). These windows were then reopened and the inhibition of vessel formation was observed in terms of number and calibre and finally compared with PBS as control. Each experiment was performed in triplicate (Yang et al. 2006).

16.2.11 Statistical Analysis

The data were subjected to ANOVA (one-way analysis of variance) to evaluate the meaning of distinction of resources of a variety of treatment groups by SPSS statistical software package. All the values are expressed as mean ± SD and $P<0.05$ was considered as statistically significant.

16.3 RESULTS AND DISCUSSION

Plants cells have a special capability towards inspire multicellular organs like organogenesis while showing to appropriate equilibrium of plant growth regulators and nutrients. This capability of plant cells to respond to developmental indication and expand into whole plants is totipotency, which is the background of micropropagation.

16.3.1 Explants Inoculation and Callus Formation from Leaf and Stem Explants

The leaf and node explants were surface sterilized and inoculated in the MS medium for callus induction supplemented with auxins NAA,2,4-D and IBA (0.5–2.5 mg/L). After 5 d of incubation for callus induction, it was transferred into light/dark (16/8 h) conditions for callus formation. On the 8 d onwards, callus formation was observed from the leaf and stem explants.

Compared to stems; the leaves showed maximum callus formation and the maximum callus formation was obtained with MS medium with NAA (2 mg/L, 80.76%) followed by 2,4-D (2.0 mg/L, 72.15%), less callus formation (38.32%) was observed in IBA (2 mg/L). The greatest callus formation was observed from stem explants from 2,4-D (2 mg/L, 61.71%), followed by NAA (2 mg/L, 42.66%), the smallest callus content (30.62%) was observed in IBA (Table 16.1) (Figure 16.1). When the hormone concentration increased, callus formation was also increased, then at a particular concentration the callus formation was reduced (Table 16.1) (Figure 16.1). The internode explants of *Verbena officinalis* were more effective than other explants (Turker et al. 2010). Asghari et al. (2012), reported callus formation in leaf explants of *Ocimum basilicum* in universal callus induction supported by combinations of auxins and cytokinins (Albarello et al. 2006). BAP and NAA are the best plant growth regulators for callus induction (Biswas et al. 2007). Callus formation in leaf explants of *Tribulus terrestris* was obtained in MS medium with different hormones and the maximum callus formation was obtained from BAP + NAA (3.0 mg/L, 90%) followed by NAA (3.0 mg/L, 84%) (Arjun 2011).

TABLE 16.1
Callus Formation on Leaf and Node Explants of *L. aspera*

Types of Hormone	Hormone Conc. (mg/L) with MS	Callus Formation (%)	
		Leaf	Node
2,4-D	0.5	38.12 ± 1.3^{e}	26.41 ± 1.3^{f}
	1.0	51.42 ± 1.2^{c}	38.24 ± 1.2^{e}
	1.5	61.21 ± 1.3^{b}	49.41 ± 1.4^{c}
	2.0	72.15 ± 2.4^{a}	61.71 ± 2.6^{b}
	2.5	60.72 ± 2.1^{b}	48.35 ± 2.6^{d}
NAA	0.5	45.21 ± 1.4^{d}	10.37 ± 0.7^{h}
	1.0	55.23 ± 1.4^{c}	22.32 ± 1.5^{f}
	1.5	69.54 ± 2.3^{b}	32.22 ± 1.4^{e}
	2.0	80.76 ± 2.8^{a}	42.66 ± 1.4^{d}
	2.5	70.11 ± 2.3^{a}	31.12 ± 1.3^{e}
IBA	0.5	9.15 ± 0.2^{h}	8.17 ± 0.4^{h}
	1.0	17.15 ± 0.4^{g}	18.45 ± 0.7^{g}
	1.5	26.52 ± 0.7^{f}	27.32 ± 1.9^{f}
	2.0	38.32 ± 1.6^{e}	30.62 ± 0.34^{e}
	2.5	26.12 ± 1.4^{f}	22.40 ± 0.34^{g}

Note: Number in parentheses shows standard deviation [SD]. Statistically significant at $p<0.05$ where $^{a}>^{b}>^{c}>^{d}>^{e}>^{f}>^{g}>^{h}$.

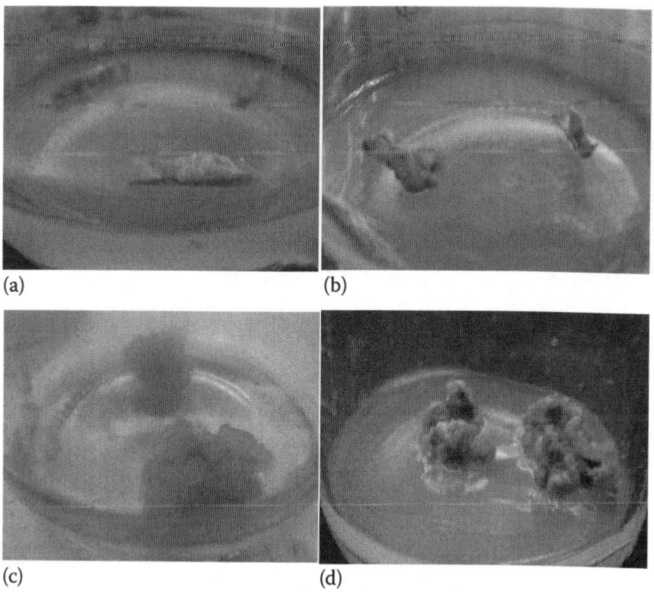

FIGURE 16.1 Biomass accumulation under *in vitro* conditions: (a and c) callus formation from stem explants; (b and d) callus formation from leaf explants.

16.3.2 Root Induction and Formation *In Vitro* Callus

The healthy developed calli were transferred to a MS medium containing dissimilar concentrations of NAA, IAA and IBA, and were kept under dark conditions. The greatest root formation was obtained on IBA (2 mg/L, 76.72%) and the number of roots also obtained per callus, per callus showed (27.43%), followed by IAA (2 mg/L, 67.22%) and the number of roots were also obtained per callus (22.62%). The least amount of root formation was observed (55.63%) on NAA (2 mg/L) and the number of roots were also obtained per callus (21.75%) (Table 16.2) (Figure 16.2).

This was further supported by Mehta et al. (2012), with the utmost root formation in MS medium with IBA observed on *Allium sativum*; the use of auxins individually or in combinations of cytokinins responsible for root formation, induction and elongation (Baksha et al. 2007; Hassan et al. 2008). In *Phyllanthus urinaria* the root formation (100%) was reported by Kalidass and Mohan (2009).

16.3.3 Antimicrobial Activity

The existing notice in the development of novel antimicrobial drugs can be partially attributed both to the rising emergence of bacterial and fungal resistance to antibiotic therapy and to newly emerging pathogens (Yaseen and Sudhakar 2010). The difference in the zone of inhibition is probably due to different solvents used for extraction and also the chemical components of the plants (Chew et al. 2012).

The antimicrobial activity by the zone of inhibition was observed and carried out for wild leaves, stems and *in vitro* calli in different solvents (hexane, ethyl acetate, chloroform, acetone and methanol) and extracts against the human pathogens *B. subtilis, S. aureus, E. coli, S. typhi, V. cholera* and *C. krusei*. In hexane extract (leaves and stems) the highest antimicrobial activity was observed on stems against *C. Krusei* (14 mm) followed by *in vitro* calli and leaf extracts showed against *E. coli*

TABLE 16.2
Root Formation *In Vitro* Callus of *L. aspera*

Types of Hormone	Hormone Conc. (mg/L) with MS	Root Formation (%)	No. of Roots Per Callus
IAA	0.5	43.14 ± 1.3^{d}	11.44 ± 0.13^{b}
	1.0	55.22 ± 1.8^{c}	16.14 ± 0.58^{b}
	2.0	67.22 ± 2.5^{b}	22.62 ± 1.99^{a}
NAA	0.5	34.51 ± 1.8^{e}	15.38 ± 0.51^{b}
	1.0	43.33 ± 2.1^{d}	16.32 ± 0.44^{b}
	2.0	55.63 ± 2.2^{c}	21.75 ± 0.82^{a}
IBA	0.5	56.67 ± 1.8^{c}	16.43 ± 0.98^{b}
	1.0	66.17 ± 1.3^{b}	21.13 ± 1.25^{a}
	2.0	76.72 ± 1.9^{a}	27.43 ± 2.34^{a}

Note: Number in parentheses shows standard deviation [SD]. Statistically significant at $p<0.05$ where $^{a}>^{b}>^{c}>^{d}$.

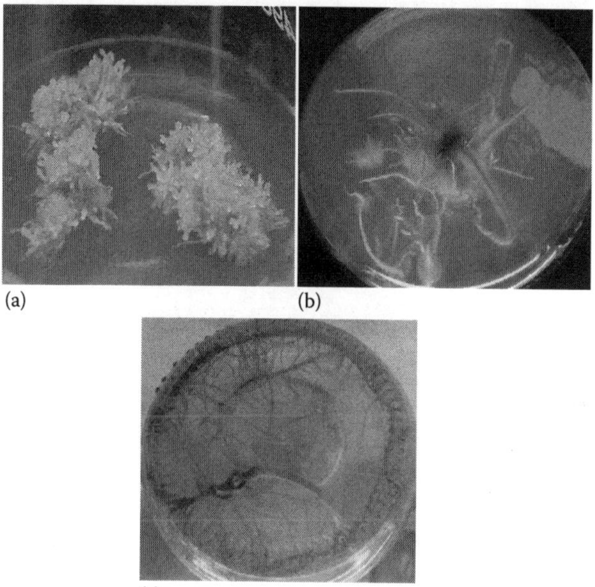

(a) (b) (c)

FIGURE 16.2 *In vitro* root induction and formation: (a) Root initiation from *in vitro* calli; (b and c) root enlargement and developed roots.

(12 mm). Ethyl acetate leaf extract showed the uppermost activity against *S. typhi* (20 mm) followed by leaf extracts against *S. aureus* (16 mm) and *E. coli* (14 mm). Acetone leaf extract showed the highest activity against *S. aureus* (16 mm), followed by stem extract (12 mm); and chloroform stem extract shows inhibition against *V. Cholera* (21 mm) followed by stem against *E. coli* (17 mm). Methanol calli extract showed the highest activity against *B. subtilis* (24 mm), *E. coli* (24 mm) and *V. cholera* (24 mm), followed by calli against *S. aureus* (23 mm) (Table 16.3). In sum, all the solvents showed antimicrobial activity and overall the methanol *in vitro* calli extract showed strong antimicrobial activity.

The ethyl acetate extract showed good activity for gram-negative bacteria in accordance with the reports of Padmakumari et al. (2012). For stem extracts, *V. cholera* of chloroform showed the maximum zone. This was further supported by Rahman et al. (2013). In contrast to these reports, Alam et al. (2011) showed that these extracts do not possess antibacterial activity. In this study, the zone of inhibition produced by commercial antibiotics was larger than those produced by extracts. It may be attributed to the fact that the plant extracts in their crude form contain a small concentration of bioactive compounds (Morshed et al. 2011).

The root extracts showed the maximum zone of inhibition against tested organisms (Chew et al. 2012; Britto et al. 2012), and reported antibacterial activity of the various extracts of Lamiaceae species against various organisms. The methanolic extracts of these plants were found to be efficient against many pathogens. The methanol and ethanol leaf extracts of *Coleus aromaticus* and *L. aspera* exhibited

TABLE 16.3
Antimicrobial Activity of Leaves, Stems and *In Vitro* Calli of Different Solvent Extracts of *L. aspera*

		Zone of Inhibition (mm)																	
		Gram +ve bacteria						Gram -ve bacteria									Fungi		
	Con	*Bacillus subtilis*			*Staphylococcus aureus*			*Escherichia coli*			*Salmonella typhi*			*Vibrio cholerae*			*Candida krusei*		
Extracts	**(µg/mL)**	Leaf	Stem	Calli	Leaf	Stem	Calli	Leaf	Stem	Calli	Leaf	Stem	Calli	Leaf	Stem	Calli	Leaf	Stem	Calli
Hexane	25	4±0.03	–	–	2 ± 0.04	–	–	–	–	–	4 ± 0.02	–	–	4 ± 0.02	–	–	4 ± 0.03	–	–
	50	6 ± 0.02	–	–	4 ± 0.03	–	–	10 ± 0.05	–	10 ± 0.02	–	–	–	6 ± 0.03	10 ± 0.01	–	6 ± 0.02	11 ± 0.02	–
	75	8 ± 0.01	12 ± 0.02	–	6 ± 0.02	11 ± 0.02	10 ± 0.02	12 ± 0.02	11 ± 0.02	12 ± 0.03	–	–	–	8 ± 0.01	11 ± 0.03	–	12 ± 0.03	14 ± 0.03	–
Ethyl acetate	25	–	–	–	4 ± 0.03	–	–	6 ±	–	–	10 ± 0.03	–	5 ± 0.03	6 ± 0.02	–	–	4 ± 0.05	–	–
	50	–	–	–	8 ± 0.01	–	–	12 ± 0.03	–	9 ± 0.02	12 ± 0.01	–	8 ± 0.02	8 ± 0.03	13 ± 0.02	10 ± 0.02	6 ± 0.01	–	–
	75	4 ± 0.04	15 ± 0.03	14 ± 0.01	16 ± 0.02	–	–	14 ± 0.03	–	10 ± 0.01	20 ± 0.03	–	10 ± 0.05	12 ±	15 ± 0.02	13 ± 0.01	10 ± 0.03	–	–
Acetone	25	–	–	–	2 ± 0.01	8 ± 0.02	–	–	–	–	6 ± 0.02	–	5 ± 0.02	10 ± 0.01	–	–	4 ± 0.02	–	–
	50	6 ± 0.05	–	–	4 ± 0.02	10 ± 0.01	–	4 ± 0.03	–	–	8 ± 0.03	–	8 ± 0.02	12 ± 0.02	–	–	6 ± 0.03	–	–
	75	8 ± 0.03	15 ± 0.02	–	16 ± 0.05	12 ± 0.01	–	6 ± 0.05	13 ± 0.02	–	10 ± 0.05	8 ± 0.01	10 ± 0.03	18 ± 0.05	–	–	12 ± 0.04	10 ± 0.03	–
Chloroform	25	–	–	–	4 ± 0.01	–	–	–	–	–	4 ± 0.02	–	9 ± 0.01	–	–	–	4 ± 0.01	–	–
	50	–	–	–	4 ± 0.02	–	–	4 ± 0.01	–	–	6 ± 0.03	–	12 ± 0.03	–	–	–	6 ± 0.03	–	–
	75	–	15 ± 0.03	–	2 ± 0.02	9 ± 0.03	–	8 ± 0.02	17 ± 0.03	–	10 ± 0.01	80.04	13 ± 0.03	6 ± 0.02	21 ± 0.03	–	–	10 ± 0.03	–
Methanol	25	–	–	17 ± 0.05	–	8 ±	12 ± 0.02	–	–	16 ± 0.03	4 ± 0.02	–	10 ± 0.01	–	–	17 ± 0.02	–	–	14 ± 0.03
	50	4 ± 0.02	–	20 ± 0.05	–	10 ± 0.03	16 ± 0.05	–	–	20 ± 0.01	6 ± 0.04	–	11 ± 0.02	–	–	20 ±	–	–	17 ± 0.04
	75	6 ± 0.05	12 ± 0.01	24 ± 0.56	6 ± 0.05	12 ± 0.05	23 ± 0.45	–	–	24 ± 0.54	8 ± 0.01	–	13 ± 0.04	–	–	24 ± 0.5	–	10 ± 0.01	20 ± 0.05

antibacterial activity against enteric pathogens, such as *Shigella* sp., *S. typhi* and *E. coli* (Shiney et al. 2012).

The *in vitro* calli extracts showed the maximum zone of inhibition for methanol extracts of *B. subtilis*, *V. cholera* and *E. coli*, followed by *S. aureus*. This was further supported by Girish et al. (2008) and Moin et al. (2012), who reported maximum inhibition for methanol extracts against *S. aureus*. The fungal species *C. krusei* also showed good activity, and this was in concordant with the reports of Shariff et al. (2006), who reported that the methanol extracts of *Rauvolfia tetraphylla* and *Physalis minima* callus showed antifungal activity. In this study, plant extracts (leaves and stems) and the *in vitro* calli extracts showed good activity for both gram-positive and gram-negative bacteria. This may be indicative for the presence of the broad spectrum of antibiotic compounds (Zain et al. 2012).

16.3.4 Antiangiogenesis

The search for plant-based antiangiogenic agents is carried out using bioassay-guided fractionation, wherein only the active extract and subsequent active fractions are pursued to minimize the efforts (Aruoma et al. 1998). The methanol leaf and stem extracts showed good inhibitory activity followed by chloroform and acetone (Figures 16.3, 16.4) (Table 16.4). This is supported by Nia et al. (2004) who reported methanol extracts of *Sphenocentrum jollyanum* that exhibited good antiangiogenic activity and acetone extracts showing angiogenic properties for *Boerhaavia diffusa* (Jadhav et al. 2011).

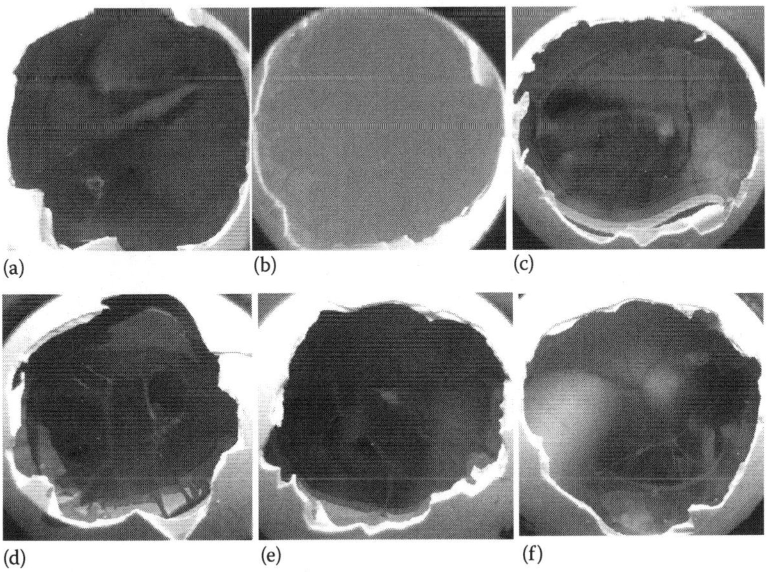

FIGURE 16.3 Antiangiogenesis of different solvent leaf crude extracts: (a) control; (b) hexane; (c) ethyl acetate; (d) acetone; (e) chloroform; (f) methanol.

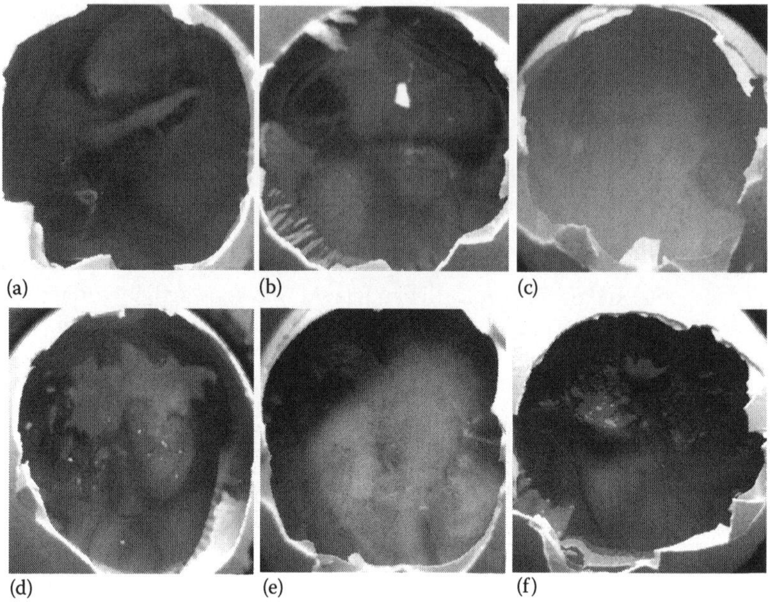

FIGURE 16.4 Antiangiogenesis of different solvent stem crude extracts: (a) control; (b) hexane; (c) ethyl acetate; (d) acetone; (e) chloroform; (f) methanol.

The *in vitro* calli extracts and the chloroform and ethyl acetate extracts showed better antiangiogenic properties (Figure 16.5). The ethyl acetate extracts of *Ardisia pyramidalis* observed better inhibitory activity (Herrera and Amor 2011), the wild plant extracts were better inhibitors compared to the *in vitro* calli extracts (Gong et al. 2013), thus making them more reliable (Figures 16.3, 16.4). This was in concordant with the reports of Moon et al. (1999), who studied the antiangiogenic activity for *Aloe vera* gel extracts and also activity for different Chinese medicinal herbs. It is apparent that no single angiogenic assay can elucidate the entire process of angiogenesis. Many technological advances have occurred recently in angiogenesis and quantification of newly-formed microvessels in laboratory animals that mimic various human diseases (Tahergorabi and Khouzai 2011).

16.4 CONCLUSION

We concluded from this work that the explants of *L. aspera* showed profuse *in vitro* biomass accumulation and root formation. The antimicrobial activity of the leaves, stems and calli extracts showed better activity and suggests that the chemical constituents of these extracts could serve as a potent source of drugs. The search and discovery of novel antiangiogenic compounds are likely to provide hope for millions of chronic disease sufferers. Since there have been no reports for micropropagation and the comparison of biological activity of *L. aspera*, this study helps the search for such compounds. In a nutshell, this work suggests that *L. aspera* should be considered as a useful source of material for human health as an antimicrobial and antiangiogenic agent.

TABLE 16.4
Antiangiogenesis Activity of Leaf, Stem and *In Vitro* Callus of Different Solvent Extracts of *L. aspera*

		Hexane			Ethyl Acetate			Acetone			Chloroform			Methanol		
Conc	PBS	L	S	C	L	S	C	L	S	C	L	S	C	L	S	C
50	9 ± 0.16	5 ± 0.13	–	4 ± 0.16	6 ± 0.14	5 ± 0.16	9 ± 0.21	6 ± 0.15	7 ± 0.15	5 ± 0.16	–	3 ± 0.16	6 ± 0.13	5 ± 0.21	6 ± 0.14	7 ± 0.21
100	11 ± 0.21	–	6 ± 0.16	4 ± 0.21	5 ± 0.15	–	–	5 ± 0.12	8 ± 0.15	7 ± 0.12	5 ± 0.13	–	4 ± 0.15	4 ± 0.16	3 ± 0.13	7 ± 0.16
150	13 ± 0.24	4 ± 0.12	–	–	8 ± 0.19	–	8 ± 0.23	3 ± 0.11	4 ± 0.15	10 ± 0.21	4 ± 0.11	3 ± 0.18	–	3 ± 0.19	4 ± 0.10	–

Note: L = Leaf, S = Stem, C = Callus.

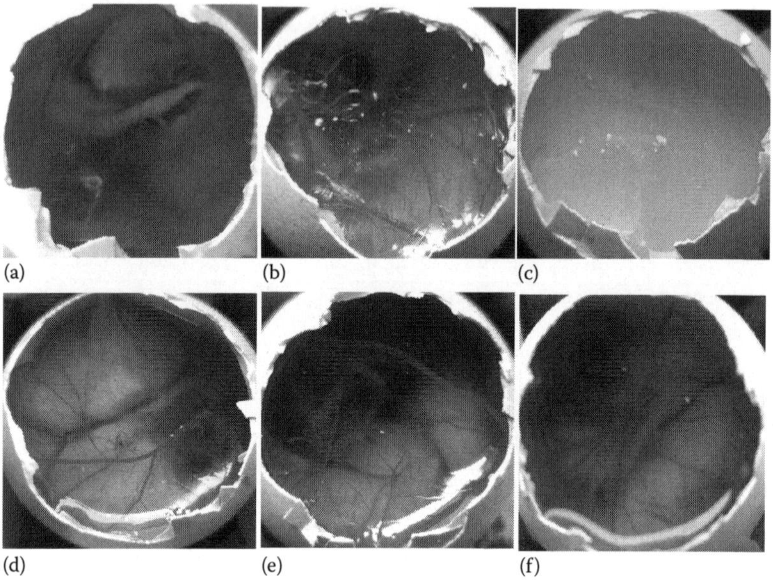

FIGURE 16.5 Antiangiogenesis of different solvent *in vitro* calli crude extracts: (a) control; (b) hexane; (c) ethyl acetate; (d) acetone; (e) chloroform; (f) methanol.

REFERENCES

Ajithkumar, D. and Seeni, S. 1998. Rapid clonal multiplication through *in vitro* axillary shoot proliferation of *Aegle marmelos* (L) Corr., A medicinal tree. *Plant Cell Reports* 17(5): 422–426.

Alam, K., Parvez, N., Yadav, S., Molvi, K., Hwisa, N., Sharif, S. M. A., Pathak, D., Murti, Y., and Zafar, R. 2011. Antimicrobial activity of leaf callus of *Bacopa Monnieri* L. *Scholars Research Library* 3(1): 287–291.

Albarello, N. E., Simo, C., Gonc, P. F., Sas, A. R., Dcastro, T. C., Gianfaldoni, M. G., Callado, C. T. H., and Mansur, E. 2006. *In vitro* propagation of *Cleome spinosa* (Capparaceae) using explants from nursery grown seeding and axenic plant *In vivo*. *In Vitro Cellular & Developmental Biology* 42(6): 601–606.

Arjun, P. 2011. Investigations on microprogation and isolationof active compounds from *Tribulus terrestris* Linn. PhD Thesis, University of Madras, Chennai, Tamil Nadu, India.

Arjun, P., Mohana Priya, S., Saranya Sivan, P. S., Krishnamoorthy, M., and Balasubramanian, K. 2012. Antioxidant and antimicrobial activity of *Nelumbo nucifera* Gaertn. leaf extracts. *Journal of Academia and Industrial Research* 1: 15–18.

Arumugam, T., Ayyanar, M., Justin, Y., Pillay, K., and Sekar, T. 2011. Phytochemical screening and antibacterial activity of leaf and callus extracts of *Centella asiatica*. *Bangladesh Journal of Pharmacology* 6(1): 55–60.

Aruoma, Okezie, I. 1998. Free radicals, oxidative stress, and antioxidants in human health and disease. *Journal of the American Oil Chemists Society* 75(2): 199–212.

Asghari, F., Hossieni, B., Hassani, A., and Shirzad, H. 2012. Effect of explants source and different hormonal combinations on direct regeneration of basil plants (*Ocimum basilicum* L.). *Australian Journal of Agricultural Engineering* 3(1): 12–17.

Badami, R. C. and Patil, K. B. 1975. Minor seed oils X: Physico-chemical characteristics and fatty acid composition of seven minor oils. *Journal of oil Technologists Association of India* 7: 82–94.

Bagavan, A., Rahuman, A. A., Kamaraj, C., and Geetha, K. 2008. Larvicidal activity of saponinfrom *Achyranthes aspera* against *Aedes aegypti* and *Culex quinquefasciatus* (Diptera: Culicidae). *Parasitology Research* 103: 223–229.

Baksha, R., Jahan, M. A. A., Khatun, R., and Munshi, J. L. 2007. *In vitro* rapid clonal propagation of *Rauvolfia serpentina* (Linn.) Benth. *Bangladesh Journal of Scientific and Industrial Research* 42(1): 37–44.

Balunas, M. J. and Kinghorn, A. D. 2005. Drug discovery from medicinal plants. *Life Sciences* 5(78): 431–441.

Biswas, A., Roy, M., Miah, M. B., and Bhadra, S. K. 2007. *In vitro* propagation of *Abrus precatorius* L.-a rare medicinal plant of Chittagong hill tracts. *Plant Tissue Culture and Biotechnology* 17(1): 59–64.

Britto, J. D. A., Sebastian, S. R., and Sujin, M. R. 2012. Antibacterial activity of selected species of Lamiaceae against human pathogens. *Indian Journal of Natural Products and Resources* 3(3): 334–342.

Brown, D. C. W., and Thorpe, T. A. 1995. Crop improvement through tissue culture. *World Journal of Microbiology and Biotechnology* 11(4): 409–415.

Chatterjee, S. K. and Majumdar, D. N. 1969. Chemical investigation of *Leucas aspera. Journal of the Institution of Chemists* 41: 98–101.

Chaudhury, N. A. and Ghosh, D. 1969. Insecticidal plants: Chemical examination of *Leucas aspera. Journal of the Indian Chemical Society* 46: 95.

Chew, A. L., Jessica J. J. A., and Sasidharan, S. 2012. Antioxidant and antibacterial activity of different parts of *Leucas aspera. Asian Pacific Journal of Tropical Biomedicine* 2(3): 176–180.

Divya, Pramod Kumar, Juli Kumari, Chanchal Kumari, Padmapriya, M., Krishnamoorthy, M., and Arjun, P. 2014. Comparative antimicrobial activity and phytochemical analysis of different extracts of potential medicinal plants of *Ocimum sanctum* Linn. and *Lantana camara* Linn. *Afro-Asian Journal of Science and Technology* 1: 053–058.

Essawi, T. and Srour, M. 2000. Screening of some Palestinian medicinal plants for antibacterial activity. *Journal of Ethnopharmacology* 70(3): 343–349.

Gani, A. 2003. Medicinal plants of Bangladesh: Chemical constituents and uses. *Asiatic Society of Bangladesh*, 215.

Girish, H. V., Sudarshana, M. S., and Rati Rao, E. 2008. *In vitro* evaluation of the efficacy of leaf and its callus extracts of *Cardiospermum halicacabum* L. on important Human pathogenic bacteria. *Advances in Biological Research* 2(1–2): 34–40.

Gong, C., Deng, S., Wu, Q., Xiang, M., Wei, X., Li, L., Gao, X., Wang, B., Sun, L., Chen, Y., Li, Y., Liu, L., Qian, Z., and Wei, Y. 2013. Improving antiangiogenesis and anti-tumor activity of curcumin by biodegradable polymeric micelles. *Biomaterials* 34: 1413–1432.

Gonzales, G. R., Quiroz, K., Carrasco, B., and Caligari, P. 2010. Plant tissue culture: Current status, opportunities and challenges. *Ciencia e investigación agrarian* 37(3): 5–30.

Goudgaon, N. M., Basavaraj, N. R., and Vijayalaxmi, A. 2003. Anti-inflammatory activity of different fractions of *Leucas aspera* Spreng. *Indian Journal of Pharmacology* 35: 397–398.

Hansen, S., Grabau, D. A., Sfrensen, F. B., Bak, M., Vach, W. and Rose, C. 2000. The prognostic value of angiogenesis by Chalkley counting in a confirmatory study design on 836 breast cancer patients. *Clinical Cancer Research* 6(1): 139–46.

Hassan, A. K. M. S. 2008. *In vitro* regeneration of a woody medicinal plant, *Pterocarpus santalinus* Linn. f., through apical and axillary shoot proliferation. *Bangladesh Journal of Life Sciences* 20(1): 29–36.

Hemendra, S. C., and Sushil Singh, K. 2011. A review of plants of genus *Leucas*. *Journal of Pharmacognosy and Phytotherapy* 3(3): 13–26.

Herrera, A. A. and Amor, E. C. 2011. Antiangiogenic activity of extracts and fractions from an endemic plant *Ardisia pyramidalis* (Cav.) Pers. From Bataan, Philippines using Duck in ovochorioallantoic membrane assay. *Journal of Medicinal Plants Research* 5(13): 2637–2646.

Hooker, J. D. 1984. *The Flora of British India*. The Muston Company, London, 690.

Hussain, A., Iqbal, A. Q., Hummera, N., and Ikram, U. 2012. *Recent Advances in Plant In Vitro Culture*. Edited by L. Annarita and M. R. R. Laura. Chapter 1. Plant Tissue Culture: Current Status and Opportunities, InTech Publisher, Croatia, EU, 1–28.

Jadhav, J., Mane, A., and Kanase, A. 2011. Antiangiogenic properties of *Boerhaavia diffusa* extracts in chick Chorioallantoic Membrane (CAM). *International Journal of Drug Development & Research* 3(4): 307–317.

Jam, M. P. and Nath, H. B. 1968. Examination of the component fatty acids of the oil from the seeds of *Leucas aspera*. *Lab Dev.* 6: 34–46.

Kalachaveedu, M., Ghosh, A., Ranjan, R., and Vedam Venkat, K. 2006. Volatile constituents of *Leucas aspera* (Willd.) *Journal of Essential Oil Research* 18: 104–115.

Kalidass, C. and Mohan, V. R. 2009. *In vitro* rapid clonal propagation of *Phyllanthus urinaria* Linn. (Euphorbiaceae) A Medicinal Plant. *Researcher* 1(4): 56–61.

Kamaraj, C., Bagavan, A., Rahuman, A. A., Zahir, A. A., Elango, G., and Pandiyan, G. 2009. Larvicidal potential of medicinal plant extracts against *Anopheles subpictus* Grassi and *Culex tritaeniorhyncus* Giles (Diptera: Culicidae). *Parasitology Research* 104: 1163–1171.

Karaman, I., Sahin, F., Gulluce, M., Ogutcu, H., Sengul, M., and Adiguzel, A. 2003. Antimicrobial activity of aqueous and methanol extracts of *Juniperus oxycedrus* L. *Journal of Ethnopharmacology* 85(2–3): 231–235.

Khaleque, A., Huq, M. E., Huq, M. S., and Mansoor, M. H. 1970. Chemical investigations on *Leucas aspera*. I. Isolation of compound-A, 3-sitosterol and et-sitosterol from the aerial parts. *Scientific Research* 7: 125–137.

Kirtikar, K. R. and Basu, B. D. 1975. *Indian Medicinal*, Vol. II, Plants New Delhi, Periodical Experts, 2019–2020.

Kirtikar, K. R. and Basu, B. D. 1991. *Indian Medicinal Plants*, Vol. III, Deharadun: Lalit Mohan Basu, 2019.

Lee, Y., Lee, D. E., Lee, H. S., Kim, K. S., Lee, W. S., Kim, S. H., and Kim, M. W. 2011. Influence of auxins, cytokinins, and nitrogen on production of rutin from callus and adventitious roots of the white mulberry tree (*Morus alba* L.). *Plant Cell, Tissue and Organ Culture* 105(1): 9–19.

Mahtabifard, A., Merritt, R. E., Yamada, R. E., Crystal, R. G., and Korst, R. J. 2003. *In vivo* gene transfer of pigment epithelium-derived factor inhibits tumor growth in syngeneic murine models of thoracic malignancies. *Journal of Thoracic and Cardiovascular Surgery* 126(1): 28–38.

Mangathayaru, K., Thirumurugan, D., Patel, P. S., Pratap, D. V., David, D. J., and Karthikeyan, J. 2006. Isolation and identification of nicotine from *Leucas aspera* (willd). *Indian Journal of Pharmaceutical Sciences* 68: 88–90.

Mehta, J., Naruka, R., Sain, M., Dwivedi, A., Sharma, D., and Mirza, J. 2012. An efficient protocol for clonal micropropagation of *Mentha piperita* L. (Pipperment). *Asian Journal of Plant Science and Research* 2(4): 518–523.

Misra, T. N., Singh, R. S., Pandey, H. S., and Singh, S. 1995. A novel phenolic compound from *Leucas aspera* Spreng. *Indian Journal of Chemistry*, Section B, 34: 1108–1110.

Misra, T. N., Singh, R. S., Prasad, C., and Singh, S. 1992. Two aliphatic ketols from *Leucas aspera*. *Phytochemistry* 32: 199–201.

Moin, S., Babu, S. S., and Mahalakshmipriya, A. 2012. *In vitro* callus production and antibacterial activity of *Barleria lupulina* Lind l. *Asia Pacific Journal of Molecular Biology and Biotechnology* 20(2): 59–64.

Moon, E. J., Lee, Y. M., Lee, O. H., Lee, M. J., Lee, S. K., Chung, M. H., Park, Y. I., Sung, C. K., Choi, J. S., and Kim, K. W. 1999. A novel angiogenic factor derived from *Aloe vera* gel: β-Sitosterol, a plant sterol. *Angiogenesis* 3(2): 117–123.

Morshed, M. A., Uddin, A., Rahman, A., Hasan, T., Roy, S., Amin, A. A. and Islam, R. 2011. *In vitro* antimicrobial and cytotoxicity screening of *Terminalia arjuna* ethanol extract. *International Journal of Biosciences* 1(2): 31–38.

Nagarajan, S., Arjun, P., Raaman, N., and Mohan Das. T. 2010. Regioselective facile one-pot Friedländer synthesis of sugar-based heterocyclic biomolecules. *Carbohydrate Research* 345: 1988–1997.

Nagarajan, S., Mohan Das, T., Arjun, P., Raaman, N. 2009. Design, synthesis and gelatin studies of 4,6-O-butylidene-α,β-unsaturated-β-C-glycosidic ketones: Application toplant tissue culture. *Journal of Materials Chemistry* 19(26): 4587–4596.

Nia, R., Paper, D. H., Essien, E. E., Iyadi, K. C., Bassey, A. I. L., Antai, A. B., and Franz, G. 2004. Evaluation of the antioxidant and antiangiogenic effects of *Sphenocentrum jollyanum* Pierre. *African Journal of Biomedical Research* 7(3): 129–132.

Padmakumari, P., Reshma M., Sulochana M., Vijaya, N., Preethi, V., and Abbulu, K. 2012. Evaluation of antibacterial, analgesic and antipyretic activity of *Leucas aspera* Spreng. *Research Journal of Pharmacognosy and Phytochemistry* 4(3): 186–190.

Pradhan, B., Chakraborty, D., and Subba, G. 1990. A triterpenoid lactone from *Leucas aspera*. *Phytochemistry* 29: 1693–1695.

Prajapati, M. S., Patel, J. B., Modi, K., and Shah, M. B. 2010. *Leucas aspera*: A review. *Pharmacognosy Review* 4(7): 85–87.

Quigley, J. P. and Armstrong, P. B. 1998. Tumor cell intravasationalucidated: The chick embryo opens the window. *Cell* 94(3): 281–284.

Rackova, L., Majekova, M., Kostalova, D., and Stefek, M. 2004. Antiradical and antioxidant activities of alkaloids isolated from *Mahonia aquifolium* structural aspects. *Bioorganic & Medicinal Chemistry* 12(17): 4709–4715.

Rahman, A., Rashu, R., Bhuiyan, T. R., Chowdhury, F., Khan, A. I., Islam, K., LaRocque, R. C., Ryan, E. T., Wrammert, J., Calderwood, S. B., Qadri, F., and Harris, J. B. 2013. Antibody-secreting cell responses after *Vibrio cholerae* O1 infection and oral cholera vaccination in adults in Bangladesh. *Clinical and Vaccine Immunology* 20(10): 1592–1598.

Rahman, M. S., Sadhu, S. K., and Hasan, C. M. 2007. Preliminary antinociceptive, antioxidant and cytotoxic activities of *Leucas aspera* root. *Fitoterapia* 78: 552–555.

Rajendra, K. and D'Souza. 2000. Secondary metabolite of Ayurvedic plants *in vitro*, Plant. biotechnology recent advances. Trivedi, P.C. (Ed.) Panima Publishing Corporation. New Delhi, pp: 350–358.

Ribatti, D., Conconi, M. T., and Nico, B. 2003. Angiogenic response induced by acellular brain scaffolds grafted onto the chick embryo chorioallantoic membrane. *Brain Research* 989(1): 9–15.

Sadhu, S. K., Okuyama, E., Fujimoto, H., and Ishibashi, M. 2003. Separation of *Leucas aspera*, a medicinal plant of Bangladesh, guided by prostaglandin inhibitory and antioxidant activities. *Chemical and Pharmaceutical Bulletin* 51: 595–598.

Sadhu, S. K., Okuyama, E., Fujimoto, H., and Ishibashi, M. 2006. Diterpenes from *Leucas aspera* inhibiting prostaglandin induced contractions. *Journal of Natural Products* 69: 988–994.

Sharif, M. D. M. and Banik, G. R. 2006. Status and utilization of medicinal plants in Rangamati of Bangladesh. *Research Journal of Agriculture and Biological Sciences* 2(6): 268–273.

Shariff, N. S., Sudharshana, M. S., Umesha, S., and Hari Prasad, P. 2006. Antimicrobial activity of *Rauwolfia tetraphylla* and *Physalis minima* leaf and callus extracts. *African Journal of Biotechnology* 5: 946–950.

Shiney, R. B., Ganesh, P., and Suresh Kumar, R. 2012. Phytochemical Screening of *Coleus aromaticus* and *Leucas aspera* and their antibacterial activity against Enteric Pathogens. *International Journal of Pharmaceutical & Biological Archives* 3(1): 162–166.

Siegel, R., Naishadham, D., and Jemal, A. 2012. *Cancer statistics. CA: Cancer Journal of Clinicians* 62(1): 10–29.

Tahergorabi, Z. and Khazaei, M. 2011. A review on angiogenesis and its assays. *Iranian Journal of Basic Medical Sciences* 15(6): 1110–1126.

Turker, A. U., Yucesan, B., and Gurel, E. 2010. Adventitious shoot regeneration from stem internode explants of *Verbena officinalis* L., a medicinal plant. *Turkish Journal of Biology* 34(3): 297–304.

Wu, Q. J., Gong, C. Y., Luo, S. T., Zhang, D. M., Zhang, S., Shi, H. S., and Yang, L. 2012. AAV-mediated human PEDF inhibits tumor growth and metastasis in murine colorectal peritoneal carcinomatosis model. *BMC Cancer* 12(1): 129.

Yang, X., Luo, P., Yang, B., and He, Q. 2006. Antiangiogenesis response of endothelial cells to the antitumour drug 10-methoxy-9-nitrocamptothecin. *Pharmacological Research* 54(5): 334–340.

Yaseen, G. and Sudhakar, J. 2010. Design, synthesis, and antimicrobial activity of 2-mercaptobenzimidazole derivatives. *International Journal of Pharma and Bio Sciences* 1(4): 281–286.

Zain, M. E., Awaad, A. S., Al-Outhman, M. R., and El-Meligy, R. M. 2012. Antimicrobial activities of Saudi Arabian desert plants. *Phytopharmacology* 2(1): 106–113.

Zhang, J., Zhang, Y., Shan, Y., Li, N., Maa, W., and He, L. 2010. Synthesis and preliminary biological evaluation of novel taspine derivatives as anticancer agents. *European Journal of Medicinal Chemistry* 45(7): 2798–2805.

Zijuan, Y., Cejia, L., Lan, X., and Yinan, Z. 2009. Phenolic alkaloids as a new class of antioxidants in *Portulaca oleracea. Phytotherapy Research* 23(7): 1032–1035.

17 Antioxidant Potential of Cup Saucer Plant *Breynia retusa* (Dennst.) Alston

Prabu Jaganathan, Murugan Rajan, Saikumar Sathyanarayanan, Iniyavan Murugaiyan, Geetha Sree Subburaj Jeyalakshmi and Parimelazhagan Thangaraj

CONTENTS

17.1 Introduction 318
17.2 Materials and Methods 318
17.2.1 Chemicals 318
17.2.2 Collection and Identification of Plant Materials 319
17.2.3 Preparation of Plant Extracts 319
17.2.4 Quantification of Total Phenolic, Flavonoid, Vitamin C and E Contents 319
17.2.5 *In Vitro* Antioxidant Assays 319
17.2.5.1 DPPH• Scavenging Assay 319
17.2.5.2 ABTS•+ Scavenging Assay 320
17.2.5.3 Phosphomolybdenum Assay 320
17.2.5.4 Ferric Reducing Antioxidant Power (FRAP) Assay 320
17.2.5.5 Metal Ion Chelating Assay 320
17.2.5.6 Nitric Oxide and Superoxide Anion Radical Scavenging Assays 321
17.2.5.7 Lipid Peroxidation Inhibition Assay 321
17.2.6 Statistical Analyses 321
17.3 Results and Discussion 321
17.3.1 Extractability of Stems 321
17.3.2 Total Phenolic and Flavonoid Contents 322
17.3.3 Vitamin C and E Contents 322
17.3.4 *In Vitro* Antioxidant Assays 323
17.3.4.1 Radical Scavenging Activity Using DPPH• Method 323
17.3.4.2 Scavenging Ability on ABTS•+ 323
17.3.4.3 Phosphomolybdenum Assay 324

17.3.4.4 Ferric Reducing Antioxidant Power (FRAP) Assay..........324
17.3.4.5 Metal Ion Chelating Assay..........325
17.3.4.6 Nitric Oxide Radical Scavenging Activity..........325
17.3.4.7 Superoxide Anion Radical Scavenging Activity..........326
17.3.4.8 Lipid Peroxidation Inhibition Assay..........326
17.4 Conclusion..........327
Acknowledgements..........327
References..........327

17.1 INTRODUCTION

Plants contain a wide variety of free radical scavenging molecules, such as phenolics, flavonoids, vitamins and endogenous metabolites. These plant-derived antioxidants have been shown to function as single and triplet oxygen quenchers, peroxide decomposers, enzyme inhibitors and synergists (Kivits et al. 1997). Electron acceptors, such as molecular oxygen, react easily with free radicals to become radicals themselves and they are also referred to as reactive oxygen species (ROS). ROSs, such as superoxide anions, hydrogen peroxide, hydroxyl, nitric oxide and peroxy nitrite radicals, play an important role in oxidative stress related to the pathogenesis of various important diseases (Halliwell and Gutteridge 1999; Finkel and Holbrook 2000). In healthy individuals, the production of free radicals is balanced by the antioxidative defense system; however, oxidative stress is generated when equilibrium favors free radical generation as a result of a depletion of antioxidant levels. The oxidation of lipids, DNA, proteins, carbohydrates and other biological molecules by toxic ROSs may cause DNA mutation and serve to damage target cells or tissues, and this often results in cell death. Moreover, the knowledge and application of such potential antioxidant activity in reducing oxidative stresses *in vivo* has prompted many investigators to search for potent and cost-effective antioxidants from various plant sources (Liu and Ng 2000; Hu and Kitts 2000; Wang et al. 2004). *Breynia retusa* is commonly known as the cup saucer plant and belongs to the family Euphorbiaceae. The juice of the stem is used in conjunctivitis (Pullaiah 2006). The leaves and fruit of *B. retusa* is a rich source of total protein, total carbohydrates, vitamins and polyphenolic compounds, which possess antioxidant and antidiabetic activity (Murugan et al. 2016). These studies can be taken as a strong platform to implement antioxidant potential in *B. retusa*. Hence, the present investigation on stems was undertaken to study the total phenolic, vitamin contents and antioxidant properties.

17.2 MATERIALS AND METHODS

17.2.1 CHEMICALS

All the chemicals used in the study were of analytical grade; 2,2-diphenyl-1-picrylhydrazyl (DPPH), 2,2 azino-bis(3-ethylbenzothiazoline)-6-sulfonic acid diammonium salt (ABTS) and 6-hydroxy-2,5,7,8-tetramethylchroman-2-carboxylic acid (Trolox) were purchased from Sigma chemicals Co (St. Louis, MO, United States). All the other reagents were obtained from Himedia Laboratories (Mumbai, Maharashtra, India).

17.2.2 Collection and Identification of Plant Materials

The fresh material of *B. retusa* were collected from Kotagiri hills in the Nilgiris District, Tamil Nadu, India during the month of October, 2014. The taxonomic identity of the plant was confirmed by Botanical Survey of India, Southern Circle, Coimbatore, Tamil Nadu, India. (No. BSI/SRC/5/23/2015/Tech/531). The stem parts were washed under running tap water to remove the surface pollutants and were air-dried under shade. The dried samples were powdered and used for further studies.

17.2.3 Preparation of Plant Extracts

The powdered stems of *B. retusa* (100 g) were packed in small thimbles and extracted successively with organic solvents (400 mL) such as petroleum ether, chloroform, ethyl acetate and methanol in increasing order of polarity using a Soxhlet apparatus. The different solvent extracts were concentrated by a rotary vacuum evaporator (Yamato RE300, Japan) and then air-dried. The dried extract obtained with each solvent was weighed and the percentage yield was calculated using Equation 17.1.

$$\text{Extract yield percentage} = [\text{A/B}] \times 100 \tag{17.1}$$

Where

A = Amount of crude extract,

B = Amount of sample.

The extracts obtained were used for the assessment of various analyses (1 mg/mL of respective organic solvents).

17.2.4 Quantification of Total Phenolic, Flavonoid, Vitamin C and E Contents

The total phenolic content was determined according to the method described by Siddhuraju and Becker (2003) and the results were expressed as gallic acid equivalents (GAE). Total flavonoids in the extracts were estimated as rutin equivalents according to the method of Zhishen et al. (1999). The vitamin C content in stem extracts was estimated based on the method of Sadasivam and Manikam (2008) and the results were expressed as milligrams of ascorbic acid equivalents per gram extract. The vitamin E content was determined by Prieto et al. (1999), method and tocopherol were used as standards.

17.2.5 *In Vitro* Antioxidant Assays

17.2.5.1 DPPH• Scavenging Assay

The DPPH radical was used to measure the free radical scavenging activity of plant extracts by the method of Blois (1958). Different concentrations of stem extracts were taken, and 3 mL of a 0.1 mM methanolic solution of DPPH was added to the aliquots of different extracts and standards. DPPH solution (5 mL) along with methanol (100 μL) was used as a negative control. All the reaction mixtures were incubated for 20 min at 27°C. DPPH radical inhibition by the plant samples was measured at 517 nm against the blank (methanol). The results were expressed as inhibitory concentration at 50 percentage of DPPH radical scavenging by plant extracts (μg/mL).

17.2.5.2 ABTS$^{\bullet+}$ Scavenging Assay

The total antioxidant activity of *B. retusa* extracts was determined by ABTS radical cation scavenging assay by the method established by Re et al. (1999). Stable radical 7 mM ABTS aqueous solution and 2.4 mM potassium persulfate were used to produce ABTS radical cation in the dark at 12–16 h. Prior to assay, ABTS solution was diluted in ethanol (1:89 v/v) to give an absorbance of 0.700 ± 0.02 at 734 nm. Triplicates of 10 μL samples and aliquots of Trolox (concentration 0–15 μM) was added to 1 mL of diluted ABTS solution. The reaction mixture was incubated at 30°C exactly 30 min and the absorbance was measured at 734 nm against the ethanol (blank). The results were expressed as μM Trolox equivalents antioxidant capacity (TEAC)/g extract.

17.2.5.3 Phosphomolybdenum Assay

The total antioxidant activity of samples was determined by the green phosphomolybdenum assay by the method of Prieto et al. (1999). The standard (1 mmol/L ascorbic acid in DMSO) and 250 μL of sample were added to 3 mL of reagent solution (0.6 mol/L sulphuric acid, 28 mM sodiumphosphate and 4 mM ammonium molybdate). The reaction mixture was incubated at 95°C for 90 min. The absorbance of the mixture was measured at 695 nm against the reagent blank (reaction mixture). The results reported are mean values and expressed as grams of ascorbic acid equivalent (AAE/100 g extract).

17.2.5.4 Ferric Reducing Antioxidant Power (FRAP) Assay

The ferric reducing ability of different extracts of stem was estimated by the method of Pulido et al. (2000). The FRAP reagent was prepared by mixing 2.5 mL of 10 mmol/L TPTZ in 40 mM HCl, 2.5 mL of 20 mM $FeCl_3$ $6H_2O$ and 25 mL of 0.3 M acetate buffer (pH 3.6). The FRAP reagent (900 μL) was mixed with 10 μL of aliquots of plant extracts and incubated at 37°C. After incubation, the ferric reducing ability of plant extracts was measured at 595 nm. The results were expressed as μM Fe (II) equivalents/mg extract.

17.2.5.5 Metal Ion Chelating Assay

The chelating activity of *B. retusa* was determined by the method of Dinis et al. (1994). Samples (500 μL) were added to 100 μL solution of 2 mM $FeCl_2$. The reaction was initiated by the addition of 400 μL of 5 mmol/L ferrozine and incubated at room temperature for 10 min. Absorbance of the samples was then measured spectrophotometrically at 562 nm against the blank (deionized water). The metal chelating capacities of the extracts were expressed as mg EDTA equivalents/g extract.

17.2.5.6 Nitric Oxide and Superoxide Anion Radical Scavenging Assays

The biological radicals such as nitric oxide (Sreejayan and Rao 1997) and superoxide anion radical (Beauchamp and Fridovich 1971) scavenging activity was estimated in the stem extracts according to the standard procedures.

17.2.5.7 Lipid Peroxidation Inhibition Assay

A modified thiobarbituric acid-reactive substance (TBARS) assay was used to measure the lipid peroxide formed using goat liver homogenates (Ohkawa et al. 1979). Malondialdehyde (MDA), a secondary end product of the oxidation of polyunsaturated fatty acids, reacts with two molecules of TBA yielding a pinkish-red chromogen. Liver homogenate (500 μL of 10%, v/v inphosphate-buffered saline pH 7.4) and 500 μL of sample were added to a test tube and made up to 1.0 mL with distilled water. Then, 50 μL of FeSO4 (0.075 M) and 20 μL of L-ascorbic acid (0.1 M) were added and incubated for 1 h at 37°C to induce lipid peroxidation. Thereafter, 0.2 mL of EDTA (0.1 mol/L) and 1.5 mL of TBA reagent (1.5 g TBA, 60 g TCA and 5.2 mL 70% $HClO_4$ in 800 mL of distilled water) were added in each sample and heated for 15 min at 100°C. After cooling, samples were centrifuged for 10 min at 3000 × g and absorbance of supernatant was measured at 532 nm. Inhibition (%) of lipid peroxidation was calculated using Equation 17.2:

$$\%\text{Inhibition} = \left[(\text{Control OD} - \text{Sample OD})/\text{Control OD}\right] \times 100 \tag{17.2}$$

17.2.6 Statistical Analyses

All the experiments were done in triplicate and the results were expressed as mean ± SD. The data were statistically analysed using one-way ANOVA by Duncan's test for all the studies. Mean values were considered statistically significant when $p < 0.05$.

17.3 RESULTS AND DISCUSSION

17.3.1 Extractability of Stems

Soxhlet extraction is a standard method for the extraction of bioactive compounds from plant sources. In the present study, petroleum ether, chloroform, ethylacetate and methanol have been used to extract the lipophilic compounds (oils and fatty acids), pigments (chlorophyll) and polyphenolics (phenolics and flavonoids). The extract yield percentage of *B. retusa* was shown in Table 17.1. The results showed that methanol extract (19.2 g/100 g sample) had a higher extract yield than other solvent extracts. Methanol extraction of medicinal plants generally yielded more components than other solvent extractions. It is worth mentioning that methanol solvent extraction may allow more hydrogen bonding with phenolic compounds (Murugan and Parimelazhagan 2014).

TABLE 17.1
Extract Yield, Total Phenolic, Flavonoid, Vitamin C and E Contents of *B. retusa*

S. No	Solvents	Extract Yield (g/100 g sample)	Total Phenolics (g GAE/100 g extract)	Flavonoids (g RE/100 g extract)	Vitamin C (g AAE/100 g extract)	Vitamin E (g TE/100 g extract)
1	Petroleum ether	2.10	1.16 ± 0.14^{c}	2.67 ± 1.36^{c}	0.20 ± 0.06^{c}	1.45 ± 0.71^{c}
2	Chloroform	3.00	1.10 ± 0.28^{c}	8.39 ± 1.80^{c}	1.67 ± 0.66^{b}	5.15 ± 2.03^{c}
3	Ethyl acetate	3.90	35.78 ± 1.91^{b}	208.94 ± 5.03^{a}	2.55 ± 0.20^{b}	15.15 ± 0.71^{b}
4	Methanol	19.20	103.86 ± 0.73^{a}	37.00 ± 9.26^{b}	4.32 ± 1.10^{a}	73.07 ± 4.38^{a}

Note: Values are means of three independent analyses ± standard deviation (n = 3). Mean values followed by different superscript letters $^{a} > ^{b} > ^{c}$ indicate significant statistical differences at $p < 0.05$.
Abbreviations: AAE = Ascorbic acid equivalents; GAE = Gallic acid equivalents; RE = Rutin equivalents; TE = α-tocopherol equivalents.

17.3.2 Total Phenolic and Flavonoid Contents

The amount of total phenolics and flavonoids in the stem extracts was analysed and shown in Table 17.1. The total phenolic contents were found to be higher in methanol extract of stems (103.86 g GAE/100 g). Moreover, ethyl acetate extract was also found to have higher flavonoid content (208.94 g RE/100 g extract) when compared to the other extracts. Hence, ethyl acetate and methanol extract was found to be more efficient solvent for extracting the phenolics and flavonoids from stems (Murugan et al. 2016). Polyphenols are the major plant compounds with antioxidant activity. This activity is believed to be mainly due to their redox properties, which can play an important role in adsorbing and neutralizing free radicals, quenching singlet and triplet oxygen or decomposing peroxides (Osawa 1994).

17.3.3 Vitamin C and E Contents

Quantitative assay results revealed the maximum amount of vitamin C was found in the *B. retusa* stem methanol extract (4.32 g AAE/100 g extract) whereas it was lower in the petroleum ether extract (0.20 g AAE/100 g extract). Vitamin E content was highest in the stem methanol extract (73.07 g TE/100 g extract). Similarly, *B. retusa* leaves and fruit also showed good vitamin contents and they may act as antioxidants (Murugan et al. 2016). From the results, the methanol solvent was the best solvent extraction for obtaining vitamin C and E contents.

17.3.4 *In Vitro* Antioxidant Assays

17.3.4.1 Radical Scavenging Activity Using DPPH• Method

Relatively stable DPPH radicals have been widely used to test the ability of compounds to act as free radical scavengers or hydrogen donors and thus to evaluate antioxidant activity (Jao and Ko 2002). The reduction capacity of DPPH radicals was determined by the decrease in its absorbance at 517 nm (Duh 1998). The results of DPPH assay were expressed in IC_{50} values. The lower value of IC_{50} indicates a higher antioxidant activity. The *B. retusa* stem methanol extract (20.04 μg/mL) showed better DPPH radical scavenging activity compared to other solvent extracts, shown in Figure 17.1. From this, it is clear that methanol extract showed better radical scavenging activity by reducing the stable DPPH radicals to a yellowish diphenylpicrylhydrazine derivative. They also exhibited good antioxidant activity in the traditional food antioxidant quercetin (4.70 μg/mL) and synthetic antioxidant BHT (31.12 μg/mL), which were used as reference compounds. Similarly, Sudhanshu et al. (2012) has also been reported to have anti-DPPH· activity in *B. retusa* extracts.

17.3.4.2 Scavenging Ability on ABTS•+

The TEAC (Trolox equivalents antioxidant capacity) was measured using the improved ABTS•+ radical decolourisation assay, one of the most commonly employed methods for antioxidant capacity, which measures the ability of a compound to scavenge ABTS•+ radicals (Awika et al. 2003). The result of ABTS cation radical scavenging activity of stem extracts of *B. retusa* is shown in Table 17.2. The ethyl acetate extract of stems showed higher cation radical scavenging activity (8396.95 μM TEAC/g extract) compared to that of other solvent extracts. Hagerman et al. (1998) reported that the high molecular weight phenolics have more ability to quench free radicals (ABTS•+). This is may be due to the presence phenolic contents in the *B. retusa* extracts.

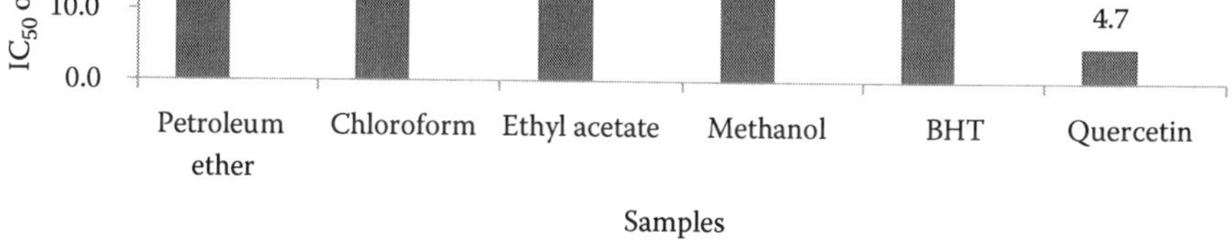

FIGURE 17.1 DPPH radical scavenging activity.

TABLE 17.2
Antioxidant Activity of *B. retusa*

S. No	Solvents	ABTS•+ Assay (μM TEACE/g extract)	Phosphomolybdenum Assay (g AAE/100 g extract)	FRAP Assay (μM Fe(II) E/mg extract)	Metal Ion Chelating Assay (mg EDTAE/g extract)
1	Petroleum ether	843.74 ± 149.26[d]	0.73 ± 0.052[d]	131.63 ± 5.57[c]	18.04 ± 0.40[b]
2	Chloroform	2544.73 ± 164.92[c]	2.12 ± 0.052[c]	2593.91 ± 17.66[b]	3.43 ± 2.20[c]
3	Ethyl acetate	8396.95 ± 152.01[a]	8.56 ± 0.46[b]	135.27 ± 10.34[c]	23.62 ± 1.51[a]
4	Methanol	4340.22 ± 136.65[b]	17.34 ± 0.08[a]	6075.66 ± 41.28[a]	24.07 ± 1.92[a]

Note: Values are means of three independent analyses ± standard deviation (n = 3). Mean values followed by different superscript letters [a] > [b] > [c] > [d] indicate significant statistical difference at $p < 0.05$.

Abbreviations: AAE = Ascorbic acid equivalents; EDTAE = Ethylenediaminetetraacetic acid; Fe (II) E = Ferrous equivalents; TEACE = Trolox equivalents antioxidant capacity equivalents.

17.3.4.3 Phosphomolybdenum Assay

The phosphomolybdneum method is based on the reduction of Mo(VI) and to Mo(V) by the antioxidant compound and the formation of green phosphate/Mo(V) complex with the maximal absorption at 695 nm. A better antioxidant capacity was shown by methanol extract (17.34 g AAE/100 g extract) compared to other solvent extracts in the order of: ethyl acetate > chloroform > petroleum ether (Table 17.2). The assay is successfully used to quantify vitamin E in seeds and being simple and independent of other antioxidant assays commonly employed, it was decided to extend its application to plant extracts (Prieto et al. 1999). The antioxidant capacity observed from the stem extracts can be correlated with its free radical scavenging activity equivalent to that of natural antioxidant ascorbic acids.

17.3.4.4 Ferric Reducing Antioxidant Power (FRAP) Assay

This simple and reliable test measures the reducing potential of an antioxidant reacting with a ferric 2,4,6-tripyridyl-S-triazine(Fe (III)-TPTZ) complex and producing a coloured ferrous 2,4,6-tripyridyl-S-triazine(Fe (II)-TPTZ) complex by a reductant at low pH, was adopted. This complex can be monitored at 593 nm (Loo et al. 2008). A higher absorbance power indicates a higher ferric reducing power. The results show (Table 17.2) that the ferric reducing capacity was much higher in stem methanol extract (6075.66 μM Fe (II) E/mg extract) and lower in petroleum ether extract (131.63 μM Fe (II) E/mg extract). Yen and Duh (1993) and Siddhuraju and Becker (2003) have reported that the reducing power of bioactive compounds, mainly low and high molecular phenolics, is associated with antioxidant activity, specifically free radical scavenging. Thus, the ferric reducing power of the different extracts of *B. retusa* reveals that there are compounds in methanolic extracts which have a high affinity to the ferrous ions and thereby scavenge them through redox reactions.

17.3.4.5 Metal Ion Chelating Assay

Iron is essential for life because it is required for oxygen transportation, respiration and the activity of many enzymes. In complex systems, such as food preparation, various different mechanisms may contribute to the oxidative processes, such as Fenton reaction, where transitioning metal ions play a vital role. Different reactive oxygen species might be generated and various target structures, such as lipids, proteins and carbohydrates, can be affected. Therefore, it is important to characterize the extracts by estimating iron chelating activity (Halliwell 1995). The method of metal chelating activity is based on chelating of Fe^{2+} ions by the reagent ferrozine, which is the quantitative formation of a complex with Fe^{2+} ions (Dinis et al. 1994). The metal chelating activity of different solvent extracts of *B. retusa* stems is shown in Table 17.2. Among the different extracts, the methanol extract of stems (24.07 mg EDTAE/g extract) showed better scavenging ability compared to other solvent extracts. Chelating agents may serve as secondary antioxidants because they reduce the redox potential, thereby stabilizing the oxidized form of the metal ions (Rajesh et al. 2008). It can be also suggested that phenolic compounds and flavonoids present in *B. retusa* may inhibit the interaction between metal and lipid through formation of insoluble metal complexes with ferrous ion.

17.3.4.6 Nitric Oxide Radical Scavenging Activity

The scavenging activity of various extracts on nitric oxide were analysed and the percentage of scavenging activity is shown in Figure 17.2. Nitric oxide radicals generated from sodium nitroprusside reacts with oxygen to form nitrite. Nitric oxide is also implicated in inflammation, cancer and other pathological conditions (Marletta 1989; Moncada et al. 1991). At a concentration of 500 μg, greater percentages of scavenging activity were observed in chloroform (31.18%) versus methanol extract (23.27%). The standards of BHA (40.5%) and rutin (14.8%) showed good nitric oxide radical scavenging activity. From the results, *B. retusa* is a potent scavenger of nitric oxide. From this, we conclude that the therapeutic properties of *B. retusa* may be useful for treating radical-related inflammation and cancer.

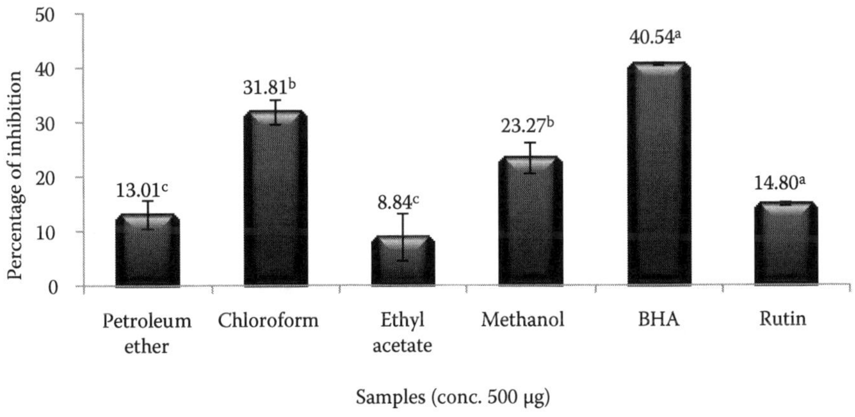

FIGURE 17.2 Nitric oxide radical scavenging activity. Values are mean of triplicate determination (n = 3) ± standard deviation; statistically significant at $p < 0.05$ where $^a > ^b > ^c$.

17.3.4.7 Superoxide Anion Radical Scavenging Activity

Superoxide anion radicals are a reactive oxygen species that can be generated in the human body by auto oxidative processes and are very harmful to cellular components. ROSs have a casual relationship with oxidative stress and the superoxide radical scavenging assay is based on the capacity of plant extracts to inhibit formazan formation by scavenging the superoxide radicals generated in riboflavin-light NBT systems (Halliwell and Gutteridge 2002). The result of superoxide anion scavenging activity of stems is shown in Figure 17.3. The extracts were found to be an efficient scavenger of superoxide radical generated in riboflavin-NBT-light systems *in vitro*. The scavenging activity of methanol extract was found to be 36.79% compared to standard ascorbic acid (55.97%). From this assay, using different extracts of *B. retusa*, it is noted that the inhibition of the formation of blue formazan and the percentage of inhibition are directly proportional to the concentration of plant extracts. The probable mechanism of scavenging superoxide anion radicals by *B. retusa* can be due to the active principles in the stem extracts which may eliminate the superoxide anion radicals which are generated through the photo-illumination process.

17.3.4.8 Lipid Peroxidation Inhibition Assay

The inhibition of Fe^{2+}-induced lipid peroxidation was assayed by the TBARS formation assay. Malonaldehyde is a naturally occurring product of lipid peroxidation and prostaglandin biosynthesis that is mutagenic and carcinogenic (Marnett 1999). Malonaldehyde is, in many instances, the most abundant individual aldehyde resulting from lipid peroxidation, and its determination by thiobarbituric acid is one of the most common assays in lipid peroxidation studies. *In vitro* malonaldehyde can alter proteins, DNA, RNA and many other biomolecules (Schauenstein et al. 1977). As shown in Figure 17.4, *B. retusa* extracts exhibited a strong inhibitory effect on lipid peroxidation and the inhibitory effect was concentration dependent. The methanol extract (49.78%) showed a higher percentage of inhibition at 500 μg concentration.

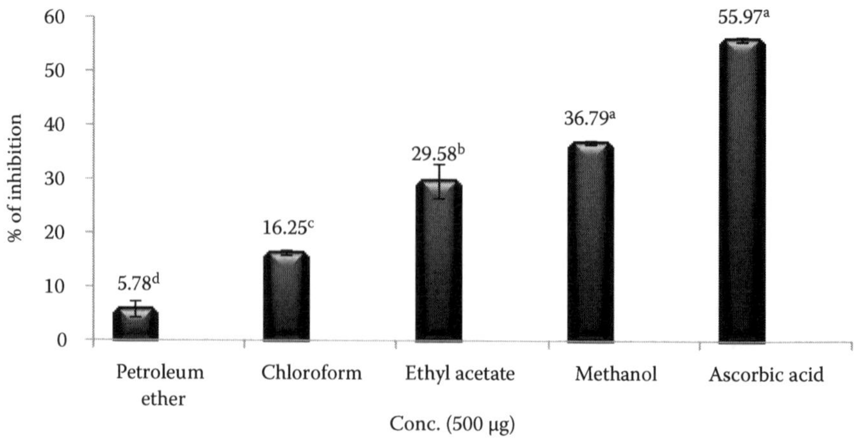

FIGURE 17.3 Superoxide anion radical scavenging activity. Values are mean of triplicate determination (n = 3) ± standard deviation; statistically significant at $p < 0.05$ where $^{a} > ^{b} > ^{c} > ^{d}$.

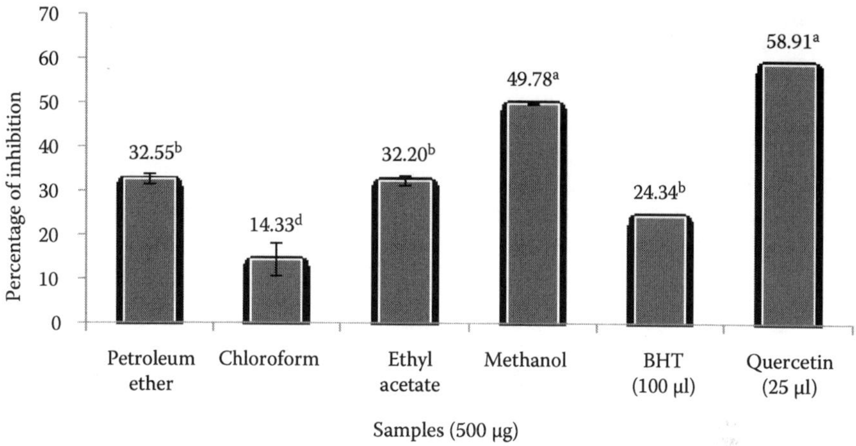

FIGURE 17.4 Lipid peroxidation inhibition assay. Values are mean of triplicate determination (n = 3) ± standard deviation; statistically significant at $p < 0.05$ where $^{a} > ^{b} > ^{c} > ^{d}$.

On the other hand, quercetin (58.91%) exhibited a higher percentage of inhibition than BHT (24.34%). From this study, it is found that plant extracts inhibit the production of malonaldehyde from the lipid peroxidation. *B. retusa* showed an efficient dose-dependent inhibition of lipid peroxidation with a percentage of inhibition compared to butylated hydroxytoluene.

17.4 CONCLUSION

Chapter 17 confirms the antioxidant potential of *B. retusa* stems, as indicated by high phenolic and vitamin contents that have the ability to scavenging different free radicals. Hence, stem extracts could be used as an easy, accessible source of natural antioxidants in the pharmaceutical and medical industries. Further detailed investigation is underway to determine the molecular mechanisms involved in antioxidant activity and determine the efficacy by *in vivo* models.

ACKNOWLEDGEMENTS

The authors are thankful to Dr. V. Narmatha Bai, the Head of the Department of Botany at Bharathiar University, Coimbatore, Tamil Nadu, India.

REFERENCES

AOAC. 1990. Official Methods of analysis, 12th ed., Washington, DC.

Awika, J. M., Rooney, L. W., Wu, X., Prior, R. L., and Cisneros-Zevallos, L. 2003. Screening methods to measure antioxidant activity of sorghum (*Sorghum bicolor*) and sorghum products. *Journal of Agricultural and Food Chemistry* 51: 6657–6662.

Beauchamp, C. and Fridovich, I. 1971. Superoxide dismutase: Improved assays and an assay applicable to acrylamide gels. *Analytical Biochemistry* 44: 276–277.

Blois, M. S. 1958. Antioxidants determination by the use of a stable free radical. *Nature* 4617: 1199–1200.

Dinis, T. C., Madeira, V. M., and Almeida, L. M. 1994. Action of phenolic derivatives (acetaminophen, salicylate, and 5-aminosalicylate) as inhibitors of membrane lipid peroxidation and as peroxyl radical scavengers. *Archives of Biochemistry and Biophysics* 315: 161–169.

Duh, P. D. 1998. Antioxidant activity of burdock (*Arctium lappa* Linne): its scavenging effect on free-radical and active oxygen. *Journal of the American Oil Chemists' Society (JAOCS)* 75: 455–461.

Finkel, T. and Holbrook, N. J. 2000. Oxidants, oxidative stress and the biology of aging. *Nature* 408: 239–247.

Hagerman, A. E., Riedl, K. M., Jones, G. A., Sovik, K. N., Ritchard, N. T., Hartzfeld, P. W., and Riechel, T. L. 1998. High molecular weight plant polyphenolics (tannins) as biological antioxidants. *Journal of Agricultural and Food Chemistry* 46: 1887–1892.

Halliwell, B. and Gutteridge, J. M. C. 1999. *Free Radicals in Biology and Medicine*, 3rd edition, Oxford University Press, Oxford, UK.

Halliwell, B., and Gutteridge, J. M. C. 2002. *Free Radicals in Biology and Medicine.* Oxford University Press.

Halliwell, B., Murcia, M. A., Chirico, S., and Aruoma, O. I. 1995. Free radicals and antioxidants in food and *in vivo*: What they do and how they work. *Critical Reviews in Food Science and Nutrition* 35: 7–20.

Hu, C. and Kitts, D. D. 2000. Studies on the antioxidant activity of *Echinacea* root extract. *Journal of Agricultural and Food Chemistry* 48: 1466–1472.

Jao, C. H., and Ko, W. C. 2002. 1, 1 Diphenyl-2-picrylhydrazyl (DPPH) radical scavenging by protein hydrolysaes from tuna cooking juice. *Fisheries Science* 68: 430–435.

Kivits, G. A. A., Vam Der Sman, F. J. P., and Tijburg, L. B. M. 1997. Analysis of catechin from green and black tea in humans: A specific and sensitive colorimetric assay of total catechins in biological fluids. *International Journal of Food Sciences and Nutrition* 48: 387–392.

Liu, F., and Ng, T. B. 2000. Antioxidative and free radical scavenging activities of selected medicinal herbs. *Life Sciences* 66: 725–735.

Loo, A. Y., Jain, K., and Darah, I. 2008. Antioxidant activity of compounds isolated from the pyroligneous acid, *Rhizophora apiculata. Food Chemistry* 107: 1151–1160.

Marletta, M. A. 1989. Nitric oxide: Biosynthesis and biological significance. *Trends in Biochemical Sciences* 14: 488–492.

Marnett, L. J. 1999. Lipid peroxidation-DNA damage by malondialdehyde. *Mutation Research – Fundamental and Molecular Mechanisms of Mutagenesis* 424: 83–95.

Moncada, S. R., Palmer, M. J., and Higgs, E. 1991. Nitric oxide: Physiology, pathophysiology, and pharmacology. *Pharmacological Reviews* 43: 109–142.

Murugan, R., Prabu, J., Chandran, R., Sajeesh, T., Iniyavan, M., and Parimelazhagan, T. 2016. Nutritional composition, in vitro antioxidant and anti-diabetic potentials of *Breynia retusa* (Dennst.). *Alston Food Science and Human Wellness* 5: 30–38.

Murugan, R. and Parimelazhagan, T. 2014. Comparative evaluations of different extraction methods for antioxidant and anti-inflammatory properties from *Osbeckia parvifolia* Arn. – An *in vitro* approach. *Journal of King Saud University – Science* 26: 267–275.

Ohkawa, H., Ohishi, N., and Yagi, K. 1979. Assay for lipid peroxides in animal tissues by thiobarbituric acid reaction. *Analytical Biochemistry* 95: 351–358.

Osawa, T. 1994. Novel natural antioxidants for utilization in food and biological systems. *Postharvest Biochemistry of Plant Food-Materials in the Tropics*, Eds. I. Uritani, V. V. Garcia, and E. M. Mendoza, 241–251. Japan Scientific Societies Press, Tokyo, Japan.

Prieto, P., Pined, M., and Aguilar, M. 1999. Spectrophotometric quantity of antioxidant capacity through the formation of a phosphomolybdenum complex: Specific application to the determination of vitamin E. *Analytical Biochemistry* 269: 337–341.

Pulido, R., Bravo, L., and Sauro-Calixto, F. 2000. Antioxidant activity of dietary polyphenols as determined by a modified ferric reducing/antioxidant power assay. *Journal of Agricultural and Food Chemistry* 48: 3396–3402.

Pullaiah, T. 2006. *Encyclopaedia of World Medicinal plants. Vol. 1*, Regency Publication, New Delhi, ISBN: 81-89233-42-4, 223.

Rajesh, M., Anusuya, N., Siddhuraju, P., and Manian, S. 2008. The antioxidant activity and free radical scavenging potential of two different solvent extracts of *Camellia sinensis* (L.) O. Kuntz, *Ficus bengalensis* L. and *Ficus racemosa* L. *Food Chemistry* 107: 1000–1007.

Re, R., Pellegrini, N., Proteggente, A., Pannala, A., Yang, M., and Evans, R. C. 1999. Antioxidant activity applying an improved ABTS radical cation decolourization assay. *Free Radical Biology and Medicine* 26: 1231–1237.

Sadasivam, S., and Manikam, A. 2008. *Biochemical Methods*, 3rd ed., New Age International (P) Limited Publishers, Chennai.

Schauenstein, E., Esterbauer, H., and Zollner, H. 1977. Aldehydes in biological systems: Their natural occurrence and biological activities, Pion. 3.

Siddhuraju, P. and Becker, K. 2003. Studies on antioxidant activities of Mucuna seed (*Mucuna pruriens* var. *utilis*) extracts and certain non-protein amino/iminoacids through *in vitro* models. *Journal of Agricultural and Food Chemistry* 51: 2144–2155.

Sreejayan, N. and Rao, M. N. A. 1997. Nitric oxide scavenging by curcuminoids. *Journal of Pharmacy and Pharmacology* 49: 105–107.

Sudhanshu, R., Nidhi, M., and Ekta Sandhya M. 2012. Antioxidant activity of *Rivea hypocratreiformis, Breynia retusa, Woodfordia fruticosa* used as traditional medicine. *International Journal of Pharmaceutical and Phytopharmacological Research* 1: 347–349.

Sudhanshu, T. and Ajay, S. 2004. Piscicidal and anti-acetylcholinesterase activity of *Euphorbia royleana* stem bark extracts against freshwater common predatory fish *Channa punctatus. Environmental Toxicology and Pharmacology* 18: 47–53.

Wang, S., Wu, J. H., Cheng, S. S., Lo, C. P., Chang, H. N., Shyur, L. F., and Chang, S. T. 2004. Antioxidant activity of extracts from *Calocedrus formosana* leaf, bark and heartwood. *Journal of Wood Science* 50: 422–426.

Wang, S. Y., Wu, J. H., Shyur, L. F., Kuo, Y. H., and Chang, S. T. 2002. Antioxidant activity of abietane-type diterpenes from heartwood of *Taiwania cryptomerioides* Hayata. *Holzforschung* 56: 487–492.

Yen, G. C., and Duh, P. D. 1993. Antioxidative properties of methanolic extracts from peanut hulls. *Journal of the American Oil Chemists' Society* 70: 383–386.

Zhishen, J., Mengcheng, T., and Jianming, W. 1999. The determination of flavonoid contents in mulberry and their scavenging effects on superoxide radicals. *Food Chemistry* 64: 555–559.

18 The Promising Antiradical Potential of *Castanospermum australe* A. Cunn. and C. Fraser ex Hook.

Sajeesh Thankarajan, Rahul Chandran, Murugan Rajan, Saikumar Sathyanarayanan and Parimelazhagan Thangaraj

CONTENTS

18.1 Introduction 332
18.2 Materials and Methods 333
  18.2.1 Collection and Identification of Plant Material 333
  18.2.2 Chemicals 333
  18.2.3 Preparation of Plant Extracts 334
  18.2.4 Extract Recovery Percent 334
  18.2.5 Quantification Assays 334
    18.2.5.1 Quantification of Total Phenolics 334
    18.2.5.2 Quantification of Tannins 334
    18.2.5.3 Quantification of Total Flavonoids 335
  18.2.6 *In Vitro* Antioxidant Assays 335
    18.2.6.1 $DPPH^{\bullet}$ Scavenging Activity 335
    18.2.6.2 $ABTS^{\bullet +}$ Scavenging Activity 336
    18.2.6.3 Phosphomolybdenum Assay 336
    18.2.6.4 Ferric Reducing Antioxidant Power (FRAP) Assay 336
    18.2.6.5 Metal Chelating Activity 337
    18.2.6.6 Hydroxyl Radical Scavenging Activity 337
    18.2.6.7 Superoxide Radical Scavenging Activity 337
    18.2.6.8 Hydrogen Peroxide Scavenging Activity 338
    18.2.6.9 Nitric Oxide Scavenging Activity 338
  18.2.7 Statistical Analyses 339
18.3 Results and Discussion 339
  18.3.1 Extract Recovery Percent 339
  18.3.2 Quantification Assays 339

18.3.2.1 Quantification of Total Phenolics 339
18.3.2.2 Quantification of Tannins 341
18.3.2.3 Quantification of Flavonoids 341
18.3.3 *In Vitro* Antioxidant Assays 342
18.3.3.1 DPPH$^•$ Scavenging Activity 342
18.3.3.2 ABTS$^{•+}$ Scavenging Activity 343
18.3.3.3 Phosphomolybdenum Assay 345
18.3.3.4 Ferric Reducing Antioxidant Power Assay 345
18.3.3.5 Metal Chelating Activity 347
18.3.3.6 Hydroxyl Radical Scavenging Activity 347
18.3.3.7 Superoxide Radical Scavenging Activity 348
18.3.3.8 Hydrogen Peroxide Scavenging Activity 349
18.3.3.9 Nitric Oxide Scavenging Activity 350
18.4 Conclusion 351
References 351

18.1 INTRODUCTION

The use of plants in traditional medicine for treating various ailments remains an integral part of the culture and traditions of a majority of the world's population. In addition, factors such as the availability, affordability and accessibility of medicinal plants have led to their high demand and usage (Aremu 2011). Secondary metabolites such as alkaloids, iridoids and phenolics, are generally produced by plants for their defense mechanisms and have been implicated in the therapeutic properties of most medicinal plants (Kliebenstein 2004). Plants therefore provide an invaluable resource useful in the development of therapeutic compounds (Gurib-Fakim 2006).

Naturally, there is a dynamic balance between the number of free radicals generated in the body and antioxidants to quench or scavenge them and protect the body against their deleterious effects (Nose 2000; Finkel and Holbrook 2000). Free radicals are fundamental to any biochemical process and represent an essential part of aerobic life and our metabolism. They are continuously produced by the body's normal use of oxygen through respiration and some cell-mediated immune functions. They are also found or generated through environmental pollutants, cigarette smoke, automobile exhaust fumes, radiation, air pollutants, pesticides, etc. (Halliwell et al. 1992; Jacob 1994; Li and Trush 1994). However, the amount of these protective antioxidant principles present under normal physiological conditions are sufficient only to cope with the physiological rate of free radical generation. Therefore, it is obvious that any additional burden of free radicals, either from the environment or produced within the body, can alter the pro-oxidant and antioxidant balance, leading to oxidative stress. In humans, the overproduction of ROS and RNS can result in tissue injury, which has been implicated in disease progression and oxidative damage of nucleic acids, proteins and lipids (Middleton et al. 2000). When there is a lack of antioxidants to quench the excess reactive free radicals, cardiovascular, cancer, neurodegenerative, Alzheimer's and inflammatory diseases may develop in the body (Krishnaiah et al. 2010).

Among synthetic antioxidants, those most frequently used in food industries are butylated hydroxyanisole (BHA), butylated hydroxytoluene (BHT), propyl gallate (PG) and tert-butyl hydroquinone (TBHQ). Reports revealed that BHA and BHT could be toxic, and the higher manufacturing costs and lower efficiency of natural antioxidants, such as tocopherols, together with the increasing consciousness of consumers with regard to food additive safety, created a need for identifying alternative natural and probably safer sources of food antioxidants (Sherwin 1990; Wanasundara and Shahidi 1998). Moreover, antioxidant activity also depends on the type and polarity of the extracting solvent, the isolation procedures, the purity of active compounds, as well as the test system and substrate to be protected by the antioxidant (Meyer et al. 1998).

The genus *Castanospermum* belongs to the family Fabaceae and has only one species, *Castanospermum australe*, commonly referred to as the black bean or the Moreton Bay chestnut. It is a medicinal as well as toxic plant which is indigenous to Australia, but cultivated in Pakistan. The seeds are toxic, but become edible when carefully prepared by pounding into flour, leaching with water and roasting. Its seeds have been utilized following extensive preparation as a food by Aborigines and contain alkaloids which have been shown to have anti-HIV and anticancer properties. *Castanospermum australe* has proved to be a fruitful source of alkaloids of the polyhydroxy indolizidine and pyrrolizidine classes, having antineoplastic and antiretroviral properties. It is evident from the fact that the various constituents of *Castanospermum australe* e.g. alkaloid, saponin and flavonoids may exhibit biological activity to a various extent, and could be considered as promising compounds for clinical utilization. The present study focused on a method to quantify the total amount of phenolics, tannins, flavonoids and *in vitro* antioxidant activities of leaf, bark and seed extracts of *C. australe*.

18.2 MATERIALS AND METHODS

18.2.1 Collection and Identification of Plant Material

Fresh leaves and bark were collected during the month of January 2014 and the seeds during the month of June 2014 from the Thiruvananthapuram district of Kerala, India. The taxonomic identity of the plant was confirmed from the Botanical survey of India, Southern Circle, Coimbatore, Tamil Nadu. The fresh plant materials were washed under running tap water to remove the surface pollutants and were air-dried under shade. Then they were separately homogenized into a fine powder using a mixer and were used for further studies.

18.2.2 Chemicals

2,2-diphenyl-1-picryl-hydrazyl (DPPH•), potassium persulfate, 2,2'-azinobis (3-ethylbenzothiozoline)-6-sulfonic acid disodium salt, 6-hydroxy-2,5,7,8-tetramethylchroman-2-carboxylic acid (Trolox), butylated hydroxy toluene (BHT), rutin, gallic acid, ferrous chloride, ferric chloride, hydrogen peroxide, ferrous ammonium sulfate, ethylene diamine tetra-acetic acid (EDTA) disodium salt, N- (1-naphthyl) ethylene diamine dihydrochloride, sodium nitroprusside, riboflavin, acetyl salicylate, pentazocine, indomethacin, carrageenan, carboxy methyl cellulose etc. were purchased from Himedia, Mumbai; SRL, Mumbai and Sigma Aldrich, Bengaluru. All the chemicals and solvents used were of analytical grade.

18.2.3 Preparation of Plant Extracts

The plant parts were extracted by the hot percolation method. The powdered plant materials such as leaves, bark and seeds were also packed in small thimbles and separately extracted successively with organic solvents such as petroleum ether, benzene and ethanol in the increasing order of their polarity using a Soxhlet apparatus. Each time before extracting with the next solvent, the thimble was air-dried. Finally, the material was macerated using hot water with constant stirring for 24 h and the water extract was also filtered using a Whatman No. 1 filter paper. The different solvent extracts were concentrated by rotary vacuum evaporator and then air-dried.

18.2.4 Extract Recovery Percent

The amount of extract recovered after successive extraction was weighed and the percentage yield was calculated by Equation 18.1:

$$\text{Extract Recovery Percent} = [\text{A/B}] \times 100 \tag{18.1}$$

where

A = Amount of extract (g),

B = Amount of plant sample (g)

18.2.5 Quantification Assays

18.2.5.1 Quantification of Total Phenolics

The total phenolics of the different plant extracts were determined according to the method described by Makkar (2003). In this method 50 μL of different plant extracts were taken into a series of test tubes and made up to 1 mL with distilled water. A test tube with 1 mL of distilled water served as the blank. Then, 500 μL of Folin–Ciocalteau Phenol reagent (1 N) was added to all the test tubes, including the blank. After 5 min, 2.5 mL of sodium carbonate solution (20%) was added to all the test tubes. The test tubes were vortexed well to mix the contents and incubated in dark for 40 min. The formation of blue color in the incubated test tubes indicated the presence of phenolics. Soon after incubation, the absorbance was read at 725 nm against the reagent blank. Gallic acid standard was also prepared and the results were expressed as gallic acid equivalents (GAE). The analyses were performed in triplicate.

18.2.5.2 Quantification of Tannins

The total phenolics contain both tannin and non-tannin phenolics. The amount of total tannins was calculated by subtracting the total non-tannin phenolics from total phenolics. For the determination of total non-tannin phenolics (Makkar 2003), 500 μL of each plant sample was incubated with 100 mg of polyvinyl polypyrrolidone (PVPP) and 500 μL of distilled water taken in a 2 mL eppendorf tube for 4 h at 4°C. After incubation, the eppendorf tubes were centrifuged at 4000 rpm for 10 min at 4°C. The supernatant contains only the non-tannin phenolics since the tannins would have been precipitated along with PVPP. The supernatant was collected and the non-tannin phenolics were determined by the same method described for the quantification of total phenolics. The analyses were also performed in triplicate, and the results were expressed in gallic acid equivalents. From these two results, the tannin content of the plant samples were calculated as follows:

$$\text{Tannins} = \text{Total phenolics}/\text{Non-tannin phenolics}$$

18.2.5.3 Quantification of Total Flavonoids

The flavonoid contents of all the extracts were quantified according to the method described by Zhishen et al. (1999). About 500 μL of all the plant extracts were taken in different test tubes and 2 mL of distilled water was added to each test tube. A test tube containing 2.5 mL of distilled water served as blank. Then, 150 μL of 5% $NaNO_2$ was added to all the test tubes, followed by incubation at room temperature for 6 min. After incubation, 150 μL of 10% $AlCl_3$ was added to all the test tubes including the blank. All the test tubes were incubated for 6 min at room temperature. Then 2 mL of 4% NaOH was added to all the test tubes which were then made up to 5 mL using distilled water. The contents in all the test tubes were vortexed well and they were allowed to stand for 15 min at room temperature. The pink colour developed due to the presence of flavonoids was read spectrophotometrically at 510 nm. Rutin was used as the standard for the quantification of flavonoids. All the experiments were done in triplicate and the results were expressed in rutin equivalents (RE).

18.2.6 *In Vitro* Antioxidant Assays

18.2.6.1 DPPH• Scavenging Activity

The antioxidant activity of the extract was determined in terms of hydrogen donating or radical scavenging ability using the stable radical DPPH, according to the method of Braca et al. (1958). Sample extracts at various concentrations were taken and the volume was adjusted to 100 μL with methanol. About 3 mL of a 0.004% methanolic solution of DPPH was added to the aliquots of samples and standards (BHT and rutin) and shaken vigorously. Negative control was prepared by adding 100 μL of methanol in 3 mL of methanolic DPPH solution. The tubes were allowed to stand for 30 min at 27°C. The absorbance of the samples and control were measured at 517 nm against the methanol blank. Radical scavenging activity of the samples was expressed as IC_{50} which is the concentration of the sample required to inhibit 50% of DPPH• concentration.

18.2.6.2 ABTS$^{•+}$ Scavenging Activity

The total antioxidant activity of the samples was measured by ABTS radical cation decolourisation assay according to the method of Re et al. (1999). ABTS$^{•+}$ was produced by reacting 7 mM ABTS aqueous solution with 2.4 mM potassium persulfate in the dark for 12–16 h at room temperature. Prior to assay, this solution was diluted in ethanol (about 1:89 v/v) and equilibrated at 25°C to give an absorbance of 0.700 ± 0.02 at 734 nm. About 1 mL of diluted ABTS solution was added to about 30 µL sample solution and 10 µL of Trolox (final concentration 0–15 µM) in ethanol. A test tube containing 1 mL of diluted ABTS solution and 30 µL of ethanol served as the negative control. All the test tubes were vortexed well and incubated exactly for 30 min at room temperature. After incubation, the absorbance of samples and standards (BHT and rutin) were measured at 734 nm against the ethanol blank. The results were expressed as the concentration of Trolox having equivalent antioxidant activity expressed as µM/g sample extracts.

18.2.6.3 Phosphomolybdenum Assay

The antioxidant activity of samples was evaluated by the green phosphomolybdenum complex formation according to the method of Prieto et al. (1999). An aliquot of 100 µL of samples and standards (BHT and rutin) were taken into a series of test tubes and were made up to 300 µL with methanol. About 300 µL methanol taken in a test tube was considered as the blank. All the test tubes were added with 3 mL of reagent solution (0.6 M sulphuric acid, 28 mM sodium phosphate and 4 mM ammonium molybdate) and vortexed well to mix the contents. The mouths of the test tubes were covered with foil and incubated in a water bath at 95°C for 90 min. After the samples were cooled to room temperature, the absorbance of the mixture was measured at 695 nm against the reagent blank. Ascorbic acid was used as the reference standard, and the results were expressed as milligrams of ascorbic acid equivalents per gram extract.

18.2.6.4 Ferric Reducing Antioxidant Power (FRAP) Assay

The antioxidant capacities of different extracts of samples were estimated according to the procedure described by Pulido et al. (2000). FRAP reagent (900 µL), prepared freshly and incubated at 37°C, was mixed with 90 µL of distilled water and 30 µL of test sample or methanol (blank). BHT and rutin were used as the standards. All the test tubes were incubated at 37°C for 30 min in a water bath. The final dilution of the test sample in the reaction mixture was 1/34. The FRAP reagent was prepared by mixing 2.5 mL of 20 mM TPTZ in 40 mM HCl, 2.5 mL of 20 mM $FeCl_3$. $6H_2O$ and 25 mL of 0.3 M acetate buffer (pH 3.6). At the end of incubation, the absorbance of the blue colour developed was read immediately at 593 nm against the reagent blank. Methanolic solutions of known $FeSO_4.7H_2O$ concentration ranging from 500 to 4000 µM were used for the preparation of the calibration curve. The parameter's equivalent concentration was expressed as the concentration of antioxidant having a ferric-TPTZ reducing ability equivalent to that of 1 mM $FeSO_4.7H_2O$.

18.2.6.5 Metal Chelating Activity

The chelating of ferrous ions by various extracts of *C. australe* was estimated by the method of Dinis et al. (1994). Initially, about 100 μL the extract samples were added to 50 μL of 2 mM $FeCl_2$ solution. Then the reaction was initiated by the addition of 200 μL of 5 mM ferrozine and the test tubes were vortexed well and left standing at room temperature for 10 min. The reaction mixture containing deionized water in place of sample was considered as the negative control. Absorbance of the solution was then measured spectrophotometrically at 562 nm against the blank (deionized water). EDTA was used as the standard metal chelating agent and the results were expressed as mg EDTA equivalents/g extract.

18.2.6.6 Hydroxyl Radical Scavenging Activity

The scavenging activities of solvent extracts of *C. australe* on hydroxyl radicals were measured according to the method of Klein et al. (1991). About 100 μL sample extracts and standards (BHT and rutin) were added with 1 mL of iron-EDTA solution (0.13% ferrous ammonium sulphate and 0.26% EDTA), 0.5 mL of EDTA solution (0.018%) and 1 mL of DMSO (0.85% v/v in 0.1 M phosphate buffer, pH-7.4). The reaction was initiated by adding 0.5 mL of ascorbic acid (0.22%) and incubated at 80–90°C for 15 min in a water bath. After incubation, the reaction was terminated by the addition of 1 mL of ice-cold TCA (17.5% w/v). About 3 mL of Nash reagent (75.0 g of ammonium acetate, 3 mL of glacial acetic acid and 2 mL of acetyl acetone were mixed and raised to 1 L with distilled water) was added and left at room temperature for 15 min. The reaction mixture without sample was used as a control. The intensity of the colour formed was measured spectroscopically at 412 nm against the phosphate buffer blank. The percentage hydroxyl radical scavenging activity is calculated by using Equation 18.2:

$$\%\ \text{Hydroxyl radical scavenging activity} = \left[(A_0 - A_1)/A_0\right] \times 100 \tag{18.2}$$

where,

A_0 is the absorbance of the control and

A_1 is the absorbance of the extract/standard.

18.2.6.7 Superoxide Radical Scavenging Activity

The assay was based on the capacity of various extracts to inhibit formazan formation by scavenging the superoxide radicals generated in riboflavin-light-NBT system (Beauchamp and Fridovich 1971). About 3 mL of reaction mixture containing 50 mM sodium phosphate buffer (pH-7.6), 20 μg riboflavin, 12 mM EDTA and 0.1 mg NBT was added to 100 μL sample solution, BHT and rutin. The reaction was started by illuminating the reaction mixture with samples for 90 seconds. The illuminated reaction mixture without sample was used as the negative control. Immediately after illumination, the absorbance was measured at 590 nm against the blank (unilluminated reaction mixture without plant sample). The scavenging activity on superoxide anion generation was calculated using Equation 18.3:

$$\text{Scavenging activity (\%)} = \left[(A_0 - A_1)/A_0\right] \times 100 \quad (18.3)$$

where,

A_0 is the absorbance of the control and
A_1 is the absorbance of the extract/standard.

18.2.6.8 Hydrogen Peroxide Scavenging Activity

The ability of the extracts to scavenge hydrogen peroxide was determined according to the method of Ruch et al. (1989). A solution of 2 mM hydrogen peroxide was prepared in 0.2 M phosphate buffer (pH-7.4) and its concentration was determined spectrophotometrically from absorption at 230 nm with molar absorbtivity 81 M^{-1} cm. About 600 μL hydrogen peroxide solution was added to plant samples (100 μg) and standards (BHT and rutin) and was then made up to 4 mL with phosphate buffer. The identical reaction mixture without the sample was taken as a negative control. All the test tubes were incubated for 10 min at room temperature. Absorbance of hydrogen peroxide at 230 nm was measured against the blank (phosphate buffer). The scavenging activity (%) was calculated as (Equation 18.4):

$$\text{Scavenging activity (\%)} = \left[(A_0 - A_1)/A_0\right] \times 100 \quad (18.4)$$

where,

A_0 is the absorbance of the control and
A_1 is the absorbance of the extract/standard.

18.2.6.9 Nitric Oxide Scavenging Activity

The procedure is based on the method (Sreejayan and Rao 1997) where sodium nitroprusside in aqueous solution at physiological pH spontaneously generates nitric oxide, which interacts with oxygen to produce nitrite mions that can be estimated using Greiss reagents. Scavengers of nitric oxide compete with oxygen leading to reduced production of nitrite ions. For the experiment, 3 mL of sodium nitroprusside (10 mM) in phosphate-buffered saline (0.2 M, pH 7.4) was mixed with 100 μL sample solution of various extracts and standards (BHT and rutin) and incubated at room temperature for 150 min. After the incubation period, 0.5 mL of Griess reagent (1% sulfanilamide, 2% H_3PO_4 and 0.1% N-(1-naphthyl) ethylene diamine dihydrochloride) was added to all the test tubes. The same reaction mixture without the sample was used as the negative control. The absorbance of the chromophore formed was read at 546 nm against the blank (phosphate buffer). The scavenging activity (%) was calculated as (Equation 18.5):

$$\text{Scavenging activity (\%)} = 18.5\left[(A_0 - A_1)/A_0\right] \times 100 \quad (18.5)$$

where,

A_0 is the absorbance of the control and
A_1 is the absorbance of the extract/standard.

18.2.7 Statistical Analyses

All the experiments were done in triplicate and the results were expressed as Mean ± SD. The data were statistically analysed using one-way ANOVA followed by Duncan's test for antioxidant studies. Mean values were considered statistically significant when $P < 0.05$.

18.3 RESULTS AND DISCUSSION

18.3.1 Extract Recovery Percent

The maximum percentage yield of leaf, bark and seed extracts of *C. austral* was obtained in ethanol for all the parts. Among the different parts used, extract recovery percent was found to be higher in bark ethanol (22.5%). In the case of leaf ethanol, the extract yield was 15.3 g/100 g of dried powder, respectively. The yield percentage in petroleum ether and benzene for all the used parts was found to be very low compared to polar solvents.

18.3.2 Quantification Assays

18.3.2.1 Quantification of Total Phenolics

The amount of total phenolics of the different extracts of leaves, bark and seeds of *C. australe* were analysed and are presented in Table 18.1. The total phenolics were found to be higher in the ethanol extract of bark (162.38 mg GAE/g extract) obtained by Soxhlet extraction. Among all the extracts obtained from the ethanol extracts of leaves, bark and seeds, ethanol extract of bark showed a higher phenolic content compared to all the other solvent extracts. The Soxhlet extraction was found to be effective in the extraction of phenolics from the different parts of *C. australe.* The seed extracts showed a very low content of phenolics in hot percolation methods (46.15 g GAE/g extract, respectively). The total phenolics of the different parts followed the trend, bark > leaf > seed. The phenolic contents of ethanol extracts of leaf were 119.21 mg GAE/g extract for Soxhlet extraction. The hot water extract of leaves and bark showed phenolic content of 106.19 and 100.48 mg GAE/g extract, respectively.

The multiple hydroxyl groups in the chemical structure of polyphenols make them ideal for free radical scavenging reactions and as metal chelating agents (Lopez et al. 2003). The arrangement of the hydroxyl groups in the phenolic molecule is also important for antioxidant reactions (Moure et al. 2001). The higher amount of phenolics in the ethanol extracts of leaves and bark could be due to the higher solubility of phenolics and other aroma compounds. It has been already investigated in many plant species that the total amount of phenolics could be significantly contributed to the antioxidant capacity of that species (Arunachalam et al. 2011). Therefore, the

TABLE 18.1
Total Phenolic Tannin Flavonoid Content of *C. australe*

Solvents	Phenolic (mg GAE/g extract)			Tannin Content (mg GAE/g extract)			Flavonoid Content (mg RE/g extract)		
	Leaf	Bark	Seed	Leaf	Bark	Seed	Leaf	Bark	Seed
Pet ether	42.86 ± 2.38	60.00 ± 3.60	30.64 ± 1.24	15.08 ± 1.44	10.32 ± 2.12	12.44 ± 1.89	23.24 ± 1.33	24.14 ± 2.03	9.11 ± 2.14
Benzene	53.33 ± 4.15	72.86 ± 4.07	36.28 ± 1.35	18.24 ± 2.14	14.63 ± 2.24	13.97 ± 2.17	28.68 ± 1.81	26.37 ± 1.70	12.89 ± 2.34
Ethanol	119.21 ± 3.82*	162.38 ± 3.33*	46.15 ± 2.40	10.59 ± 1.15	26.98 ± 1.08*	9.51 ± 1.55	70.22 ± 1.68*	83.33 ± 1.15*	31.78 ± 0.38
Water	106.19 ± 2.65*	100.48 ± 3.60*	35.38 ± 6.04	19.52 ± 2.11*	21.11 ± 1.97*	15.13 ± 1.80	66.44 ± 0.38*	63.78 ± 2.04*	36.44 ± 1.39

Note: Values are mean of triplicate determination (n = 3) ± standard deviation, GAE = Gallic Acid Equivalents.
*Statistically significant at $P < 0.05$.

higher amount of phenolics in all the parts of *C. australe* can be taken as a good indication of its higher antioxidant capacity.

18.3.2.2 Quantification of Tannins

The solvent extracts obtained from Soxhlet of leaves, bark and seeds of *C. australe* were analysed for their tannin content and the results are shown in Table 18.1. The tannin contents were found to be higher in ethanol extracts of bark prepared by the hot extraction method and were 32.44 mg GAE/g extract. The tannin content was also found to be very low in the ethanol extract of seeds obtained from hot (9.51 mg GAE/g extract) percolation methods. The amount of total phenolics and total tannins revealed that the proportion of free phenolics was higher in ethanol extracts of leaves and bark compared to seeds. The tannin content of ethanol extracts of different parts followed the trend bark > leaf > seed for hot percolation methods.

The greater amount of tannins in the extracts of bark of *C. australe* can be due to the higher polymerization of existing polyphenolic compounds. Recently, it has been reported that the high molecular weight phenolics, such as tannins, have more ability to quench/scavenge free radicals. Tannins are constituents of several drugs because of their astringent property. They are used in the treatment of haemorrhoids, diarrhoea, dysentery and leucorrhoea, and as a useful medicine for the throat (Allport 1970). Since the tannin content was low in seeds, there will not be any unfavourable reaction associated with tannin and anti-nutritional factors. Apart from that, the high amounts of tannins in bark and leaves of *C. australe* may enhance the free radical scavenging activity of its extracts.

18.3.2.3 Quantification of Flavonoids

The flavonoid content in different parts of *C. australe* were analysed and are presented in Table 18.1. Among the extracted samples, the ethanol extract of bark was found to have appreciable amount of flavonoid content (83.33 mg RE/g extract). The leaf water extracts showed higher flavonoid content in hot percolation methods. However, the seed extracts didn't show any appreciable amount of flavonoid content. The flavonoid content of ethanol and water extracts prepared from Soxhlet apparatus were 70.22 and 66.44 mg RE/g extract, respectively. The flavonoid content of all the parts of *C. australe* was found to be in the order of bark > leaf > seed in most of the extracts prepared by the Soxhlet apparatus.

Flavonoids are one of the most diverse and important groups of natural phenolics. In the food processing industry, flavonoids have been shown to inhibit heat or chemical-initiated lipid peroxidation as well as chelating metallic and super oxide ions (Kim et al. 1990). Tkahama et al. (1984) reported that flavonoids are potent inhibitors of molecular oxygen (O_2), thus acting as free radical scavengers (antioxidant). Flavonoids also scavenge other free radicals such as OH and NO_2 (Bors et al. 1990). Flavonoids suppress the effects of active oxygen species (H_2O_2 and O^{2-}) in many other vulnerable biological systems (Nakayama et al. 1993). Flavonoids are used as natural antioxidants in food, medicinal and nonnutritive plant materials due to their ability to inhibit and scavenge reactive oxygen species (Kim et al. 1990; Larson 1988). Therefore, the quantification of the flavonoid content of plant parts is important, since it can be correlated with the anticancer and radical scavenging activity of the plant. Since *C. australe* possess good flavonoid content in its bark and leaves, it could be assumed that it can have a higher free radical scavenging activity, which involves the transfer of electron or hydrogen atoms from flavonoids to free radicals. Moreover, the higher number of flavonoids in the Soxhlet-extracted samples suggests that hot percolation is more effective in extracting the flavonoids from the leaves and bark of *C. australe* rather than simple maceration.

18.3.3 *In Vitro* Antioxidant Assays

18.3.3.1 DPPH• Scavenging Activity

The DPPH radical scavenging activities of different extracts of leaves, bark and seeds of *C. australe* are shown in Figure 18.1. Among all the parts analysed, the ethanol extracts of leaves (96.90 µg/mL) and bark (158.74 µg/mL) showed better IC_{50} values for DPPH· radical scavenging activities compared to other solvent extracts. Among extracts obtained by Soxhlet extraction, the DPPH radical scavenging activities followed the trend as leaf > bark > seed. The IC_{50} of ethanol extracts of seeds have shown that they have less free radical scavenging activity compared to leaf and bark. The IC_{50} of standard natural antioxidant rutin was found to be much better than that of plant extracts whereas that of synthetic antioxidant BHT was found to be 52.91 µg/mL and was little comparable with the activity of ethanol leaf extract obtained by the hot percolation method.

The DPPH radical has been widely used to test the ability of compounds as free radical scavengers or hydrogen donors to evaluate the antioxidant activity of plant extracts and foods (Soares et al. 1997). The antiradical scavenging activity of different extracts of *C. australe* would be related to the nature of phenolics, thus contributing to their electron transfer/hydrogen donating ability (Blios 1958). The enhanced activity of ethanol extract may be due to the polar nature and extracting ability of phenolic compounds from the parts under study. On the other hand, the DPPH radical scavenging efficiency of extracts from *C. australe* might have also been partly attributed to Millard reaction products other than phenolic constituents because they also effectively participate as radical scavengers (Brand-Williams et al. 1995). Extracts obtained from the cold percolation technique have shown lower activity than those obtained from hot percolation method. It may be because of failure to extract thermo labile compounds which play a major role in antioxidant activity. Even though the

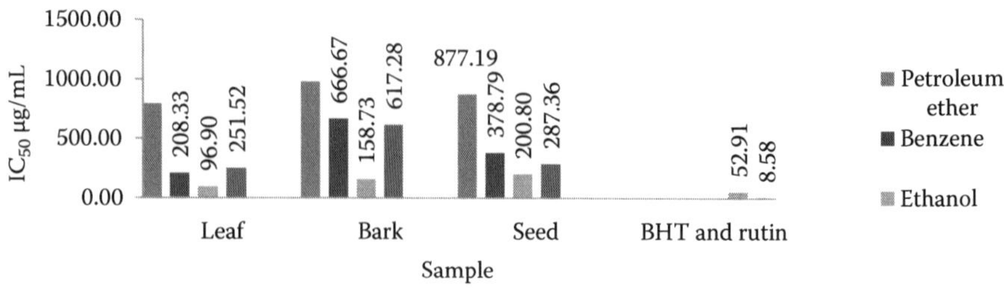

FIGURE 18.1 DPPH radical scavenging activities of *C. australe*.

DPPH radical scavenging activities of different extracts of the plant seemed to be below average, it is not evidence for plant has reduced amount of radical scavenging activities. Because, however DPPH is stable and long-lasting nitrogen radical, unlike radicals present in living organisms, it has no similarity to the highly reactive and transient peroxyl radicals that are involved in lipid peroxidation (Huang et al. 2005). Thus, antioxidants that react quickly with peroxyl radicals may react slowly or may be inert to the DPPH radical. Steric accessibility is a major determinant of the reaction mechanism; hence small molecules have higher apparent antioxidant capacity due to their better access to the DPPH radical site (Prior et al. 2005). Therefore, further antioxidant assays should be carried out to assess the antiradical ability of the extracts.

18.3.3.2 ABTS$^{\bullet+}$ Scavenging Activity

The results of ABTS$^{\cdot+}$ cation radical scavenging activities of different extracts of leaves, bark and seeds of *C. australe* are presented in Table 18.2. The ethanol extracts of leaf and bark obtained from the Soxhlet apparatus showed higher ABTS cation radical scavenging activities (6728.56 and 6184.12 µM Trolox equivalents/g extract, respectively) as compared to that of other solvent extracts. However, the seed was shown to have lower total antioxidant activity in its solvent extracts. The ABTS scavenging activity of leaf ethanol extract was found to have comparable activity with the standard synthetic antioxidant BHT (8423.95 µM Trolox equivalents/g extract). As in previous results, the extracts obtained through Soxhlet extraction were found to perform better than that of maceration. The ethanol and water extracts of *C. australe* obtained by Soxhlet extraction shown to have total antioxidant activity ranging from 3128.68 to 6728.56 µM Trolox equivalents/g extract. The higher scavenging activity of bark was observed in its ethanol fraction and was 5386.5 µM Trolox equivalents/g extract. In the case of hot water, bark exhibited higher activity (5542.13 µM Trolox equivalents/g extract) than leaves and seeds.

The total antioxidant activity of *C. australe* extracts with a higher phenolic content seems to be efficient for functioning as potential nutraceuticals or antioxidants when they are ingested along with nutrients. The ABTS radical is soluble in both aqueous and organic solvents, is not affected by ionic strength, and can be used to measure the antioxidant capacity of hydrophilic and lipophilic compounds in test samples (Arno 2000; Roginsky and Lissi 2005). The radical is suitable for evaluating the antioxidant capacity of phenolics due to their comparatively lower redox potentials (0.68 V). Many phenolic compounds can thus react with the ABTS radicals because of this thermodynamic property (Osman et al. 2006). Apart from these, Hagerman et al. (1998) have reported that the high molecular weight phenolics (tannins) has more ability to quench free radicals (ABTS$^{\cdot+}$) and that effectiveness depends on the molecular weight, the number of aromatic rings and nature of hydroxyl group substitution versus those of specific groups. On the other hand, the formation of tannin protein complexes, both in soluble and insoluble complexes as the result of food processing, have also been shown to be potential free radical scavenger and radical sinks (Riedl and Hagerman 2001). This study indicated that the extracts of *C. australe* have a strong hydrogen donating ability and could serve as free radical scavengers by acting as primary antioxidants.

TABLE 18.2
ABTS Cation Radical Scavenging Activities and Phosphomolybdenum Assay of *C. australe*

	ABTS Cation Radical Scavenging Activities of *C. australe* (μM TE/g extract)			Phosphomolybdenum Assay of *C. australe* (mg AAE/g extract)		
Solvents	**Leaf**	**Bark**	**Seed**	**Leaf**	**Bark**	**Seed**
Pet. ether	2613.37 ± 95.26	2757.55 ± 91.87	2458.22 ± 102.33	108.15 ± 10.96	117.04 ± 7.98	130.37 ± 3.53
Benzene	3269.45 ± 120.31	3459.97 ± 89.66	2994.31 ± 96.51	126.81 ± 9.32	107.04 ± 9.40	150.74 ± 5.13
Ethanol	6728.56 ± 105.39*	6184.12 ± 95.21*	4831.28 ± 94.23	297.22 ± 12.29*	254.63 ± 3.70*	123.33 ± 4.44
Water	5126.75 ± 116.42*	5542.13 ± 112.51*	3128.68 ± 103.32	139.26 ± 4.49	122.96 ± 9.47	54.44 ± 2.42
BHT		8423.95 ± 96.32*			451.85 ± 32.78*	
Rutin		9942.69 ± 109.84*			305.93 ± 5.25*	

Note: Values are mean of triplicate determination (n = 3) ± standard deviation, TE = Trolox Equivalents.
*Statistically significant at $P < 0.05$.

18.3.3.3 Phosphomolybdenum Assay

The antioxidant activity of different solvent extracts of leaves, bark and seeds of *C. australe* were analysed by the phophomolybdenum assay and shown in Table 18.2. Among the different parts used, the ethanol extract of leaves showed higher activity compared to the extracts of bark and seeds. Among the Soxhlet-extracted samples, ethanol extracts of leaves and bark showed higher antioxidant activity by the reduction of phosphomolybdenum complex and were 297.22 and 254.63 mg AAE/g extract, respectively. The seeds of *C. australe* showed lower phosphomolybdenum reduction for all its extracts. The antioxidant activities of extracts obtained by Soxhlet extraction showed the same trend as leaf > bark > seed.

Phosphomolybdenum assay was successfully used to determine the ability of extracts to reduce Mo(VI) to Mo(V) and subsequent formation of green phosphate/Mo(V) complex at an acid pH. It was also used to quantify vitamin E in seeds, and being simpler and independent of other antioxidant measurements commonly employed, it was decided to extend its application to plant extracts (Prieto et al. 1999). Moreover, it is a quantitative one, since the antioxidant activity is expressed as the number of equivalents of ascorbic acid. Thus, the total antioxidant capacity observed for the extracts of *C. australe* can be correlated with their free radical scavenging activity equivalent to that of natural antioxidant ascorbic acid. The reduction of Mo(VI) to Mo(V) by the leaf and bark extracts of *C. australe* may be due to the electron transfer or hydrogen ion transfer by the bioactive compounds, specifically phenolics and flavonoids present in the respective parts.

18.3.3.4 Ferric Reducing Antioxidant Power Assay

Antioxidant potential of leaves, bark and seeds of *C. australe* were estimated from their ability to reduce TPTZ-Fe(III) complex to TPTZ Fe(II) and are given in Table 18.3. Among the different solvent extracts obtained from Soxhlet extraction, ethanol extracts of leaves (46.64 mM Fe (II)/mg extract) and seeds (41.69 mM Fe (II)/mg extract) showed higher ferric reducing power than bark.

In the case of hot water extracts of different parts of *C. australe,* leaves showed comparable ferric-reducing antioxidant activity. The standard antioxidants BHT and rutin was found to show antioxidant power which was far better from the plant extracts. The macerated seed samples were observed for their lower antioxidant power which was even lower than that of petroleum ether and benzene extracts for leaf. It was also found that the ferric reducing power of *C. australe* followed the trend of leaf > seed > bark.

Antioxidants can be explained as reductants, and the inactivation of oxidants by reductants can be described as redox reactions in which one reaction species (oxidant) is reduced at the expense of the oxidation of another antioxidant. The ferric-reducing antioxidant power assay measures the antioxidant effect of any substance in the reaction medium as a reducing ability. Yen and Duh (1993) and Siddhuraju et al. (2002) have reported that the reducing power of bioactive compounds, mainly low and high molecular phenolics, was associated with antioxidant activity, specifically free radical scavenging. It has also been proved that the potential antioxidants, through *in vitro* ferric reducing antioxidant power assay, increased the total antioxidant capacity of blood plasma (Serafini et al. 2003). Results have revealed

TABLE 18.3
FRAP Assay and Metal Chelating Activities of *C. australe*

Solvents	FRAP Assay of *C. australe* (mM Fe (II)/mg extract)			Metal Chelating Activities of *C. australe* (mg EDTA equivalents/g extract)		
	Leaf	**Bark**	**Seed**	**Leaf**	**Bark**	**Seed**
Pet. Ether	21.20 ± 0.62	15.55 ± 0.56	15.32 ± 0.77	31.03 ± 3.27	16.97 ± 2.83	12.90 ± 1.95
Benzene	29.43 ± 0.87	20.54 ± 2.11	18.36 ± 1.00	33.41 ± 1.82	26.82 ± 2.01	23.07 ± 2.98
Ethanol	46.64 ± 2.37*	34.43 ± 0.46*	41.68 ± 0.78*	44.21 ± 1.49*	27.66 ± 1.50	29.27 ± 1.70
Water	40.89 ± 1.43*	29.82 ± 4.29	10.26 ± 0.79	117.55 ± 2.21*	52.95 ± 2.56*	30.37 ± 3.05
BHT		181.92 ± 4.03*			–	
Rutin		433.19 ± 25.91*			–	

Note: Values are mean of triplicate determination (n = 3) ± standard deviation.
*Statistically significant at $P < 0.05$.

that the extracts with higher amount of phenolics have higher reducing power. However, extracts from the cold percolation technique have shown less activity than that of extracts obtained from hot percolation, because of its low ability to extract phenolics from plant parts. Thus, the ferric reducing power of different extracts of *C. australe* reveals that there are compounds in the ethanolic extracts which have high affinity to ferrous ions and thereby quench/scavenge them through redox reactions.

18.3.3.5 Metal Chelating Activity

The metal chelating activities of leaves, bark and seeds of *C. australe* were calculated and are presented in Table 18.3. Among the different solvent extracts of the parts, hot water showed better chelating ability in all the parts compared to the other solvents. The higher metal chelating activity was shown by the hot water extract of leaves (117.55 mg EDTA equivalents/g extract). The metal chelating abilities of different parts of *C. australe*, followed the trend of leaf > bark > seed for hot extraction methods. The metal chelating activities of water and ethanol extract of bark prepared through Soxhlet extraction and maceration was 52.95 and 27.66 mg EDTA equivalents/g extract, respectively. The petroleum ether and benzene extracts of leaf also showed comparable activity and were 31.03 and 33.41 mg EDTA equivalents/g extract, respectively. On the other hand, the seed extracts didn't show any chelating effect on metal.

Iron is an essential metal for normal physiology, but if it undergoes the Fenton reaction, this reduced metal may from highly harmful hydroxyl radicals and thereby contributing to oxidative stress (Hippeli and Elstner 1999). Metal chelating capacity was significant as they reduced the concentration of the catalysing transition metal in lipid peroxidation (Duh et al. 1999). It was already reported that chelating agents which form σ- bonds with a metal are effective as secondary antioxidants because they reduce the redox potential, thereby stabilizing the oxidized form of the metal ion (Gordon 1990). Antioxidants inhibit interaction between metals and lipids through the formation of insoluble metal complexes with ferrous ion. Hence, the metal chelating assay of *C. australe* reveals that the water extract for all the parts in both extraction methods have shown effective activity, suggesting that its action as antioxidant may be related to its sequestering of Fe^{2+} ions that may otherwise catalyse Fenton type reactions or participate in metal catalysed hydroperoxide decomposition reactions.

18.3.3.6 Hydroxyl Radical Scavenging Activity

The scavenging activities of different plant extracts on hydroxyl radicals were evaluated and are shown in Figure 18.2. Among the extracts prepared, the higher hydroxyl radical scavenging activity was shown by the ethanol extract of leaf (60.86%). In case of bark and seeds, the ethanol extracts obtained from a Soxhlet apparatus showed higher activity was noted as 57.71% and 51.29% for bark and seeds, respectively. But, the standards, synthetic antioxidant BHT and the natural antioxidant rutin were found to have good scavenging activities (87.23% and 98.11%) even at 50 μg/mL. Among the different parts of *C. australe*, the scavenging activities of ranged from 35.24% to 66.21%.

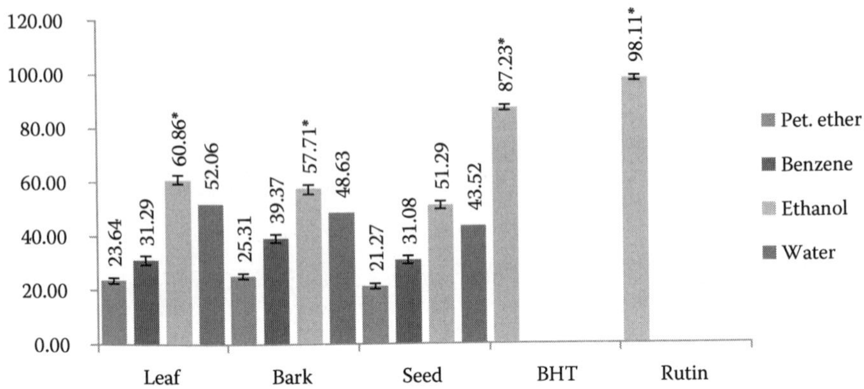

FIGURE 18.2 Hydroxyl radical scavenging activities of *C. australe*. Values are mean of triplicate determination (n = 3) ± standard deviation; statistically significant at $p < 0.05$.

Hydroxyl radical can be formed from superoxide and hydrogen peroxide in the presence of metal ions, such as copper and iron. Among the oxygen radicals, hydroxyl radicals are the most reactive. and induce severe damage to adjacent biomolecules (Sakanaka et al. 2005). Hydroxyl radicals can react with lipid, polypeptides, proteins and DNA, especially thiamine and guanosine (Siddhuraju and Becker 2007). When a hydroxyl radical reacts with aromatic compounds, it can add across a double bond, resulting in hydroxyl cyclohexadienyl radical. The resulting radical can undergo further reactions, such as reactions with oxygen to give a peroxyl radical or decompose to phenoxyl type radicals by water elimination (Lee et al. 2004). Hagerman et al. (1998) have also reported that high molecular weight, and the proximity of many aromatic rings and hydroxyl groups are more important for the free radical scavenging by tannins than their specific functional groups. Thus, the hydroxyl radical scavenging of *C. australe* ethanolic extracts can be directly related to the amount of phenolics and tannins present in the sample, since the potential scavenging abilities of phenolic substances might be due to the active hydrogen donor ability of hydroxyl substitution.

18.3.3.7 Superoxide Radical Scavenging Activity

The superoxide anion radical scavenging activities of leaves, bark and seeds of *C. australe* are shown in Figure 18.3. The extracts were found to be an efficient scavenger of superoxide radicals generated in riboflavin-NBT-light system *in vitro*. The ethanol extracts made by the Soxhlet extraction of all the parts showed higher superoxide radical scavenging activities compared to other solvent extracts at a concentration of 100 µg/mL. Among them, ethanol extract of leaves obtained by the hot percolation method was higher (69.84%). In the case of bark and seeds, the higher activities were shown by the ethanol extracts from Soxhlet extraction and were 56.28% and 33.12%, respectively. The superoxide radical scavenging activities of different parts of *C. australe* followed the trend of leaf > bark > seed. The scavenging activities of hot water extracts of leaves and bark were 48.31% and 41.36%, respectively.

Superoxide radicals are known to be a very harmful species to cellular components, and as a precursor of more reactive oxygen species (Halliwell and Gutteridge

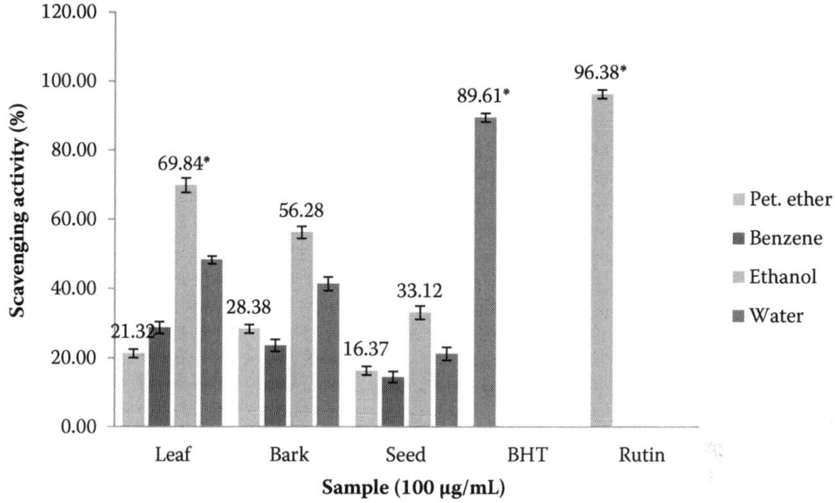

FIGURE 18.3 Superoxide radical scavenging activities of *C. australe*. Values are mean of triplicate determination (n = 3) ± standard deviation; statistically significant at $p < 0.05$.

1985). Scavenging superoxide radicals is necessary because they acts as the precursor for other major ROSs, like hydrogen peroxide, hydroxyl and singlet oxygen (Lee et al. 2004). Numerous biological reactions generate superoxide radicals, which is a highly toxic species. Although they cannot directly initiate lipid oxidation, superoxide radical anions are potent precursors of highly reactive species, such as hydroxyl radicals, and thus the study of scavenging of this radical is important (Kannat et al. 2007).

Since, the ethanol extract obtained by the Soxhlet extraction of leaves and bark of *C. australe* showed an appreciable percentage of scavenging activity against superoxide radicals, it can be used against adverse effects caused by superoxide radicals in the body. The active principles in the plant extracts may eliminate the radicals by its reduction to attain the octant stage, or through the formation of water molecules.

18.3.3.8 Hydrogen Peroxide Scavenging Activity

The ability to scavenge hydrogen peroxide by various extracts of *C. australe* is shown in Figure 18.4. The higher percentages of scavenging activities were found at a concentration of 100 μg/mL in the ethanol and water extracts of bark (76.23% and 65.74%, respectively) prepared by hot percolation method. But, BHT and rutin used as standards were found to have good scavenging activity (72.39% and 91.62%) even at a concentration of 50 μg/mL. The ethanol and water extracts of leaf made through hot percolation method have shown hydrogen peroxide scavenging activities as 59.63% and 52.81%, respectively. The seed extracts could not make any significant scavenging activity against hydrogen peroxide. The scavenging activities of ethanol and water extracts obtained from hot percolation methods for all the parts ranged from 19.61% to 76.23%. The scavenging activities followed the trend of bark > leaf > seed for the hot percolation method.

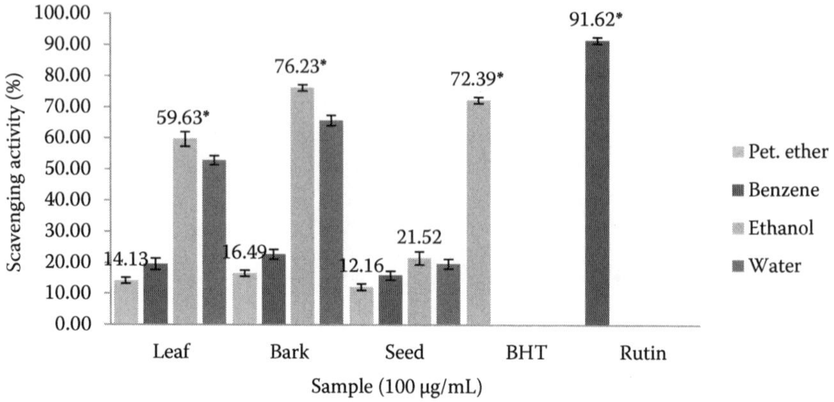

FIGURE 18.4 Hydrogen peroxide scavenging activities of *C. australe*. Values are mean of triplicate determination (n = 3) ± standard deviation; statistically significant at $p < 0.05$.

Hydrogen peroxide is not very reactive, but sometimes it is toxic to cells because it may give rise to hydroxyl radicals in the cells (Halliwell 1991). Therefore, removing H_2O_2 is very important for the antioxidant defence in cell or food systems. Dietary polyphenols have also been shown to protect mammalian and bacterial cells from cytotoxicity induced by hydrogen peroxide, especially compounds with the orthodihydroxy phenolic structure, quercetin, catechin, gallic acid ester and caffeic acid ester (Nakayama 1994). The results can be correlated to the ability of polar organic solvents to extract phenolics, tannins and flavonoids of bark, which showed maximum scavenging of hydrogen peroxide when compared to other parts. The lower percentage of the activity of samples obtained by maceration reveals that the active components responsible for hydrogen peroxide reduction may be heat-stable compounds, which can be extracted by the hot percolation methods such as Soxhlet extraction. Therefore, the ethanol extracts of bark and leaf *C. australe* can be used as a potent hydrogen peroxide scavenger in body systems.

18.3.3.9 Nitric Oxide Scavenging Activity

The nitric oxide scavenging activities of various sample extracts of *C. australe* were analysed and the percentages of scavenging activities are shown in Figure 18.5. The scavenging activities of Soxhlet-extracted ethanol samples of leaf and bark were 49.16% and 57.68%, respectively. In case of seeds, a higher scavenging activity was observed in the hot water extract, which was 41.32%. The scavenging activities of the standards BHT and rutin were found to be of 91.13% and 95.87%, respectively at a concentration of 50 µg/mL. The hydrogen peroxide scavenging activities of the extracts of petroleum ether and benzene for all the parts were found to be very low when compared to ethanol and water.

Nitric oxide or reactive nitrogen species, formed during their reaction with oxygen or with superoxides, such as NO_2, N_2O_4, N_3O_4, NO_3^- and NO_2^+, are very reactive. These compounds are responsible for altering the structural and functional behaviour of many cellular components. The plant products may have the property

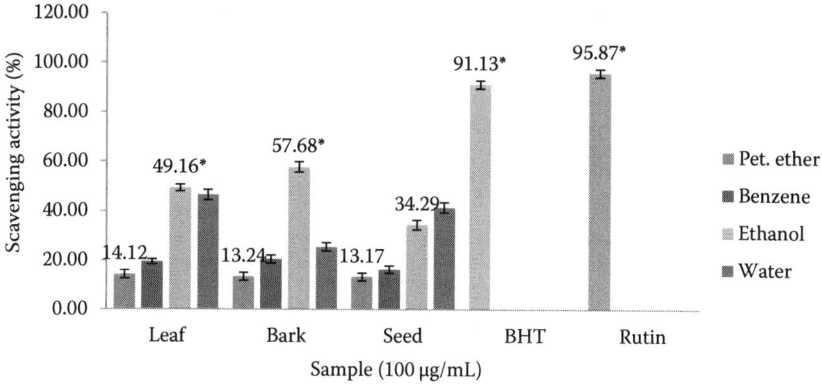

FIGURE 18.5 Nitric oxide scavenging activities of *C. australe.* Values are mean of triplicate determination (n = 3) ± standard deviation; statistically significant at $p < 0.05$.

to counteract the effect of NO formation, and in turn may be of considerable interest in preventing the ill effects of excessive NO generation in the human body. Further, the scavenging activity may also help to arrest the chain of reactions initiated by excess generation of NO that are detrimental to the human health. Nitric acid is also implicated in inflammation, cancer and other pathological conditions (Moncada et al. 1991). Since the ethanol extracts of leaf and bark of *C. australe* showed higher scavenging activity, it is clear that it can be used for reducing the deleterious effects caused by the reactive nitrogen species in the human body. Thus, it is clear that the heat-liable nitric oxide scavenger present in the bark and leaf of *C. australe* becomes inactive during the hot percolation extraction.

18.4 CONCLUSION

In conclusion, these findings justify that *Castanospermum australe* can be a valuable natural antioxidant source, which could provide potential nutraceuticals for human health. Further, detailed exploration, chemical studies and screening for medicinal properties will provide cost-effective and reliable sources of medicine for the welfare of humanity.

REFERENCES

Allport, Noel. 1970. Drugs containing tannins. *Chemistry and Pharmacy of Vegetable Drugs.*

Aremu, A. O., Amoo, S. O., Ndhlala, A. R., Finnie, J. F., and van Staden, J. 2011. Antioxidant activity, acetylcholinesterase inhibition, iridoid content and mutagenic evaluation of *Leucosideasericea. Food and Chemical Toxicology* 49(5): 1122–1128.

Arno, M. B. 2000. Some methodological problems in the determination of antioxidant acitivity using chromogen radicals: a practical case. *Trends in Food Science & Technology* 11: 419–421.

Arunachalam, K., Saravanan, S., and Parimelazhagan, T. 2011. Nutritional analysis and antioxidant activity of Palmyrah (*Borassus flabellifer* L.) seed embryo for potential use as food source. *Food Science and Biotechnology* 20(1): 143–149.

Beauchamp, C. and Fridovich, I. 1971. Superoxide dismutase: Improved assays and an assay applicable to acrylamide gels. *Analytical Biochemistry* 44: 276–277.

Blios, M. S. 1958. Antioxidants determination by the use of a stable free radical. *Nature* 4617: 1199–1200.

Bors, W., Heller, W., Michel, C., and Saran, M. 1990. Flavonoids as antioxidants: determination of radical-scavenging efficiencies. *Methods Enzymology.* 186: 343–355.

Braca, A., Tommasi, N. D., Bari, L. D., Pizza, C., Politi, M., and Morelli, I. 2001. Antioxidant principles from *Bauhinia terapotensis. Journal of Natural Products* 64: 892–895.

Brand-Williams, W., Cuvelier, M. E., and Berset, C. 1995. Use of free radical method to evaluate antioxidant activity. *Lebensm Wiss Technol Journal* 28: 25–30.

Dinis, T. C. P., Madeira, V. M. C., and Almeida, L. M. 1994. Action of phenolic derivatives (acetoaminophen, salycilate and 5-aminosalycilate) as inhibitors of membrane lipid peroxidation and as peroxyl radical scavengers. *Archives of Biochemistry and Biophysics* 315: 161–169.

Duh, P. D., Tu, Y. Y., and Yen, G. C. 1999. Antioxidant activity of water extract of Harng Jyur *Chrysanthemum morifolium* Ramat *Lebensm Wiss Technol Journal* 32: 269–277.

Finkel, T. and Holbrook, N. J. 2000. Oxidants, oxidative stress and the biology of ageing. *Nature* 408: 239–247.

Gordon, M. H. 1990. The mechanism of antioxidant action *in vitro. Food Antioxidants*, Ed. B. J. F. Hudson, London: Elsevier Applied Science, 1–18.

Gurib-Fakim, A. 2006. Medicinal plants: Traditions of yesterday and drugs of tomorrow. *Molecular Aspects of Medicine* 27: 1–93.

Hagerman, A. E., Riedl, K. M., Jones, G. A., Sovik, K. N., Ritchard, N. T., Hartzfeld, P. W. and Riechel, T. L., 1998. High molecular weight plant polyphenolics (Tannins) as biological antioxidants. *J. Agric. Food Chem.* 46: 1887–1892.

Halliwell, B. 1991. Reactive oxygen species in living systems: Source, Biochemistry and role in human disease. *American Journal of Medicine* 91: 14–22.

Halliwell, B. and Gutteridge, J. M. C. 1985. In free radicals , aging and disease. *Free radicals Biology and medicine.* (2nd ed) Oxford Calrendron Press. pp: 279–315.

Halliwell, B., Gutteridge, J. M. C., and Cross, C. E. 1992. Free radicals, antioxidants, and human disease: where are we now? *Journal of Laboratory and Clinical Medicine* 119: 598–620.

Hippeli, S. and Elstner, E. F. 1999. Transition metal ion-catalyzed oxygen activation during pathogenic processes. *FEBS Letters* 443: 1–7.

Huang, D., Ou, B., and Prior, R. L. 2005. The chemistry behind antioxidant capacity assays. *Journal of Agricultural and Food Chemistry* 53: 1841–1856.

Jacob, R. A. 1994. Nutrition, health, and antioxidants. *InForm* 5: 1271–1275.

Kannat, S. R., Chander, R., and Sharma, A. 2007. Antioxidant potential of mint (*Mentha spicata* L.) in radiation-processed lamb meat. *Food Chemistry* 100: 451–458.

Kim, J. Y., Germolec, D. R., and Luster, M. I. 1990. Panax ginseng as a potential immunomodulator, studies in mice. *Immunopharmacology and Immunotoxicology* 12(2): 257–76.

Klein, S. M., Cohen, G., and Cederbaun, A. I. 1981. Production of formaldehyde during metabolism of dimethyl sulphoxide by hydroxyl radical generating system. *Biochemistry* 20: 6006–6012.

Kliebenstein, D. J. 2004. Secondary metabolites and plant/environment interactions: A view through *Arabidopsis thaliana* tinged glasses. *Plant, Cell & Environment* 27(488): 675–684.

Krishnaiah, D., Sarbatly, R., and Nithyanandan, R. 2010. A review of the antioxidant potential of medicinal plant species. *Food and Bioproducts Processing* 10(1016): 04–08.

Larson, R. A. 1988. The anti-oxidants of higher plants. *Phytochemistry* 27(4): 969–978.

Lee, J., Koo, N., and Min, B. D. 2004. Reactive oxygen species, aging and antioxidative nutraceuticals. *Comprehensive Reviews in Food Science and Food Safety* 3: 21–33.

Li, Y. and Trush, M. 1994. Reactive oxygen-dependent DNA damage resulting from the oxidation of phenolic. *Cancer Research* 54: 1895s–1898s.

Lopez, M., Martínez, F., del Valle, C., Ferrit, M., and Luque, R., 2003. Study of phenolic compounds as natural antioxidants by a fluorescence method. *Talanta* 60: 609–616.

Makkar, H. P. S. 2003. *Quantification of Tannins in Tree and Shrub Foliage: A Laboratory Manual.* Dondrecht, The Netherlands: Kluwer Academic Publishers.

Meyer, A. S., Heinonen, M., and Frankel, E. N. 1998. Antioxidant interactions of catechin, cyanidin, cafeic acid, quercetin, and ellagic acid on human LDL oxidation. *Food Chemistry* 61: 71–75.

Middleton, J. E., Kandaswami, C., and Theoharides, T. C. 2000. The effects of plant flavonoids on mammalian cells: Implication for inflammations, heart disease and cancer. *Pharmacological Reviews* 52: 673–751.

Moncada, A., Palmer, R. M. J., and Higgs, E. A. 1991. Nitric oxide physiology, pathophysiology and pharmacology. *Pharmacological Reviews* 43: 109–142.

Moure, A., Cruz, J. M., Franco, D., Domingueza, J. M., Sineiro, J., Domíngueza, H., Nunez, M. J., and Parajo, J. C. 2001. Natural antioxidants from residual sources. *Food Chemistry* 72: 145–171.

Nakayama, T. 1994. Suppression of hydroperoxide-induced cytotoxicity by polyphenols. *Cancer Research* 54: 1991–1993.

Nakayama, T., Yamaden, M., Osawa, T., and Kawakishi, S. 1993. Suppression of active oxygen-induced cytotoxicity by flavonoids. *Biochemical Pharmacology* 45: 265–267.

Nose, K. 2000. Role of reactive oxygen species in the regulation of physiological functions. *Biological and Pharmaceutical Bulletin* 23: 897–903.

Osman, A. M., Wong, K. K. Y., Hill, S. J., and Fernyhough, A. 2006. Isolation and the characterization of the degradation products of the mediator ABTS-derived radicals formed upon reaction with polyphenols. *Biochemical and Biophysical Research Communications* 340: 597–603.

Oyaizu, M. 1986. Studies on product of browing reaction prepared from glucosamine. *Japanese Journal of Nutrition and Dietetics* 44: 307–315.

Prieto, P., Pineda, M., and Aguilar, M. 1999. Spectophotometric quantitative of antioxidant capacity through the formation of a phosphomolybdenum complex: Specific application to the determination of vitamin E. *Analytical Biochemistry* 269: 337–341.

Prior, R. L., Wu, X., and Schaich, K. 2005. Standardized methods for the determination of antioxidant capacity and phenolics in food and dietary supplements. *Journal of Agricultural and Food Chemistry* 53: 4290–4302.

Pulido, R., Bravo, L., and Sauro-Calixto, F. 2000. Antioxidant activity of dietary polyphenols as determined by a modified ferric reducing/antioxidant power assay. *Journal of Agricultural and Food Chemistry* 48: 3396–3402.

Re, R., Pellegrini, N., Proteggente, A., Pannala, A., Yang, M., and Rice Evans, C. 1999. Antioxidant activity applying an improved ABTS radical cation decolourization assay. *Free Radical Biology and Medicine* 26: 1231–1237.

Riedl, K. and Hagerman, A. E. 2001. Tannin-protein complexes as radical scavengers and radical sinks. *Journal of Agricultural and Food Chemistry* 49: 4917–4923.

Roginsky, V. and Lissi, E. A. 2005. Review of methods to determine chain-breaking antioxidant activity in food. *Food Chemistry* 92: 235–254.

Ruch, R. J., Cheng, S. J., and Klaunig, J. E. 1989. Prevention of cytotoxicity and inhibition of intracellular communication by antioxidant catechins isolated from Chinese green tea. *Carcinogenesis* 10: 1003–1008.

Sakanaka, S., Tachibana, Y., and Okada, Y. 2005. Preparation and antioxidant properties of extracts of Japanese persimmon leaf tea (kakinoha-cha). *Food Chemistry* 89: 569–575.

Serafini, M., Bugianesi, R., Manani, G., Valturna, S., Santis, S. D., and Crozier, A. 2003. Plasma antioxidants from chocolate. *Nature* 28; 424(6952): 1013.

Sherwin, E. R. 1990. Antioxidants. *Food Antioxidants*. Ed. A. L. Branen, P. M. Davidson, and S. Salminen. New York: Marcel Dekker Inc.

Siddhuraju, P. and Becker, K. 2007. The antioxidant and free radical scavenging activities of processed cowpea (*Vigna unguiculata* (L.) Waip.) seed extracts. *Food Chemistry* 101: 10–19.

Siddhuraju, P., Mohan, P. S., and Becker, K. 2002. Studies on the antioxidant activity of Indian Laburnum (*Cassia fistula* L.): A preliminary assessment of crude extracts from stem bark, leaves, flowers and fruit pulp. *Food Chemistry* 79: 61–67

Soares, J. R., Dins, T. C. P., Cunha, A. P., and Ameida, L. M. 1997. Antioxidant activity of some extracts of *Thymus zygis*. *Free Radical Research* 26: 469–478.

Sreejayan, N., and Rao, M. N. A. 1997. Nitric oxide scavenging by curcuminoids. *Journal of Pharmacy and Pharmacology* 49: 105–107.

Tkahama, U., Youngman, R. J. and Elatner, E. F. 1984. Flavonoids: Quenches molecular oxygen species. *Phytobiochem. Phytobiophys.* 17: 175.

Wanasundara, U. N., and Shahidi, F. 1998. Antioxidant and pro-oxidant activity of green tea extracts in marine oils. *Food Chemistry* 63: 335–342.

Yen, G. C., and Duh, P. D. 1993. Antioxidative properties of methanolic extracts from peanut hulls. *Journal of the American Oil Chemists' Society* 70: 383–386.

Zhishen, J., Mengecheng, T., and Jianming, W. 1999. The determination of flavonoid contents on mulberry and their scavenging effects on superoxide radical. *Food Chemistry* 64: 555–559.

19 *In Vitro* Plant Regeneration, Comparative Biochemical and Antioxidant Potential of Calli and Seeds of *Sesbania grandiflora* (L.) Poiret

*Krishnamoorthy Vinothini,
Masilamani Sri Devi, Sudharshan Sekar,
Blassan P. George, Heidi Abrahamse,
Bettine van Vuuren and Arjun Pandian*

CONTENTS

19.1 Introduction 356
19.1.1 Aim and Objectives of This Study 357
19.1.2 Plant Description and Distribution 358
19.1.3 Medicinal Uses of *S. grandiflora* 358
19.1.4 Secondary Metabolites of *S. grandiflora* 358
19.2 Materials and Methods 359
19.2.1 Plant Collection and Surface Sterilization of Explants 359
19.2.2 Preparation of Tissue Culture Media 359
19.2.3 Callus Induction 359
19.2.4 Shoot Induction and Elongation 360
19.2.5 Root Induction and Acclimatization 360
19.2.6 Organic Solvent Extraction of Plant Materials 360
19.2.7 Proline Estimation 361
19.2.8 Total Phenolic Content (TPC) Analysis 361
19.2.9 Dot-Blot Rapid Assay with DPPH 361
19.2.10 DPPH Free Radical Scavenging Activity 361
19.2.11 Ferric Reducing Antioxidant Power (FRAP) Assay 362
19.2.12 Statistical Analysis 362

19.3 Results and Discussion 362
19.3.1 *In Vitro* Seed Germination........ 362
19.3.2 Callus Formation, Growth and Development........ 364
19.3.3 Shoot, Root Induction and Elongation........ 367
19.3.4 Proline Estimation 370
19.3.5 Total Phenolic Content (TPC)........ 370
19.3.6 FRAP (Ferric Ion Reducing Antioxidant Power) Assay........ 371
19.3.7 Dot-Blot DPPH Assay........ 371
19.3.8 DPPH Free Radical Scavenging Activity 371
19.3.9 Principle Component Analysis (PCA)........ 372
19.4 Conclusion 374
Acknowledgements........ 374
References........ 375

19.1 INTRODUCTION

Since the prehistoric era, plants have forever been closely related with our daily life, moreover as food products or for their therapeutic values. Throughout the world, especially in the developing countries, the curative power of the plants is documented by both the scientific community and everyday enthusiasts (Adaikan et al. 2001; Arjun 2011). Due to the depletion of habitats and ruthless collection, medicinal plants are on the verge of extinction. The demand for plant-based therapeutics is increasing in both developed and developing countries, because natural products are easily biodegradable, produce minimal environmental hazards, contain no adverse side effects and easily available at affordable prices (Arjun 2011). Four thousand years ago, the therapeutic knowledge of the Indian subcontinent was termed as Ayurveda. Ayurveda remains an important system of medicine and drug therapy in India. It is nearly estimated, that of the 17,000 species discovered, nearly 3000 species are used in this medicinal field. The pharmacological properties of various Ayurvedic crude drugs sustain its therapeutic claims (van Acker et al. 1995; Arjun 2011).

Natural products have always represented a significant, though often underappreciated, resource for the development of new medicine. There are about 121 prescription drugs in modern medicine that are derived from higher plants (Adaikan et al. 2001). These plant metabolites, according to their composition, are grouped as alkaloids, glycosides, corticosteroids, essential oils, etc. A large proportion of the drugs used in modern medicine are either directly isolated from plants or synthetically modified from a lead compound of natural origin (Arjun 2011). An increasing number of compounds are being isolated from cell cultures, including at least ten groups of phenylpropanoids, ten groups of alkaloids, five groups of terpenoids and three groups of quinines. Thousands of complex molecules were invented from plants during their development to serve the purpose of resistance against plant pathogens and predators. These compounds belong to a group collectively known as secondary metabolites. Their molecules are known to play a vital role in the adaptation of plants to their environment, but also represent an important source of pharmaceuticals (Rao and Ravishankar 2002).

Plant tissue culture is a technique of culturing plant cells, tissues and organs on synthetic medium under an aseptic environment and controlled conditions of light, temperature and humidity. There has been an increasing interest in developing *in vitro* propagation techniques for establishing multipurpose clones of selected plants from within highly variable natural populations (Sinha 2000). Plant tissue culture technology holds great promise for micropropagation, conservation and enhancement of the natural levels of valuable secondary plant products and to meet pharmaceutical demands (Harisaranraj et al. 2009). Accumulation of phytohormones to the culture medium redirects the growth and differentiation of somatic cells (Skoog and Miller 1957; Arjun 2011). Novel cell production and isolation in cultured plant cells can arise in two diverse developmental pathways of organogenesis or somatic embryogenesis (Arjun 2011).

Antioxidants are substances that, when present in low concentrations compared to those of an oxidisable substrate, cause significantly delays or prevent the oxidation of a particular substance (Halliwell and Gutteridge 1999). The generation of free radicals or reactive oxygen species (ROS) during metabolism and other activities beyond the antioxidant capacity of a biological system gives rise to oxidative stress (Zima et al. 2001). Oxidative stress plays a role in heart disease, neurodegenerative disease, cancer and the aging process (Astley 2003). Plants containing chemical compounds (such as ascorbic acid, carotenoids and phenolic compounds) exhibit antioxidant properties. The addition of antioxidants to foods would help the human body reduce the losses from oxidation disturbance (Arjun 2011). Proline acts as a signalling molecule and influences defence pathways, regulates complex metabolic and developmental processes and offers additional opportunities for plant improvement (Laszlo and Arnould 2009). In addition to its role in protein synthesis and the plant cells' response to environmental stresses, circumstantial evidence suggests that proline may also play a role in development as a metabolite and as a signal molecule (Mattioli et al. 2009).

Sesbania is a folk remedy for catarrh, bruises, eye problems, dysentery, fevers, smallpox, headaches, sore throat, sores and stonatitis. It is also used for the treatment of bronchitis, anaemia, inflammation, leprosy, gout, opthalmia, and rheumatism and as a potent antidote for tobacco and smoking-related diseases (Ghani 1998). The leaf juice of *S. grandiflora* was deemed safe for oral consumption and exhibited no gross behavioural changes, except for an increase in urination. Its leaf juices are used as antiurolithiatic activity against calcium oxalate stones and free radical scavenging properties (Doddola et al. 2008). Shareef et al. (2011) reported that its seed oils have a very good potential for edibles and industrial purposes, as well for maintaining nutritional balance like other commonly used vegetable oils.

19.1.1 Aim and Objectives of This Study

Based on its medicinal properties, there is no comparative study report of *in vitro* calli and *S. grandiflora*. An *in vitro* tissue culture technique was selected, the effects were standardized, and a suitable medium for multiple shoots production from *in vitro* seedling explants was chosen to increase the biomass yield of calli on plant growth regulators (PGRs) and different explants. The total phenol, proline and antioxidant potential of wild plant seeds and *in vitro* calli was calculated through dot-blot assay and free radical scavenging by DPPH method activity of various solvent extracts of seed and calli.

19.1.2 Plant Description and Distribution

The plant *Sesbania grandiflora* belongs to Fabaceae family. It is a loosely branching tree (15 m tall), its leaves are pinnately compound (30 cm long), with many pairs of leaflets, and it is oblong to elliptical in shape (12.4 × 5.15 mm dimensions). Its stems are tomentose and unarmed. Its roots normally heavily nodulated with large nodules. Its flowers are large clusters that are white-yellowish, rose-pink or red in colour and calyx-shaped (15.22 mm long). Its standard dimensions are 10.5 × 6 cm, and its pods are long (20 to 60 cm) and thin (6 to 9 mm), wide (15–40 mm), and flat, with pale colour seeds that are elliptical and red brown (Hooker 1978; Orwa et al. 2009). *S. grandiflora* is considered to be native of many Southeast Asian countries, and the use of perennial *Sesbania* species has largely been restricted to South and Southeast Asia (Abbs and Rexin 2013).

19.1.3 Medicinal Uses of *S. grandiflora*

The whole plant is traditionally and medicinally very important. Its leaves are acrid, bitter, sweet and cooling, which are used for tonics, aperients and diuretics. Its juice is used as snuff in coryza, and to treat headaches, influenza, coughs, colds, fevers and epilepsy. Made into a paste, it can be externally applied for wounds, ulcers, rheumatic, and inflamed joints, its leaf juice is known for its antihelminthic activity, and it is also used for the treatment of biliousness, fever, gout, itchiness, leprosy, epilepsy, night blindness and antipyretic (Avalaskar et al. 2011). Its bark is astringent, cooling, tonic, bitter, febrifuge, and anthelmintic. Pounded bark can be externally applied for scabies; bark juice can be used for dyspepsia, diarrhoea, gastralgia, abdominal colic and gastrointestinal disorders. Its flowers are cooling, bitter, astringent, acrid and emollient, and can be used for laxatives and antipyretics. Its juices can be applied on the eyes for night blindness, and also be used for leucorrhoea, fevers, periodic fever, small pox, poisoning cases, biliousness and debility. Ripe pods are used for brain tonic and as a memory promoter. The sweet flowering buds can be pickled and eaten, and used as a laxative, alexiteric or flatulent colic and for treating anaemia and emaciation (Prescott et al. 2002; Ramesh et al. 2007). The *S. grandiflora* leaves have bactericidal activity, and its extracts have been used for the treatment of nasal catarrh and fevers, and are also being used for anxiolytic and anticonvulsive effects (Ramesh et al. 2007).

19.1.4 Secondary Metabolites of *S. grandiflora*

The active metabolites responsible for these activities were proteins, alkaloids, flavonoids, tannins, saponins, keampferol, glycosides, diterpenoids, steroids, triterpenes, gums, phenylalanine, mucilages, valine, linoleic acid, nicotinic acid, β-sitosterol, carbohydrates and anthraquinone (Ghani 1998; Paranjpe 2001; Avalaskar et al. 2011).

19.2 MATERIALS AND METHODS

Methanol, acetone, hexane, chloroform and ethyl acetate, 2,2-Diphenyl-1-picrylhydrazyl (DPPH), TPTZ (2,4,6-tripyridyl-s-triazine), gallic acid, quercetin, ferrous sulphate and analytical grade chemicals supplied by Hi-Media, Merck and Sigma Chemicals were used.

19.2.1 Plant Collection and Surface Sterilization of Explants

The wild plant seeds of *S. grandiflora* were collected from Cuddalore, Tamil Nadu, India. Seeds were surface sterilized to produce contamination-free plants that can be maintained under aseptic conditions. Seeds were checked for viability and washed thoroughly under running tap water (15 min) without damage to the tissues. In order to avoid the interaction of microbes such as bacteria and fungi in plant tissue culture, seeds were then transferred to a beaker containing sodium hypochlorite (3%) with rapid shaking for 15–30 min and changing the solution at 10 min intervals. Then, the seeds were washed with distilled water and treated with 0.3% mercuric chloride for 2–5 min. The surface sterilized seeds were washed with sterile distilled water five times. The excess water on the seeds was removed by using sterile tissue paper before culture. The seeds were transferred to a semi-solid medium under aseptic conditions in a laminar flow chamber (Arjun 2011).

19.2.2 Preparation of Tissue Culture Media

MS (Murashige and Skoog 1962) Media used six individual stock solutions of macro, micro, minor, iron and vitamins, which were prepared and stored. The iron stock was stored in a black bottle to prevent photolysis of chemicals. All the stock solutions were stored in refrigerator and used within one year. Meso-inositol, cytokinin and auxin stock solutions were freshly prepared and used for a month. For preparation of medium, all the six stock solutions were mixed thoroughly with required amounts of sterile distilled water. Sucrose (3%, 30 g/L), 0.1% meso-inositol (100 mg/L) and the required amount of plant growth hormones were added to the medium and buffered by 1N HCl or 1N NaOH to adjust the medium pH to 5.6–5.8 before autoclaving. The medium was solidified by adding agar (0.8%, 8 g/L). Sterile distilled water was used to make the final volume. The medium was poured into culture vials and autoclaved at a pressure of 15 lbs for 15 min at 121°C. A photoperiod (16/8 h) light/dark condition (25 ± 2°C) under a cool white light (2000 lux) fluorescent tube was provided.

19.2.3 Callus Induction

Surface sterilized leaf explants were cut into small pieces (0.5–1.0 cm) barring the cut ends and transferred to MS basal medium. Three different auxins [(indole-3-acetic acid (IAA), α-naphthalene acetic acid (NAA) and 2,4-dichlorophenoxyacetic acid (2,4-D)] cytokinins and 6-benzylaminopurine (BAP) at the same concentration (0.5 mg/L) were used in the callus induction. Leaf explants cultured on the MS medium without any growth regulators were used as controls for callus induction. All treatments were conducted thrice with replicates in each treatment.

19.2.4 Shoot Induction and Elongation

The organogenic calli were obtained after four weeks of culture and were further inoculated into the fresh MS medium supplemented with auxins (IAA, NAA and 2,4-D), cytokinins (BAP) and gibberellins (GA_3) at different concentrations alone (0.1–3 mg/L) and in various combinations (0.2–3 mg/L) to study their ability for shoot induction and regeneration. The well-developed shoots were subcultured on the fresh MS medium every 28 d. All cultures were maintained (25 ± 2°C) with relative humidity (60%) under fluorescent lights with light/dark (16/8 h) cycles. The length of individual shoots was measured and recorded after six weeks of culture and the mean shoot length was determined. All treatments were conducted thrice with replicates in each treatment.

19.2.5 Root Induction and Acclimatization

In vitro generated shoots were excised from cultures and transferred to half-strength MS basal medium supplemented with auxins, cytokinins and gibberellins (GA_3) alone and different concentrations (0.5–3.0 mg/L) for the induction and development of roots. The cultures were incubated in the dark (two weeks). The cultures were subcultured every 28 d onto the fresh half-strength MS medium supplemented with plant growth regulators. The length of individual roots was measured and recorded after six weeks of culture and the mean root length was determined. All treatments were conducted thrice with replicates in each treatment.

Regenerated plantlets were removed from the culture medium. The plantlets were washed with sterile distilled water and transferred to plastic cups containing sterilized vermiculite and soil (2:1 w/w) mixture. Initially, the cups were covered with polythene bags to prevent excess transpiration and also to maintain high relative humidity (80%). The potted plants were kept under photoperiod (16/8 h) for two weeks and half-strength MS salt solution was poured regularly. After the two-week period, the polythene bags were removed and the potted plants were exposed to direct light and sterile liquid medium was poured. Then, they were transferred to room temperature (two weeks), after which the plants were transferred to greenhouse slowly.

19.2.6 Organic Solvent Extraction of Plant Materials

The seeds and *in vitro* calli of *S. grandiflora* were shade-dried; the shade-dried samples were powdered, the powdered samples were cold-macerated with different solvents (hexane, ethyl acetate, chloroform, acetone and methanol) for 3 d with occasional stirring. Then all the extracts were filtered through Whatman filter paper, the same process repeated twice, the pooled solvent removed at low temperature (40°C–60°C) below abridged pressure in the rotary evaporator, the evaporated samples were kept in the refrigerator condition for further analysis (Arjun et al. 2012a).

19.2.7 Proline Estimation

The proline estimation method described by Malaisamy and Mohan (2014) was implemented with some modifications. Wild plant seeds and fresh and dried *in vitro* calli (calli derived from internode, cotyledon explants), were taken and homogenized with cold ethanol (2 mL, 40%) and used in the mortar and pestle. The mortar contained a tiny quantity of well-washed and disinfected sand. Agitated (10 min) and homogenized samples were filtered through Whatman filter paper, then aliquot (2 mL) was used for the quantification of proline. The absorbances were noticed (528 nm), the results were obtained and those results were expressed in proline (μM) per mg of fresh and dried forms, then absorbance values compared with the standard curve of proline.

19.2.8 Total Phenolic Content (TPC) Analysis

The total phenolic content was determined using the modified Folin–Ciocalteau method described by Arjun et al. (2015) with a few modifications. Fresh plant samples (0.05 g) were homogenized with the solvents of acetone and water (1:1, v/v) for 15 min (4°C), homogenized aliquot samples (9 μL) was taken and those samples was mixed with Folin–Ciocalteau reagent (109 μL). Three min of equilibrium time under room temperature went by (25°C) and then Na_2CO_3 (180 μL, 7.5%, w/v) solution was added. These mixed solutions were allowed to stand for five min at 50°C and then keep at 25°C for two min and absorbance was measured at 760 nm (Zenyth 200rt Microplate Reader UK-Biochrom Ltd). Total phenolic content was calculated using the standard curve of gallic acid and expressed as mg gallic acid equivalents (mg GAE/g).

19.2.9 Dot-Blot Rapid Assay with DPPH

The rapid screening of antioxidants was performed by Sivaraj (2011). Aliquots of different solvent extracts (3 μL) were spotted carefully on TLC plates. Sheets bearing the dry spots were placed upside down (10 sec) in 2,2-diphenyl-1-picrylhydrazyl (DPPH) (0.4 mM) solution and the layer was dried. The stained silica layer revealed a purple background with yellow spots, which showed radical scavenging capacity. The odd electron in the DPPH free radical gave a strong absorption maximum at 517 nm and was purple in colour. The color turned from purple to yellow as the molar absorption of DPPH radical at 517 nm was reduced from 9660 to 1640, when the odd electron of DPPH radical became paired with hydrogen from a free radical scavenging antioxidant to form the reduced DPPH-H. The resulting decolourisation was stoichiometric with respect to number of electrons captured.

19.2.10 DPPH Free Radical Scavenging Activity

The seeds and *in vitro* calli of different solvents like hexane, ethyl acetate, acetone, chloroform and methanol extracts were subjected to free radical scavenging activity by the DPPH method. The DPPH free radical scavenging activity was determined based on Nagarajan et al. (2012). The DPPH solution (0.1 mM, M.W. 394.32) was prepared in methanol, the solvent extracts were prepared in methanol at different concentrations (50–250 μg/mL) and the DPPH solution (1.0 mL) was added (1.0 mL). The mixtures were incubated in dark conditions at room temperature for 30 min and absorbance was measured (517 nm). For the comparison quercetin standard was used indifferent concentrations (2–10 μg/mL). The potential to scavenge the DPPH free radical was calculated using the following equation; $Ab_{control}$-Ab_{sample}/$Ab_{control}$X100. Whereas $Ab_{control}$ = Absorbance of control, Ab_{sample} = Absorbance of sample.

19.2.11 Ferric Reducing Antioxidant Power (FRAP) Assay

The modified method of Xu and Chang (2007) was used to find out the ferric reducing ability of samples. Briefly, acetate buffer (300 mM, pH 3.6) and weigh sodium acetate trihydrate (3.1 g) were added into the glacial acetic acid (16 mL) and made up to 1 L with distilled water. TPTZ (2,4,6-tripyridyl-s-triazine; M.W. 312.34), TPTZ (10 mM) dissolved in HCl (40 mM: M.W. 36.46) and $FeCl_3.6H_2O$ (20 mM; M.W. 270.30) were used to make the FRAP reagent. The working FRAP reagent was prepared by mixing a, b and c (10:1:1). The FRAP reagent (150 μL) was read at 600 nm and the samples were added (20 μL). They were incubated at room temperature under dark condition for eight min and read at 600 nm. The results were expressed in mM Fe (II)/g fresh and dry mass.

19.2.12 Statistical Analysis

The results were subjected to one-way analysis of variance (ANOVA) to evaluate the important of differentiation by means of a variety of conduct different groups, using the SPSS numerical software package, the values are accessible as mean ± S.D and $P<0.05$. SIMCA software version 13.0 31 was used for the comparison studies.

19.3 RESULTS AND DISCUSSION

Worldwide, climate change and use of lands for intensive farming and development are threatening the plant genetic diversity. Hence, the conservation of biodiversity is very important (Krishnan et al. 2011).

19.3.1 *In Vitro* Seed Germination

Under *in vitro* conditions, seed germination efficiency was obtained on the same concentration (0.1 mg/L) in different PGRs (2,4-D, IAA, NAA and BAP), seed germination efficiency was observed for 10, 20 and 30 d (Table 19.1). From the 10 d, maximum seed germination was observed for BAP (36.15%) followed by IAA (34.21%), less germination was observed for 2,4-D (21.83%). On 20 d, the highest content was observed on BAP (84.32%) followed by IAA (82.44%) and

TABLE 19.1
In vitro* Seed Germination Efficiency of *S. grandiflora

Concentrations of Hormones (mg L^{-1})	Germination Efficiency (%)		
	10 d	20 d	30 d
BAP (0.1)	36.15 ± 3.1[d]	84.32 ± 3.2[a]	78.07 ± 4.1[b]
NAA (0.1)	28.12 ± 3.5[e]	71.33 ± 3.8[b]	78.11 ± 2.4[b]
IAA (0.1)	34.21 ± 0.7[d]	82.44 ± 3.3[a]	53 ± 1.8[c]
2,4-D (0.1)	21.83 ± 1.5[e]	78.28 ± 3.3[b]	70.83 ± 2.4[b]

Note: Number in parentheses shows standard deviation (SD). Statistically significant at $p<0.05$ where $^a>^b>^c>^d>^e$.

2,4-D (78.28%), less germination was observed on NAA (71.33%). On 30 d, seed germination for BAP and NAA showed 78% followed by 2,4-D (70.83) and IAA (53%) (Table 19.1; Figures 19.1 and 19.2).

In vitro seed germination (GA_3 at 1.0 mg/L concentration) of *Genipa americana* was reported by Costa et al. (2002), 80% seed germination was observed in *Annona glabra* under *in vitro* conditions using GA_3 (2 mg/L) (Deccetti 2000). The *S. grandiflora* germination of seeds and the establishment of seedlings are less susceptible to environmental stress conditions through plant tissue cultures to produce more plants within a short period. Every four weeks, *in vitro* germinated plantlets were subcultured. The *in vitro* explant parts of cotyledons, nodes and internodes were collected from seedlings for more biomass production.

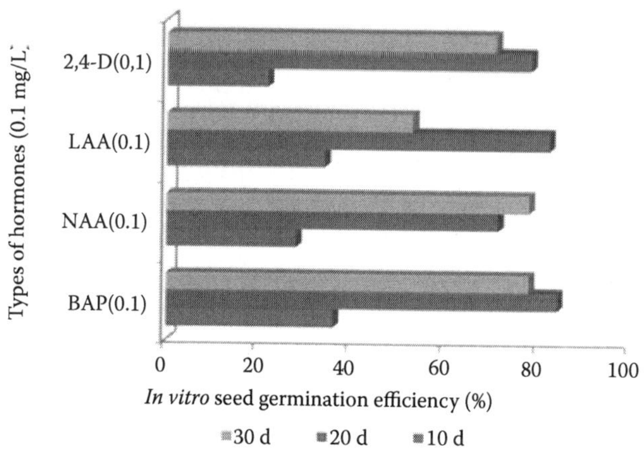

FIGURE 19.1 *In vitro* seed germination efficiency of *S. grandiflora*.

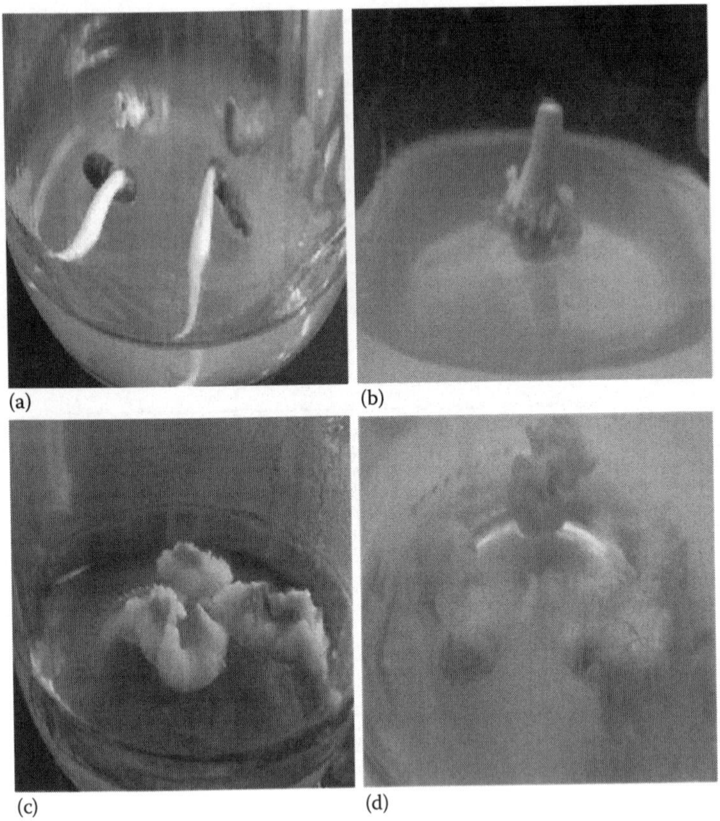

FIGURE 19.2 *In vitro* seed germination and callus formation of *S. grandiflora*. (a) Seed germination; (b) callus formation on node; (c) callus formation from in vitro cotyledon; (d) callus formation on internode.

19.3.2 Callus Formation, Growth and Development

The *in vitro* callus formation in cotyledon, node and internode explants were obtained though MS medium at the same concentrations (0.5 mg/L) supplemented with different PGRs (2,4-D, NAA and IAA). After one week of incubation, the callus formation was first noticed in cotyledon, node and internode explants. On the 10 d, maximum callus formation was observed in internodes on 2,4-D (52.3%), followed by nodes (48.4%); and fewer callus formation was observed from nodes on NAA (31.3%). Twenty-day maximum calli was obtained from internodes (78.9%) for 2,4-D and from NAA (68%) and fewer callus formation was observed in IAA (57.7%). On the 30 d, callus content was obtained from internodes with 2,4-D (95.6%), followed by nodes (91.6%) (Table 19.2; Figures 19.1, 19.3 and 19.4).

In the present study, callus growth was observed for different days up to 30 d. Different types of calli were observed and MS basal medium supplemented with

TABLE 19.2

Callus Formation from Cotyledons (C), Node (N) and Internodes (In) of *S. grandiflora*

Concentrations of	10 d			20 d			30 d		
Hormones mg L^{-1}	C	N	In	C	N	In	C	N	In
2,4-D (0.5)	33.4 ± 2.1^{f}	49.4 ± 2.8^{d}	53.3 ± 3.8^{d}	59.6 ± 3.7^{e}	65.9 ± 4.7^{d}	79.9 ± 4.7^{c}	78.9 ± 2.7^{c}	92.6 ± 2.3^{a}	96.6 ± 3.3^{a}
IAA (0.5)	35.6 ± 2.6^{f}	32.3 ± 2.2^{f}	45.3 ± 3.1^{e}	60.2 ± 3.2^{d}	58.7 ± 3.3^{e}	68.7 ± 4.4^{d}	86.2 ± 3.2^{b}	83.4 ± 4.2^{b}	83.7 ± 5.0^{b}
NAA (0.5)	42.3 ± 1.4^{e}	45.2 ± 2.1^{e}	44.5 ± 2.8^{e}	66.6 ± 1.2^{d}	64.8 ± 1.3^{d}	69.9 ± 1.0^{d}	87.7 ± 1.8^{b}	88.1 ± 1.3^{b}	89.4 ± 1.2^{b}
BAP (0.5)	38.1 ± 67.8^{f}	36.9 ± 2.3^{f}	40.6 ± 3.3^{e}	67.8 ± 2.1^{d}	61.3 ± 2.9^{d}	75.2 ± 2.3^{c}	89.1 ± 2.3^{b}	91.1 ± 2.4^{a}	90 ± 2.3^{a}

Note: Number in parentheses shows standard deviation (SD). Statistically significant at $p<0.05$ where $^{a}>^{b}>^{c}>^{d}>^{e}>^{f}$.

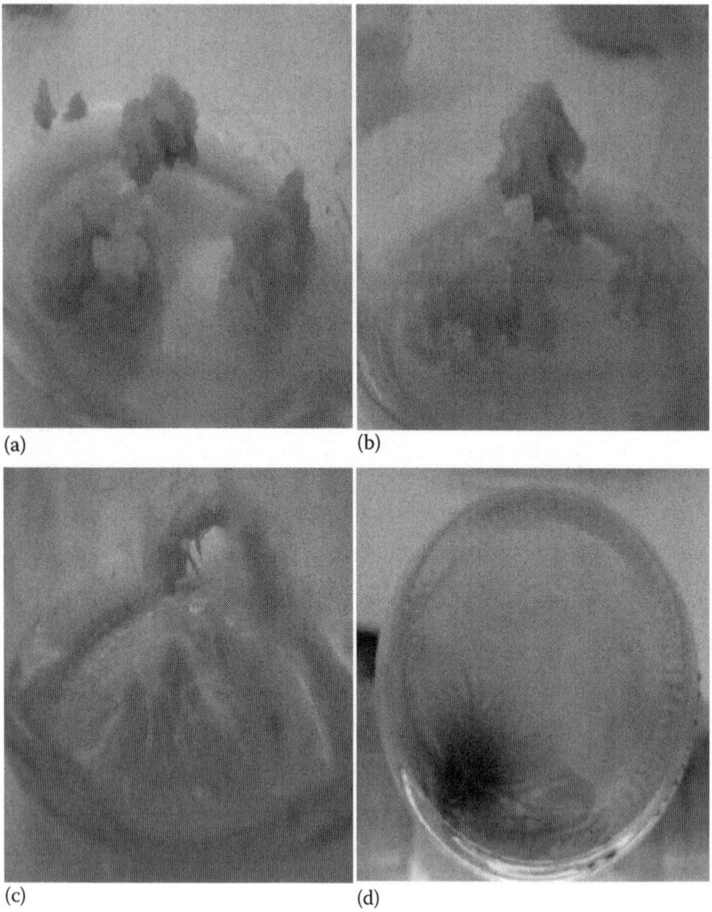

FIGURE 19.3 Callus, organogenic callus, and multiple root formations: (a) friable callus, (b) organogenic callus; (c) multiple roots from callus and (d) multiple roots from node.

2,4-D (3 mg/L) showed good friable callus, BAP (1 mg/L) showed compact callus (Figure 19.4) and IAA+BAP (3+3 mg/L) showed organogenic calli. The callus grown in MS medium was greenish in color, granular and fast growing. Every three to four weeks, the calli were subcultured for further proliferation, multiple shoots, root formation and induction of organogenesis (Figure 19.5).

Callus formation from the basal-cut ends of nodes and shoot-tip explants of *Leptadenia reticulata* on MS medium enriched with BA or KIN is related to the report on *Gymnema sylvestre* and *Holostemma ada-kodien* (Komalavalli and Rao 2000; Martin 2002). The friable callus developed at the basal cut ends of shoot-tip and node explants on MS medium supplemented with BA alone or in combination with NAA, IAA and IBA, was reported by Martin (2004).

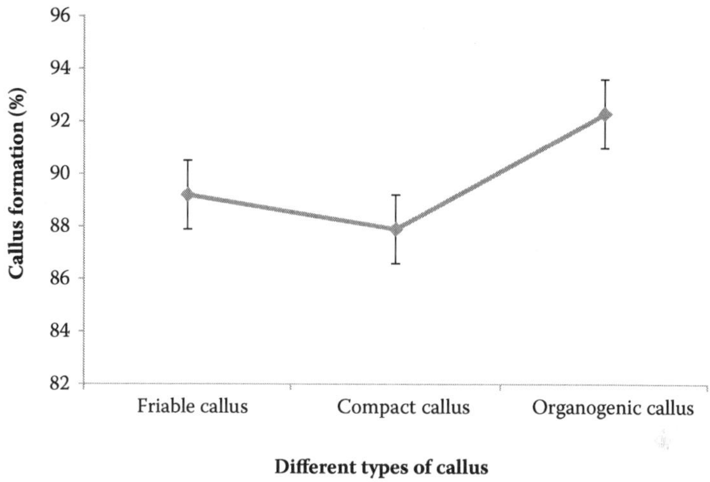

FIGURE 19.4 Different types of callus formation of *S. grandiflora*.

FIGURE 19.5 Multiple shoots and multiple root formation (a, b, c).

19.3.3 Shoot, Root Induction and Elongation

Different PGRs (2,4-D, IAA, NAA and BAP) at a concentration of 0.1 mg/L induced shoot initiation and maximum shoot initiation was observed in BAP (95.2%), followed by NAA (72.6%). The highest content of shoot formation was achieved on IBA (3 mg/L, 75.6%) and multiple shoot formation (63.7%) was observed, followed by NAA+BAP (0.4+0.2 mg/L), which showed 58.9% shoot and 53.3% multiple shoot formation (Table 19.3; Figures 19.4 and 19.6). Auxin and cytokine alone (2,4-D, IAA, NAA, BAP and IBA) and combined (NAA+BAP) shoot elongation was measured (cm). The highest shoot elongation was obtained on NAA+BAP (0.4 + 0.2 mg/L) with

TABLE 19.3
Shoot Induction, Elongation, Multiple Shoot Formation from Callus of *S. grandiflora*

Concentrations of Hormones (mg L^{-1})	Shoot Induction	Shoot Formation	Multiple Shoot Formation	Shoot Elongation (cm)
BAP (0.1)	95.2 ± 3.2^{a}			8.50 ± 1.3^{c}
NAA (0.1)	72.6 ± 3.3^{b}			13.21 ± 2.3^{b}
IAA (0.1)	70.3 ± 2.8^{b}			9.0 ± 1.4^{c}
2,4-D (0.1)	55.3 ± 1.93^{c}			7.21 ± 1.1^{c}
IBA (1.0)		53.3 ± 1.4^{c}	41.3 ± 1.2^{d}	
IBA (2.0)		63.4 ± 1.3^{b}	50.4 ± 2.9^{c}	
IBA (3.0)		75.6 ± 3.5^{a}	63.7 ± 2.7^{b}	
IAA+BAP (0.2+0.4)		21.1 ± 2.7^{f}	31.7 ± 2.4^{e}	6.32 ± 1.5^{c}
IAA+BAP (0.4+0.2)		32.2 ± 2.5^{e}	41.4 ± 2.5^{d}	9.31 ± 2.5^{c}
NAA+BAP (0.2+0.4)		45.7 ± 2.2^{d}	43.5 ± 2.4^{d}	13.13 ± 1.2^{b}
NAA+BAP (0.4+0.2)		58.9 ± 2.3^{c}	53.3 ± 2.3^{c}	20.42 ± 2.3^{a}

Note: Number in parentheses shows standard deviation (SD). Statistically significant at $p<0.05$ where $^{a}>^{b}>^{c}>^{d}>^{e}>^{f}$.

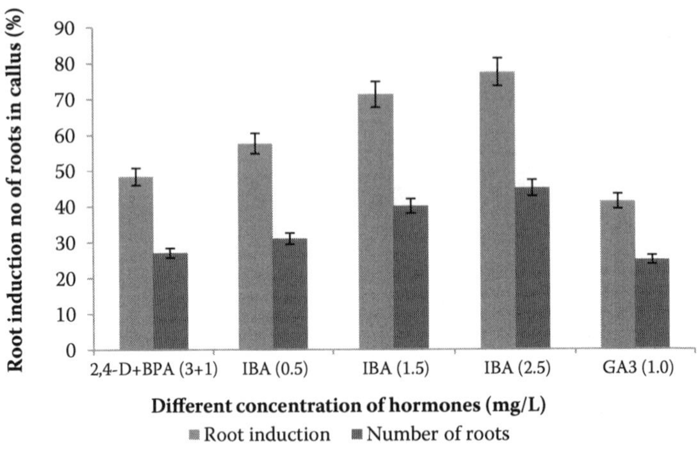

FIGURE 19.6 Root induction and number of root formation in callus of *S. grandiflora*.

highest elongated shoots (20.42 cm), followed by NAA with 13.21 cm respectively (Table 19.3; Figure 19.6).

The direct shoot regeneration in cotyledonary explants of *T. terrestris* was achieved and reported by Ali et al. (1997). Raghu et al. (2010) reported that BA increased multiplication but with a lower rate of shoot elongation, the maximum number (6–7 shoots/node) was obtained in Woody Plant Medium supplemented with BA (4.0 mg/L).

In this study, the results obtained in the root formation indirectly from organogenic callus by different plant growth regulators can be used in root-specific compound isolation and conservation of the plant through tissue culture methods. Adventitious root formations and root inductions are responsible due to auxins in the media; within 3 d, the root induction was observed in *S. grandiflora* (Table 19.4; Figure 19.7). Topmost root induction was observed on IBA (2.5 mg/L) (77.3%), the number of roots was observed (45%) followed by 2,4-D+BAP (3+1 mg/L) with shoot induction (48.3%) and the number of roots was obtained (27%), respectively (Table 19.4; Figures 19.4–19.7). Adventitious root formation in many plant species induced by auxins and silver nitrate on leaf explants of *Beta vulgaris* (sugar beet) were reported by Gurel and Gurel (1998).

Plant regeneration through organogenesis callus cultures derived from petioles, mature leaves, roots and stems of young seedlings of *Psoralea corylifolia* were reported by Saxen et al. (1997) and *Zingiber officinale in vitro* organogenesis through callus culture was reported by Rout and Das (1997). The *in vitro* regeneration also

TABLE 19.4
Root Induction Number from *In Vitro* Callus of *S. grandiflora*

Concentrations of Hormones (mg L^{-1})	Root Induction	No. of Roots (%)
2,4-D+BPA (3+1)	48.3 ± 1.3^{c}	27 ± 1.2^{e}
IBA (0.5)	57.6 ± 2.1^{b}	31 ± 1.4^{d}
IBA (1.5)	71.2 ± 2.9^{a}	40 ± 1.1^{c}
IBA (2.5)	77.3 ± 2.7^{a}	45 ± 2.8^{c}
GA_3 (1.0)	41.2 ± 2.2[illegible]	25 ± 2.1^{e}

Note: Number in parentheses shows standard deviation (SD). Statistically significant at $p<0.05$ where $^{a}>^{b}>^{c}>^{d}>^{e}$.

(a) (b)

FIGURE 19.7 Plantlet under lab conditions and harvested callus: (a) lab conditions and (b) harvested callus kept for drying.

reported by Pande et al. (2002) and the efficient systems of *Solanum laciniatum* and *Echinacea pallid in vitro* regeneration established from explants on medium supplemented with NAA and BAP were reported by Sharma and Rajam (1995). The successful plant regeneration from leaf and stem derived callus of *Centella asiatica* on MS medium supplemented with BA (4.0 mg/L), KIN (2.0 mg/L), NAA (0.25 mg/L), and adenine sulfate (20 mg/L) were reported by Patra and Rai (1998). The *in vitro* regeneration efficient systems of *E. pallida* and *S. laciniatum* established on MS medium were supplemented with BAP and NAA (Sharma and Rajam 1995). In the current study, *S. grandiflora* shoots had well-developed adventitious roots with BAP and NAA–supplemented medium, the plantlets regenerated were transferred to plastic cups containing sterilized vermiculite and soil (2:1 w/w). The potted plants were reserved under the (16/8 h) light/dark photoperiod conditions (Figure 19.7).

19.3.4 Proline Estimation

When the plants were exposed to stressful conditions and metabolites, amino acids accumulated. These amino acids traditionally considered as precursors and constituents of proteins and play an important role in plant metabolism and development. Besides acting as an excellent osmolytes, they play three major roles during stress conditions (as a metal chelator, as an antioxidative defence molecule and as a signalling molecule) (Hayat 2012). Based on adequate knowledge and the importance of proline, *S. grandiflora*'s seeds, fresh and dried calli derived from cotyledon, nodes and internodes have been estimated for the presence of proline. The maximum proline content (112.91 mg/g) was observed from the dried calli (internode and node), followed by calli (cotyledon) (63.71 mg/g), respectively (Table 19.5).

19.3.5 Total Phenolic Content (TPC)

The total phenolic content was estimated on fresh and dried callus (node and internode), fresh and dried callus (cotyledon) and seeds. Phenolic content was in the range of 2.12–17.15 mg/g, the highest content of TPC was observed on seeds (17.15 mg/g) followed by dried calli (node and internode) (15.22 mg/g) (Table 19.5). Compared

TABLE 19.5
Quantification of Proline, Phenol and FRAP assay of *In Vitro* Calli and Seeds

Types of Samples	Proline (mg/g)	Phenol (mg GAE/g)	FRAP mg (Fe(II)/g)
Fresh callus (node & internode)	13.32 ± 2.2^{g}	3.23 ± 0.6^{h}	139 ± 3.2^{c}
Dried callus (node & internode)	112.91 ± 4.1^{d}	15.22 ± 2.1^{g}	236 ± 4.1^{a}
Fresh callus (Cotyledon)	11.41 ± 1.1^{h}	2.12 ± 2.0^{h}	109 ± 3.2^{d}
Dried callus (Cotyledon)	63.71 ± 2.6^{e}	8.52 ± 1.2^{h}	169 ± 4.2^{b}
Seeds	12.37 ± 1.8^{g}	17.15 ± 2.4^{g}	105 ± 2.1^{d}

Note: Values are mean of triplicate determination (n=3) ± standard deviation, Statistically significant at $p<0.05$ where $^{a}>^{b}>^{c}>^{d}>^{e}>^{f}>^{g}>^{h}$.

between fresh and dried calli, the dried calli showed the maximum amount of phenolic contents in *S. grandiflora*. The TPC was calculated using a standard curve gallic acid was expressed as mg of gallic acid equivalents (GAE 100 g^{-1}) of fresh weight and dry weight of the samples. According to Sarkar et al. (2012) that the phenolic contents in plant *S. grandiflora* extracts were obtained (48.2 mcg/mg).

Phenolics and flavonoids are the common natural compounds in plants with a wide range of physiological properties and possess the ability to reduce oxidative damage associated with many diseases, including cardiovascular diseases, cancer, atherosclerosis, diabetes, hepatitis, immune deficiency diseases, arthritis and aging (Arjun 2011).

19.3.6 FRAP (Ferric Ion Reducing Antioxidant Power) Assay

The FRAP assay monitors the reaction of Fe^{2+} with TPTZ (2,4,6-Tripyridyl-s-Triazine), the mixture's violet-blue colour was formed and absorbed at 593 nm (Berker et al. 2007). In the present study, the highest ferric reduction was estimated in dried calli (node and internode) (236 mg/g), followed by dried calli (cotyledon) (169 mg/g), respectively. Ferrous sulphate was used as an internal standard and was calculated FRAP mg (Fe(II)/g) (Table 19.5). Antioxidants, non-enzymatic life forms against reactive O, N-species, are essential for the daily life of humans, most of the antioxidant compounds are introduced through the diet. Due to these established methods, the total antioxidant capacity of food the plant extracts, isolated compounds and food samples was directly measured (Ou et al. 2002).

19.3.7 Dot-Blot DPPH Assay

The solvent extracts in TLC plate sprayed DPPH (0.4 mM) solution, stained the silica layer and exposed the purple colour background with yellow spots, which showed radical scavenging capacity in the dot-blot assay. All the extracts showed antioxidant activity in this method (Figure 19.8). The technique is characteristically based on the inhibition and accumulation of oxidized commodities, while the production of free radical scavenging is inhibited by adding antioxidants (Sivaraj 2011).

19.3.8 DPPH Free Radical Scavenging Activity

In biological systems, oxidative damage produced through ROS is the main cause for several degenerative diseases (Wang et al. 2013). It involves preventing and protecting the cells from ROS, therefore plants and plant products are very important sources of natural antioxidants. The consumption of polyphenolic compounds, flavonoids, carotenoids and all secondary metabolites reduce the risk of cancer and age-related disorders (Perez-Vizcaino and Duarte 2010).

In the current study, the scavenging activity of *S. grandiflora* ranged 10.23% to 97.15%, the highest scavenging was observed on chloroform seed extract (97.15%), followed by methanol seed extracts (94.51%), ethyl acetate seed extract (93.66%), acetone seed extract (93.45%), chloroform calli extract (92.42%), methanol calli extract (92.21%) and hexane calli extract (91.25%), respectively. IC_{50} values of hexane calli extract showed 70.66 μg/mL followed by methanol calli (71.31 μg/mL) and

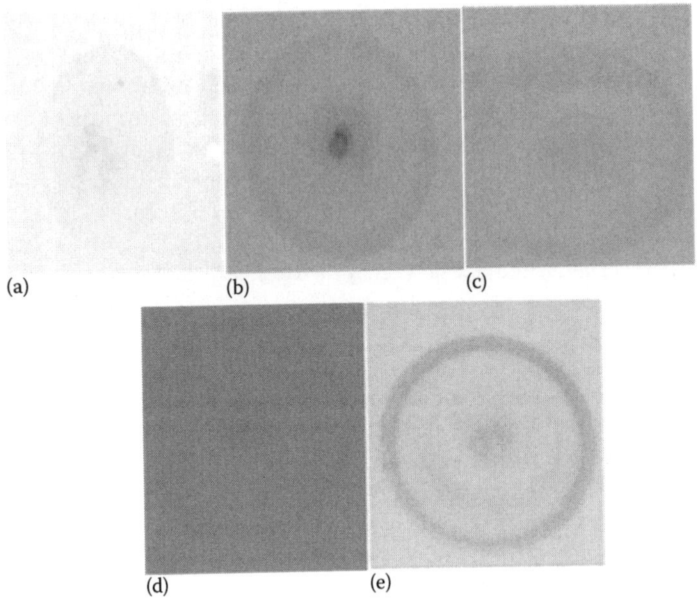

FIGURE 19.8 DPPH dot-blot rapid assay method of different organic solvent extracts of *S. grandiflora*: (a) hexane, (b) ethyl acetate; (c) chloroform; (d) acetone and (e) methanol.

chloroform seeds (71.75 µg/mL), respectively. Both wild plant seeds and *in vitro* calli showed potential antioxidant activity. Ascorbic acid was used as the standard (2–10 µg/mL, 18.38% to 91.32%). The seed extracts showed the highest scavenging activity and a lower value of IC_{50} indicated highest antioxidant activity (Table 19.6).

The aqueous ethanol extracts of *Jatropha dioica, Flourensia cernua, Eucalyptus camaldulensis* and *Turnera diffusa* extracts tested against antioxidant activity among four plants. *J. dioica* showed highest scavenging (Jorge et al. 2015). Methanolic leaf extracts of *Nelumbo nucifera* showed brawny antioxidant (94.97%, IC_{50}, 230.62 µg/mL) activity, (Arjun et al. 2012b). *Marrubium vulgare* aqueous extracts, *Thymus numidicus* methanol leaf extracts and *Senecio candicans in vitro* callus and *in vivo* leaf extracts showed more potent antioxidant activity in a similar manner (Vanderjagt et al. 2002; Hariprasath et al. 2015).

19.3.9 Principle Component Analysis (PCA)

PCA is a functional, arithmetical method that has found relevance in fields such as face recognition and image compression and is a common technique for discovery patterns in data of lofty dimension. It covers SD (standard deviation), covariance, eigenvectors and eigen-values. The scores t1, t2, etc., are new variables summarizing the X-variables. The scores are orthogonal, i.e. completely independent of each other. The score t1 (first component) explains the largest variation of the X space, followed by t2 etc. Hence the scatter plot of t1 vs. t2 is a window in the X space, displaying how the X observations are situated with respect to each other. This plot shows the

TABLE 19.6
DPPH Scavenging Activity of Dried *In Vitro* Calli and Wild Plant Seeds of *S. grandiflora*

	Hexane		Ethyl Acetate		Chloroform		Acetone		Methanol	
Con. (µg/mL)	Calli	Seed	Calli	Seed	Calli	Seed	Calli	Seed	Calli	Seed
50	41.18 ± 0.6^{f}	13.56 ± 0.3^{i}	10.23 ± 1.2^{i}	38.52 ± 1.4^{g}	43.85 ± 1.5^{f}	41.39 ± 1.6^{f}	14.15 ± 0.9^{i}	37.99 ± 1.8^{g}	43.2 ± 1.4^{f}	38.5 ± 1.7^{g}
100	57.45 ± 1.3^{e}	20.19 ± 0.2^{h}	22.22 ± 2.4^{h}	49.18 ± 2.1^{f}	57.32 ± 1.9^{e}	56.6 ± 2.1^{e}	37.14 ± 1.1^{g}	49.7 ± 2.1^{f}	57.11 ± 1.8^{e}	49.38 ± 2.2^{f}
150	69.31 ± 2.4^{c}	32.11 ± 0.5^{g}	46.47 ± 2.0^{f}	61.45 ± 2.2^{c}	69.53 ± 2.1^{c}	79.47 ± 2.3^{c}	63.04 ± 1.6^{d}	62.86 ± 2.4^{d}	70.13 ± 2.2^{c}	60.83 ± 2.8^{d}
200	81.67 ± 3.1^{b}	41.39 ± 0.2^{f}	68.03 ± 2.1^{d}	79.85 ± 3.1^{c}	82.13 ± 2.4^{b}	92.05 ± 3.1^{a}	69.39 ± 2.1^{d}	78.49 ± 2.6^{c}	81.71 ± 2.7^{b}	72.61 ± 3.1^{c}
250	91.25 ± 3.8^{a}	53.13 ± 0.8^{e}	82.43 ± 2.6^{b}	93.66 ± 3.6^{a}	92.42 ± 2.6^{a}	97.15 ± 3.2^{a}	85.81 ± 2.8^{b}	93.45 ± 3.7^{b}	92.21 ± 3.2^{a}	94.51 ± 3.5^{a}
IC_{50} values	70.66 ± 0.9	240.89 ± 5.4	158.59 ± 1.3	99.78 ± 2.3	73.51 ± 2.1	71.75 ± 0.08	137.14 ± 0.9	98.77 ± 1.6	71.31 ± 0.62	104.42 ± 2.7

Note: Statistically significant at $p<0.05$ where $^{a}>^{b}>^{c}>^{d}>^{e}$.

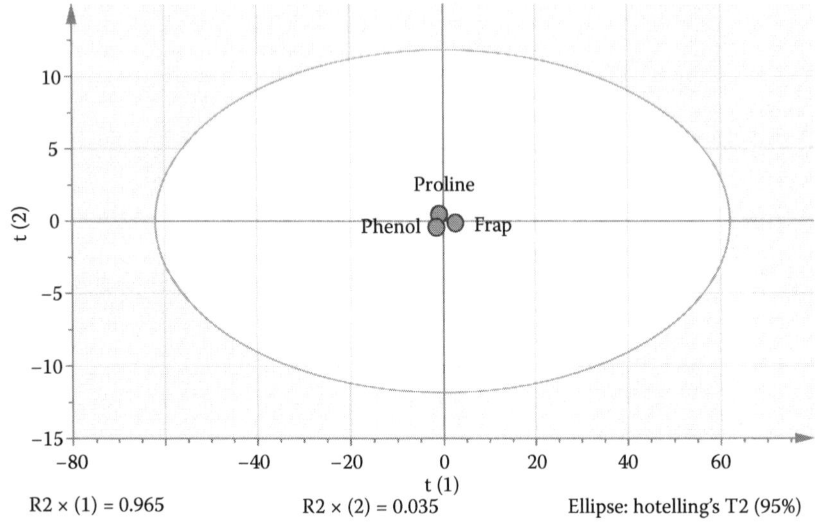

FIGURE 19.9 Principle component analysis of *S. grandiflora.*

possible presence of outliers, groups, similarities and other patterns in the data. The score plot is a map of the observations. In this two-dimensional score plot, SIMCA draws the tolerance ellipse based on Hotelling's T2 model. PCA further confirmed all phytochemical analysis methods interlinked with one another (Figure 19.9).

19.4 CONCLUSION

The *in vitro* regeneration established a reliable protocol for the application of genetic engineering techniques to enhance the production of medicinally important metabolites. The present study is the first report on efficient *in vitro* regeneration of *S. grandiflora*. Plant tissue culture has often been considered to be a suitable system for large-scale production. This can be applied for the production of secondary metabolites calli and cell culture. A large amount of *in vitro* calli harvested were compared with the wild plants' seeds, phenolic content, proline, FRAP, dot-blot assay and antioxidant activity. The study focused on plant tissue culture techniques that extracted the bioactive compounds for medical purposes and the pharmaceutical industry and to conserve the natural environment of endemic, rare, threatened and endangered medicinal plants.

ACKNOWLEDGEMENTS

The authors would like to thank, University Grants Commission (UGC) – Science for Meritorious Fellowship, Government of India, New Delhi for the financial support.

REFERENCES

Abbs, F. R. and Rexin, N. A. 2013 Phytochemical study on *Sesbania grandiflora. Journal of Chemical and Pharmaceutical Research* 5(2): 196–201.

Adaikan, P. G., Gauthman, K., and Prasad, R. N. V. 2001. History of herbal medicine with an insight on the pharmacological properties of *Tribulus terrestris. The Aging Male* 4: 163–169.

Ahmed, K. A. 2005. A proposed taxonomic system of the genus *Tribulus* in Egypt by means of numerical taxonomy. *Arab University Journal of Agricultural Sciences* 13: 219–231.

Ali, G., Mughal, M. H., Srivastava, P. S., and Iqbal, M. 1997. Micropropagation of *Tribulus terrestris* L., an important medicinal plant. *Journal of Plant Biology* 40: 202–205.

Arjun, P. 2011. Investigations on microprogation and isolation of active compounds from *Tribulus terrestris* Linn. PhD thesis, Faculty of Sciences, University of Madras, Chennai, Tamil Nadu, India.

Arjun, P., Mohana Priya, S., Saranya Sivan, P. S., Krishnamoorthy, M., and Balasubramanian, K. 2012a. Antioxidant and antimicrobial activity of *Nelumbo nucifera* Gaertn. leaf extracts. *Journal of Academia and Industrial Research* 1: 15–18.

Arjun, P., Semwal, D. K., Ruchi, B. S., Mishra, S. P., Vijayan, A. B., and Krishnamoorthy, M. 2015. Quality Retention and Shelf-Life Improvement of Fresh-cut Apple, Papaya, Carrot and Cucumber by Chitosan-Soy Based Edible Coating. *Current Nutrition & Food Science* 11(4): 282–291.

Arjun, P., Sivan, P. S., Mohana Priya, S., Krishnamoorthy, M., and Balasubramanian, K. 2012b. Phytochemical analysis and anticancer activity of *Nelumbo nucifera* extracts. *Journal of Academia and Industrial Research* 1: 81–85.

Astley, S. B. 2003. Dietary antioxidants – Past, present and future. *Trends in Food Science and Technology* 14: 93–98.

Avalaskar, A., Itankar, P. R., Joshi, V. S., Agrawal, M., and Vyas, J. 2011. Phytochemical and TLC studies of ethanolic extract of *Sesbania grandiflora* (Fabaceae). *International Journal of Pharma Tech Research* 3: 1346–1349.

Berker, K. I., Guclu, K., Tor, I., and Apak, R. 2007. Comparative evaluation of Fe (III) reducing power-based antioxidant capacity assays in the presence of phenanthroline, bathophenanthroline, tripyridyltriazine (FRAP), and ferricyanide reagents. *Talanta* 72(3): 1157–1165.

Costa, M. A. P. C., Carmo, D. O., Souza, F. V. D., Magalhaes, G. L., and Hansen, D. S. 2012. Efeito de diferentes concentrações de GA_3 no alongamento de brotações in vitro de jenipapo (*Genipa americana*). *Confresso Brasileiro De Fruiticultura.* 17, Belém, Anais.

Deccetti, S. F. C. 2000. Propagacao *in vitro* de *Annona glabra* L. Lavras: Universidade Federal de Lavras, Dissertacao (Mestrado). 101.

Doddola, S., Pasupulati, H., Koganti, B., and Koganti, V. S. R. G. 2008. Evaluation of *Sesbania grandiflora* for antiurolithiatic and antioxidant properties. *Journal of Natural Medicines* 62(3): 300–307.

Ghani, A. 1998. Medicinal Plants of Bangladesh; Published by Asiatic Society of Bangladesh.

Gurel, E. and Gurel, S. 1998. Plant Regeneration from unfertilized ovaries of sugar beet (*Beta vulgaris* L.) cultured *In Vitro. Traditional Journal of Botany* 22: 233–238.

Halliwell, B. and Gutteridge, J. M. C. 1999. Free radicals, 'reactive species' and toxicology. B. Halliwell, J. M. C. Gutteridge. (Eds.) *Free Radicals in Biology and Medicine.* Oxford University Press, New York. 544–616.

Hariprasath, L., Jegadeesh, R., Arjun, P., and Raaman, N. 2015. *In vitro* propagation of *Senecio candicans* DC and comparative antioxidant properties of aqueous extracts of the *in vivo* plant and *in vitro* derived callus. *South African Journal of Botany* 98: 134–141.

Harisaranraj, R., Suresh, K., Babu, S. S., and Achudhan, V. V. 2009. Phytochemical-based strategies for pathogen control and antioxidant capacities of *Rauwolfia serpentina* extracts. *Recent Research in Science and Technology* 1(2): 67–73.

Hayat, S., Hayat, Q., Alyemeni, M. N., Wani, A. S., Pichtel, J., and Ahmad, A. 2012. Role of proline under changing environments: A review. *Plant Signalling and Behaviour* 7: 1456–1466.

Hooker, J. D. 1978. Flora of British India, International Book distributors, Dehradun, Indian Reprint II. 115.

Imen, B. E. H. A., Maher, C., Radhia, B., Ikbal, C., Mohamed, B., and Fethia, H. S. 2014. Chemical composition and antioxidant, antibacterial, allelopathic and insecticidal activities of essential oil of *Thymus algeriensis* Boiss. *Industrial Crops and Products* 77: 631–639.

Jorge, E. W. P., Juan, C. C. E., Raul, R. H., Maria, L. C. L., Lluvia, I. L., Guadalupe, V. N. M., and Cristobal, N. A. 2015. Total phenolic content, *in vitro* antioxidant activity and chemical composition of plant extracts from semiarid Mexican region. *Asian Pacific journal of Tropical Medicine* 8(2): 104–111.

Komalavalli, N. and Rao, M. V. 2000. *In vitro* micropropagation of *Gymnema sylvestre* – A multipurpose medicinal plant. *Plant Cell Tissue and Organ Culture* 61: 97–105.

Krishnan, P. N., Decrus, S. W., and Radha, R. K. 2011. Conservation of medicinal plants of Western Ghats, India and its sustainable utilization through *in vitro* technology. *In Vitro Cellular and Developmental Biology* 47: 110–122.

Laszlo, S. and Arnould, S. 2009. Proline: A multifunctional amino acid. *Trends in Plant Science* 15(2): 89–97.

Malaisamy, M. and Mohan, N. 2014. Impact of Ozone on Morphological, Physiological, and Biochemical Changes in Cow Pea (*Vigna unguiculata* [L.] Walp.). *Ozone: Science and Engineering* 36: 36–42.

Martin, K. 2002. Rapid propagation of Holostemma ada-kodien Schult, a rare medicinal plant, through axillary bud multiplication and indirect organogenesis. *Plant Cell Reports* 21: 112–117.

Martin, K. P. 2004. Benzyladenine induced somatic embryogenesis and plant regeneration of *Leptadenia reticulata*. *Biologia Plantarum* 48: 285–288.

Mattioli, R., Costantino, P., and Trovato, M. 2009. Proline accumulation in plants. *Plant Signalling and Behaviour* 4(11): 1016–1018.

Murashige, T. and Skoog, F. 1962. A revised medium for rapid growth and bioassay with tobacco tissue cultures. *Physiologia Plantarum* 15: 473–497.

Nagarajan, S., Arjun, P., Raaman, N., Shah, S., Elizabeth Sobhia, M., and Mohan Das, T. 2012. Selective synthesis of sugar-based β-lactum derivatives: Docking studies and its biological evaluation. *Tetrahydron* 68: 3037–3045.

Nagarajan, S., Mohan Das, T., Arjun, P., and Raaman, N. 2009. Design, synthesis and gelation studies of 4,6-O-butylidene-α,β-unsaturated-β-C-glycosidic ketones: Application to plant tissue culture. *Journal of Material Chemistry* 19: 4587–4596.

Orwa, C., Mutua, A., Kindt, R., Jamnadass, R., and Simons, A. 2009. Agroforestree Database: A tree references and selection fugide version 4. World Agroforestry Centre, Kenya.

Ou, B., Huang, D., Hampsch-Woodill, M., Flanagan, J. A., and Deemer, E. K. 2002. Analysis of antioxidant activities of common vegetables employing oxygen radical absorbance capacity (ORAC) and ferric reducing antioxidant power (FRAP) assays: A comparative study. *Journal of Agricultural and Food Chemistry* 50: 3122.

Pande, D., Malik, S., Bora, M., and Srivastava, P. 2002. Rapid protocol for *in vitro* micropropagation of *Lepidium sativum* Linn and enhancement in the yield of lepidine. *In Vitro Cellular and Developmental Biology* 38: 451–455.

Paranjpe, P. 2001. Indian Medicinal Plants, Caukhamba Sanskrit Pratishthan, New Delhi Edition. 1, 3.

Patra, A. and Rai, B. 1998. Successful plant regeneration from callus cultures of *Centella asiatica* (Linn.) urban. *Plant Growth Regulators* 24: 13–16.

Perez-Vizcaino, F. and Duarte, J. 2010. Flavonols and cardiovascular disease. *Molecular Aspects of Medicine* 31(6): 478–494.

Prescott, L. M., Harley, J. P., and Klein, D. A. 2002. *Microbiology*. 5th International Edition, New York: Tata McGraw-Hill, 106–133.

Raghu, A. V., Geetha, S. P., Martin, G., Balachandran, I., and Mohanan, K. V. 2010. Micropropagation of *Tribulus terrestris* Linn. *Indian Journal of Natural Products and Resources* 1: 232–235.

Ramesh, T., Mahesh, R., and Begum, V. H. 2007. Effect of *Susbania grandiflora* on lung antioxidant defense system in cigarette smoke exposed rats. *International Journal of Biological Chemistry* 1(3): 141–148.

Rao, R. and Ravishankar, G. A. 2002. Plant cell cultures: Chemical factories of secondary metabolites. *Biotechnology Advances* 20: 101–153.

Rout, G. R. and Das, P. 1997. *In vitro* organogenesis in ginger (*Zingiber officinale* Rosc.). *Journal of Herbs, Spices and Medicinal Plants* 4: 41–51.

Sarkar, B. K., Mehta, J., Sharma, G., and Soni, P. 2012. Investigation on the antioxidant activity of *Sesbania grandiflora* flower extract. *Journal of Harmonized Research in Pharmacy* 1(1): 1–4.

Saxen, C., Palai, S. K., Samantaray, S., Rout, G. R., and Das, P. 1997. Plant regeneration from callus cultures of *Psoralea corylifolia* Linn. *Plant Growth Regulation* 22: 13–17.

Shareef, H., Rizwani, G. H., Zia-ul-Haq, M., Ahmad, S., and Zahid, H. 2011. Tocopherol and phytosterol profile of *Sesbania grandiflora* (Linn.) seed oil, *Journal of Medicinal Plants Research* 6(18): 3478–3481.

Sharma, P. and Rajam, V. M. 1995. Genotype, explant and position effects on organogenesis and somatic embryogenesis in eggplant (*Solanum melongena* L.). *Journal of Experimental Botany* 46: 135–141.

Sinha, R. K., Majumdar, K., and Sinha, S. 2000. *In vitro* bud breaks in axillary nodal segments of mature trees of *Acacia nilotica*. *Indian Journal of Plant Physiology* 34: 21–24.

Sivaraj, C. 2011. Isolation of compounds of antioxidant potential from *Manilkara zapota* Linn. PhD thesis, Faculty of Sciences, University of Madras, Chennai, Tamil Nadu, India.

Skoog, F. and Miller, C. O. 1957. Chemical regulation of growth and organ formation in plant tissue cultures *in vitro*. *Symposia of the Society for Experimental Biology* 11: 118–131.

Van Acker, S. A., Tromp, M. N. J., Heanen, G. R. M., Vander, W. J. F., and Bast, A. 1995. Flavonoids as scavenges of nitric oxide radicals. *Biochemical and Biophysical Research Communications* 214: 755–759.

Vanderjagt, T. J., Ghattas, R., Vanderjagt, D. J., Crossey. M., and Glew, R. H. 2002. Comparison of the total antioxidant content of 30 widely used medicinal plants of New Mexico. *Life Sciences* 70: 1035–1040.

Wang, C. H., Wu, S. B., Wu, Y. T., and Wei, Y. H. 2013. Oxidative stress response elicited by mitochondrial dysfunction: Implication in the pathophysiology of aging. *Experimental Biology and Medicine* 238(5): 450–460.

Xu, B. J., and Chang, S. K. C. 2007. A comparative study on phenolic profiles and antioxidant activities of legumes as affected by extraction solvents. *Journal of Food Science* 72: 159–166.

Zima, T. S., Fialova, L., Mestek, O., Janebova, M., Crkovska, J., Malbohan, I., Stıpek, S., Mikulıkova, L., and Popov, P. 2001. Oxidative stress, metabolism of ethanol and alcohol-related diseases. *Journal of Biomedical Science* 8: 59–70.

Index

A

Abrus precatorius, 93
Abscisic acid (ABA), 229
Acetylcholine, 162
Acetylcholinesterase, 165
AD; *see also* Alzheimer's disease, 133–136, 154
Adhatoda vasica, 107
Aegle marmelos, 93
Aerva lanata, 93
Aesculus turbinate, 120
Alkaloid, 27, 358
Aloe vera, 93
Alzheimer's disease, 162, 251
Amchi system, 240, 245, 251, 253
Amlich, see Badriphal, 242
Ammi majus, 60
Amyloid precursor protein, 163
Analgesic, 286
Andrographis paniculata, 93
Annona muricata, 93
Annona squamosa, 93
Anthocyanidins, 23
Anthocyanins, 264
Anthraquinone, 358
Antiallergic activity, 199
Antibacterial, 281
Anticarcinogenic, 281
Antidepressant, 286
Antidyspeptic, 262
Anti-inflammatory, 196–199, 281, 286
Antimicrobial, 198
Antiobesity, 115, 120, 123
Antiplasmodial, 198
Anti-RSV virus, 198
Antispasmodic, 262
Antitumour, 199
Aphrodisiac, 286
Aristolochia, 80
- *albida*, 82
- *bracteolata*, 82
- *elegans*, 82
- *indica*, 82
- *odoratissima*, 82
- *radix*, 82
- *shimadai*, 82

Aromatherapy, 181
Artocarpus heterophyllus, 93
Ashwagandha, 182
Asparagus recemosus, 93
Azadirachta indica, 93

B

Benincasa hispida, 93
Berries, 33
Betalain, 27
BHA; *see also* Butylated hydroxyanisole, 333, 335–337
BHT; *see also* Butylated hydroxytoluene, 333
Biophytum sensitivum, 93
Boerhavia diffusa, 93
Brassica oleracia, see Cabbage, 2
Breynia retusa
- antioxidant, 317–321, 323–326
- flavonoid, 319, 322
- lipid peroxidation, 321, 326
- total phenolic, 319, 322
- vitamin C and E, 319, 322

C

Cabbage
- ascorbic acid estimation, 4, 6–7
- carotene estimation, 6
- chlorophyll estimation, 5, 8
- genotypes, 4
- green, 2
- iodine estimation, 5, 10–11
- medicinal properties, 2
- red, 2
- savoy, 2
- β-carotene estimation, 4, 7–8

Camellia sinensis, 118
Capsaicinoids, 28
Carbohydrates, 358
Carotenoids, 20
Castanospermum austral
- *in vitro* antioxidant assays, 335, 342
  - ABTS, 336, 343
  - DPPH, 335, 342
  - FRAP; *see also* Ferric reducing antioxidant power assay, 336, 345
  - metal chelating activity, 337, 347
  - phosphomolybdenum, 336, 345
- tannins, 334, 341
- total flavonoids, 335, 341
- total phenolics, 334, 339

Catechins, 23
Catharanthus roseus, 93, 166
Cathranthine, 166
Celastrus paniculatus, 166
Centella asiatica, 93, 178

Chalcone, 180
Cholinergic, 179
Cholinergic neurons, 164
Cholinesterase, 165
Citrus, 32
CNS; *see also* Central nervous system diseases, 103
Constipation, 90
Cruciferous vegetables, 28
Curcuma longa, 178
Curcumin, 178
CVS; *see also* Cardiovascular disorders, 103

D

Dactylorhiza hatagirea, 249
Delphinine, 267
Diarrhoea, 90
Dietary fibre, 25
Diterpenoids, 358
DNA; *see also* Deoxyribonucleic acid, 318
DNS; *see also* 3,5-Dinitrosalicylic acid, 105
DPPH; *see also* 2,2-Diphenyl-1-picrylhydrazyl, 318–319, 323
Dr Alois Alzheimer, 162
Dwarf umbrella tree, 195

E

Embelic acid, 267
Epiphytic dwarf shrubs, 194
Euphorbia royleana Boiss, 178

F

Falcarinol, 199
Ficus amplissima, 101, 103
Flatulence, 90
Flavonoids, 22, 358
Flaxseed, *see* Linseed, 30
Fusarium graminarium, 288
  culture, 288
  extraction of toxin, 288

G

Galantamine, 179
Galanthus caucasicus, 179
Gallic acids, 23
Garlic, 35
Ginkgo biloba, 179
Ginsenoside, 181, 267
Glabridin, 180
Glucosinolates, 24
Glutamatergic, 179
Glycoside hydrolases, 78
Glycosides, 358
Glycyrrhiza glabra, 180
Glyderinine, 180
Grapes, 34
Gums, 358

H

Heteroblasty, 194
Histamine, 200–201
Homoblastic, 194
Huperzia serrata, 180
Huperzine A, 180
Hyaluronidases, 78
Hyperglycaemia, 113
Hyperinsulinaemia, 113
Hyponatremia, 90

I

Inflammation, 162
Inula racemosa, 252–253
Isobruceine, 267
Isoflavones, 23

J

Jyothismati oil, 166

K

Kabra, 245
Keampferol, 358

L

Lagerstroemia speciosa, 115
Leaf constants, 273
Leucas aspera, 298–310
  antiangiogenesis, 303, 309–310
  antimicrobial activity, 302, 306–308
  medicinal uses, 299
  plant description, 299
  secondary metabolites, 300
Leucoderma, 49
  aetiology, 52
  available treatments, 54
  clinical features, 50
  incidence, 50
  types, 51
Lichochalocone A, 180
Licorice, 180
Licorice isoflavones, 180
Lignans, 25
Limonoids, 21
Linoleic acid, 358
Lukoskin, 47–69
Lupane type triterpenoid saponins, 199

Lutein, 21
Lycopene, 21

M

Malondialdehyde, 321
Malus domestica, 119
Manu, 252
Marsh orchid, 249
Melissa officinalis, 181
Mucilages, 358
Muntingia calabura, 261–267
  botanical information, 261
  HPLC, 265–267
  HPTC, 264
  medicinal value, 262
  phytochemical investigations, 262–264

N

Neurodegenerative, 133
Neurofibrillary tangles, 163
Nicotinic acid, 358

O

Oats, 29
Organogenic callus, 302
Organosulphur compounds, 24

P

Panax ginseng, 181
Panax quinquefolium, 122
PG; *see also* Propyl gallate, 333
PGRs; *see also* Plant growth regulators, 357
  BAP; *see also* 6-Benzylaminopurine, 359
  cytokinins, 359
  2,4-D; *see also* 2,4-Dichlorophenoxyacetic acid, 359
  IAA; *see also* Indole-3-acetic acid, 359
  NAA; *see also* α-Naphthalene acetic acid, 359
Phaseolus vulgar, 115
Phenolic compounds, *see* Polyphenols, 21
Phenylalanine, 358
Phospholipase, 78
Phytoestrogens, 26
Phytonutrients, 18–19
  classification, 19
Phytostanols, 26
Piper, 80
  longum, 82
  nigrum, 82
  peltatum, 82
  umbellatum, 82
Plant hormones, 229
Plasmodium falciparum, 198
p-Nitrophenol, 104
pNPG; *see also* p-Nitrophenylglucopyranoside, 104
Pogostemon mollis, 287–289
  antifungal activity, 287, 288
  antimycotoxigenic activity, 287, 289
  collection and identification, 287
  extraction, 287
Poncirus trifoliate, 149
Psoralea corylifolia, 60
Pteridophytes, 223
Pterolobium hexapetalum, 269–281
  collection and identification, 272
  pharmacognosy studies, 272–274, 276–280

R

RH-3, 243
Root induction, 302, 306
Rosmarinic acid, 182
Rutin, 335–337

S

Salacia reticulate, 119
Salvia officinalis, 181
Saponins, 358
Scolopendra, 74
Sea buckthorn, 240
Sea buckthorn oil, 242–243
Serpentine, 166
Sesbania grandiflora, 355
Shilajit, 178–179
Shuang-Qi gout capsule, 198
Soybeans, 31
Steroids, 358
Stilbenes, 23
Stomatal
  index, 273
  number, 273
Symphytine, 267

T

Tabersonine, 166
Tannins, 358
Taraxacum officinale, 119
TBHQ; *see also* Tert-butyl hydroquinone, 333
Tea, 33
Terpenes, 25
Tinospora cordifolia, 151
Tocopherols, 241–242
Tocotrienols, 26, 241
Tomatoes, 30
Triglyceride, 113

Triterpenes, 358
Trolox; *see also* 6-Hydroxy-2,5,7,8-tetramethylchroman-2-carboxylic acid, 318
Type I diabetes mellitus, 89
Type II diabetes mellitus, 89
Tyrosinase enzyme, 64

U

Umbellate inflorescences, 194
Umbrella plant, 195
Urticadioica, 153

V

Valine, 358
Vegetables, 1
Vein-Islet number, 273
Vindoline, 166
Vinpocetine, 166
Vitis vinifera, 119
Vivipary, 223–224

W

Withania somnifera, 154, 182

X

Xanthophylls, 230

Z

Zearalenone, 288
  identification and quantification, 290
Zeaxanthin, 21
α-Amylase, 101, 105, 107
α-Glucosidase, 101, 104, 107
β-Lactoglobulin, 115
β-Sitosterol, 358
γ-Glutamyl transpeptidases, 78
ε-Polylysine, 115